D0609535

SEDIMENTOLOGY AND GEOCHEMISTRY OF DOLOSTONES

Based on a Symposium

Sponsored by the Society of Economic Paleontologists and Mineralogists

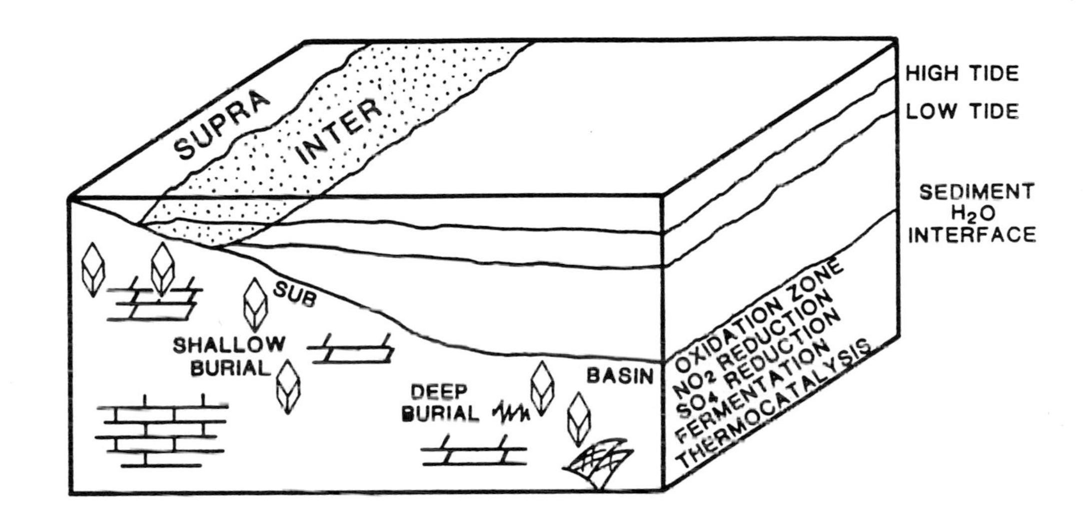

Edited by

Vijai Shukla and Paul A. Baker

Texaco U.S.A., E & P Technology Division, 3901 Briarpark, Houston, Texas 77042, and Department of Geology, Duke University, Box 6729 College Station, Durham, North Carolina 27708

Barbara H. Lidz, Editor of Special Publications

Special Publication No. 43

Tulsa, Oklahoma, U.S.A. ISBN: 0-918985-77-3 *November 1988*

A PUBLICATION OF
THE SOCIETY OF ECONOMIC PALEONTOLOGISTS AND MINERALOGISTS

ISBN: 0-918985-77-3

P.O. Box 4756
Tulsa, Oklahoma 74131

Printed in the United States of America

PREFACE

During the first half of 1984, both editors had independently approached SEPM to organize a symposium on dolostones. In December 1984, O. H. Pilkey, Jr., formally invited the editors to join forces and co-organize a dolostone symposium for the SEPM Third Annual Midyear Meeting held in Raleigh, North Carolina, during September 26–28, 1986.

The symposium, entitled *"SEDIMENTOLOGY AND GEOCHEMISTRY OF DOLOSTONES,"* was an enormous success. Twenty-three papers were presented in an all-day session; in addition, four related papers were presented in a poster session. In May 1986 (before the actual symposium), and in anticipation of the popularity of the symposium, we formally proposed to the SEPM Special Publications Committee that the proceedings be published as a Special Publication. In November 1986 the SEPM Council gave its imprimatur.

The present volume consists of 18 papers of which one (Sass and Bein) was invited for this publication and another (Holail and others) was part of a poster session in 1986. The remaining 16 papers represent data presented at the original symposium.

A publication such as this represents a synergistic effort by many people. First, we thank the authors for their contributions and for bearing with our editorial vicissitudes, demands, and deadlines. Reviews are essential for the presentation of good science, and we express our gratitude to the following reviewers for their efforts: D. A. Budd, E. Busenburg, R. R. Carlton, P. W. Choquette, W. C. Dawson, S. L. Dorobek, G. M. Friedman, J. M. Gregg, P. M. Harris, G. M. Harwood, J. D. Humphrey, J. Kaldi, M. Kastner, D. L. Kissling, O. C. Kopp, L. S. Land, M. Magaritz, R. K. Matthews, P. I. Nabelek, C. W. Naeser, R. A. Palmer, N. E. Pingitore, R. K. Stoessel, P. K. Swart, M. E. Tucker, J. Veizer, and B. H. Wilkinson. We thank O. H. Pilkey, Jr., the SEPM Publications Committee and its chairperson B. H. Lidz, and the SEPM Council for helping us bring this publication to fruition. Sara Wolpin, Brooklyn, New York, drafted the emblem for this publication.

Finally, one of us (VS) expresses profound thanks to the managers at Texaco's research facility. D. A. Bennett, T. L. Burnett, E. L. Stoudt, and J. W. Wood were enthusiastic and unhesitatingly supportive. They provided the company's time and resources for the organization of the original symposium and, subsequently, this publication.

Vijai Shukla
Paul A. Baker

CONTENTS

SECTION VII: DOLOMITE ORIGINS: CASE HISTORIES

INTRODUCTION

The need has always existed for understanding the processes of dolomitization and the origin of thick sequences of dolostones. This need becomes even more critical because pre-Cretaceous dolostones commonly host economically important deposits of natural resources and fossil fuels, e.g., petroleum reservoirs, mineral deposits, and groundwater aquifers.

In the past, two SEPM Research Symposia dealt with dolomitization of limestones and various models of dolomitization. These symposia culminated in the publication of SEPM Special Publications 13 and 28. Special Publication 13 (*DOLOMITIZATION AND LIMESTONE DIAGENESIS,* L. C. Pray and R. C. Murray, eds., 1965) was seminal in characterizing penecontemporaneous dolomite in arid and humid climates. "Supratidal" and "sabkha" became synonymous with dolomite. In the early 1970s as inorganic geochemistry was increasingly applied to dolomite studies, it became apparent that all dolomites were not supratidal nor sabkha-related. This awareness, in combination with chemohydrology of carbonate aquifers, led to the second panacea—the mixed-water (or "dorag") model. This is a chemohydrological model drawing on earlier works on kinetics in brackish-water zones. Special Publication 28 (*CONCEPTS AND MODELS OF DOLOMITIZATION,* D. H. Zenger, J. B. Dunham, and R. L. Ethington, eds., 1980) marked a major step in recognizing numerous types of dolomites and many models of dolomitization. In addition, papers emphasized usage of geochemistry and studies on kinetics of dolomitization.

It is pertinent to inquire what direction dolomite research has taken in the last eight years or so. How have the "concepts and models" described in Special Publication 28 been applied to sedimentary dolomites? Have we gained new insights and awareness into the processes of dolomitization? Have the fabric and geochemical approaches been fruitful, and has there been serious integration of geochemistry and sedimentology of dolostones? Finally, almost a decade has passed since the research on which Special Publication 28 was based, and therefore is it time to take stock, once again?

The present publication attempts to answer these questions. The wide scope of topics, both analytical and conceptual, has been arranged into seven sections, each emphasizing a specific theme:

Section I: TECHNIQUES AND EXPERIMENTAL STUDIES

1. Lumsden and Lloyd—ESR spectra of dolomites.
2. Swart—Nuclear-track mapping of dolomites.
3. Morrow and Ricketts—Dolomite and sulfate inhibition.

Section II: ORGANOGENIC DOLOMITES

4. Burns and Others—Drakes Bay dolomites.
5. Compton—Monterey dolomites.

Section III: DOLOMITES IN MVT DEPOSITS

6. Gregg—Dolomites and clay diagenesis.
7. Buelter and Guillmette—Viburnum trend dolomites.

Section IV: ROCK-WATER INTERACTIONS DURING DOLOMITIZATION

8. Banner and others—Evidence from stable and radiogenic isotopes.
9. Machel—Fluid flow direction.

Section V: GEOCHEMISTRY OF DOLOMITE TEXTURES AND FABRICS

10. Cander and others—Dolomite textures and fabrics.
11. Shukla—Dolomite fabrics.

Section VI: DOLOMITE DIAGENESIS

12. Zenger and Dunham—Burial dolomitization or burial neomorphism?
13. Moore and others—Dolomite recrystallization.
14. Holail and others—Dolomitization and dedolomitization.
15. Fischer—Diagenesis, texture, and porosity.

Section VII: DOLOMITE ORIGINS: CASE HISTORIES

16. Sass and Bein—Dolomites and salinity.
17. Mullins and others—Deep-water dolomites.
18. Ruppel and Cander—Seawater and brines.

Papers in each section are preceded by a brief introduction by the editors. Collectively, the studies present a vignette of close integration of sedimentologic and geochemical characteristics of dolomites and dolostones. Every publication has a principal message and ours is: whereas enormous progress has been made in dolostone research since 1965, nevertheless the subject is ripe for further study and much work needs to be done; we have come a long way but still have miles to go.

We hope this publication will be an appropriate and timely complement to the two earlier SEPM Special Publications (Numbers 13 and 28) on dolostones.

VIJAI SHUKLA　　PAUL A. BAKER

SECTION I
TECHNIQUES AND EXPERIMENTAL STUDIES

SECTION I: INTRODUCTION
TECHNIQUES AND EXPERIMENTAL STUDIES

The three papers in this section describe two techniques and one experimental study.

Lumsden and Lloyd describe Electron Spin Resonance (ESR) techniques, which can differentiate Mn^{2+} partitioning between Ca^{2+} and Mg^{2+} sites in dolomite. They also evaluate this partitioning in terms of age and stoichiometry of dolomite.

Swart describes nuclear-track mapping techniques to distinguish various dolomitization events. Dolomites with high U concentration are interpreted as replacements of aragonite and Mg-calcite precursors, whereas dolomites low in U resulted from replacement of calcite.

Morrow and Ricketts describe experiments that evaluated the influence of sulfate on dolomitization of calcite in saline solutions. These experiments confirmed the general conclusions of the landmark paper by P. A. Baker and M. Kastner on inhibition of dolomitization by sulfate; however, Morrow and Ricketts show that sulfate inhibition acts by retarding the dissolution of calcite, rather than preventing dolomite precipitation.

AN UPDATE OF ESR SPECTROSCOPY STUDIES OF DOLOMITE ORIGIN

DAVID N. LUMSDEN AND ROGER V. LLOYD

Department of Geological Sciences and Department of Chemistry Memphis State University Memphis, Tennessee 38152

ABSTRACT: Electron Spin Resonance (ESR) spectroscopy can determine the absolute amounts of Mn(II) in the Ca and Mg sites in dolomite and in associated calcite. The ESR spectra of Mn(II) in dolomite can be qualitatively divided into three types that have little overlap. Type 1 spectra have sharp peaks, and the partitioning of Mn into cation sites can be determined. Such spectra are common in stoichiometric and nonstoichiometric dolomites from both lithified and unconsolidated deposits (14 of 20 modern dolomites; 13 of 27 deep-marine dolomites). There is no apparent relation between Mn partitioning ratios and the absolute amount of Mn, the presence of a free radical center peak, or the total amount of dolomite in the sample. Modern dolomite, deep-marine dolomite, and nonstoichiometric Phanerozoic dolomites have average Mn partitioning ratios of 2, 5, and 6, respectively, suggesting that the ratios are not age dependent. Stoichiometric dolomites have an average partitioning ratio of approximately 30; thus, ratios and stoichiometry may be related. Type 2 spectra were observed in six of 20 modern dolomites and in 10 of 27 deep-marine dolomites. These spectra have broad peaks, and the Mg and Ca sites cannot be individually resolved. Because they are not found in older lithifield Phanerozoic dolomites, type 2 spectra may be related to lattice disorder. Type 3 spectra, observed in four deep-marine dolomites, do not have interpretable Mn peaks. A center peak assignable to radiation damage and/or free radicals may be present, independent of the Mn spectra. Age and thermal history data can be obtained from this peak.

INTRODUCTION

Modes of dolomitization cannot be distinguished by current techniques in any but the most diagenetically simple settings (Machel and Mountjoy, 1986; Hardie, 1987; Zenger and Dunham, this volume). Electron Spin Resonance (ESR) spectroscopy can often discriminate among dolomite samples that otherwise appear identical and thus provides data that may be useful in assessing dolomite origins. This paper describes the application of ESR to the analysis of sedimentary dolomite. We compare results for dolomites from unconsolidated sediments from a variety of environments (supratidal, freshwater lake, saline lakes, and deep marine), and contrast these results with data obtained for lithified Phanerozoic carbonates.

We use ESR to obtain the Mn(II) distribution ratio between the Mg and Ca sites in dolomite and to determine the partitioning of the Mn(II) between dolomite and associated calcite. Although partitioning in the latter case can, in principle, be obtained by selective dissolution, in practice this is difficult to do. The former distribution ratio cannot be determined by any other technique.

CHARACTERISTICS OF ELECTRON SPIN RESONANCE

Electron Spin Resonance measures the behavior of unpaired electrons in a magnetic field. Two classes of substances that have unpaired electrons are free radicals and transition metal ions. Divalent Mn is a common transition metal trace element in carbonates, and its electron structure typically gives a sharp ESR signal at room temperature. The specifics of the signal are controlled by the "environment" of the Mn ion. Carbonates provide several environments, each with distinct symmetry and distortion and, therefore, a unique ESR spectrum.

ESR spectra are presented as first derivatives (Fig. 1). The ESR spectrum of Mn(II) consists of peaks arising from five electronic transitions in the ion. Each of these is further split into six peaks as a result of nuclear hyperfine interactions with ^{55}Mn (Fig. 2). In room temperature spectra of powders, we observe only the $\pm^1/_2$ electronic transition split into six peaks. The site symmetry of the Mn(II) ions influences the details of the spectra. In particular, distortion from octahedral symmetry (as measured by the zero field splitting parameter) causes peaks to broaden.

The amount of Mn in a given cation site is proportional to the area of the peak representing the site. Dolomite has two distinctly different crystallographic sites that are hosts for Mn: (1) Ca sites (octahedral) and (2) Mg sites (distorted octahedral). The Ca site of calcite has a distortion intermediate between the Ca and Mg sites in dolomite. The signals from Mn in each of the three sites (two in dolomite, one in calcite) are, in favorable cases, distinguishable in ESR spectra (Fig. 2). Thus, the relative partitioning of Mn among the three sites can be determined. The signals are best distinguished in the high-field peak, which is universally used for analysis.

In many cases, the spectra show additional peaks, not due to the presence of Mn, that appear in the "center" of the Mn spectra (referred to as the "center peak"; see Fig. 2, peak 7). The center peak can be a complex of peaks that result from the effect of ionizing radiation on the sample. It is attributed to odd-electron species, such as CO_3^-, or to radicals from humic acids. The same radiation damage sites are used in thermoluminescence measurements (Aitken, 1985). The ionizing radiation includes alpha and beta particles generated by the decay of radionuclides in the carbonate lattice (U, Th, K) and, in some cases, by cosmic radiation (Hennig and Grun, 1983). We have shown empirically that a portion of the center peak is temperature sensitive and can function as a paleothermometer (Lloyd and Lumsden, 1987).

Thus, there are several ESR characteristics that can be used to study dolomite origin, including the partitioning ratios of Mn between Ca and Mg sites and between dolomite and calcite, and the size and structure of the radiation damage signal.

Dolomite does not obey Goldschmidt's rule; that is, Mn (radius 0.83 A; Shannon and Prewitt, 1969) goes into the smaller Mg sites (radius 0.72 A) rather than the larger Ca sites (radius 1.0 A). This puzzling observation (Wildeman, 1970), which has recently been confirmed (Shepherd and Graham, 1984; Prissok and Lehmann, 1986), may be due to lattice distortion overriding the effects of ionic charge and radius (Rosenberg and Foit, 1979). The mole ratio of

Sedimentology and Geochemistry of Dolostones, SEPM Special Publication No. 43

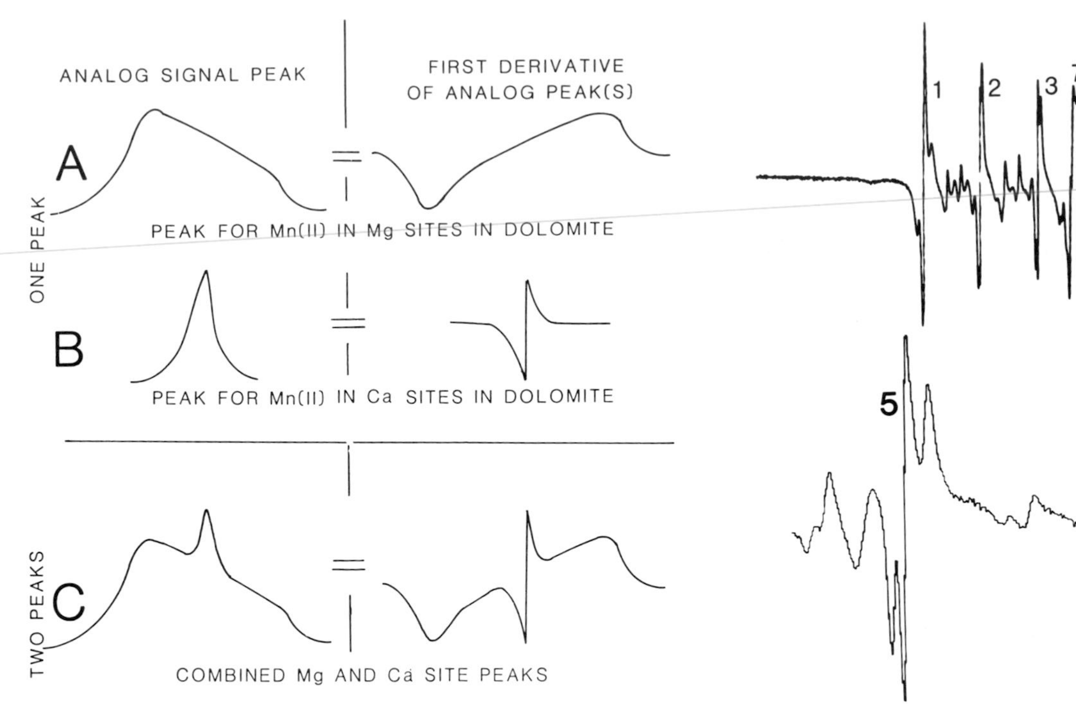

FIG. 1.—The relation between an analog signal and a first derivative signal. (A) A broad-analog peak (left) and its first derivative equivalent (right). (B) The same situation for a narrow-analog peak. (C) The effect of combination of a broad and a narrow peak, the typical situation for Mn in dolomite. The addition of a third peak due to the presence of calcite adds more complexity (see Fig. 2B).

FIG. 2.—Example ESR patterns. (A) Typical full scan of a type 1 pattern showing a center peak (seven) in a background of six peaks from the Mn hyperfine interaction. (B) High-field peak (peak six in A) used for analysis. Peaks for the Ca and Mn sites in dolomite are indicated by solid and dashed lines, respectively, and the Ca site in calcite is indicated by dotted line. Mn partitioning ratio for a given sample of dolomite is the area of the Mn signal from the Mg site divided by that from the Ca site. Modified from Lumsden and Lloyd (1984).

Mn:Ca as a measure of relative concentration (Viezer, 1983) indicates that manganese is approximately 10^5 times more abundant in certain kinds of dolomite (average = 270 ppm; Weber, 1964) than in sea water (0.002 ppm).

METHOD

The focus of this paper is on dolomite from young unconsolidated sediments, where the environment of deposition and geochemical parameters are reasonably well known (Table 1), and deep-marine dolomites (Table 3).

We use a Varian E-4 ESR spectrometer linked to a microcomputer to digitize, store, and analyze spectra. Powders prepared for X-ray diffraction are ideal for ESR; 0.1 g is more than sufficient. Samples must have no more than approximately 15% calcite in the carbonate fraction. There is no significant interference from other common sedimentary minerals, although iron in the dolomite may cause the type 3 spectra discussed later. Determining the relative peak areas by double numerical integration of the first derivative spectra is the most reliable technique. This compensates for differences in peak widths and shapes. We use a standard numerical integration program based on

TABLE 1.—CHARACTERISTICS OF SELECTED DOLOMITE-BEARING UNCONSOLIDATED LAKE AND SHORE-ZONE SEDIMENTS

Location	Climate	Environment	Temperature (°C)	Gypsum	Salinity	Mg/Ca in Water	Dolomite (percent)	Calcite in Dolomite	Crystal Size (μm)	Age (yrs)	Dolomite Type
BELIZE (1)	Warm, humid, dry	Supra-tidal		y			70	57–58	0.5–3.0	850–3,000	replacement cement
ANDROS (2)	Warm, humid, dry	Supra-tidal		y	5–6	40	20	56–62	1–4	150–2,200	replacement cement
COORONG LAGOON (3)	Warm, dry	Lake	28		0.5–7	20–100			0.5–20	300–2,000	
DEEP SPRINGS LAKE (4)	Warm, dry	Lake, playa				4–16		52–56	1–2	<3,500	precipitated
ELK LAKE (5)	Cool, moist	Varved lake	4	n	<1	7.7	10–30	50–51		<10,000	

Salinity = salinity where sea water is 1.
References: (1) Mazzullo and others, 1987; (2) Shinn and others, 1965; (3) von der Borch, 1965; (4) Clayton and others, 1968; (5) W. E. Dean, pers. commun., 1987.

TABLE 2.—DATA FOR MODERN UNCONSOLIDATED SAMPLES

Sample No.	Pattern Type	Mn Ratio	Center Peak**	Nonstoichiometric percent $CaCO_3$	Percent Dol	ppm Mn	Remarks
BZF31A2	1	NR	Y	55	78	84	Upper crust
BZF31A3	1	1	Y	57	86	46	Upper crust
BZF31A4	1	4	Y	57	64	49	Upper crust
BZF31B5	1	NR	S	58	35		Between upper and middle
BZF31C6	1	2	Y	58	62	85	Middle crust
BZF31C8	1	2	Y	58	69	26	Middle crust
BZF31C9	1	NR	Y	58	63	22	Middle crust
BZF31D10	1	1	N	57	97	238	Between middle and lower
BZF31E11	1	1	N	57	97	146	Lower crust
BZF31E12	1	2	N	58	79	205	Lower crust
BZF31E13	1	2	N	58	84	314	Lower crust
C1	1	5	Y	59	30		
C2	2	—	Y	53	100		Halite, magnesite
A-1	2	—		61	60		HMC, LMC, aragonite
EL-1	1	210	S	50	60		HMC, LMC
DS-1/8	1*	2	N	55	100		
DS-9/16	2*	2	N	52	100		
DS-17/22	2*	2	N	52	100		
DS-23/28	2*	2	N	54	100		
DS-28/33	2	—	N	54	65		
Average		2.2 ± 1.1		56 ± 2.5			Excluding EL-1

ESR patterns are coded: 1 = Type 1 patterns; 2 = Type 2 patterns. Percent dolomite is the percent in the carbonate fraction, not for the bulk sample. Samples BZF are from Belize, C from Coorong, A from Andros, EL from Elk Lake, and DS from Deep Springs Lake. HMC = high-mg calcite; LMC = low-mg calcite.

*These patterns are on the border of types 1 and 2. The partitioning ratio could be determined only for sample DS-1/8. The balance of the DS samples are similar and have close to the same ratio.

**Y = yes, center peak present; S = strong center peak; N = no center peak.

NR = no ratio, Mn content is so low that a reasonably reliable peak ratio could not be determined.

Newton's and Simpson's rules to compute areas of the high-field peaks. The error is inversely proportional to the absolute amount of Mn in the dolomite; we estimate it to be ±10% in most cases.

RESULTS

The ESR spectra of dolomite can be divided into three types that show little overlap (Fig. 3). Type 1 spectra have well-defined peaks from which the Mn peak areas of the Ca and Mg sites in dolomite can be determined. In type 2 spectra, the peaks are broad and the amount of Mn in the various sites cannot be separately determined. Type 3 spectra show one very broad peak that amounts to a change in background level.

We have established a calibration between peak areas and total amount of Mn in dolomite and calcite based on NBS standards; this allows us to obtain quantitative estimates of the amount of Mn in the three lattice sites. Detection down to 25 ppm for the sum of all sites is possible under favorable conditions. Excess calcite can be removed by selective dissolution. The technique is free of interference from other phases or from other chemical forms of Mn. We can also observe ESR peaks that are attributable to radiation damage of the carbonate structure. The size of this peak increases with the age of the sample because of radiation damage but eventually reaches a maximum due to equilibrium with chemical and/or thermal annealing. Thus, its development is a function not only of time but also of diagenetic changes.

Modern Unconsolidated Shore-Zone and Lake Dolomite

Ambergris Cay, Belize.—

Samples were collected from three crusts separated by thin intervals of ooze in a 60-cm core (Mazzullo and others, 1987; Table 1). Isotopic data suggest precipitation from near-normal marine water, although meteoric and hybrid fluids may have influenced the bottom samples. The deposits have a high proportion of dolomite (average = 70%), with the balance of the minerals consisting of high-Mg calcite (HMC), some low-Mg calcite (LMC), and traces of aragonite. Dolomite mosaics of 0.5- to 2.5-μm subhedral crystals replace high-Mg calcite in micrite and allochems, and dolomite cement is composed of well-formed 1- to 3-μm rhombs. The dolomite is markedly nonstoichiometric (57 to 58% $CaCO_3$), and the X-ray diffraction peak for the (01.5) order reflection is very broad.

All samples give type 1 ESR spectra, typically with a partitioning ratio of 1 to 2. No pattern of change in partitioning with depth is apparent (Table 2). The bottom four samples lack a center peak and have five to 10 times more Mn than shallower samples. They also have a lighter carbon isotope signature than shallow samples (J. Gregg, pers. commun., 1987).

Andros Island, Bahamas.—

Dolomite crusts of similar origin and age are found on Andros Island and in the lower Florida Keys (Atwood and Bubb, 1970; Shinn, 1968; Shinn and others, 1965; Shinn and others, 1969). Sediment derived from storm tides is eventually dolomitized in the supratidal environment. Dolomite typically forms 10 to 30% of the total sample and is most abundant (80%) in well-lithified crusts on the surface. Small crystals of dolomite (1 to 3 μm) replace calcium carbonate in pellets, skeletal grains, and micrite, and 1- to 4-μm crystals line cavities. The dolomite is very nonstoichiometric (55 to 62% $CaCO_3$) and order peaks are weak to absent.

The one Andros sample available has a type 2 spectrum.

TABLE 3.—DATA FOR DEEP-MARINE DOLOMITE

Sample No.	Percent Ca	Percent Dol	Crystal Size (μm)	Age (Ma)	Mn		Center Peak	ppm Mn	Comment
					Pattern Type	Ratio			
15/142/7/2/60	57	20	4	1	2	—	S	760	Mud
15/154/14/cc	57	10	5	1	2	—	W	4080	Mud
15/154/16/cc	55	85	4	1	2	—	N	3260	Stiff
10/95/20/cc	53	35	6	1	1	7	S/C		Hard, limestone
10/95/21/cc	55	90	6	1	1	5	S/C	<30	Crumbly, chalky
10/95/22/cc	53	60	11	1	1	3	S/C	<40	Crumbly, chalky
77/535/50/1/125	55	18	8	—	1	4	Y	390	Crumbly, claystone
77/536/22/1/80	57	95	—	100	1	9	S/C	120	Crystalline, dolostone
77/537/11/1/25	53	40	14	140	1	10	S	530	Crumbly
77/537/11/1/120	54	38	14	140	1	12	—		Crumbly
77/540/36/1/65	54	20	5	95	1	2	S	<260	Crumbly, claystone
77/540/37/1/39	55	40	5	95	1	1.5	S	<70	Crumbly, claystone
77/540/43/1/25	56	20	4	110	1	1	S		Crumbly, claystone
2/12/3/cc	55	25	7	40	2	—	N		Mud
10/86/10/cc	57	90	5	100	1	1:1:1	S		Dolomite sand
11/103/2/5/142	56	10	5	8	2	—	W		Mud
13/129/2/1/128	56	60	5	10	1	3	N		Crumbly
14/138/6/cc	58	80	10	93	2	—	W		Crystalline
39/357/24/5/90	58	10	33	40	2*	—	W		Crumbly
41/367/22/2/62	58	50	15	93	2	—	Y		Crumbly
42/373/2/1/130	56	75	8	4	3	—	N		Crumbly, red
75/530/84/2/117	59	60	8	87	2	—	W		Crumbly, red
75/530/85/4/123	58	65	10	87	2	—	N		Crumbly, red
47/398/103/1/42	—	5	—	108	3	—	N		Crystalline, nodule
50/416/5/1/10	55	30	6	51	1	0.8	S		Crumbly, mudstone
82/558/26/1/102	57	5	18	25	3	—	N		Crumbly, coccolithic
82/558/26/3/89	—	5	22	30	3	—	N		Crumbly, coccolithic
Average	56 ± 1.7					4.6 ± 3.8			

Sample number is according to DSDP code: leg/site/core/section/interval. The column headed percent Ca refers to the amount of Ca in the dolomite lattice (nonstoichiometry). Percent dolomite is the percent in the total sample, not the percent in the carbonate fraction. Center peaks are coded as: Y = yes, peak present; N = no peak; S = strong peak; W = weak peak; C = complex. The 1:1:1 ratio is for the three sites, Ca and Mg sites in dolomite and calcite.

The strong center peak may be due to cosmic-ray damage because this sample was from a surface crust.

Coorong Lagoon, Australia.—

The dolomite-bearing coastal-margin lakes of Coorong Lagoon have widely varying but generally high salinities and are often dry for part of the year (von der Borch, 1965; DeDeckker and Geddes, 1980). Dolomite crystal size is in the micron range; submicron to 1-μm spherular dolomite was observed (von der Borch and Jones, 1976).

Samples from the various lakes in the complex differ in their mineralogy. Sample C-1, from Lake 15, contains quartz, calcite, and high-Mg calcite as well as dolomite. The dolomite is nonstoichiometric (60% $CaCO_3$), without order peaks. Sample C-2 is from the "yogurt" that precipitates in "Magnesite Lake." The sediment contains magnesite and halite as well as nonstoichiometric dolomite (54% $CaCO_3$) and shows order peaks. The Mn ratio of sample C-1 is 5, whereas C-2 has a type 2 spectrum. Both samples have a strong center peak.

Elk Lake, Minnesota.—

Elk Lake is a marl lake in which diagenetic dolomite forms from HMC when the Mg:Ca ratio of the water is 7 to 12 (W. E. Dean, pers. commun., 1987). Dolomite content varies but generally increases downcore back to 10 ka. The single sample used here was at 30 cm. Elk Lake is an unusual carbonate-secreting lake in that it contains calcite, high-Mg calcite, rhodochrosite ($MnCO_3$), and dolomite, but no aragonite. Quartz and feldspar are also present. The dolomite is stoichiometric (50.6% $CaCO_3$). Interference obscures the order peak at (01.5).

The Mn(II) ESR signal is well developed, with a partitioning ratio of 210, the highest obtained thus far. Such high ratios are typical of stoichiometric dolomite (Lumsden and Lloyd, 1984). The presence of rhodochrosite and the

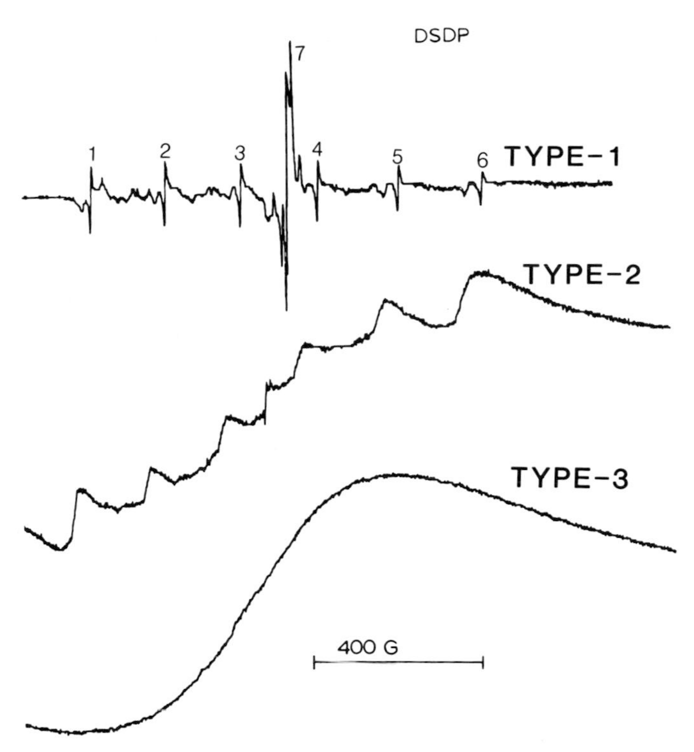

FIG. 3.—Examples of pattern types. Type 1 patterns show six Mn(II) peaks in which Ca, Mg, and calcite lines are visible and the ratios can be determined. A strong, complex, center peak is also present. Type 2 patterns have six peaks, but the separate sites cannot be resolved and they usually cannot be used to give Mn(II) distributions. A weak center peak is also present. Type 3 patterns show only a single broad peak that amounts to a change in background level.

very high Mn partitioning ratio may be related. A strong center peak signal is also present.

Deep Springs Lake, California.—

The saline Deep Springs Lake has been studied intensively in efforts to determine the origin of its dolomite (Jones, 1965). Direct precipitation was suggested (Peterson and others, 1966) but disputed (Clayton and others 1968). We studied splits of five samples collected at intervals along a piston core described in Clayton and others (1968). Dolomite crystals are typically less than 2 μm. The dolomite is moderately nonstoichiometric (52 to 56% $CaCO_3$) with broad-order peaks. Less than 1% of the dolomite is detrital (Peterson and others, 1963).

All samples produced similar spectra, transitional from type 1 to type 2. The top sample was sufficiently well resolved for determination of a ratio (approximately 2). Low absolute amounts of Mn in the dolomite contribute to the poor resolution of the peaks and to a relatively high error. There is no center peak in any sample.

Summary.—

Numerous studies of modern supratidal and lake dolomites suggest that, on the whole, modern unconsolidated sediments contain 1- to 2-μm euhedral dolomite rhombs in the voids among other carbonate grains (Graf and others, 1961; Curtis and others, 1963; Illing and others, 1965; Deffeyes and others, 1965; Muller and others, 1972; Hsu and Kelts, 1978; Muller and Wagner, 1978; Patterson and Kinsman, 1982). The rhombs are commonly less than 5,000 years old and may have formed in one generation. Supratidal samples were taken from areas of seasonal variation in environmental conditions, but warm brines occurred at least part of the time. Lake dolomites are found in both fresh water and evaporite situations, the latter being far more common. Almost all modern dolomites are very nonstoichiometric (55 to 60% $CaCO_3$, average 56%). Order peaks typically are very broad to absent in X-ray diffraction patterns.

The one freshwater sample is stoichiometric and has the highest Mn ratio yet observed. It is associated with $MnCO_3$, and this may in some way be related to the extreme Mn ratio. Thirteen modern saline-related samples have type 1 spectra and six have type 2. Their center peaks range from absent to strong.

Deep-Marine Dolomite

A set of 27 deep-marine samples was selected to represent a diversity of dolomite character (Table 3). Dolomite in deep-marine sediments is similar in appearance to supratidal and lake dolomite but is seldom less than 1 ma. It typically forms modest percentages of the sediment (1 to 25% in most samples), is of micron size (average 6 μm), and forms euhedral rhombs dispersed in the often low-carbonate sediment. Deep-marine dolomite is usually very nonstoichiometric (55 to 60% $CaCO_3$, average 56%). In X-ray diffraction patterns, order peaks vary from sharp to somewhat broadened.

All three types of Mn signal were observed: 13 type 1, 10 type 2, and four type 3. The type 3 spectra come from samples in which the total dolomite concentration is very low, iron is present, or the dolomite is in a concretion. The type 1 spectra have an average partitioning ratio of approximately 5, and three samples had ratios slightly less than 1. We are unable to find any pattern of relation between center-peak signal intensity and other parameters.

Lithified Phanerozoic Dolostones

Dolomite in lithified Phanerozoic rocks typically forms mosaics of interlocking crystals two to three orders of magnitude larger than those in modern and deep-marine sediments. It follows that the ESR signal gives bulk analysis that integrates the development of the dolomite through all phases of its formation, most of which must have taken place during diagenesis. Both sedimentary and metamorphic dolomites have type 1 spectra (Lumsden and Lloyd, 1984; Lloyd and others, 1985). The center peak varies from strong to weak and may be related to paleothermal events (Lloyd and Lumsden, 1987).

DISCUSSION

Type 1 dolomite spectra were observed for the freshwater lake sample, for 13 of 19 evaporite-related modern samples, for 13 of 27 deep-marine samples, for lithified sediments, and for metamorphic dolomite. In summary, type 1 spectra are characteristic of well-crystallized dolomite, both stoichiometric and nonstoichiometric, formed under a variety of conditions. The fact that the highest partitioning ratio was observed in dolomite associated with rhodochrosite suggests a relation to Mn abundance; however, a plot of partitioning ratios vs. the absolute amount of Mn (data in Tables 2 and 3) does not confirm this, nor is there a relation to the total amount of dolomite in the sample. There is a hint of a relation between crystal size and Mn ratio, but the data are few (Table 3).

In previous work (Lumsden and Lloyd, 1984), we suggested, on the basis of Mn partitioning data, that lithified Phanerozoic dolomites form two groups. One group (group B) consists of stoichiometric dolomites with a variable partitioning ratio (average 30 ± 23; Fig. 4). The second group (group A) has a variable $CaCO_3$ content but a relatively low and consistent Mn partitioning ratio (average 6 ± 2; Fig. 4). The contrast between these groups suggests that differences in their geochemistry of formation lead to differences in their trace-element incorporation and bulk composition. Data for both modern and deep-marine dolomites studied here plot in the field of group A. The modern dolomites have a partitioning ratio that averages 2 ± 1 (exclusive of the 210 ratio for the stoichiometric Elk Lake sample, which plots with group B), whereas deep-marine dolomite averages 5 ± 4 (Fig. 4). Both sets are nonstoichiometric, with averages of 56% $CaCO_3$. A scan of histograms of the Mn partitioning data and the results of an analysis of variance (0.05 significance level) indicate that deep-marine dolomites have a broader range and statistically higher mean value than modern dolomites but are similar to group B dolomites. It is tempting to distinguish between modern and deep-marine dolomites on the basis of their different Mn partitioning ratios, but the small number of samples precludes a firm conclusion.

Kretz (1982) proposed, on theoretical grounds, that dolomite should have a Mn partitioning ratio of 1.5. Our observations suggest that a ratio near 5 is the stable value toward which dolomites adjust with time and/or temperature, as shown by the results obtained for the nonstoichiometric Phanerozoic sedimentary dolomites (Fig. 4) and for metamorphic dolomite (Lloyd and others, 1985). In contrast, stoichiometric (group B) dolomites have a mode of origin that involves a nonthermodynamic, possibly kinetically controlled, mechanism of trace-element incorporation.

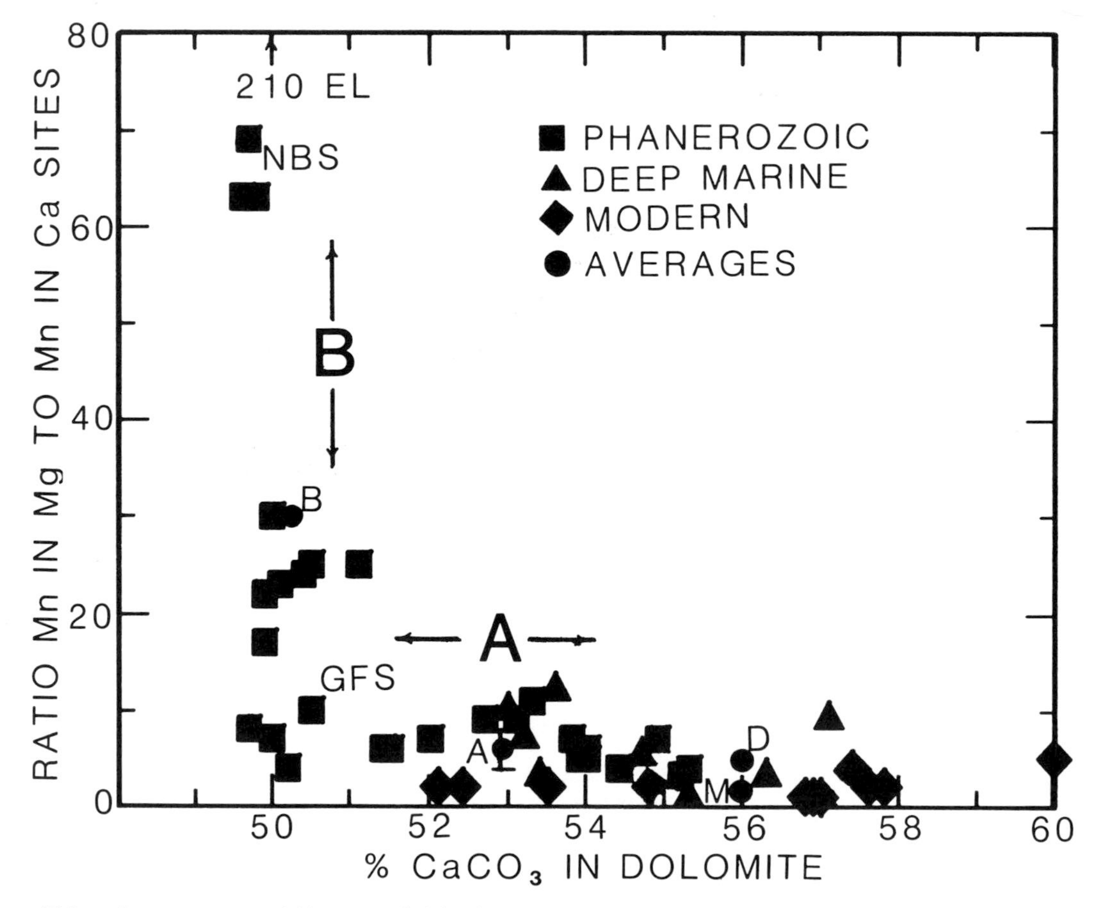

FIG. 4.—Plot of Mn ratios versus nonstoichiometry of dolomites. Phanerozoic sedimentary dolomites form two groups (Lumsden and Lloyd, 1984). Group B samples come from evaporite-related formations, are stoichiometric, and have a variable Mn partitioning ratio (15 samples, including four common to both groups, range 5 to 70, average 30, point B). Group A samples come from units of nonsaline origin, are nonstoichiometric, and have low Mn partitioning ratios (19 samples, including four common to both groups, average 6, point A). Both deep-marine dolomites and modern samples plot with group A. The former have an average Mn partitioning ratio of 5 (point D); the latter have a partitioning ratio of 2 (point M). The Elk Lake sample (EL) plots off the graph. Points for NBS 88a and G. Frederic Smith (GFS) dolomite standards are indicated.

Type 2 spectra occur in saline-lake samples, saline supratidal samples, and many deep-marine samples. The cause of the peak broadening in type 2 spectra is uncertain. These spectra were not observed, however, in either lithified Phanerozoic dolostones or the well-crystallized and stoichiometric sample from Elk Lake. This suggests that they are functions of lattice disorder. Perhaps there is a range of lattice sites with a variety of distortions leading to a broad "average" peak. Another possibility is that the abudance of minute crystal domains observed in young dolomite (Reeder and Wenk, 1979) causes broadening by mutual interference along crystallite boundaries. Minor amounts of iron may influence the peaks.

Type 3 spectra were found only in deep-marine samples; one was from a concretion, others had little dolomite (5% or less) or associated iron-bearing minerals. We speculate that they result from the molecular-scale magnetic influences of a high iron content.

The size of at least a portion of the center ESR peak is a function of the amount of radiation damage. This portion increases in size with increasing age of the dolomite. It is necessary to know the total amount of radionuclides (U, Th, K) so that the "annual dose" can be compared to the "total dose" to determine age. The radiation damage sites, however, anneal with time and temperature and ultimately reach an equilibrium number (Aitken, 1985). Other factors (exposure to cosmic radiation, variable temperatures, variable rates of annealing in variable solutions, and so on) cloud the issue, so that each geologic situation presents unique relations between peak size and age. Center-peak size shows no correlation with stoichiometry, crystal size, or Mn(II) ratio. We previously noted (Lloyd and Lumsden, 1987) a decrease in the size of the temperature-sensitive portion of the center peak with approach to an intrusion. This suggests that the center peak may be used as a paleothermometer. Thus, when the amount of radionuclides incorporated in the dolomite is known, it may be possible to determine when the intrusive event occurred.

We suggest two future lines of research. Investigation of the relation between organic radicals and the character of the center peak could provide evidence about the influence of organic processes on the origin of dolomite. Laboratory synthesis of dolomite using various combinations of time, temperature, and Mn concentrations should determine what relation, if any, exists between Mn concentrations in the ambient environment and its partitioning into the cation sites of dolomite.

CONCLUSIONS

The ESR Mn spectra for dolomite can be divided into three types that exhibit little overlap. Type 1 are well-defined spectra in which the absolute amount of Mn(II) in the two cation sites (Ca and Mg), and therefore the partitioning of Mn between the sites, can be determined. If a modest amount of calcite is also present (less than 15%), the partitioning ratio between all three cation sites can be determined. These spectra are found in well-ordered dolomite, both stoichiometric and nonstoichiometric, in modern sediments, deep-marine sediments, well-lithified Phanerozoic sediments, and metamorphic rocks. In the case of manganese, there appears to be a stable, perhaps thermodynamically controlled, cation site ratio in dolomite of approximately 5. This value is obtained by a combination of sufficient time and/or temperature. There is no apparent relation between the total amount of Mn in the dolomite and Mn partitioning ratios. We suggest that there are at least two origins for dolomite, reflected in both their stoichiometry and their trace-element distribution. It is possible that the difference in Mn partitioning ratios between modern and deep-marine dolomites (2 and 5, respectively) is real, but more data are needed for proof.

A center peak attributable to radiation damage of the carbonate lattice may be present, independent of the Mn spectrum. The size of this peak depends on a dynamic interaction between time (creation) and annealing (destruction).

ACKNOWLEDGMENTS

We could not have done this study without the samples sent to us by Jay Gregg and Sal Mazzullo (Belize), Gene Shinn (Andros), R. M. Forester and Chris von der Borch (Coorong), Blair Jones (Deep Springs Lake), and Walter Dean (Elk Lake). A Memphis State University Grant for Faculty Research and a grant from the Southern Regional Education Board supported travel to the Deep Sea Drilling Program repository.

REFERENCES

AITKEN, M. J., 1985, Thermoluminescence Dating: Academic Press, New York, 359 p.

ATWOOD, D. K., AND BUBB, J. N., 1970, Distribution of dolomite in a tidal flat environment: Journal of Geology, v. 78, p. 499–505.

CLAYTON, R. N., JONES, B. F., AND BERNER, R. A., 1968, Isotope studies of dolomite formation under sedimentary conditions: Geochimica et Cosmochimica Acta, v. 32, p. 415–432.

CURTIS, R., EVANS, G., KINSMAN, D. J., AND SHEARMAN, D. J., 1963, Association of dolomite and anhydrite in the Recent sediments of the Persian Gulf: Nature, v. 197, p. 679–680.

DEDECKKER, P., AND GEDDES, M. C., 1980, Seasonal fauna of ephemeral saline lakes near the Coorong Lagoon, South Australia: Australian Journal of Marine and Freshwater Research, v. 31, p. 677–699.

DEFFEYES, K. S., LUCIA, F. J., AND WEYL, P. K., 1965, Dolomitization of Recent and Plio-Pleistocene sediments by marine evaporite waters on Bonaire, Netherlands Antilles, *in* Pray, L. C., Murray, R. C., eds., Dolomitization and Limestone Diagenesis: A Symposium: Society of Economic Paleontologists and Mineralogists Special Publication 13, p. 71–88.

GRAF, D. L., EARDLEY, A. J., AND SHIMP, N. F., 1961, A preliminary report on magnesium carbonate formation in glacial Lake Bonneville: Journal of Geology, v. 69, p. 219–223.

HARDIE, L. A., 1987, Dolomitization: A critical view of some current views: Journal of Sedimentary Petrology, v. 57, p. 166–183.

HENNIG, G. J., AND GRUN, R., 1983, ESR dating in Quaternary geology: Quaternary Science Review, v. 2, p. 157–238.

HSU, K. J., AND KELTS, K., 1978, Late Neogene chemical sedimentation in the Black Sea, *in* Matter, A., Tucker, M. E., eds. Modern and Ancient Lake Sediments: International Association of Sedimentologists Special Publication 2, p. 129–145.

ILLING, L. V., WELLS, A. J., AND TAYLOR, J. C. M., 1965, Penecontemporary dolomite in the Persian Gulf, *in* Pray, L. C., Murray, R. C., eds., Dolomitization and Limestone Diagenesis: A Symposium: Society of Economic Paleontologists and Mineralogists Special Publication 13, p. 89–111.

JONES, B. F., 1965, The hydrology and mineralogy of Deep Springs Lake, Inyo County, California: U.S. Geological Survey Professional Paper 502-A, 56 p.

KRETZ, R., 1982, A model for the distribution of trace elements between calcite and dolomite: Geochimica et Cosmochimica Acta, v. 46, p. 1979–1981.

LLOYD, R. V., AND LUMSDEN, D. N., 1987, The influence of temperature on the radiation damage line in the ESR spectra of metamorphic dolomites: A potential paleothermometer: Chemical Geology, v. 64, p. 103–108.

———, ———, AND GREGG, J. M., 1985, Relationship between paleotemperatures of metamorphic dolomites and ESR determined Mn(II) partitioning ratios: Geochimica et Cosmochimica Acta, v. 49, p. 2565–2568.

LUMSDEN, D. N., AND LLOYD, R. L., 1984, Mn(II) partitioning between calcium and magnesium sites in studies of dolomite origin: Geochimica et Cosmochimica Acta, v. 48, p. 1861–1865.

MACHEL, H. G., AND MOUNTJOY, E. W., 1986, Chemistry and environments of dolomitization–A reappraisal: Earth Science Reviews, v. 23, p. 175–122.

MAZZULLO, S. J., REID, A. M., AND GREGG, J. M., 1987, Dolomitization of Holocene Mg-Calcite supratidal deposits, Ambergris Cay, Belize: Geological Society of America Bulletin, v. 98, p. 224–231.

MULLER, G., IRION, G., AND FORSTNER, U., 1972, Formation and diagenesis of inorganic Ca-Mg carbonates in the lacustrine environment: Naturwissenschaften, v. 59, p. 158–164.

———, AND WAGNER, F., 1978, Holocene carbonate evolution in Lake Balaton (Hungary): A response to climate and impact of man, *in* Matter, A., Tucker, M. E., eds., Modern and Ancient Lake Sediments: International Association of Sedimentologists Special Publication 2, p. 57–81.

PATTERSON, R. J., AND KINSMAN, D. J. J., 1982, Formation of diagenetic dolomite in coastal sabkha among Arabian (Persian) Gulf: American Association of Petroleum Geologists Bulletin, v. 66, p. 28–43.

PETERSON, M. N. A., BIEN, G. S., AND BERNER, R. A., 1963, Radiocarbon studies of Recent dolomite from Deep Springs Lake, California: Journal of Geophysical Research, v. 68, p. 6493–6505.

———, VON DER BORCH, C. C., AND BIEN, G. S., 1966, Growth of dolomite crystals: American Journal of Science, v. 264, p. 257–272.

PRISSOK, F., AND LEHMANN, G., 1986, An EPR study of Mn^{2+} and Fe^{3+} in dolomites: Physics and Chemistry of Minerals, v. 13, p. 331–336.

REEDER, R. J., AND WENK, H. R., 1979, Microstructures in low temperature dolomites: Geophysical Research Letters, v. 6, p. 77–80.

ROSENBERG, P. E., AND FOIT, F. F., 1979 The stability of transition metal dolomites in carbonate systems: A discussion: Geochimica et Comochimica Acta, v. 43, p. 951–955.

SHANNON, R. D., AND PREWITT, C. T., 1969, Effective ionic radii in oxides and fluorides: Acta Crystallographica, v. B25, p. 925–946.

SHEPHERD, R. A., AND GRAHAM, W. R. M., 1984, EPR of Mn^{2+} in polycrystalline dolomite: Journal of Chemical Physics, v. 81, p. 6080–6084.

SHINN, E. A., 1968, Selective dolomitization of Recent sedimentary structures: Journal of Sedimentary Petrology, v. 38, p. 612–616.

———, GINSBURG, R. N., AND LLOYD, R. M., 1965, Recent supratidal dolomite from Andros Island, Bahamas, *in* Pray, L. C., and Murray, R. C., eds., Dolomitization and Limestone Diagenesis: A Symposium: Society of Economic Paleontologists and Mineralogists Special Publication 13, p. 112–123.

———, LLOYD, R. M., AND GINSBURG, R. N., 1969, Anatomy of a modern carbonate tidal flat, Andros Island, Bahamas: Journal of Sedimentary Petrology, v. 39, p. 1202–1228.

VIEZER, J., 1983, Chemical diagenesis of carbonates: Theory and application of trace element technique, *in* Arthur, M. A., ed., Stable Isotopes in Sedimentary Geology: Society of Economic Paleontologists and Mineralogists Short Course Notes No. 10, p. 3-1 to 3-100.

VON DER BORCH, C. C., 1965, The distribution and preliminary geochemistry of modern carbonate sediments of the Coorong area, South Australia: Geochimica et Cosmochimica Acta, v. 29, p. 781–799.

———, AND JONES, J. B., 1976, Spherular modern dolomite from the Coorong area, South Australia: Sedimentology, v. 23, p. 587–591.

———, RUBIN, M., AND SKINNER, B. J., 1964, Modern dolomite from South Australia: American Journal of Science, v. 262, p. 1116–1118.

WEBER, J. N., 1964, Trace element composition of dolostones and dolomites and its bearing on the dolomite problem: Geochimica et Cosmochimica Acta, v. 28, p. 1817–1868.

WILDEMAN, T. R., 1970, The distribution of Mn^{2+} in some carbonates by electron paramagnetic resonance: Chemical Geology, v. 5 p. 167–177.

THE ELUCIDATION OF DOLOMITIZATION EVENTS USING NUCLEAR-TRACK MAPPING

PETER K. SWART

Division of Marine Geology and Geophysics, Rosenstiel School of Marine and Atmospheric Science, University of Miami, 4600 Rickenbacker Causeway Miami, Florida 33149

ABSTRACT: The concentrations and distribution of uranium and boron have been measured in dolomites and limestones from a core taken on the island of San Salvador in the Bahamas. The analyses reveal a wide range of concentrations both within and between the two predominant types of dolomite. The crystalline dolomites show unexpectedly high concentrations of U in skeletal components (2 to 7 ppm), but low values in void-filling cements (0.5 to 1 ppm). In contrast, the fabric-destructive microsucrosic dolomites are uniformly low in U (0.5 to 1 ppm) with occasional red algal fragments exhibiting concentrations as high as 1.5 ppm. Data presented here suggest that the U concentrations of the dolomites are inherited from original sedimentary and diagenetically altered components. It is suggested that the rocks that have higher concentrations of U, and in which the original fabrics are largely preserved, were dolomitized directly from the aragonite and high-Mg calcite (HMC) precursors. The U concentration is retained during dolomitization because in carbonite-rich fluids the uranyl ion (UO_2^{2+}) is complexed principally with the carbonate ion (CO_3^{2-}). As the activity of CO_3^{2-} is usually limiting in producing solutions supersaturated with respect to dolomite, CO_3^{2-} produced from the dissolution of metastable precursors is reincorporated into dolomite. In contrast, dolomites with lower U concentration formed from a low-Mg calcite (LMC) precursor which previously lost U during stabilization by meteoric waters.

Concentrations of B in the dolomites were similar (1 to 3 ppm) to values determined for modern LMC organisms (this study) and therefore suggest dolomitization from predominantly marine fluids. Comparisons with ranges reported in the literature show B concentrations in this investigation to be much lower. This is attributable to the ability of the nuclear-track technique to recognize contamination within the sample and consequently to allow it to be eliminated from the analysis.

INTRODUCTION

The study of carbonate diagenesis is often hindered by the inability to make geochemical measurements at the same scale as petrographic observations. The electron microprobe has overcome some of these problems, but as a result of its high detection limits, it is unable to measure elements at true trace concentrations (less than 100 ppm), or to make stable isotopic analyses for carbon and oxygen. Over the past decade, petrologists have commonly attempted to overcome some of these problems by resorting to increasingly more sophisticated systems of micro-sampling, or use of the ionmicroprobe and cathodoluminescence to infer concentrations of certain trace elements. In this study use is made of the fact that specific elements undergo nuclear reactions when subjected to a flux of neutrons. These can be mapped at a resolution of between 20 and 50 μm using nuclear-track detectors. In particular, ^{235}U experiences a fission reaction (n,f), producing two approximately equal size particles that create a narrow path of intense damage on an atomic scale. The damage is produced in the mineral itself and, if present, in an overlying particle detector. The damage may be made visible under an optical microscope using a chemical reagent that attacks the damaged area. By counting the number of tracks per unit area and relating the density to a standard with a known uranium concentration and 235/238 ratio, the concentration of U in the sample can be measured. The distribution and concentration of B can be mapped in a similar manner making use of a different nuclear reaction, neutron flux, and detector material. The mapping of these two elements provides direct measurements of the concentration of U and B at a resolution of approximately 20 to 30 μms.

In this paper I report the use of nuclear-track mapping techniques to study the paragenetic sequence and processes of dolomitization that have occurred in later Tertiary and Pleistocene carbonates of the Bahamas. The materials for this study are taken from a 168-m core from San Salvador in the Bahamas.

Uranium in Sea Water

The work of Langmuir (1978) and others has established that in the marine environment U exists mainly as the uranyl ion (UO_2^{2+}), complexed predominantly with CO_3^{2-} and $H(PO_4)_2^{2-}$ ions. Using stability constant data compiled by Langmuir (1978) and the model of Garrels and Thompson (1962), approximately 70% of the U in sea water can be calculated to exist as $UO_2(CO_2^{3})_2^{2-}$ and the remainder is present as either $UO_2(CO_3)_3^{4-}$ or $UO_2(HPO_4)_2^{2-}$ ions. The distribution of U species in sea water led Swart and Hubbard (1982) to suggest that calculation of a distribution coefficient for uranium relative to Ca^{2+} was inappropriate and that U is partitioned relative to carbonate rather than to Ca^{2+}.

Uranium in Carbonates

The concentration of U has been extensively investigated in certain types of modern carbonates, usually as a consequence of radiometric dating using uranium series disequilibrium techniques. In particular, scleractinian corals and molluscs have been well studies (Broecker, 1963; Lahoud and others, 1966; Swart and Hubbard, 1982). There is, however, a paucity of data on many other organisms and nonskeletal components. A summary of the reported ranges in the literature is shown in Figure 1. Of particular interest in these data are the rather wide ranges of values reported for corals and molluscs. This may be a reflection of the large number of analyses made on these organisms in different diagenetic conditions and by a variety of different workers from separate geographic localities. Swart and Hubbard (1982) reviewed these data and concluded that upon death, coral skeletons readily adsorb U and increase in concentration from normal live levels of 1.5 to 2 ppm to between 3 to 4 ppm (see also Cross and Cross, 1983; Swart, 1984). Similar processes may also occur in molluscs, but in this case the wide range of reported values (0.01 to 3.3 ppm) is a result of the presence of both aragonitic and cal-

Sedimentology and Geochemistry of Dolostones, SEPM Special Publication No. 43

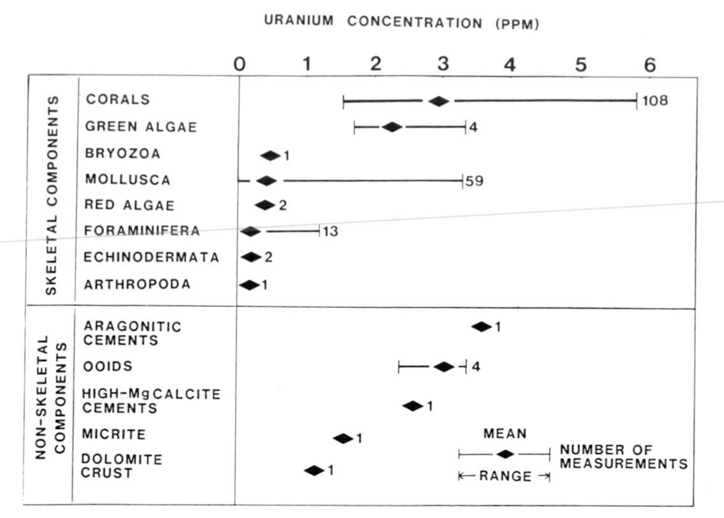

FIG. 1.—Ranges of U reported in modern carbonates. Data and references from a compilation by Chung (1988).

citic mineralogies (Chung and Swart, unpubl. data). Uranium frequently shows an association with organic material (Amiel and others, 1972; Taylor, 1979) and it has been documented (Amiel and others, 1973) that certain organic components within certain coral skeletons can be highly enriched in U.

As a general rule, biogenic and inorganic low-Mg calcite (LMC) and high-Mg calcite (HMC) contain lower U concentrations than aragonite. The exceptions are certain marine aragonitic gastropods, such as pteropods, which possess U concentrations between 0.01 and 0.05 ppm (Harrison and Swart, unpubl. data), and early diagenetic calcite cements, which frequently contain U values between 2 and 3 ppm (Gvirtzman and others, 1973).

Diagenesis.—

The behavior of U during early carbonate diagenesis has been studied by Lahoud and others (1966), Haglund and others (1969), Gvirtzman and others (1973), Chung and others (1985), and Chung and Swart (1986). These workers discovered that there was generally a decrease in the U concentration of carbonate rocks during freshwater diagenesis as metastable aragonite and HMC were dissolved and/or replaced by LMC. Early marine aragonite and HMC cements, however, are not depleted in U and frequently possess values comparable to the original sedimentary components.

In contrast to other carbonates, the U concentration of dolomites has not been extensively investigated by any analytical technique, and this study represents the first attempt to use fission-track mapping to examine dolomitization processes. One of the only studies of U in dolomites (Rodgers and others, 1982) examined rocks from the Niue Island in the South Pacific. These workers reported concentrations of between 0.3 and 10.2 ppm U and observed that dolomites contained twice the concentration of U as co-occurring calcite. Rodgers and others (1982) concluded that the high concentration of U in these rocks was unusual and that the source of U was not present at the time of initial diagenesis. Hence, U must have been introduced during dolomitization, possibly from former hydrothermal activity.

Boron in Carbonates

A wide range of B values has been reported in the literature for modern marine carbonates (see compilation in Milliman, 1974). The highest concentrations have been documented in scleractinian corals (21 to 100 ppm), whereas pelecypods and gastropods, regardless of mineralogy, have on average concentrations of less than 5 ppm (Milliman, 1974). No data are available for nonskeletal components. Larger ranges have been reported in limestones (1–240 ppm) and dolomites (30–337 ppm) (Weber, 1964; Christ and Harder, 1978). Most of the reported B measurements have been made in an attempt to determine the paleosalinity of the depositional or diagenetic environment, but in the case of limestones and dolomites, the range of data is such that no clear conclusion can be made. A significant problem in the interpretation of B in any type of rock relates to the method of analysis and removal of contamination. It is clear that adsorbed B is particularly difficult to remove from samples and that previous analyses may in some manner be contaminated. In addition, as the apparent concentration of B also seems to be related to the crystal and grain size of a component, useful comparisons can only be made on samples that are of similar textures.

PREVIOUS WORK ON SAN SALVADOR DOLOMITES

Dolomites from the 168-m San Salvador core have been described by Supko (1970, 1977) and Dawans and Swart (1988). Dolomite first appears at 34.2 m below the surface and extends to about 157 m, replacing middle Miocene to late Pliocene shallow-marine carbonates (Williams, 1985; Swart and others, 1987). Overlying the dolomite is a Pleistocene limestone interval that contains decreasing quantities of aragonite and HMC with depth (Supko, 1970). On the basis of a comparison of the Sr isotopic composition of the dolomites with the known seawater evolution curve of Burke

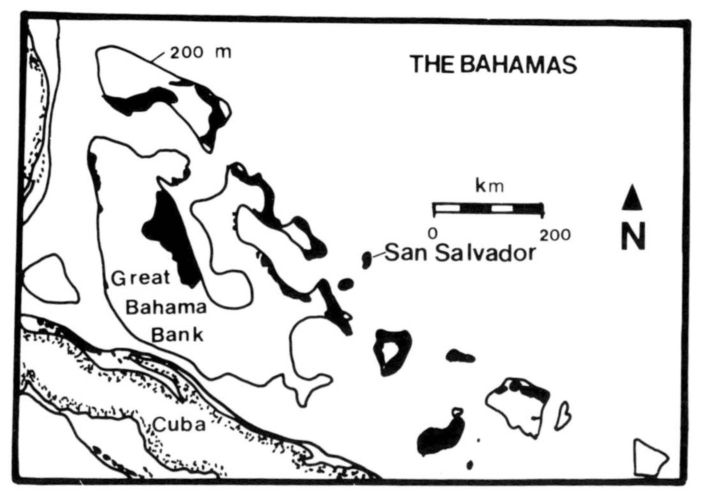

FIG. 2.—Location of the island of San Salvador.

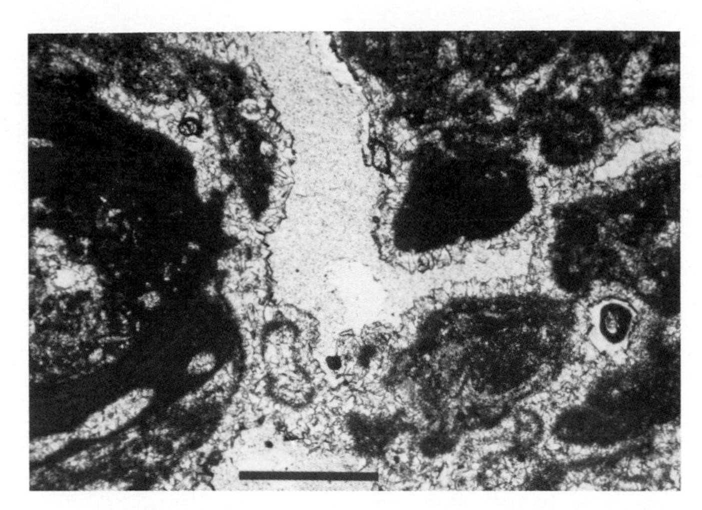

FIG. 3.—Photomicrograph of CM dolomite from a depth of 55.8 m. Petrographic characteristics of this type are a preservation of the precursor fabric and void-lining cements. Scale bar = 500 μm.

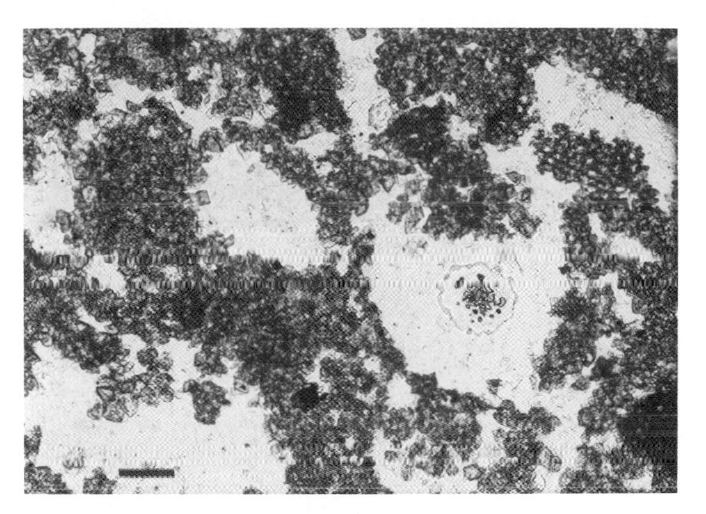

FIG. 4.—Photomicrograph of MS dolomite from a depth of 64 m. Petrographic characteristics of this type are a fine microsucrosic groundmass of euhedral rhombs with occasional red algal fragments. Scale bar = 100 μm.

and others (1982) and DePaolo (1986) and U series disequilibrium dating, a formation age of between 150 ka and 1.7 Ma has been estimated for these dolomites (Swart and others, 1987). Dawans and Swart, 1988: recognized two texturally and geochemically distinct types of dolomites in this core. One of these dolomites was termed microsucrosic (MS) because it is composed of a mosaic of 10- to 50-μm-size euhedral rhombs (Fig. 3). The other variety, termed crystalline mimetic (CM), was hard and crystalline, and mimetically preserved the precursor fabric (Fig. 4). These two dolomites grade subtly into one another through a transitional variety termed crystalline microsucrosic (CMS), which exhibits petrographic characteristics intermediary between the CM and MS types. In the Miocene portion of the core, the crystalline dolomites were placed into an additional category, crystalline non-mimetic (CNM), on the basis of non-preservation of the precursor fabric. The MS

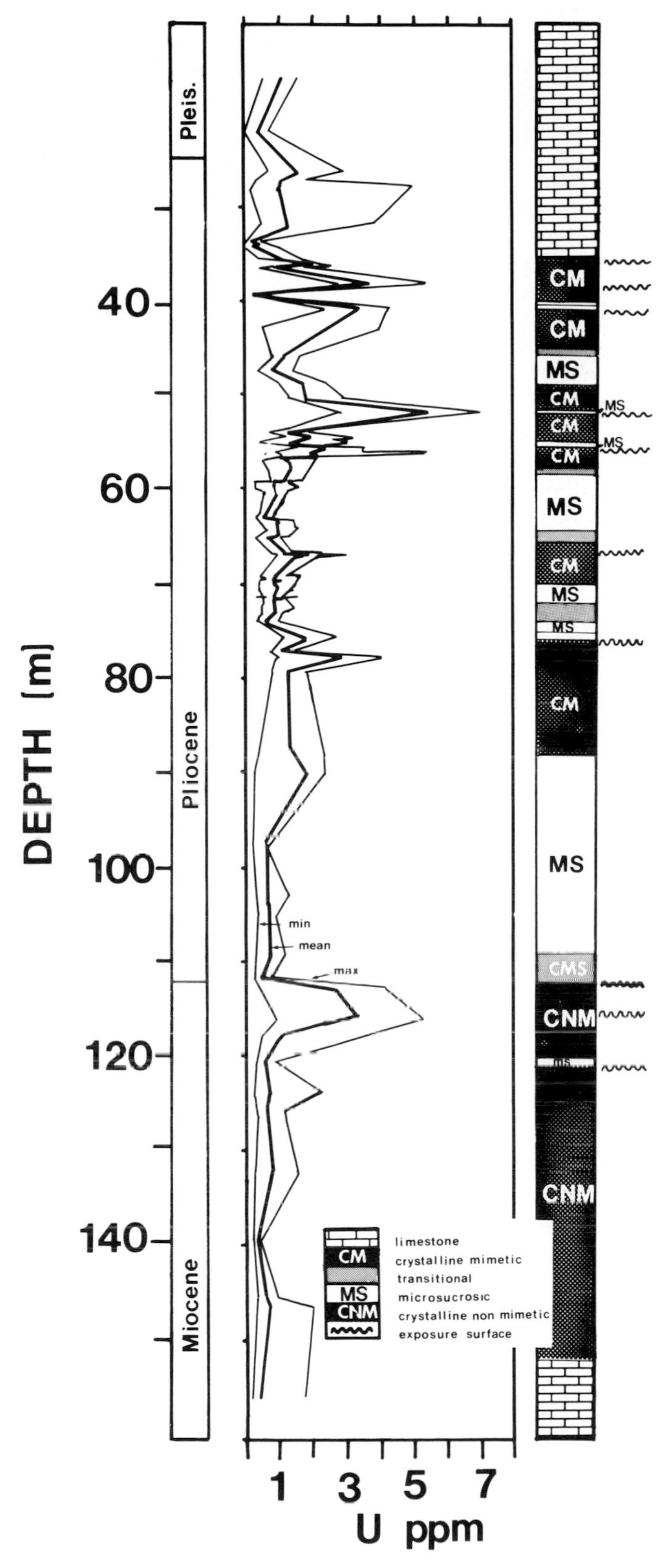

FIG. 5.—The distribution of textural types of dolomites (Dawans and Swart, 1988) and average U concentrations. The upper and lower lines represent the maximum and minimum U concentrations observed in 20 spot analyses (see text). For position of analyses, see Table 3. Stratigraphy from McNeill and others (1988).

and CM/CNM dolomites also showed differences in their major- (Mg and Ca) and minor-element (Sr, Mn, Fe, V, and Rare Earth Elements, and $\delta^{18}O$ values (Swart, unpub. data). The MS dolomites are consistently of near-stoichiometric composition (a summary of the petrographic and geochemical features of the dolomite types is shown in Table 1). The CM and MS dolomites form end members of a gradational series of textures that exhibit a series of alternations through the Pliocene and Miocene intervals of the core (Fig. 5).

The depositional textures and facies are similar to those described in late Cenozoic carbonates by Beach (1982), Pierson (1982), and Williams (1985) in other areas of the Bahamas. Each of the dolomite alternations is characterized by a shallow-upward sequence, capped either by a subaerial-exposure horizon or an erosional surface. Hence, the CM dolomites tend to replace peritidal to shallow subtidal marine carbonates underlain by coral-rich boundstones and packstones, whereas MS dolomites, at the bottom of the alternations, replace mostly lagoonal deposits composed of skeletal and nonskeletal wackestones and mudstones.

METHODS

Samples for this investigation were taken throughout the limestone and dolomite intervals of the core and embedded in polyester casting resin (Polylite Polyester Resin, 32–033 Cecra Coat). From these blocks, polished thin sections were prepared using epoxy (Epoxy Patch, Hysol Division, The Dexter Corporation) to mount the blocks to the slides. The depths from which these samples were taken are listed in Table 2. The thin sections were placed adjacent to LEXAN plastic detectors in an aluminum canister and irradiated with a total integrated thermal neutron flux of 10^{17} n/cm^2. During the irradiation, ^{235}U undergoes a fission reaction to produce fragments. As the thermal neutron fission cross section (σ_f) for ^{235}U [583 barns (b)] is significantly highly than either that of ^{238}U [4 μbarns (μb)] or ^{232}Th (2.5 μb) (Walker and others, 1983), contributions from fissioning of these elements are minor. These fragments leave tracks in the adjacent plastic detector (Fig. 6A) that can be made visible under an optical microscope by etching in a solution of 7N NaOH at a temperature of 70°C for 3 min. In order to calibrate the technique, standards were prepared from two NBS glasses, NBS-617 and NBS-615 (Carpenter and Reimer, 1974; U.S.Department of Commerce, 1981). The concentration of U can be calculated in a sample by the following formula:

$$\text{U(sample)} = \frac{\text{U(standard)} \times \text{D(sample)}}{\text{D(standard)}} \times \frac{\text{R(standard)}}{\text{R(sample)}} \quad (1)$$

TABLE 1.—TEXTURAL AND GEOCHEMICAL DESCRIPTION OF THE SAN SALVADOR DOLOMITES

Textural Type	Macroscopic Description	Matrix Appearance Under Binoculars	$MgCO_3$ mole %	$\delta^{13}C$ ‰	$\delta^{18}O$ ‰	Sr ppm
CM Crystalline mimetic	Very hard, dense, compact to porous (megaporosity is fabric selective after corals and molluscs), greyish to brownish dolomite. *Mimetic replacement of precursor fabric.*	Matrix composed of crystals tightly interlocking, allowing no visible intercrystalline pore space. Sharp edges and surfaces on breaking.	42.5 to 47.0	0.7 to 3.2	1.35 to 2.30	102 to 497
			m = 44.9 *s* = 1.2 *n* = 42	−*m* = 2.25 *s* = 0.55 *n* = 32	*m* = 1.74 *s* = .34 *n* = 32	*m* = 241 *s* = 79 *n* = 30
CNM Crystalline Non-mimetic	Very hard, dense, compact to porous (megaporosity is mainly vuggy to caveronous), greyish brown to greyish white dolomite, mottled with limestone. *Non-mimetic replacement of precursor fabric.*	Same as CM	43.0 to 45.0	0.7 to 3.4	1.40 to 2.70	198 to 224
			m = 43.8 *s* = 0.6 *n* = 15	*m* = 1.88 *s* = 0.86 *n* = 8	−*m* = 2.02 *s* = 0.45 *n* = 8	*m* = 207 *s* = 11 *n* = 5
MS Microsucrosic	Dull, earthy or chalky, soft to friable, highly porous, white dolomite. *Non-mimetic replacement of precursor fabric.*	Matrix has very fine texture and is composed of crystals (<20 μm) less effectively interlocking than in CM and CNM, joining at different angles and allowing visible pore space.	46.0 to 50.0	0.5 to 2.5	2.2 to 4.2	80 to 212
			m = 47.7 *s* = 1.03 *n* = 52	*m* = 1.97 *s* = 0.41 *n* = 40	−*m* = 2.71 *s* = .51 *n* = 40	*m* = 106 *s* = 20 *n* = 39
CMS	Transitional between CM and MS varieties. *Possess characteristics of both CM & MS types.*		45.0 to 48.2 *m* = 46.6 *s* = 1.1 *n* = 13	0.6 to 2.84 *m* = 2.03 *s* = 0.51 *n* = 12	1.6 to 2.8 *m* = 1.87 *s* = .71 *n* = 12	95 to 226 *m* = 132 *s* = 18 *n* = 7

m = mean, *s* = standard deviation, *n* = number of samples.

TABLE 2.—COMPARISON OF TRACK DENSITIES AND COMPUTED URANIUM AND BORON CONCENTRATIONS USING NBS-615 AND NBS-617 AS STANDARDS

	URANIUM			
	Density tracks/cm² × 10⁶	% 235	Calculated Value	NBS Value
NBS-615	2.836	.2792	.800 ± .04*	.823 ± .002
NBS-617	.515	.616	.074 ± .004**	.072 ± .001
	BORON			
	Density tracks/cm² × 10⁶		Calculated Value	NBS Value
NBS-615	9.000		1.24 ± .04*	1.30
NBS-617	1.446		.20 ± .01**	.20

*Value calculated using NBS-617 as a standard.
**Value calculated using NBS-615 as a standard.

In this equation D = the track density of either sample or standard, U = the concentration of uranium, and R = percent of ^{235}U. Because the two NBS standards not only have different uranium concentrations, but also different ^{235}U/^{238}U and ^{235}U/^{232}Th ratios, assessment can be made of the presence of more energetic neutrons during the irradiation. Such types of neutrons, if present, may fission ^{238}U and ^{232}Th, thereby leading to anomalous concentrations of U. Data for this comparison (Table 2) indicate that within the limits of uncertainty of the technique, the calculated concentrations of U agree between the two standards.

Determination of the track density for both standard and sample was carried out using a video camera and a CRT monitor. Two types of U determinations were made on the samples. The first type was analysis of individual components, each spot being 25 μm² in size. In the case of these analyses, the percent error is estimated to be $1/\sqrt{n}$, in which n = the number of tracks present in the area analyzed. In order to determine the concentration of a specific component within a section, a number of spots were averaged together. These data are shown in Table 3. The second type of analysis represents mean values of 20 random spots, each 2 mm apart, and is used to determine the bulk U concentration of the thin section. For all samples the mean and standard deviation values of U are presented. For analyses in which the tract distribution exhibits a Poisson distribution, the estimated error is also calculated (Table 3).

Determination of B by the nuclear-track method relies on some of the same principles as the U method, but with several important differences. First, the nuclear process utilized is the ^{10}B(n,α)^{7}Li reaction and hence the tracks produced are α tracks, not fission tracks. Although (n,α) reactions occur during the higher flux used for the fission reaction, the α tracks cannot be seen in the LEXAN plastic detector. Second, a lower neutron dosage is used (10^{13} n/cm²), which will not produce similar reactions in other elements. In order to detect α tracks, a cellulose nitrate detector is used (KODAK CN 85). The cellulose nitrate is etched in using a 2.5N NaOH solution at a temperature of 60°C for 30 min. Counting is carried out in a method analogous to that of U, and calibration is achieved using NBS-615 and NBS-617, which contain 1.3 and 0.2 ppm of B, respectively (Fig. 6B; Carpenter and Reimer, 1974).

The production of α tracks is also possible by other nuclear reactions, such as ^{6}Li(n,α)^{3}H, ^{235}U(n,f)α + f, ^{18}O(n,α)^{15}N, and ^{17}O(n,α)^{14}C. The most important of these is the ^{6}Li reaction as the cross-capture section (σ_α) for this reaction is only four times less than that for ^{10}B (941 vs. 3838 b; Walker and others, 1983). Taking into consideration the relative differences in abundance of ^{10}B and ^{6}Li compared to their more abundant isotopes, it can be expected that for a sample containing equal amounts of Li and B, approximately 10% of the α tracks would be produced by the Li reaction. Obviously, such a contribution would increase if the concentration of Li were greater, and it therefore is important that an estimate or measurement be made of Li in these samples. There have been only a few measurements of Li reported in the literature. For example, Weber (1964) measured a mean concentration of 15.2 ppm in 292 dolomite rocks (compared to between 66 and 337 ppm B in the same rocks) and a lower value of 3.4 ppm for separated dolomite minerals. Analyses on modern skeletal organisms such as corals are limited to a study by Liv-

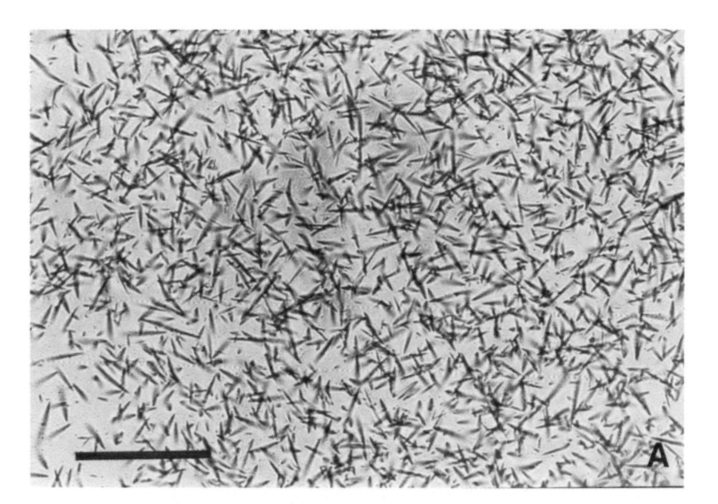

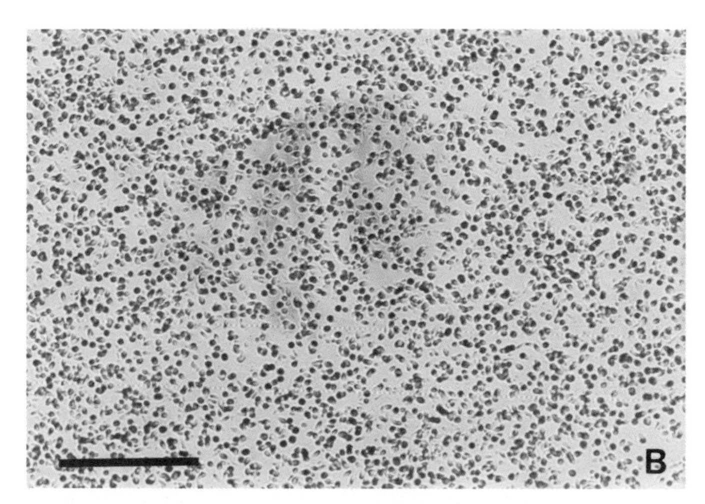

FIG. 6.—(A) Fission tracks in LEXAN plastic. Scale bar = 30 μm. (B) Alpha tracks in a cellulose nitrate detector CN 85. Scale bar = 30 μm.

TABLE 3.—URANIUM AND BORON DATA FROM THE SAN SALVADORE CORE

Depth (m)	Uranium mean (ppm)	Standard deviation	Relative Standard Error %	Textured type	Dolomite cement (ppm)		Calcite cement (ppm)		Red Algae (ppm)		Boron (ppm)	
					conc.	σ	conc.	σ	conc.	σ	conc.	σ
15.85	1.08	0.25	1.1	1								
21.74	0.39	0.16	3.3	1								
25.95	1.59	0.48	1.2	1			0.22	0.04				
26.86	1.11	0.45	2.0	1								
27.74	1.00	0.97	4.8	1			0.10	0.03				
31.83	1.32	0.73	2.4	1								
33.51	0.22	0.09	4.5	1								
33.72	0.42	0.13	2.3	1			0.14					
33.93	0.27	0.19	7.3	1			0.11					
35.43	1.07	0.31	1.4	c			0.18		0.62	0.08		
35.67	1.34	0.47	1.6	c								
35.98	2.27	0.27	0.4	c			0.04					
36.28	1.11	0.32	1.5	c								
37.87	3.85	0.65	0.4	c	0.14	0.04						
39.02	0.14	0.08	8.1	1								
40.55	3.43	0.42	0.3	c	0.80	0.05						
42.62	2.51	0.85	1.1	c	0.47	0.07						
45.79	1.17	0.27	1.1	s								
47.26	0.83	0.22	1.5	m								
48.72	1.67	0.34	0.8	c								
50.29	1.81	0.35	0.8	c								
51.74	5.81	1.16	0.6	c	0.66	0.05						
53.38	1.79	0.30	0.7	c								
53.72	1.23	0.33	1.2	c								
54.15	1.55	0.50	1.3	c					0.98	0.09	1.93	0.12
54.21	1.95	0.46	0.8	c	0.66	0.10						
54.48	1.70	0.32	0.9	c								
54.85	1.25	0.77	2.7	c							1.30	0.03
55.15	1.29	0.29	1.0	c							1.83	0.05
55.49	2.42	0.40	0.6	c	0.44	0.06			0.40	0.06		
55.79	2.04	0.57	1.0	c							1.80	0.05
56.07	1.84	1.03	2.1	c	0.59	0.06			1.32	0.12		
56.34	2.15	0.54	0.9	c								
56.52	0.99	0.18	1.0	s								
56.77	1.33	0.42	1.4	s								
57.53	1.35	0.38	1.3	s					1.93	0.09		
58.93	1.14	0.22	1.0	m								
59.05	0.87	0.23	1.4	m					0.80	0.04		
59.82	0.88	0.33	2.0	m					1.12	0.06		
60.09	0.96	0.19	1.0	m					1.11	0.07		
62.80	0.54	0.12	1.6	m					0.57	0.03		
63.11	0.93	0.29	1.6	m					1.46	0.08	1.23	0.04
64.02	1.00	0.20	1.0	m					0.89	0.03		
64.63	0.94	0.24	1.3	m					1.50	0.07		
64.94	0.62	0.20	2.0	m								
66.46	1.14	0.18	0.7	s					1.08	0.05		
66.71	2.08	0.50	0.8	c	0.60	0.06						
66.92	1.48	0.29	0.8	c	0.73	0.07						
68.99	0.85	0.18	1.3	m					0.84	0.04		
69.12	1.09	0.31	1.5	m					0.91	0.04		
69.42	0.83	0.33	2.3	m					1.86	0.13		
70.34	0.93	0.27	1.6	m					1.16	0.10	1.52	0.09
71.01	0.88	0.16	1.1	m							1.80	0.08
71.07	0.81	0.15	1.1	m								
71.10	0.93	0.21	1.3	m							1.45	0.05
71.16	0.80	0.19	1.3	m								
71.17	1.04	0.19	1.0	m							1.83	0.07
71.19	1.03	0.24	1.2	m							1.45	0.06
71.20	0.83	0.13	1.0	m							2.06	0.09
72.47	0.90	0.22	1.4	m								
72.90	0.78	0.17	1.4	m								
73.66	0.57	0.18	2.2	m								
75.37	1.91	0.39	0.8	c	1.27				2.12		1.50	0.05
76.83	0.99	0.16	0.9	c					0.90			
77.44	3.16	0.82	0.8	c	0.68							
78.96	1.22	0.26	1.0	c	1.06				0.97			
86.65	1.34	0.55	1.8	c					0.89			
89.73	1.71	0.57	1.4	c	0.81							
97.62	0.49	0.14	2.2	m					0.39			
102.44	0.64	0.22	2.3	m					0.92			
105.15	0.56	0.16	2.1	m					0.71			
109.24	0.67	0.18	1.6	m								
111.37	0.44	0.12	2.1	m					0.39			
112.47	2.64	1.22	1.5	n	0.62				3.55			
115.85	3.42	1.05	0.9	n	0.48							
117.35	1.16	0.57	2.3	n	0.35							

TABLE 3.—*Continued*

Depth (m)	Uranium mean (ppm)	Standard deviation	Relative Standard Error %	Textured type	Dolomite cement (ppm)		Calcite cement (ppm)		Red Algae (ppm)		Boron (ppm)	
					conc.	σ	conc.	σ	conc.	σ	conc.	σ
120.43	0.52	0.18	2.4	n	0.25				0.62	0.08		
123.48	0.71	0.49	4.5	n	0.60							
125.52	0.62	0.23	2.3	n			0.40				1.83	0.12
131.86	0.82	0.38	2.7	n	0.33							
137.20				n							2.75	0.18
138.11	0.51	0.15	2.0	n	0.34						2.20	0.13
142.23				n							2.08	0.09
145.12	0.60	0.18	2.1	n	0.39							
146.04	0.72	0.55	4.5	n	0.40						2.33	0.15
155.49	0.44	0.33	6.1	l	0.60		0.32				2.25	0.10

Relative standard error = $[(\sigma/\sqrt{n} \div \text{mean}] \times 100$, in which n = number of tracks counted and σ = standard deviation of field of views (Naeser and others, 1979). All bulk analyses were tested for a Poisson distribution using the Kolmogorov-Smirnov test. Only samples from 27.7, 54.8, and 146 m did not exhibit a Poisson distribution at the 99.9% confidence limits.

ingston and Thompson (1971), in which Li concentrations were only 1 ppm, approximately an order of magnitude less than B values reported by the same workers. The utility of such information as applied to this investigation, however, is suspect, because the B values measured in modern organisms during this investigation are generally much lower than those previously reported. Nevertheless, Li concentrations probably do not exceed those of B in carbonate rocks and, hence, contributions to the α track density from ^{6}Li should not exceed 10%. In this study, which is similar to that by Truscott and Shaw (1984), no correction has been applied for possible contribution for ^{6}Li(n,α)^{3}H.

In order to ascertain whether there was any association between uranium and organic material in the carbonates, the percent carbonate and organic matter was determined in a selection of samples from between 60- and 80-m depths. The carbonate determinations were performed by treating the sample with HCl in a LECO filtering crucible, then drying and weighing. The weight loss is reported as weight percent carbonate. The carbonate-free residue is then burned in a LECO carbon analyzer to determine weight percent organic carbon.

RESULTS

Uranium

The San Salvador dolomites exhibit a range of U concentrations varying between 0.5 and 6 ppm (Table 3, Fig. 5). In general, the CM dolomites are consistently enriched in U and extremely heterogeneous, compared to the MS and CMS dolomites (see Table 4). Dolomitized grains, such as peloids and ooids, which exhibit preservation of original fabrics, contain U concentrations (2 to 3 ppm) comparable to, or in excess of, known concentrations in modern analogues. Edges of grains are frequently enriched in U, perhaps a consequence of micritization. Void-filling cements in the same rocks, however, are consistently low, between 0.5 and 1 ppm (Fig. 7). In contrast, the MS dolomites are comparatively homogeneous, possessing bulk concentrations between 0.5 and 1 ppm. The only exception to this uniform distribution in the MS dolomites is red algal fragments, which occasionally exhibit U concentrations as high as 1.5 ppm (Fig. 8). The difference between the CM and MS dolomites is most pronounced in the upper portion of the dolomitized interval and diminishes with depth. The

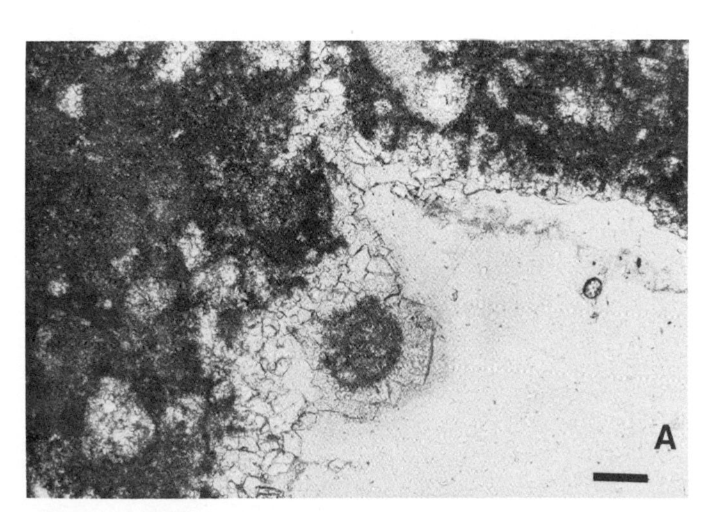

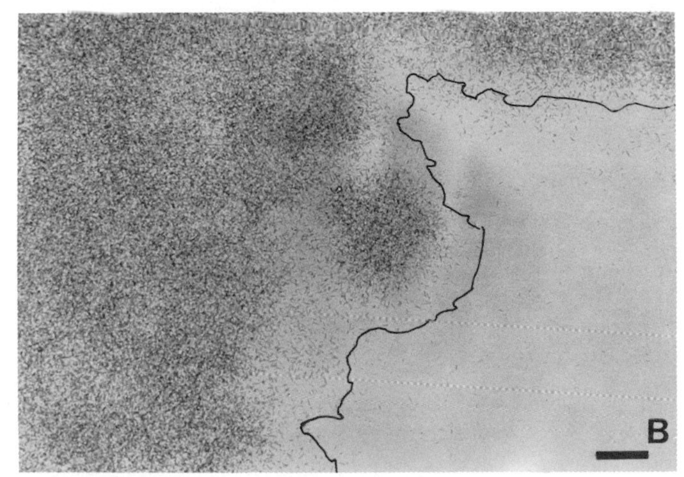

FIG. 7.—Photomicrograph of CM dolomite from a depth of 51.7 m (A) showing void-lining dolomite cement and (B) fission-track point corresponding to view in (A). Note low concentration of tracks in cements. Scale bar = 100 μm.

TABLE 4.—AVERAGE U AND B IN VARIOUS TEXTURAL TYPES, MINERALOGIES, AND SEDIMENTARY COMPONENTS

Type	Bulk	Cement	Red Algae	Boron
CM	2.00(0.99)	0.69(0.27)	1.04(0.48)	1.61(.22)
CMS	1.26(0.18)	—	1.33(0.43)	1.93(–)
MS	0.82(0.18)	—	0.98(0.40)	1.62(.26)
CNM	1.17(0.96)	0.42(0.12)	2.09(1.47)	2.24(.30)
L	0.73(0.48)	0.18(0.08)	—	2.25(–)

Values in parentheses refer to standard deviation of analyses.

TABLE 5.—CONCENTRATION OF ORGANIC MATERIAL IN SOME DOLOMITES FROM SAN SALVADOR

Sample Depth (*m*)	TOC (Wt. percent)	Carbonate (Wt. percent)
55.53	.02/.05*	100.5/98.1*
56.97	.02	100.2
57.03	.05	99.4
59.25	.04	100.1
61.23	.08	100.1
63.86	.08/.06*	99.5/98.9*

TOC = total organic carbon.
*Replicate analyses.

concentration of U reaches a maximum within the bulk rock and individual sedimentary components coincident with exposure horizons and, in fact, such surfaces can be easily identified by plotting bulk U concentration against depth. The CNM dolomites, which are present only in the Miocene section of the core, exhibit a distinctly bimodal distribution of uranium concentrations, one with a mean of 0.5 ppm and the other 2.5 ppm.

The limestone section above 33 m exhibits a gradual decrease in mean U concentration with depth toward the dolomite–limestone interface at which bulk limestone values are slightly higher than 0.1 ppm. High concentrations of U in the limestone section coincide with exposure surfaces. Limestones below the dolomitized interval display U concentrations less than 0.4 ppm.

Percent Organic Material

The percent organic material in the dolomites is low, between 0.02 and 0.08 weight percent. The measured values are reported in Table 5.

Boron

The analysis of B by the nuclear-track method was complicated as B appeared to be readily adsorbed onto crystal surfaces and trapped as relict sea water within small pores. This produced high track densities along crystal faces and in dolomites with high porosities or with fine grain sizes. The polyester casting resin used in this study did not produce high concentrations of tracks as has been observed in samples impregnated in epoxy (Truscott and Shaw, 1984), but because epoxy was used to attach the samples to the slide, contributions were visible in samples that were slightly wedged, thereby allowing the epoxy holding the sample to the slide to come into contact with the detector.

In the Pliocene section of the core, no significant differences in B are observed between the CM and MS dolomites. In fact, B concentrations are uniformly low (1.2 to 2.1 ppm), with the exception of micritic infill of pores, which occurred close to exposure horizons (3–4 ppm). In the Miocene portion of the core, which contains coarser grained dolomites, B concentrations are higher (1.8 to 2.8 ppm; Fig. 10). Partially dolomitized limestones beneath the main dolomitized interval contain as much as 2.8 ppm B, whereas the calcite possesses less than 0.05 ppm (Fig. 11).

Samples of modern organisms were also analyzed in order to establish concentrations in some marine carbonates using the α-track method. These data, shown in Table 5, reveal considerably lower concentrations than reported previously for modern marine carbonates (Milliman, 1974).

DISCUSSION

Uranium

A significant observation made in this investigation is the extremely high and heterogeneous concentrations of U present in many of the rocks which, when analyzed by Supko (1977) and Dawans and Swart (1988), were reported to be 100%

A

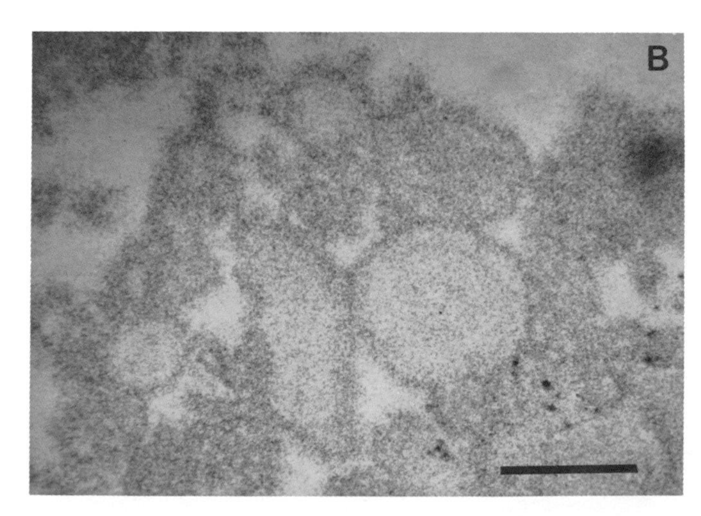

FIG. 8.—Photomicrograph (A) and fission-track print (B) of a sample of CNM dolomites from a depth of 112.4 m. Higher track density can be seen in micritic rim surrounding sediment grains. Scale bar = 500 μm.

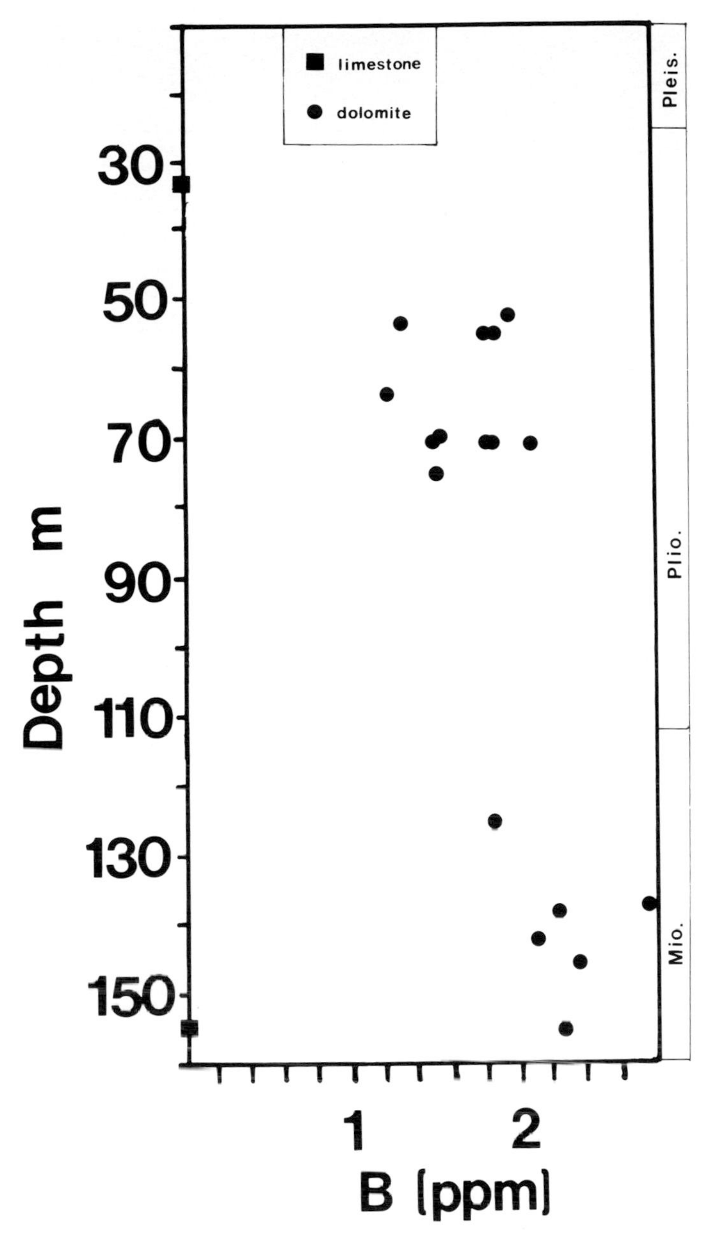

FIG. 9.—Change in the concentration of B with depth.

dolomite. These concentrations are surprising for two reasons. First, as dolomite is a hexagonal carbonate, it might be expected to have relatively low concentrations of the larger radii elements, such as U. Second, the concentrations of U in some of the dolomitized components are in many instances higher than measured in modern analogues.

One possible explanation of the U concentrations of these dolomites may lie in the fact that U is often complexed with the CO_3^{2-} ion in waters of a pH and Eh near that of sea water (Langmuir, 1978). As the concentrations of HCO_3^- and CO_3^{2-} ions in sea water are extremely low (10^{-4} and 10^{-5} m, respectively; Stumm and Brauner, 1975), it is probable that during dolomitization the CO_3^{2-} ion is conserved, especially during fabric-preserving reactions (Morrow, 1982). Therefore, practically all the U released during the dissolution of the precursor carbonates is in the complexed carbonate form, and the U concentration of the original components is inherited by the dolomitized sediment. Such an argument implies that the U concentration of the precursors was not depleted prior to dolomitization and that their mineralogy was still largely aragonite and HMC.

Reduced U concentrations could be produced either (1) under conditions of greater fluid supply, or (2) through the dolomitization of a precursor already low in U, such as LMC. An enhanced supply of dolomitizing fluids would also increase the amounts of allochthonous HCO_3^- and CO_3^{2-} ions and reduce the $UO_2(CO_3)_3^{2-}/CO_3^{2-}$ ratio and the U concentration of the dolomite. This hypothesis, however, is not in agreement with the observations and conclusions of Dawans and Swart (1988), who propose that, because the MS dolomites were fabric destructive, possessed high porosity, and were of near-ideal chemical composition, they formed under conditions of lower fluid flow and, hence, reduced the allochthonous CO_3^{2-} ion input (similar to equation 2):

$$2CaCO_3 + Mg^{2+} = CaMg(CO_3)_2 + Ca^{2+}. \quad (2)$$

Dolomitization with an excess supply of CO_3^{2-} ions would proceed according to a reaction similar to equation (3), would result in a net reduction in the porosity of 76% (Morrow, 1982), and may be responsible for the formation of the CM dolomites:

$$CaCO_3 + CO_3^{2-} + Mg^{2+} = CaMg(CO_3)_2. \quad (3)$$

These rocks have most of their inter- and intraparticle porosity filled with dolomite cement. This cement could have formed either according to equation (2), producing an U concentration similar to that of the precursor, or equation (3), in which case the U concentration would be partially dictated by the $UO_2(CO_3)_2^{2-}/CO_3^{2}$ ratio of the dolomitizing fluid and an appropriate distribution coefficient. An alternative scenario for the difference between the MS and CM dolomites is that the precursor of the MS dolomite had already been largely converted to LMC by fresh water prior to dolomitization, whereas the precursors of the CM dolomite had been leached, but only partially stabilized to LMC. During dolomitization, the diagenetically altered precursors of both the textural types retained their U concentrations. The differential reactivity of the calcitic and aragonitic components resulted in the production of dolomites with varying mole percent $MgCO_3$ and fabrics. This scenario concurs both with the interpretation of Dawans and Swart (1988), and the fact that the limestones immediately overlying the dolomitized interval, which have now been completely converted to LMC, contain concentrations of U lower than 0.5 ppm.

Cements.—

An often controversial question is whether dolomite cements such as those observed in the San Salvador core (Fig. 7) are replaced calcite, and therefore not true cements, or primary dolomite. Using the difference in U concentration between the calcite and dolomite cements (<0.1 ppm vs. 0.5 to 1 ppm), it might be speculated that the dolomites are not replaced calcite, but rather primary in origin. If the

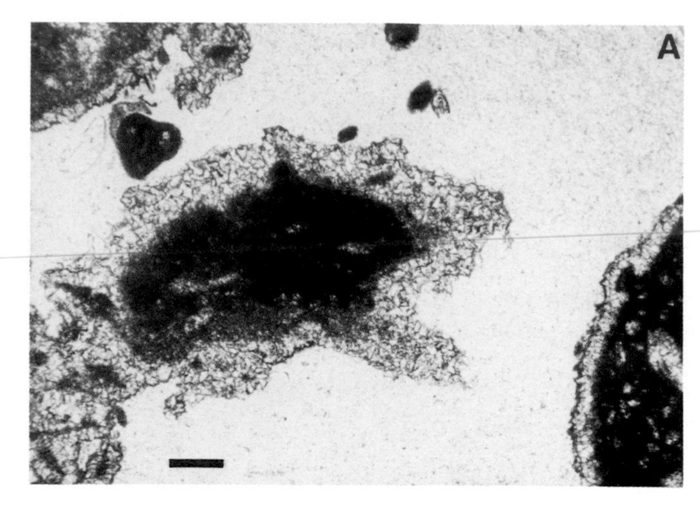

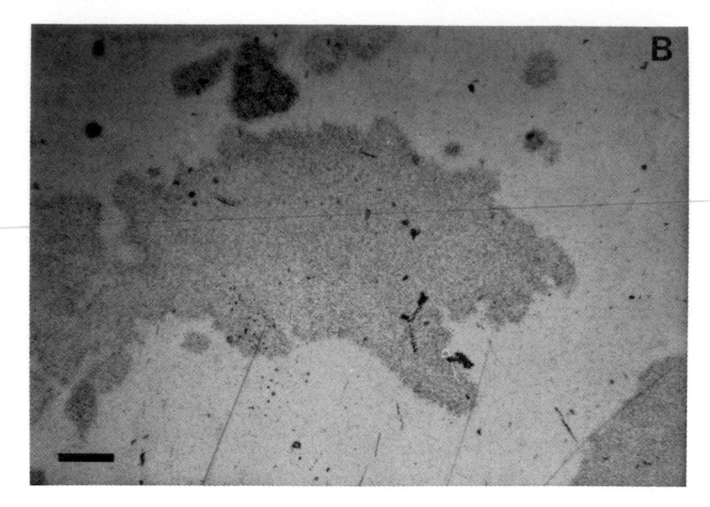

FIG. 10.—(A) Photomicrograph of CM dolomite (55.8 m) and (B) distribution of α tracks as shown in (A). Note homogeneous distribution of tracks. Scale bar = 200 μm.

dolomite is replacive, then concentrations of U might be expected to be similar or even lower than the original calcite. This is not the case, however. The U concentration of the cements, therefore, probably indicate primary precipitation from a seawater-like medium.

Carbon-isotopic composition.—

If the carbonate ions now present in the dolomite were derived mainly from the precursor, then there should be only a minimal change in the carbon-isotopic composition of the CM dolomites as compared to those of modern analogs, but a larger shift to more depleted isotopic values in the MS rocks. In fact, carbon-isotopic data on these dolomites (Dawans and Swart, 1988) show that the $\delta^{13}C$ of both types of dolomite lies between +1.5 and +3‰ (PDB), with the MS dolomites being only slightly depleted; comparisons with bulk $\delta^{13}C$ values of modern reef carbonates, such as in the study of Weber and Woodhead (1969), indicate $\delta^{13}C$ values for such sediments of between +1 and +3‰. The absence of a larger $\delta^{13}C$ depletion in the MS dolomites, which might be expected if the MS dolomites had been stabilized to LMC prior to dolomitization, indicates that either the paragenetic sequence suggested above is incorrect, or that CO_3^{2-} ions supplied by the dolomitizing solution are in isotopic equilibrium with a larger reservoir of HCO_3^- and therefore not strongly influenced by the carbon-isotopic composition of the carbonate dissolved from the precursor.

Organic material.—

The association between organic material and U, which is frequently documented in the literature (Taylor, 1979), led to speculation as to whether the differential U concentrations that were visible in the dolomites were related to varying amounts of organic matter; however, no preferential associations between U and percent organic material were observed in the dolomites examined in this investigation (Table 4). Concentrations of organic material were

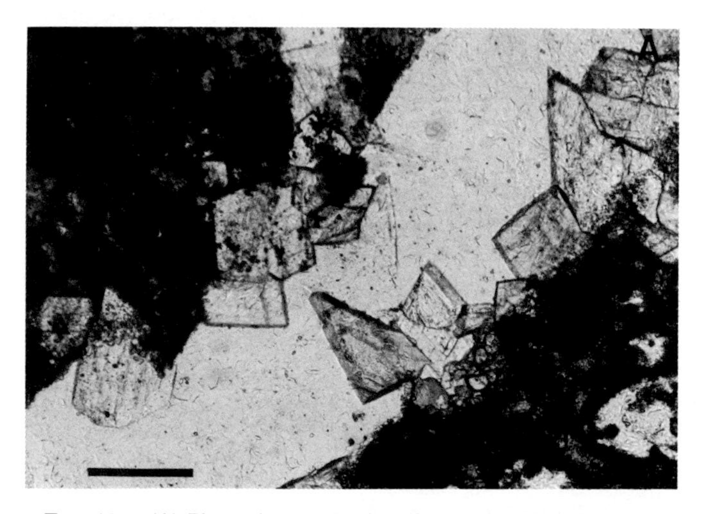

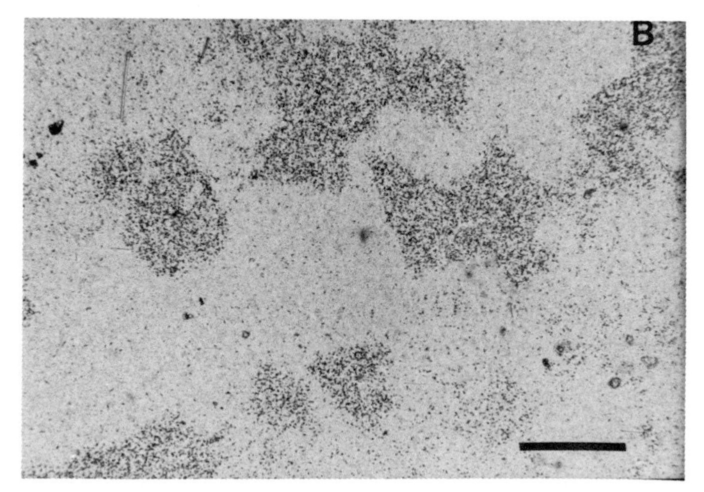

FIG. 11.—(A) Photomicrograph of partially dolomitized limestone near bottom of core (155.45 m). Section has been stained with Alizarin Red and the limestone appears darker. (B) Alpha-track distribution shows boron concentrated in dolomite lining fracture. Scale bar = 250 μm.

uniformly low in all the samples analyzed and could have resulted entirely from surface contamination during handling of the sample.

Association with exposure surfaces.—

A second source of the high U concentrations may be provided by the abundant exposure surfaces throughout the core (Fig. 5). These surfaces possibly act as sinks of U during deposition, either as a result of the accumulation of uranium-rich windblown dust, or U uptake by biologically associated activities. Dolomites slightly above and below all exposure surfaces are enriched in U compared to the rest of the core, and the U enrichment provides an easily recognizable indicator of exposure surfaces, even when petrographic and sedimentologic features are ambiguous (Fig. 5). For example, at the top of the Miocene section, two exposure horizons were identified by Dawans and Swart (1988) on the basis of changes in facies and dolomite texture, but without direct sedimentologic features. Both these surfaces were found in this study to correspond to high concentrations of U, suggesting that they were relict exposure surfaces.

Comparisons with modern analogs.—

As stated in the introduction, concentrations of U in modern "live" organisms appear to be lower than in "dead" organisms. In corals, there is an increase from values of between 1.5 and 2.0 ppm in live individuals to concentrations in excess of 3 to 4 ppm upon removal of the protective organic tissues. This phenomenon has also been noticed in molluscs and may be a general feature of skeletal carbonates. For example, the concentration of U in the red algal (0.5 to 3.5 ppm) components from San Salvador is considerably higher than that found in modern examples (0.5 ppm), as measured by Gvirtzman and others, (1973) and Swart and Hubbard (1982). One explanation is that upon death, U is adsorbed onto the skeleton, precipitated in small skeletal voids, and/or concentrated within organic components in the skeleton. Eventually, the U concentration reaches a value typical of the red algae seen in the core, and upon dolomitization, this value is preserved. An alternative explanation is that, since the red algae are typically the first components to be stabilized to LMC (Land, 1967), this transformation occurs in an environment which contains abundant U leached from metastable aragonitic components or derived from exposure surfaces. Hence, the neomorphosed skeletons contain higher concentrations of U, which is again preserved during dolomitization. An association with organic material is considered unlikely because the percentage of organic material in all dolomites examined is extremely low (Table 4).

Comparisons with previous studies.—

The range of U concentrations measured in the San Salvador dolomites is comparable to that determined by Rodgers and others (1982) for rocks of similar age from the Island of Niue in the South Pacific. More recent work by Aharon and others (1987) on the same rocks has revealed many other geochemical resemblances to San Salvador, including Sr, O, and C isotopic values, suggesting perhaps a similar paragenetic sequence and mechanism of dolomitization. Rodgers and others (1982) concluded on the basis of the differences between co-occurring calcite and dolomites that the high uranium concentrations in the Niue dolomites were introduced by the dolomitizing fluids rather than inherited from the carbonate precursors. Although similar differences exist in San Salvador, the data presented in this paper show that dolomites with very different U concentrations are closely associated and probably resulted from the same episode of dolomitization (Swart and others, 1987). If U is supplied by the dolomitizing fluids, then it is likely that all the rocks should have similar U concentrations and certainly should not vary by an order of magnitude, as was observed in this study. The low U concentrations of the limestones may have resulted from a later diagenetic event.

TABLE 6.—CONCENTRATION OF BORON IN SOME MARINE CALCAREOUS ORGANISMS

Organism	Locality	B ppm (±sd))
Coral (*Acropora sp.*)	Great Barrier Reef	11.5 ± .81
Halimeda	Great Barrier Reef	5.1 ± .54
Red algae	Great Barrier Reef	3.4 ± .43
Ooids	Bahamas	4.9 ± .52
Oyster	Florida	3.0 ± .23
Barnacles	Florida	1.3 ± .09
Calcite cements	Miami Limestone	0.8 ± .12

Boron

The measurement of B in ancient and modern carbonate rocks is fraught with difficulties both in terms of the analytical technique and the interpretation. As mentioned earlier, reported B values in carbonates extend from 1 to 240 ppm. Values measured in this investigation, however, on both modern skeletal components and the rocks from the San Salvador core, are considerably lower. The low values in this study reflect the ability of the nuclear-mapping technique to isolate contamination and, hence, exclude it from the analysis. The concentrations of B measured in modern calcareous organisms range from 1.3 ppm in barnacles to 11.5 ppm in scleractinian corals (Table 5). Although these data suggest slightly lower concentrations of B in LMC organisms, the fine crystal structure of aragonite may allow relict sea water to be trapped between adjacent crystals. On the other hand, carbonates that formed on freshwater-dominated environments exhibit extremely low B values, less than 0.8 ppm.

The relatively high concentrations of B in the dolomites from San Salvador suggest formation from a fluid that has a B content similar to that of a marine composition. The absence of any difference in B concentration between the MS and CM types further infers that the dissimilarity of textures between the CM and MS dolomites cannot be explained through major changes in the salinity of the dolomitizing fluids.

The concentration of B is higher in the Miocene CNM dolomites (Fig. 9) than in the overlying Pliocene rocks. The higher B content is apparently related to the nature of the dolomitizing fluids, because samples that contain both limestone and dolomites show very low concentrations of B in the limestones (0.05 ppm), while exhibiting high values in the dolomites (3 ppm). This increase in B cannot be related to an artifact, such as crystal size, because the rhombs in this type of dolomite are frequently 200 to 500 μm in size (Fig. 11), whereas the limestone precursor is generally finer grained. Although no conclusive explanation can be offered at present for the higher B values in the CNM dolomites, it might be speculated that, as a separate episode of dolomitization was proposed for the Miocene (Swart and others, 1987), the fluids involved also possessed a slightly higher salinity.

CONCLUSIONS

(1) Uranium concentrations in the San Salvador dolomites show distinct differences between the CM and MS textural types (2.0 vs. 0.8 ppm). In the CM dolomites, texturally preserved components contain concentrations at least as high as modern analogs, whereas the cements possess considerably reduced levels but still higher than calcitic cements in the upper portion of the core. The MS dolomites are more uniform in their U concentration (0.5 to 1.1 ppm) with occasional red algal fragments possessing concentrations as high as 1.5 ppm.

(2) The CNM dolomites in the Miocene portion of the core contain a bimodal pattern of U distribution, one between 0.5 and 0.8 ppm and the other between 2.5 and 3 ppm. These patterns relate to preservation of the original U concentrations of the precursor.

(3) The patterns of U distribution suggest that the CM dolomites were produced through direct dolomitization of a uranium-rich metastable precursor, whereas the MS rocks were depleted of U during calcitization before dolomitization. Comparison of the U concentration of the dolomitic and calcitic cements indicates that the dolomites are not replacement, but rather direct precipitates.

(4) No preferential association was observed between dolomites with higher U concentrations and their organic matter content, indicating that organic material is not the host of U in these rocks.

(5) High B concentrations (1 to 3 ppm) compared to low values in the overlying limestones (0.1 to 0.05 ppm) indicate that dominantly marine fluids were the dolomitizing agents.

ACKNOWLEDGMENTS

The author thanks Gong Chung and Steve Harrison for help with nuclear-track analyses. This research was supported by National Science Foundation Grants EAR-8419359, EAR-8676088, and the Petroleum Research Fund (18022-AC2). Additional support was provided by Dr. R. N. Ginsburg and the Industrial Associates of the Comparative Sedimentology Laboratory. Analyses of total organic matter were provided courtesy of Marathon Oil and P. D. Crevello. Irradiations were provided by Georgia Institute of Technology under the University Reactor Sharing scheme. I especially acknowledge the assistance of Dr. A. Karem in this regard. The manuscript benefited from reviews by Drs. D. E. Fisher, C. W. Naeser, and V. Shukla. This is a contribution from the Stable Isotope Laboratory and the Comparative Sedimentology Laboratory in the Division of Marine Geology and Geophysics at the Rosenstiel School of Marine and Atmospheric Sciences.

REFERENCES

AHARON, P. SOCKI, R. A., AND CHAN L., 1987, Dolomitization of atolls by sea water convection flow: Test of a hypothesis at Niue, South Pacific: Journal of Geology, v. 95, p. 187–203.

AMIEL, A. J., MILLER, D. S., AND FRIEDMAN, G. M., 1972, Uranium distribution in carbonate sediments of a hypersaline pool, Gulf of Elat, Red Sea, Israel: Journal of Earth Sciences, v. 21, p. 187–191.

———, ———, AND ———, 1973, Incorporation of uranium in modern corals: Sedimentology, v. 20, p. 523–528.

BEACH, D. K., 1982, Depositional and diagenetic history of Plio-Pleistocene carbonates of northwestern Great Bahamas Bank: Evolution of a carbonate platform: Unpublished Ph.D. Dissertation, University of Miami, Miami, Florida, 452 p.

BROECKER, W. S., 1963, A preliminary evaluation of uranium series disequilibrium as a tool for absolute age measurement on marine carbonates: Journal of Geophysical Research, v. 688, p. 2817–2834.

BURKE, W. H., DENISON, R. E., HEATHERINGTON, E. A., KOEPNICK, R. B., NELSON, H. F., AND OTTO, J. B., 1982, Variation of seawater $^{87}Sr/^{86}Sr$ throughout Phanerozoic time: Geology, v. 10, p. 516–519.

CARPENTER, B. S., AND REIMER, G. M., 1974, Calibrated glass standards for fission track use: NBS Special Publication 260–49, U.S. Department of Commerce, U.S. Government Printing Office, 16 p.

CHRIST, C. L., AND HARDER, H., 1978, Boron, *in* Wedepohl, K. H., ed., Handbook of Geochemistry: Springer-Verlag, Berlin, p. 5.A.1 to 5.0.3.

CHUNG, G. S., 1988, The application of the nuclear track mapping of uranium to the study of diagenesis in carbonate rocks: Unpublished Ph.D. Dissertation, University of Miami, Miami, Florida, 256 p.

———, EVANS C., AND SWART, P. K., 1985, Uranium as an indicator of diagenesis and water flow in the Pleistocene Miami Limestone: Geological Society of America, Program with Abstracts, v. 17, p. 546.

———, AND SWART, P. K., 1986, Uranium as a diagenetic indicator in carbonate systems: Geological Society of America, Program with Abstracts, v. 18, p. 564.

CROSS, T. S., and CROSS, B. W., 1983, U, Sr, and Mg in Holocene and Pleistocene corals *A. palmata* and *M. annularis:* Journal of Sedimentary Petrology, v. 53, p. 587–594.

DAWANS, J. M. AND SWART, P.K. 1988, Textural and geochemical alternations in Late Cenozoic Bahamian dolomites, Sedimentology v. 35, 385–403.

DEPAOLO, D. J., 1986, Detailed record of the Neogene Sr isotopic evolution of sea water from DSDP Site 590B: Geology, v. 14, p. 103–106.

GARRELS, R. M., AND THOMPSON, M. C., 1962, A chemical model for seawater at 20°C and at one atmosphere pressure: American Journal of Science, v. 260, p. 57–66.

GVIRTZMAN, G., FRIEDMAN, G. M., AND MILLER, D. S., 1973, Control and distribution of uranium in coral reefs during diagenesis: Journal of Sedimentary Petrology, v. 43, p. 983–997.

HAGLUND, D. S., FRIEDMAN, G. M., AND MILLER, D. S., 1969, The effect of freshwater on the redistribution of uranium in carbonate sediments: Journal of Sedimentary Petrology, v. 39, p. 1283–1296.

LAHOUD, J. A., MILLER, D. S., AND FRIEDMAN, G. M., 1966, Relationship between depositional environment and uranium concentration of Molluskan shells: Journal of Sedimentary Petrology, v. 36, p. 541–547.

LAND, L. S., 1967, Diagenesis of skeletal carbonates: Journal of Sedimentary Petrology, v. 37, 6. 914–930.

LANGMUIR, D., 1978, Uranium solution-mineral equilibria at low temperature with applications to sedimentary ore deposits: Geochimica et Cosmochimica Acta, v. 42, p. 547–569.

LIVINGSTON, H. D., AND THOMPSON, G., 1971, Trace element concentrations in some modern corals: Limnology and Oceanography, v. 16, p. 786–796.

MCNEILL, D. F., GINSBURG, R. N., CHANG, S. B. R., AND KIRSCHVINK, J. L., 1988, Magnetostratigraphic dating of shallow-water carbonates: Geology, v. 16, p. 8–12.

MILLIMAN, J. D., 1974, Marine Carbonates: Springer-Verlag, Berlin, 375 p.
MORROW, D. W., 1982, Diagenesis, 1. Dolomite. The chemistry of dolomitization and dolomite precipitation: Geoscience, v. 9, p. 5–13.
NAESER, C. W., GLEADOW, A. J. W., AND WAGNER, G. A., 1979, Standardization of fission-track data reports: Nuclear Tracks, v. 3, p. 133–136.
PIERSON, B. J., 1982, Cyclic sedimentation, limestone diagenesis and dolomitization in upper Cenozoic carbonates of the southeastern Bahamas. Unpublished Ph.D. Dissertation, University of Miami, Miami, Florida, 312 p.
RODGERS, K. A., EASTON, A. J., AND DOWNES, C. J., 1982, The chemistry of carbonate rocks of Niue Island, South Pacific: Journal of Geology, v. 90, p. 645–662.
SUPKO, P. R., 1970, Depositional and diagenetic features in subsurface Bahamian rocks: Unpublished Ph.D. Dissertation, University of Miami, Miami, Florida, 168 p.
———, 1977, Subsurface dolomites, San Salvador, Bahamas: Journal of Sedimentary Petrology, v. 47, p. 1063–1077.
STUMM, W., AND BRAUNER, P. A., 1975, Chemical speciation, *in* Riley, J. P., and Skirrow, G., eds., Chemical Oceanography: Academic Press, London, p. 173–239.
SWART, P. K., 1984, U, Sr, and Mg in Holocene and Pleistocene corals—Discussion: Journal of Sedimentary Petrology, v. 54, p. 326–329.
———, AND HUBBARD, J. A. E. B., 1982, Uranium in scleractinian corals: Coral Reefs, v. 1, p. 13–19.
———, RUIZ, J., AND HOLMES, C., 1987, Use of strontium isotopes to constrain the timing and mode of dolomitization of upper Cenozoic sediments in a core from San Salvador, Bahamas: Geology, v. 15, p. 262–265.
TAYLOR, G. H., 1979, Biogeochemistry of uranium minerals, *in* Trudinger, P. A., and Swaine, D. J., Biogeochemical cycling of mineral forming elements: Studies in Environmental Science, v. 2, p. 485–514.
TRUSCOTT, M. G., AND SHAW, D. M., 1984, Boron in chert and Precambrian siliceous iron formations: Geochimica et Cosmochimica Acta, v. 48, p. 2313–2320.
UNITED STATES DEPARTMENT OF COMMERCE, 1981, NBS Standard Reference Materials Catalog, 1981–1983 Edition: NBS Special Publication 260, U.S. Government Printing Office, Washington, D.C., 114 p.
WALKER, F. W., MILLER, D. G., AND FEINER, F., 1983, Chart of the Nuclides: General Electric Company.
WEBER, J. N., 1964, Trace element composition of dolostones and dolomites problem: Geochimica et Cosmochimica Acta, v. 28, p. 1817–1868.
———, AND WOODHEAD, P. M. J., 1969, Factors affecting the carbon and oxygen isotopic composition of marine carbonate sediments, II. Heron Island, Great Barrier Reef, Australia: Geochimica et Cosmochimica Acta, v. 33, p. 19–38.
WILLIAMS, S. C., 1985, Stratigraphy, facies evolution and diagenesis of late Cenozoic limestones and dolomites, Little Bahama Bank, Bahamas: Unpublished Ph.D. Dissertation, University of Miami, Miami, Florida, 217 p.

EXPERIMENTAL INVESTIGATION OF SULFATE INHIBITION OF DOLOMITE AND ITS MINERAL ANALOGUES[1]

DAVID W. MORROW AND BRIAN D. RICKETTS

Institute of Sedimentary and Petroleum Geology, Geological Survey of Canada, 3303-33rd Street N.W. Calgary, Alberta T2L 2A7

ABSTRACT: Time series experiments relating to the dolomitization of calcite at 215° to 225°C in saline solutions of near-seawater salinity were conducted to ascertain the influence of sulfate and carbonate in solution on the rate of calcite dolomitization. A concentration of about 0.004M sulfate in solution prevented the dolomitization of calcite. At concentrations of less than 0.004M, dolomitization proceeded at a slower rate than in experiments where no sulfate was present. The final concentration of sulfate was controlled by the precipitation of anhydrite.

The presence of sulfate in solution did not prevent the direct precipitation of dolomite in experiments in which the solid reactants were carbonate minerals other than calcite ($BaCO_3$ and $2PbCO_3 \cdot PbOH$). Also, the presence of sulfate in the calcite dolomitization experiments slowed the rate of calcite dissolution from 3 days in sulfate-free solutions to 6 or 7 days in sulfate-bearing solutions. These observations indicate that sulfate in solution may inhibit dolomitization primarily by retarding the rate of calcite dissolution, rather than by inhibiting the direct precipitation of dolomite from solution.

The rate of calcite dolomitization was greater in solutions with higher carbonate/bicarbonate concentrations. This provides some confirmation for hypotheses regarding the importance of carbonate in solution as a kinetic factor that expedites dolomitization.

INTRODUCTION

Dolomite, an important rock-forming mineral, has worldwide importance as a host rock for hydrocarbons and for base-metal deposits. In spite of the economic significance of this mineral, our understanding of its origin in terms of geologic processes is limited by uncertainties in our understanding of the chemical controls on the precipitation of dolomite from solution. The chemical controls on dolomite precipitation at near-surface conditions of temperature and pressure have been the subject of continued interest, because of the inability of experimentalists to obtain unequivocal precipitates of dolomite at temperatures less than 100°C (Lippman, 1973; Land, 1980) in spite of the fact that dolomite is greatly oversaturated in most natural solutions (Carpenter, 1980). Consequently, there has been a determined search for chemical variables that retard the precipitation of dolomite. Such variables, which affect the rate of a chemical reaction, are commonly referred to as kinetic factors.

On the basis of experiments by Baker and Kastner (1981), Kastner (1983, 1984) has inferred that the sulfate concentration of natural solutions is the major kinetic factor in governing whether dolomitization will occur in a particular geologic setting. An interesting aspect of this kinetic factor is that it was first identified in high-temperature experiments (>100°C) and subsequently inferred to affect dolomite precipitation at lower temperatures. The efficacy of dissolved sulfate in preventing dolomitization at 200°C from solutions with as little as 0.004M initial SO_4^{2-} certainly implies that similar concentrations of dissolved sulfate would suffice to prevent dolomite precipitation at temperatures less than 100°C. Objections to the sulfate inhibition theory are based on documented occurrences of modern dolomite formation in environments containing appreciable dissolved sulfate (Hardie, 1987) and also on the precipitation of dolomite mineral analogues, such as $PbMg(CO_3)_2$, from sulfate-bearing solutions at low temperatures (Morrow and Ricketts, 1986).

The purpose of this study is to provide an independent assessment of the sulfate inhibition theory of Baker and Kastner (1981) on the basis of precipitation of dolomite analogues at higher temperatures (145° to 225°C) and on the precipitation of dolomite under experimental conditions similar to those of Baker and Kastner (1981). The influence of carbonate-ion activity at these temperatures was also investigated to test the hypothesis that dolomitization is facilitated by higher carbonate activities (Lippman, 1973). The data reported in this study provide partial confirmation for the sulfate inhibition theory and provide some indication of the mechanism of sulfate inhibition. Preliminary results indicate that carbonate-ion activity also influences dolomitization.

METHODS

Two sets of experimental runs were performed at temperatures ranging from 145° to 225°C in Teflon-lined stainless steel bombs of 25-ml capacity at near-equilibrium vapor pressure. In the first set of runs, reagent grade solid carbonates, including $BaCO_3$, $2PbCO_3.PbOH$, and $SrCO_3$ (BDH Chemicals Ltd. and Baker and Adamson Ltd.) were placed in the bombs along with 15 ml of magnesium-bearing saline aqueous solution. X-ray diffraction analysis confirmed that these carbonates were the minerals witherite, hydrocerrusite, and strontianite, respectively. These minerals, which are rhombohedral carbonates analogous to calcite, will form double-magnesian carbonates analogous to dolomite on reaction with magnesium-bearing solutions (Lippman, 1973). The experiments undertaken here (Tables 1–3) are a continuation at higher temperatures of the dolomite analogue precipitation experiments of Morrow and Ricketts (1986).

The second set of experiments involve the dolomitization of reagent grade calcite (Fisher Scientific Ltd.). In these runs 30 mg of calcite were placed in hydrothermal bombs along with 15 ml of $MgCl_2$, $CaCl_2$, and NaCl-bearing solutions. Several series of these runs were performed with solutions that contained variable concentrations of dissolved sulfate (Tables 4–7), and the remainder were per-

[1]Contribution number 31787 of the Geological Survey of Canada.

Sedimentology and Geochemistry of Dolostones, SEPM Special Publication No. 43
 ISBN 0-918985-77-3

TABLE 1.—$BaCO_3$ IN SOLUTION ABOVE 100°C

Temp. (°C)	Initial Solution Composition[1]		Final Ba/Mg or (Ca/Mg) of Solution	Solid Phases (XRD)		Run-Time in Days
145			0.21	Magnesite	−67%	1–28
	$MgCl_2$	−0.08M		Norsethite	−33%	
—	$NaHCO_3$	−0.05M				
220	Na_2CO_3	−0.025M	0.58	Magnesite	−45%	2–22
	NaCl	−0.20M		Unknown #1	−22%	
				Hydromagnesite	−16%	
				Halite	−13%	
				Silicon[2]	−4%	
220	$MgCl_2$	−0.08M	(0.69)	Dolomite	−71%	3–19
	$CaCl_2$	−0.06M		Unknown #1	−13%	
	NaCl[2]	−0.28M		Brucite	−11%	
				Halite	−5%	
145			0.05	Magnesite	−61%	4–28
	$MgSO_4$	−0.08M		Barite	−39%	
—	$NaHCO_3$	−0.05M				
220	Na_2CO_3	−0.025M	very	Magnesite	−32%	5–22
	NaCl	−0.20M	small	Brucite	−25%	
				Barite	−21%	
				Halite	−15%	
				Silicon	−7%	
220	$MgCl_2$	−0.08M	(0.71)	Dolomite	−51%	6–19
	$CaCl_2$	−0.06M		Brucite	−16%	
	NaCl[2]	−0.28M		Barite	−14%	
	Na_2SO_4	−0.004M		Unknown #1	−13%	
				Magnesite	−6%	

1. 40 mg of $BaCO_3$ was the reactant in each run (BDH Chemicals Ltd.).
2. Silicon was added to some reaction products as a standard.

formed with variable concentrations of dissolved carbonate (Tables 4, 8).

Experimental runs ranged in time from 4 to 30 days. At the end of each run, the solution and its reaction products were rapidly quenched, decanted, and separated in a high-speed centrifuge. About 5 ml of the centrifuged solution

TABLE 2.—$PbCO_3$ IN SOLUTION ABOVE 100°C

Temp. (°C)	Initial Solution Composition[1]		Final Pb/Mg or (Ca/Mg) of Solution	Solid Phases (XRD)		Run-Time in Days
145			very	$PbMg(CO_3)_2$	−57%	1–28
			small	Magnesite	−22%	
				Unknown #1	−16%	
	$MgCl_2$	−0.08M		Halite	−5%	
—	$NaHCO_3$	−0.05M				
220	Na_2CO_3	−0.025M	0.01	Unknown #1	−24%	2–22
	NaCl	−0.20M		Halite	−23%	
				Hydromagnesite	−21%	
				Magnesite	−18%	
				Unknown #2	−7%	
				Silicon[2]	−7%	
220	$MgCl_2$	−0.08M	(0.78)	Hydrocerrusite	−63%	3–19
	$CaCl_2$	−0.06M		Unknown #1	−37%	
	NaCl	−0.28M				
145			very	Magnesite	−58%	4–28
			small	$PbMg(CO_3)_2$	−31%	
				Unknown #1	−6%	
	$MgSO_4$	−0.08M		Cerrusite	−5%	
—	$NaHCO_3$	−0.05M				
220	Na_2CO_3	−0.025M	very	Halite	−39%	5–22
	NaCl	−0.20M	small	Magnesite	−38%	
				MgO_2 sulfate		
				hydrate	−14%	
				Unknown #1	−5%	
				Silicon	−4%	
220	$MgCl_2$	−0.08M	(0.79)	Halite	−47%	6–19
	$CaCl_2$	−0.06M		Anhydrite	−18%	
	NaCl	−0.28M		Dolomite	−13%	
	Na_2SO_4	−0.004M		Unknown #1	−12%	
				Hydromagnesite	−6%	
				Brucite	−4%	

1. 54 mg of $2PbCO_3.PbOH$ was added as a reactant to each run (BDH Chemicals Ltd.).
2. Silicon added to some reaction products as a standard.

TABLE 3.—$SrCO_3$ IN SOLUTION 22 DAYS AT 225°C

Initial Solution Composition[1]		Final Sr/Mg or (Ca/Mg) of Solution	Solid Phases (XRD)		Run No.
$MgCl_2$ $CaCl_2$ NaCl	−0.08M −0.06M −0.28M	0.27 and (0.77)	Brucite Halite Dolomite Unknown Strontium chloride hexahydride	−67% −12% −9% −7% −5%	1
$MgCl_2$ $NaHCO_3$ Na_2CO_3 NaCl	−0.08M −0.05M −0.025M −0.02M	0.74	Magnesite Unknown #1 Brucite Halite Unknown #2 Unknown #3 Strontium chloride hexahydride	−35% −27% −26% −4% −3% −3% −2%	2
$MgCl_2$ $CaCl_2$ NaCl Na_2SO_4	−0.08M −0.06M −0.28M −0.004M	0.27 and (0.78)	Brucite Halite Dolomite Strontium magnesium carbonate	−69% −23% −5% −3%	3
$MgSO_4$ $NaHCO_3$ Na_2CO_3 NaCl	−0.08M −0.05M −0.025M −0.20M	0.01	Strontium sulfate Magnesite Brucite	 −48% −35% −17%	4

1. 30 mg of $SrCO_3$ was added as a reactant to each run (Baker and Adamson Ltd.).

TABLE 4.—$CaCO_3$ IN SOLUTION AT 225°C FOR 16 DAYS

Run No.	Initial Solution Composition[1]		Final Ca/Mg Molar Ratio of Solution	Solid Phases (XRD)	
1	$MgCl_2$ $CaCl_2$ NaCl	−0.08M −0.06M −0.28M	1.024	Brucite Dolomite Calcite	−79% −29% −trace
2	$MgCl_2$ $NaHCO_3$ Na_2CO_3 NaCl	−0.04M −0.025M −0.0125M −0.10M	0.8047	Dolomite Brucite Magnesite	−52% −32% −16%
3	$MgCl_2$ $NaHCO_3$ Na_2CO_3 NaCl	−0.08M −0.05M −0.025M −0.20M	0.957	Brucite Magnesite	−77% −23%
4	$MgCl_2$ $CaCl_2$ NaCl Na_2SO_4	−0.08M −0.06M −0.28M −0.004M	1.076	Brucite Anhydrite Calcite Dolomite Magnesite	−41% −34% −10% −14% −10%
5	$MgSO_4$ $NaHCO_3$ Na_2CO_3 NaCl	−0.04M −0.025M −0.0125M −0.10M	0.6471	Brucite Anhydrite Magnesite Calcite	−50% −36% −12% −2%
6	$MgSO_4$ $NaHCO_3$ Na_2CO_3 NaCl	−0.08M −0.05M −0.025M −0.20M	0.143	Anhydrite Magnesite Unknown Calcite $NaCa(CO_3)_2$	−34% −27% −24%? −6% −4%

1. 30 mg $CaCO_3$ were added as a reactant to each run (Fisher Scientific Ltd.).

TABLE 5.—$CaCO_3$ IN SULFATE-FREE SOLUTION AT 215°C

Run No.	Final Ca/Mg in Solution[1]	Final pH	Solid Phases (XRD)	Time in Days
1	1.054	7.5	Brucite −59%, Dolomite −41%	3
2	1.114	7.4	Dolomite −74%, Brucite −26%	7
3	1.192	7.5	Brucite −60%, Dolomite −25%, Magnesite −15%	13
4	1.273	7.5	Brucite −91%, Dolomite −9%	20
5	1.287	7.4	Brucite −100%	26
6	1.293	7.5	Brucite −100%	28

1. 30 mg $CaCO_3$ were added as a reactant to each run (Fisher Scientific Ltd.), and the initial solution composition for all runs was $MgCl_2$—0.08M, $CaCl_2$—0.06M, and NaCl—0.28M.

TABLE 6.—$CaCO_3$ IN 0.004M SULFATE-BEARING SOLUTION AT 215°C

Run No.	Final Ca/Mg[1]	Final (SO_4^{2-})	Final pH	Solid Phases (XRD)	Time in Days
1	0.961	0.0022M	nm	Anhydrite −41%, Calcite −37%, Brucite −17%, Dolomite −5%	5
2	1.188	0.0017M	nm	Brucite −53%, Dolomite −27%, Anhydrite −18%, Magnesite −2%	9
3	1.127	0.0018M	nm	Brucite −53%, Dolomite −28%, Anhydrite −19%	14
4	1.143	0.0019M	nm	Anhydrite −41%, Brucite −34%, Dolomite −24%, Magnesite −1%	17
5	1.276	0.0022M	nm	Brucite −50%, Anhydrite −50%	22
6	1.150	0.0019M	nm	Brucite −50%, Anhydrite −50%	22
7	0.967	0.0028M	7.6	Calcite −64%, Brucite −18%, Anhydrite −18%	4
8	1.360	0.0027M	7.4	Anhydrite −64%, Dolomite −18%, Brucite −18%	7
9	1.273	0.0023M	7.5	Anhydrite −60%, Brucite −25%, Dolomite −15%	11
10	1.247	0.0026M	7.3	Anhydrite −41%, Brucite −34%, Dolomite −25%	14
11	1.539	0.0022M	7.5	Brucite −54%, Anhydrite −42%, Dolomite −4%	17
12	1.563	0.0025M	7.2	Brucite −83%, Anhydrite −16%, Dolomite −1% (a trace)	20

1. 30 mg $CaCO_3$ were added as a reactant to each run (Fisher Scientific Ltd.), and the initial solution composition for all runs was $MgCl_2$—0.08M, $CaCl_2$2—0.06M, NaCl—0.28M, and Na_2SO_4—0.004M. Runs 1 to 6 and runs 7 to 12 are two separate sets of time series runs. The final Ca/Mg is the molar solution ratio of calcium and magnesium at the end of each run. The final (SO_4^{2-}) and final pH are the sulfate concentrations and pH of the solutions at the end of each run.

TABLE 7.—$CaCO_3$ IN 0.02M SULFATE-BEARING SOLUTION AT 215°C

Run No.	Final Ca/Mg	Final (SO_4^{2-})	Final pH	Solid Phases (XRD)	Time in Days
1	0.773	0.0065M	7.2	Anhydrite −45%, Brucite −45%, Calcite −10%	4
2	0.997	0.0064M	7.5	Anhydrite −54%, Brucite −44%, Gypsum −2%	8
3	0.962	0.0037M	7.4	Anhydrite −52%, Brucite −48%	12
4	0.905	0.0031M	7.6	Anhydrite −69%, Brucite −21%, Bassanite −4%, Gypsum −4%, Magnesite −2%	18
5	1.112	0.0033M	7.4	Anhydrite −56%, Brucite −42%, Gypsum −2%	22
6	0.903	0.0043M	7.2	Anhydrite −86%, Brucite −12%, Gypsum −2%	26

1. 30 mg $CaCO_3$ were added as a reactant to each run (Fisher Scientific Ltd.), and the solution composition for all runs was $MgCl_2$—0.053M, $CaCl_2$—0.04M, NaCl—0.186M, and Na_2SO_4—0.02M.

TABLE 8.—$CaCO_3$ IN SOLUTIONS OF VARIABLE CARBONATE-ION ACTIVITY AT 215°C

Run No.	Solution Composition[1]	Final Ca/Mg	Final pH	Solid Phases (XRD)	Time in Days
1	$NaHCO_3$ absent	0.925	8.88	Calcite −65%, Brucite −23%, Dolomite −10%	2
2	$NaHCO_3$—0.0025M	0.965	7.06	Dolomite −85%, Calcite −15%	2
3	$NaHCO_3$—0.01M	1.023	6.59	Dolomite −100%	2
4	$NaHCO_3$ absent	1.089	8.30	Dolomite −54%, Brucite −46%	6
5	$NaHCO_3$—0.0025M	1.104	8.00	Dolomite −61%, Brucite −39%	6
6	$NaHCO_3$—0.01M	1.011	7.93	Dolomite −61%, Brucite −39%	6

1. 30 mg $CaCO_3$ were added as a reactant to each run (Fisher Scientific Ltd.), and the solutions in all runs contained $MgCl_2$—0.08M, $CaCl_2$—0.06M, and NaCl—0.28M with variable concentrations of $NaHCO_3$.

was decanted into sealed test tubes, and the reaction solids were pipetted onto circular microcover glass slides and weighed. The reaction solutions were analyzed where appropriate for Ca^{2+}, Mg^{2+}, Ba^{2+}, Pb^{2+}, and Sr^{2+} by atomic absorption on a Perkin Elmer #603 atomic absorption spectrophotometer. The pH of many of these solutions was measured at room temperature (21°C) on an Orion Research Microprocessor Ionalyzer 901. Regrettably, however, several days had elapsed after the completion of runs before pH measurements were performed, except for those runs that involved variable carbonate in solution, for which pH measurements were taken immediately after the reactions were quenched (Table 8). Also, where appropriate, the concentration of the sulfate anion (SO_4^{2-}) was determined on a 2120I Dionex ion chromatograph using an AS 4A column with a sodium carbonate-sodium bicarbonate eluent.

The reaction mineral solid products were analyzed by powder X-ray diffraction on a Phillips X-ray diffractometer with iron-filtered Co K alpha radiation. Mineral phases were also identified and examined by scanning electron microscopy on a Cambridge Stereoscan 150 MK2; SEM photomicrographs were taken of most mineral reaction products, and some minerals were analyzed for CaO, MgO, and SO_2 by an energy-dispersive X-ray spectrometer (KEVEX). A modified version of the SOLMNEQ program of Kharaka and Barnes (1973) was used to calculate the activities of ions in solution and the saturation states of minerals in those runs for which pH measurements were taken.

The hydrothermal bombs used in this study were not totally sealed with respect to the vapor phase, particularly at temperatures over 100°C. The rate of fluid loss was about 1 ml/wk over the 3- to 4-wk-long time-series runs (Fig. 1). The vapor phase of the reaction systems is composed mainly of water and CO_2 gas with the remainder being relict atmospheric components. Progressive loss of vapor caused the ionic strength of the solution to increase slightly during runs. More important, loss of CO_2 from the system caused the eventual dissolution of dolomite after 20 days in all time-series runs, regardless of differences in their solution compositions, and enhanced the precipitation of brucite (Fig. 2). This is a significant shortcoming to the experiments reported here; however, for the general dolomitization reaction,

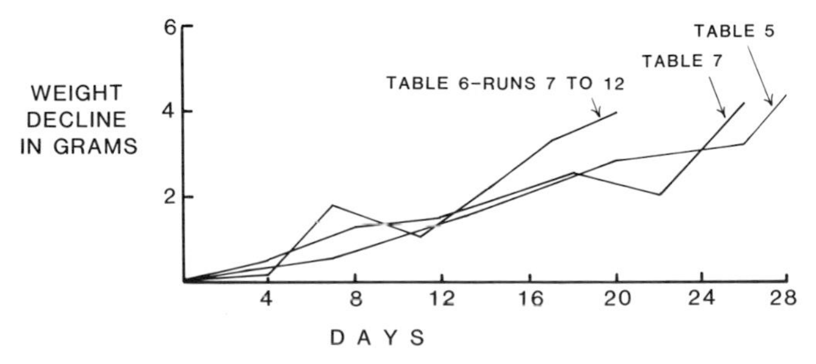

FIG. 1.—Graph showing rate of loss of water plus CO_2 vapor from teflon-lined hydrothermal bombs during time-series runs. Loss of CO_2 from system lowers saturation level of carbonate minerals in solution but does not affect relative stability of calcite with respect to dolomite.

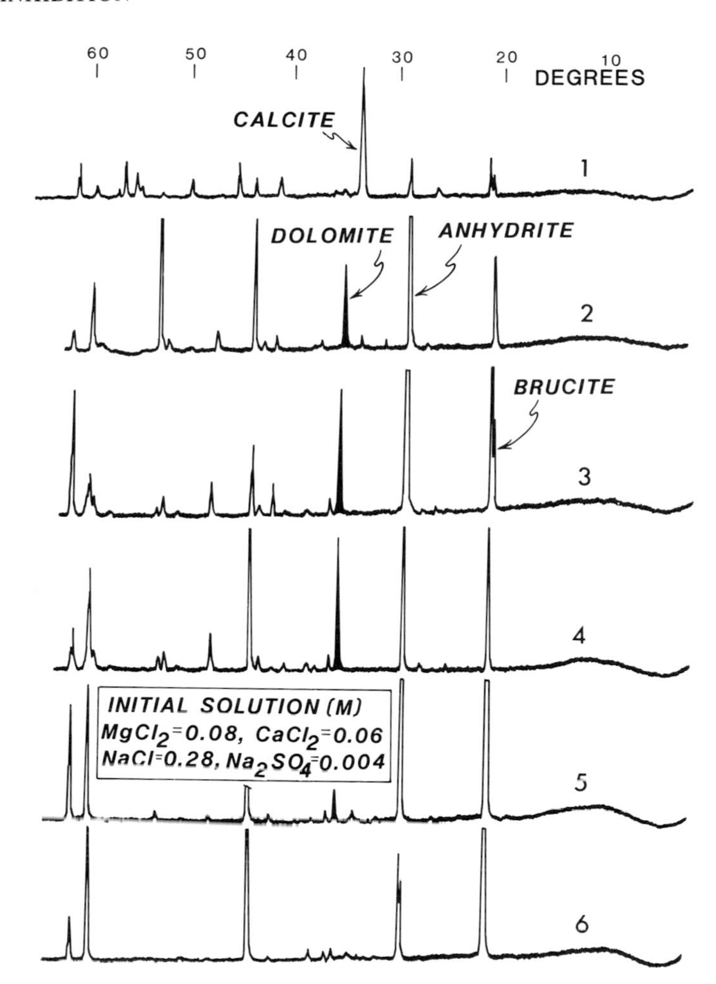

FIG. 2.—X-ray diffractograms of a time series of runs with calcite reactants and a solution sulfate concentration of 0.004M (runs 7 to 12 of Table 6 are runs 1 to 6 of this figure). Runs 1 through 6 of this figure were removed after 4, 7, 11, 14, 17, and 20 days, respectively. Calcite dissolution is delayed for 5 or 6 days.

$$(2 - X)CaCO_3 + Mg^{2+} + XCO_3^{2-} = CaMg(CO_3)_2 + (1 - X)Ca^{2+}$$

the relative thermodynamic stability of calcite with respect to dolomite is dependent only on the Ca/Mg activity ratio and is independent of the activity of dissolved carbonate species (Stoessell, 1987). This indicates that the results reported here may be interpreted on the basis of the Ca/Mg activity ratios of the reaction solutions as indicators of the relative thermodynamic stability of calcite vs. dolomite.

RESULTS

The results reported here are in two parts. The first part is an extension of our earlier work (concerning the precipitation of dolomite analogues at low temperatures; Morrow and Ricketts, 1986) to higher temperatures. The second part includes results obtained for the dolomitization of calcite under a range of conditions of sulfate and carbonate concentration in experiments of variable duration. Few of the data presented here permit unequivocal interpretation, but they do provide the beginning of a framework of data necessary to interpret the origin of dolomite under conditions similar to those of the experiments described here and earlier by Baker and Kastner (1981).

Dolomite analogues–high temperature.—

A variety of experimental runs was performed using $BaCO_3$, $PbCO_3$ (actually $2PbCO_3.PbOH$), and $SrCO_3$ at temperatures ranging from 145°C to 225°C (Tables 1–3). At 145°C the behavior of runs seeded with solid $BaCO_3$ and $PbCO_3$ paralleled that reported by Morrow and Ricketts (1986) for lower temperatures (25° to 80°C). Runs seeded with $BaCO_3$ at 145°C formed magnesite and barite in sulfate-bearing solution but formed $BaMg(CO_3)_2$ (norsethite) in sulfate-free solution (Table 1). This behavior may have a thermodynamic origin related to the low solubility of barite rather than to a kinetic inhibition effect caused by sulfate in solution. The calculated concentration of barium in solution, assuming barite solubility of about 8×10^{-8} (see Barnes, 1979, Fig. 9.25), is 0.23 μg/ml, which is consistent with the observed value of less than 0.2 μg/ml of barium in run 4 of Table 1. In run 1 of Table 1 the high final barium concentration of 950 μg/ml perhaps reflected near-equilibration with $BaMg(CO_3)_2$. The precipitation of $PbMg(CO_3)_2$ from both sulfate-free and sulfate-bearing solutions at 145°C (runs 4 and 1, Table 2) also supports a thermodynamic interpretation for the absence of $BaMg(CO_3)_2$ in sulfate-bearing runs.

At temperatures above 200°C, using the same calcium-free magnesium-bearing solutions as in the runs at 145°C (runs 2 and 5 of Tables 1 and 2), there is a marked change in the behavior of runs seeded with $BaCO_3$ and $PbCO_3$. No dolomite analogue minerals were precipitated, and instead, magnesite was the dominant magnesium-bearing phase formed. This behavior may itself be analogous to the calcite-dolomite-magnesite stability relations outlined by Rosenberg and Holland (1964), in which the magnesite stability field expands with increasing temperature at the expense of the dolomite field. It seems possible that in both the $BaCO_3$—$BaMg(CO_3)_2$—$MgCO_3$ and $PbCO_3$—$PbMg(CO_3)_2$—$MgCO_3$ systems, the magnesite field also expands with increasing temperature but at a more rapid rate than in the dolomite-magnesite system, preventing the precipitation of dolomite analogues at temperatures lower than those required to prevent dolomite precipitation.

The dolomite analogue experiments described up to this point were conducted with calcium-free solutions. The addition of calcium to the experimental solutions in the dolomite analogue experiments at a temperature of 220°C caused dolomite to form in both sulfate-free and sulfate-bearing solutions, even though the reactants used were $BaCO_3$ and $PbCO_3$ instead of $CaCO_3$ (runs 3 and 6 of Table 1, and run 6 of Table 2; Fig. 3). The initial sulfate concentration (0.004M) of the sulfate-bearing solution was the same as that found by Baker and Kastner (1981) to prevent the dolomitization of calcite. The precipitation of dolomite in the calcite-free runs in both sulfate-bearing and sulfate-free solutions may indicate that sulfate in solution does not in itself prevent the precipitation of dolomite. More complete documentation of these types of experiments, in which dolomite is precipitated in the absence of calcite and in which the concentration of sulfate in solution is determined during the runs, might yield a more definitive answer to the question of whether sulfate in solution in itself will retard or prevent the direct precipitation of dolomite from solution.

In addition to experiments using $BaCO_3$ and $PbCO_3$ as reactants, four runs were performed using $SrCO_3$ as a reactant. Previously, Morrow and Ricketts (1986) were unable to precipitate $SrMg(CO_3)_2$ at temperatures below 100°C. In the runs performed here at 225°C (Table 3), only one run formed a small amount of $SrMg(CO_3)_2$. Evidently, the precipitation of this mineral occurs only at higher temperatures (Froese, 1967). Like the other dolomite analogue runs, however, runs in which $SrCO_3$ was used as a reactant with calcium-bearing solutions formed dolomite in both sulfate-free and sulfate-bearing solutions (runs 1 and 3; Table 3). This persistent precipitation of dolomite from sulfate-bearing calcium-containing solutions in the dolomite analogue experiments led us to repeat the calcite dolomitization experiments of Baker and Kastner (1981).

Dolomite precipitation–variable sulfate in solution.—

Four sets of six runs each were performed in order to evaluate the influence of time vs. the concentration of sulfate in solution on the dolomitization reaction (Tables 5–7). The same initial solution was used with 30 mg of solid calcite reactant for all runs in an individual run set, and each run was sampled separately in time intervals of 2 to 4 days. All runs were performed at a temperature of 215°C. Three different initial solutions were used; two of these were of the same composition ($MgCl_2$—0.08M, $CaCl_2$—0.06M, NaCl—0.28M and $MgCl_2$—0.08M, $CaCl_2$—0.06M, NaCl—0.28M, Na_2SO_4—0.004M) as those used by Baker and Kastner (1981), and the third had a lower ionic strength but much greater sulfate concentration ($MgCl_2$—0.053M, $CaCl_2$—0.04M, NaCl—0.186M, Na_2SO_4—0.02M). The first two initial solutions had an ionic strength equal to that of sea water (0.7), whereas the third was about 0.45. The low ionic strength of the third solution was necessary to prevent anhydrite or gypsum precipitation. Calculations with a modified version of the SOLMNEQ program of Kharaka and Barnes (1973) indicates that these initial solutions were undersaturated with respect to all possible mineral phases at room temperature (25°C) but that both sulfate-bearing solutions were oversaturated with respect to anhydrite at 215°C.

In the time-series experiment using the sulfate-free initial solution, calcian dolomite replaced the reactant calcite completely in 3 days (Table 5). This dolomite became more stoichiometric after 7 days but diminished in amount after about 17 days until it finally disappeared shortly after 20 days (Table 5; run 1, Table 4; Appendix). Brucite was present in abundance in all runs, and magnesite appeared in the run sampled after 13 days. SEM photomicrographs (Fig. 4) show that 10-μm-size dolomite crystals initially form polycrystalline aggregates that are intergrown with large, tabular, brucite crystals. After 7 days, these dolomite crystals appear to have grown slightly, and the crystal faces are no longer smooth but are marred by irregular shallow cavities in the face centers. These cavities have rectilinear edges that appear to be parallel to crystal faces. After 20 days,

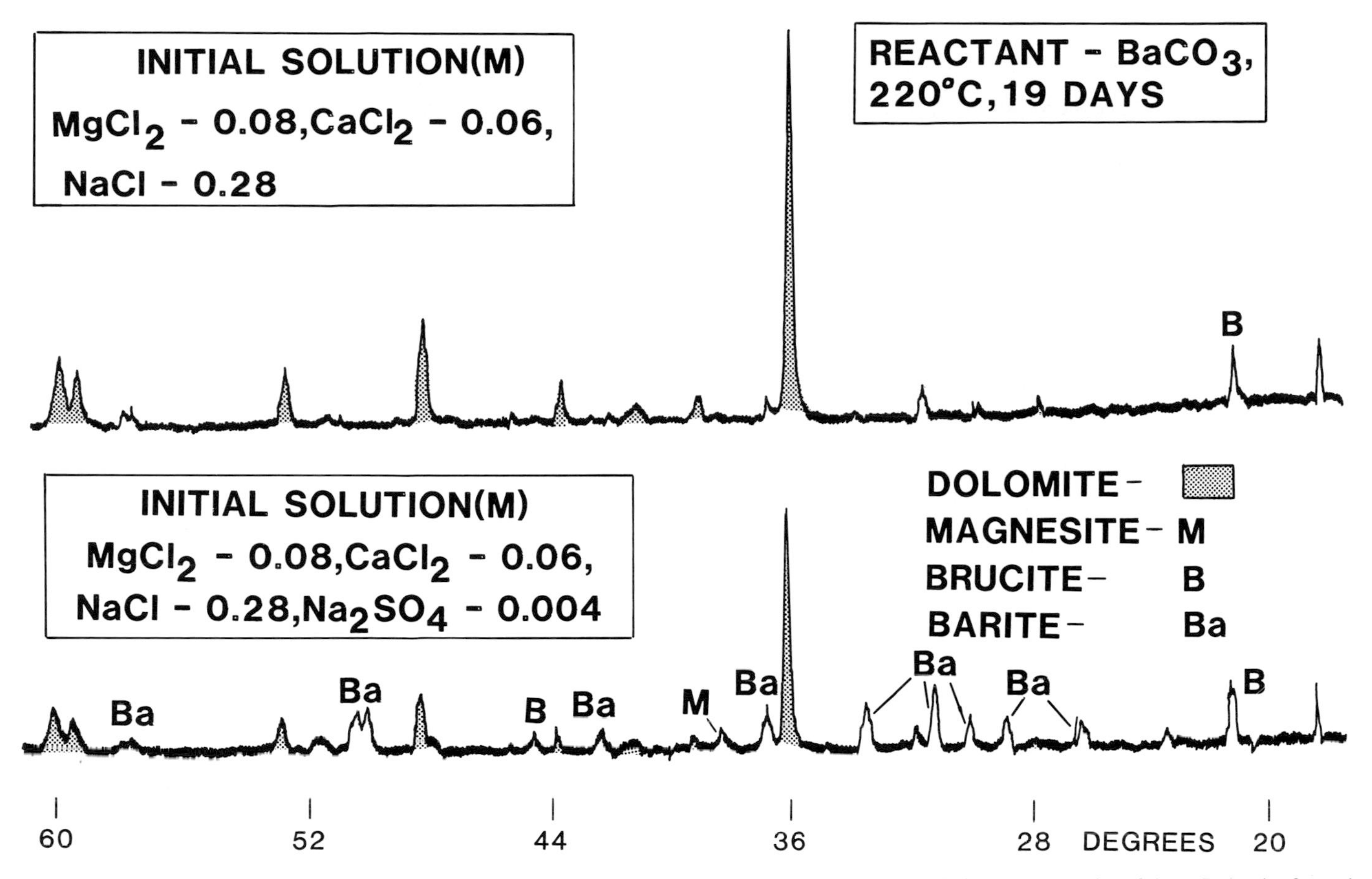

FIG. 3.—X-ray diffractograms of runs seeded with $BaCO_3$ but in sulfate-free and sulfate-bearing solutions that contain calcium. Dolomite formed in both runs, even in a solution with an initial 0.004M concentration of sulfate and in the absence of a calcite reactant.

only small rounded remnants of the formerly euhedral dolomite crystals remain among large, euhedral, brucite crystals (Fig. 4). Preferential dissolution of crystal edges and corners undoubtedly caused this "rounding" during dolomite dissolution after about 10 days. Dolomite dissolution and brucite precipitation also caused the Ca/Mg ratio to increase progressively (runs 1 to 6, Table 5).

Two time-series experiments were conducted using the sulfate-bearing initial solution composition found by Baker and Kastner (1981) to prevent the dolomitization of calcite totally. The concentration of sulfate in this solution (0.004M) is twice that found by them to inhibit dolomitization largely, although not completely. In both of these sets of six runs, calcite remained largely undolomitized for 6 to 7 days, and after 7 days the reactant calcite was totally dolomitized by an initially calcian dolomite that became stoichiometric after 11 to 12 days (Fig. 2; Table 6; Appendix). As with the sulfate-free runs, the dolomite formed in these sulfate-bearing solutions disappeared after about 20 days. After 7 days, anhydrite and brucite became abundant and remained so to the end of the runs. SEM photomicrographs show that the dolomite initially formed is very similar in crystal size and form (Fig. 5) to the first-formed dolomite in the sulfate-free runs. Similarly, these dolomites are intimately intergrown with brucite and possibly also with lathe-like anhydrite crystals. After 20 days, only the rounded remnants of former dolomite crystals remain among large and well-formed brucite and anhydrite crystals. As with the sulfate-free runs, the Ca/Mg ratio of each of these sulfate-bearing run sets rises progressively, reflecting the dissolution of dolomite and precipitation of brucite (Table 6). Magnesite occurs in small amounts in some of these runs (see also run 4 of Table 4).

A single time-series experiment was conducted using the solution with a high-sulfate concentration (0.02M). In this set of runs, calcite was partly dissolved after 5 days and had disappeared after 8 days (Table 7). Dolomite did not form at any time, and instead, anhydrite and brucite were the dominant reaction products. Small amounts of gypsum and magnesite were present in some runs. Like the other experiments, the Ca/Mg ratio of the reaction solutions increased progressively. This was caused by the precipitation of brucite.

Dolomite precipitation–variable carbonate in solution.—

A time-series experiment was conducted using solutions with variable concentrations of $NaHCO_3$. These solutions were of the same composition as the solution used for the sulfate-free runs of Table 5 with the addition of 0.0025M $NaHCO_3$ or 0.01M $NaHCO_3$ (Table 8). The intermediate 0.0025M $NaHCO_3$ concentration is about the same as the

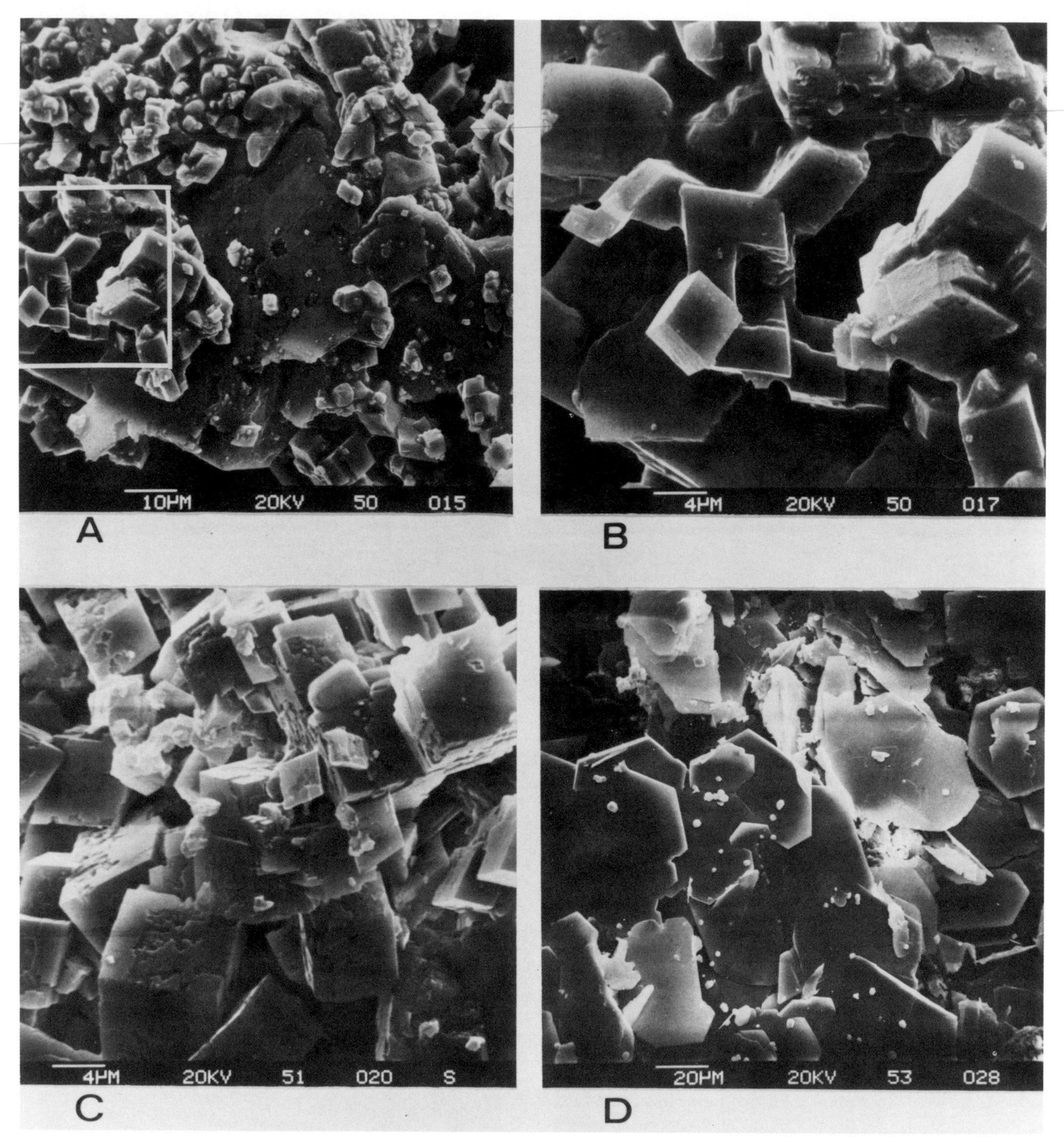

FIG. 4.—Scanning electron photomicrographs of reaction products from calcite-seeded runs in sulfate-free solutions at 215°C (see Table 5). (A) View of dolomite euhedra and crystal aggregates among large, irregular, platelike brucite crystals after 3 days. (B) Closer view of dolomite crystal aggregate shown in Figure 5A, emphasizing the smooth, well-formed crystal faces. (C) After 7 days, dolomite displayed slightly larger crystals with etched or incompletely filled crystal faces. (D) Dolomite reduced by dissolution to small rounded "nubbins" after 20 days. Note abundant euhedral brucite crystals.

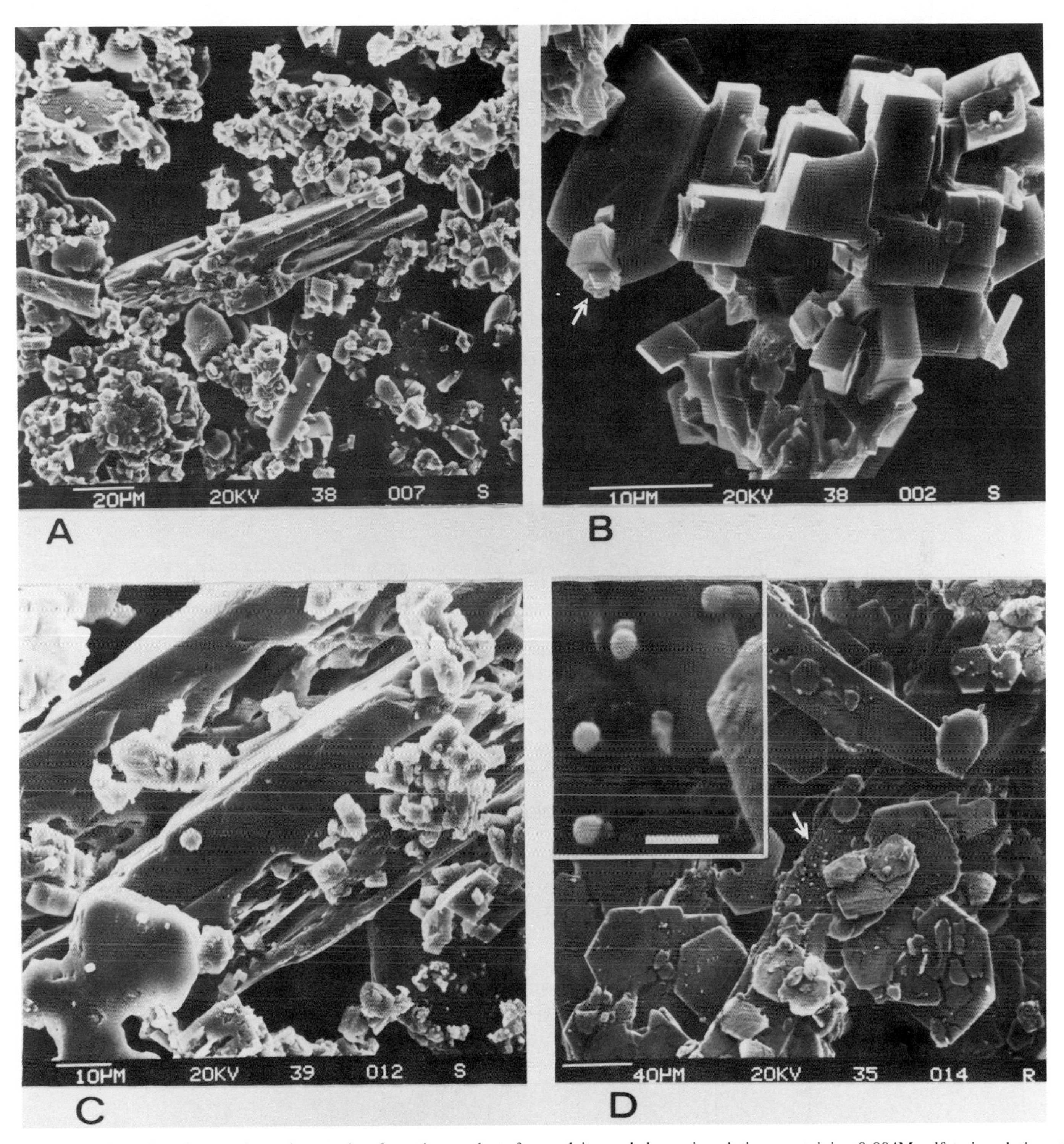

FIG. 5.—Scanning electron photomicrographs of reaction products from calcite-seeded runs in solutions containing 0.004M sulfate in solution at 215°C (see Table 6). (A) View of dolomite euhedra and crystal aggregates among larger prismatic anhydrite and tabular brucite crystals after 7 days. (B) Closer view of dolomite crystal aggregate after 7 days. Dolomite is well formed with smooth crystal faces. Arrow indicates dolomite intergrown with a platy brucite crystal. (C) Aggregates of dolomite crystals after 11 days. Anhydrite crystals are much larger than after 7 days. (D) Large and euhedral-bladed anhydrite and platy brucite crystals after 22 days. Scattered remnants of solution-rounded dolomite crystals dot surfaces of larger brucite and anhydrite crystals. A closer view of these small, rounded dolomite crystal "nubbins" is shown in the inset photomicrograph and indicated by the arrow.

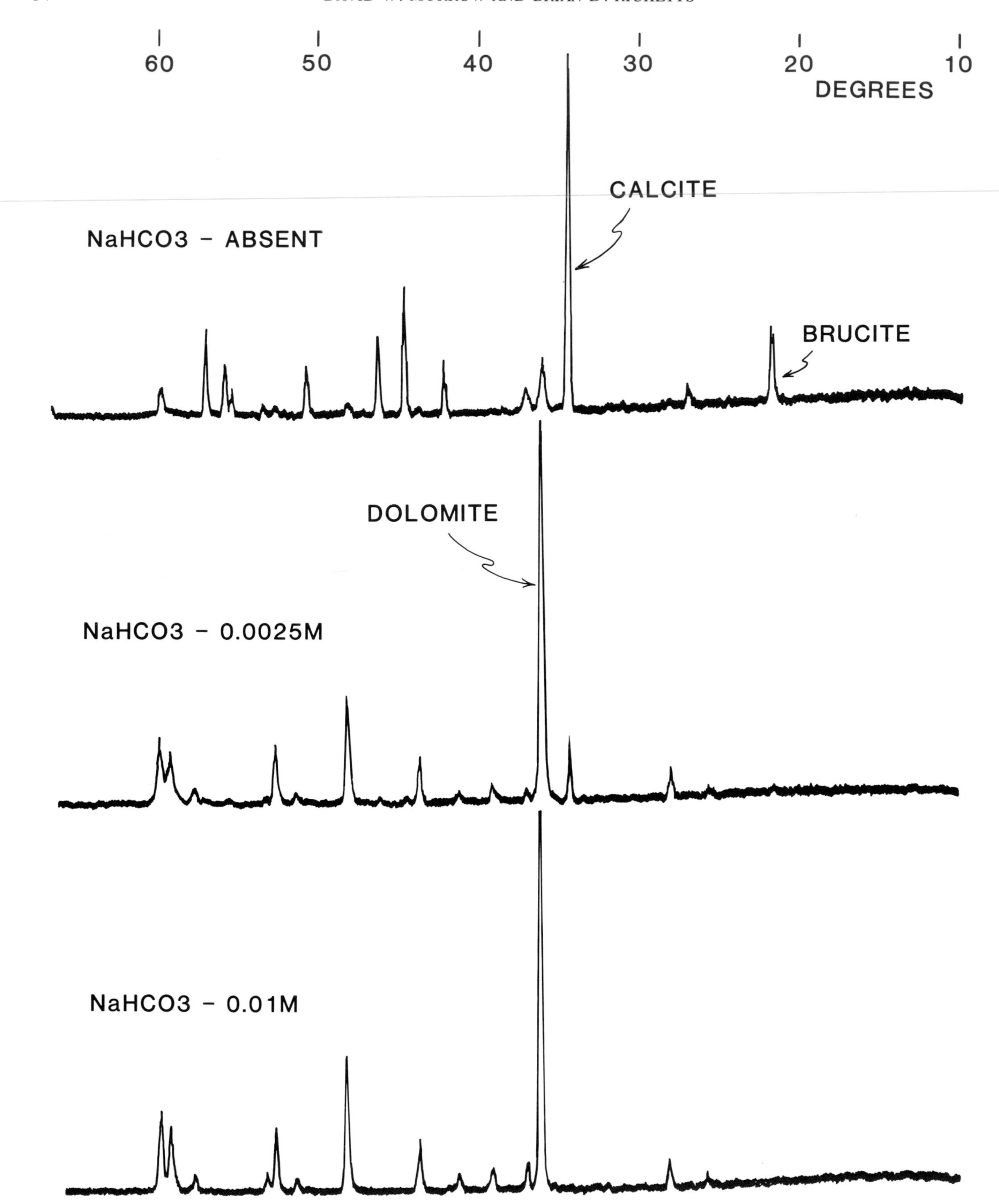

FIG. 6.—X-ray diffractograms of runs 1, 2, and 3 (Table 8), having different initial concentrations of $NaHCO_3$, and reaction times of 2 days. Dolomitization of calcite proceeded more rapidly in solutions containing high initial $NaHCO_3$ concentrations.

combined carbonate and bicarbonate concentration of sea water (Garrels and Thompson, 1962). Runs of 2 days with these variable concentrations of carbonate revealed striking differences in dolomitization rates. Calcite in runs with higher initial concentrations of $NaHCO_3$ were dolomitized much more rapidly (Fig. 6). After 6 days, however, dolomitization went to completion regardless of the initial concentration of $NaHCO_3$ (Table 8). The dolomites formed in these runs were all stoichiometric except for the dolomite formed after 3 days in the $NaHCO_3$-free solution (run 1, Table 8), which was slightly calcian (Appendix). There was a slight increase in the Ca/Mg ratio over the 6-day period of these runs (Table 8). As before, this increase may reflect the precipitation of brucite.

DISCUSSION

The results reported here differ from those reported by Baker and Kastner (1981) in that the initial concentration of sulfate required to prevent the dolomitization of calcite was higher in these experiments and may lie somewhere between 0.004M and 0.02M. The disappearance of dolomite in all time-series exeriments after 20 days may be related to the loss of CO_2 attendant on the progressive loss of the vapor phase from the hydrothermal bombs (Fig. 1). This loss of carbonate caused the dissolution of carbonate minerals such as dolomite and the subsequent precipitation of brucite. The occurrence of brucite as a high-temperature reaction product in $MgCl_2$—$CaCl_2$—$CaCO_3$ reaction systems is well known. Rosenberg and Holland (1964) infused their reaction vessels with CO_2g at a pressure of 60 bars specifically to prevent brucite precipitation. As mentioned previously, the relative thermodynamic stability of dolomite vs. calcite is not affected by changes in carbonate-ion activity. Consequently, the differences in behavior of calcite and dolomite between experimental runs in which sulfate or carbonate concentrations are independently varied probably reflect the influence of kinetic factors.

In adopting this assumption, we have ignored the effect of slight differences in the activity coefficients for the various ions in solutions of slightly different compositions and ionic strengths. These activity coefficient differences will have an effect on the saturation of mineral phases in these solutions. Calculations with SOLMNEQ indicate that mineral saturations between the sulfate-free and sulfate-bearing time series runs will vary by less than a factor of two for any particular carbonate-ion concentration. These calculations were based on the assumptions that the rates of loss of CO_2 from the system were similar for each time series run, that the system was near equilibrium at all times, and that the measured weight of solid phases after each run and the estimates of mineral percentages in the solids were accurate. Calculations of this type cannot be totally accurate, but they do indicate that the pronounced differences in the mineral reaction rates observed here are not likely to be caused by large variations in the relative degree of saturation of calcite vs. dolomite in the various time-series experiments reported here.

The final Ca/Mg activity ratios of solutions from the dolomitization runs indicate that they lie well within the combined stability fields of dolomite and magnesite (Fig. 7). Estimates were made of the actual activities of calcium and magnesium for those solutions for which pH measurements were taken using the SOLMNEQ program. The simple molar Ca/Mg ratios of the remaining solutions were plotted as activity ratios under the assumption that the activity coefficients for calcium and magnesium were approximately equal. Comparison of molar ratios with activity ratios calculated by SOLMNEQ indicated that these ratios differed by less than 50% of the molar ratio values. The dashed lines in Figure 7 are stability field boundaries extrapolated from the data of Rosenberg and Holland (1964) following the assumption that molar Ca/Mg ratios approximate activity ratios. The solution activity ratios from this study fall almost directly on their dolomite-magnesite field boundary line. This is consistent with the sporadic occurrence of magnesite in the reaction products of some of the dolomitization runs in which the Ca/Mg ratios were close to 1.0. The few runs with Ca/Mg ratios much less than 1.0 (runs 5 and 6, Table 4; Fig. 7) precipitated magnesite rather than dolomite.

The precipitation of anhydrite was an important factor in governing the solution sulfate concentration in these experiments. The initial solutions of runs 1 to 12 of Table 6 contained 0.004M sulfate and were identical in composition to solutions found by Baker and Kastner (1981) to prevent the dolomitization of calcite. Ordered stoichiometric dolomite formed in these run sets, but the precipitation of anhydrite lowered the concentration of sulfate to the range

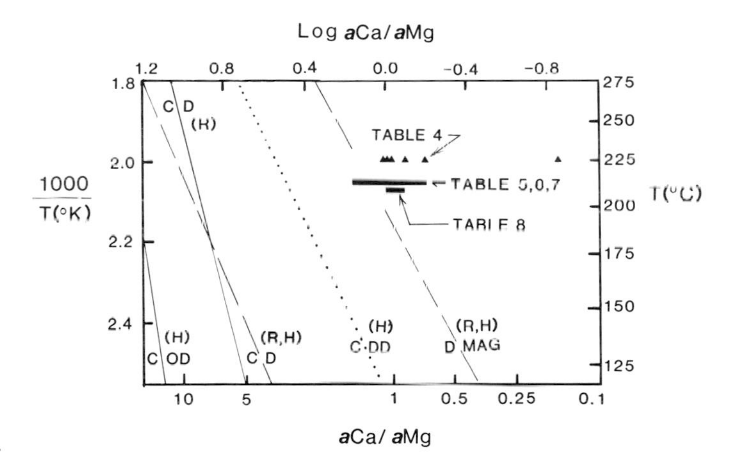

FIG. 7.—A plot of Ca/Mg activity ratios vs. the reciprocal of the temperature. Ca/Mg activity ratios for runs from Tables 5, 6, 7, and 8 were estimated by calculation from the SOLMNEQ program. Ranges of these calculated values are shown as bar graphs. The Ca/Mg solution ratios shown for runs from Table 4 are molar ratios only (solid triangles) and are assumed to approximate the actual activity ratios. Also shown are some published stability field boundaries between calcite and dolomite and between dolomite and magnesite. The boundaries between calcite and ordered dolomite (C—OD) from Helgeson and others (1978), (H) and between calcite and dolomite, (C—D) from Robie and others (1979), (R) are shown as solid lines. The dotted boundary is the boundary between calcite and disordered dolomite, (C—DD) from Helgeson and others (1978), (H). The dashed lines are the stability boundaries between calcite and dolomite, (C—D) and between calcite and magnesite (D—MAG) extrapolated from the data of Rosenberg and Holland (1964), (R,H). The final Ca/Mg ratios of this study are close to those at the dolomite-magnesite boundary.

0.0017M to 0.0029M (Table 6). The solutions of runs 1 to 6 of Table 7, in which dolomite did not form but was oversaturated (Fig. 7), contained 0.02M dissolved sulfate initially. Precipitation of anhydrite lowered the sulfate concentrations of this run set to the range 0.0031M to 0.0065M. This concentration range is essentially at or above the sulfate concentration found by Baker and Kastner (1981) to inhibit dolomitization. Consequently, these experiments may be interpreted to provide independent verification of the results of Baker and Kastner (1981), although it is not evident why anhydrite was not observed as reaction products in their experiments. As mentioned previously, calculations by SOLMNEQ indicate that anhydrite is slightly oversaturated at 215°C in all the sulfate-bearing solutions used in these time-series experiments, even without the contribution of calcium from the calcite reactant. Dissolution of the calcite reactant caused an increase in anhydrite saturation and promoted its precipitation.

A striking difference between all sulfate-free and sulfate-bearing runs concerns the length of time that the reactant calcite remained incompletely dissolved. In sulfate-free runs, calcite dissolved completely in about 3 days (run 1, Table 5; Table 8), but in all sulfate-bearing runs some calcite remained undissolved for as long as 7 days after the start of runs (Table 6; Fig. 2) or even after 16 days in some sulfate-bearing runs (Table 4). This might indicate that dissolved sulfate inhibited the dissolution of calcite and delayed the precipitation of dolomite in the 0.004M sulfate runs. Busenberg and Plummer (1985) found that sulfate in solution influenced the rate of calcite precipitation, so it might be expected that sulfate would also influence the rate of calcite dissolution.

Alternatively, the survival of calcite for longer time periods in the sulfate-bearing runs may be attributed to a delay in the dolomitization reaction because of a sulfate inhibition of dolomite precipitation. This does not explain, however, the disappearance of calcite in the 0.02M sulfate reaction where no dolomite formed (Table 7). Perhaps the disappearance of calcite in the high-sulfate system was controlled by the progressive precipitation of anhydrite. The declining concentration of sulfate in solution as the reaction proceeded (Table 7) indicates that anhydrite continued to precipitate throughout the reaction. This is consistent with the much lower Ca/Mg solution ratios in the 0.02M sulfate runs than in the 0.004M sulfate runs (Tables 6, 7). This explanation is also consistent with the slightly more rapid disappearance of calcite in the 0.02M sulfate runs than in the 0.004M runs.

It is of interest to compare the behavior of calcite in the 0.02M runs with the analogous high-sulfate runs in which $PbCO_3$ was the reactant. $PbMg(CO_3)_2$ formed readily at temperatures below 150°C from solutions with 0.08M sulfate in solution, but no anglesite ($PbSO_4$) precipitated (run 4, Table 2). This may indicate that the precipitation of anhydrite in the runs seeded with calcite influenced the dolomitization process. Possibly, anhydrite precipitated rapidly in the immediate vicinity of calcite crystals that were undergoing dissolution. This may have caused the saturation of dolomite to remain low and thus prevented dolomitization of the calcite during its dissolution. This is a possible alternate explanation to the kinetic inhibition mechanism inferred by Baker and Kastner (1981) to prevent dolomitization in sulfate-bearing solutions, although they did not report anhydrite as a reaction product in their experiments.

This explanation is also consistent with the behavior of $BaCO_3$ in sulfate-free vs. sulfate-bearing solutions. The precipitation of barite ($BaSO_4$), as previously mentioned, caused the barium contents of solutions from sulfate-bearing runs to be reduced to extremely low amounts (Table 1) and obviously prevented the precipitation of $BaMg(CO_3)_2$. The direct precipitation of dolomite in sulfate-bearing runs seeded with $BaCO_3$ also may indicate that dissolved sulfate does not greatly inhibit the precipitation of dolomite. The precipitation of barite, however, probably reduced the sulfate content of the solution to low levels. Similarly, the precipitation of dolomite in sulfate-bearing runs seeded with $2PbCO_3.PbOH$ is consistent with this hypothesis (run 6, Table 2). In this case, the lead sulfate (anglesite) did not form, and the sulfate concentration of the solution was probably controlled by the more soluble anhydrite.

In the case of calcite in sulfate-bearing solutions, the data are consistent with a two-stage process. During calcite dissolution, contemporaneous anhydrite precipitation lowered the calcium concentrations and limited the saturation of dolomite to low levels, such as during the initial 7-day period of the 0.02M sulfate run (Table 7). After the disappearance of calcite, it is possible that dolomite could not precipitate because of the absence of any solid carbonate phases that could provide nucleation sites. In other words, in sulfate-bearing solutions, the dolomitization of calcite in these types of experiments is dependent both on the saturation level of dolomite in solution and on the presence of nucleation sites in the form of solid carbonate phases. Also, if sulfate in solution retarded the dissolution of calcite directly by forming surface coatings of $CaSO_4$ on calcite crystals, then these crystals would not serve as readily as nuclei for the growth of dolomite crystals.

The alternate hypothesis for the dependence of dolomitization on the concentration of sulfate in solution involves the inhibition of spontaneous dolomite nucleation and precipitation by sulfate in solution. The evidence presented here at least suggests that this may not be accurate and that the sulfate inhibition mechanism preventing dolomitization involves primarily the interaction of sulfate with calcite. If this is correct, then dolomites formed in nature by direct precipitation from solution, such as has been inferred for some deep-sea dolomites (Baker and Burns, 1985), could have precipitated from sulfate-bearing solutions, whereas dolomites formed by the *in situ* dolomitization of precursor calcite require a sulfate-free environment. It would be of interest to analyze natural dolomites for their sulfate content, as has been done for some natural calcites (Busenberg and Plummer, 1985).

The effect of variations in the carbonate activity of solutions on the dolomitization reaction has been seldom investigated, although some authors have hypothesized that higher carbonate concentrations will increase the rate of dolomite precipitation from oversaturated solutions (Lippman, 1973). Oomori and others (1983) found that the precipitation of disordered protodolomite from evaporated-seawater brines at 33°C was favored at higher carbonate concentrations. The results shown in Figure 6 and in Table 8 are consistent with these previous hypotheses and experimental data. Dolomitization of calcite was much more rapid in solutions containing $NaHCO_3$ than in solutions with no added $NaHCO_3$ at 215°C. Dolomites formed in these runs with variable carbonate activity were stoichiometric and moderately well ordered (Appendix).

CONCLUSIONS AND FUTURE WORK

The primary focus of this study has been to investigate the data and conclusions of Baker and Kastner (1981) relating to the influence of dissolved sulfate on the precipitation of dolomite at 200°C. These results have been extrapolated to lower temperatures in an effort to explain natural occurrences of low-temperature dolomites (Kastner, 1983, 1984). The work of Baker and Kastner (1981), however, may be applied more directly to the interpretation of the origin of ancient examples of high-temperature dolomites (typically 150° to 250°C precipitational temperatures) that are more common than is generally appreciated (Hardie, 1987). Replication of initial experimental results, such as those of Baker and Kastner (1981), is an important stage in the development of any scientific theory. It is our hope that this study contributes to this goal.

The main conclusions of this study may be summarized as follows.

(1) The existence of a sulfate inhibition effect on dolomitization is confirmed in these experiments.
(2) Sulfate in solution inhibited dolomitization in these experiments primarily by retarding the rate of calcite dissolution and by causing the precipitation of anhydrite. Anhydrite precipitation may inhibit dolomitization by two means: by diminishing calcium-ion activity during calcite dissolution, and possibly by coating calcite crystals with a thin $CaSO_4$ layer, which prevented the nucleation of dolomite on calcite crystal surfaces. After the disappearance of calcite, the saturation of dolomite in solution may not have been great enough for spontaneous precipitation.
(3) There is an indication from the precipitation of dolomite in sulfate-bearing solutions seeded with carbonate minerals other than calcite that sulfate in solution does not strongly inhibit the direct precipitation of dolomite from solution.
(4) The rate of dolomitization of calcite is much more rapid in solutions with higher carbonate and bicarbonate activities.

There is considerable scope for future work in experiments of this type. More could be done to document the early stages of calcite replacement by dolomites in solutions with variable sulfate concentration and to document the suggestion that direct precipitation of dolomite from sulfate-bearing solutions is possible, even at sulfate concentrations greater than 0.004M. Future work could also be focused on the kinetic effect that higher carbonate concentrations have on the rate of the dolomitization reaction.

ACKNOWLEDGMENTS

We thank Shell Oil of Canada, Ltd., for providing us with solution sulfate analyses. Pat Longhurst, Bradley Gorham, and Jennifer Wong at Shell Oil and at the Institute of Sedimentary and Petroleum Geology in Calgary assisted with the acquisition of the chemical data reported here. Many people suggested improvements to the manuscript. Foremost among these were Ronald K. Stoessel, Terry Gordon, Duncan F. Sibley, and Paul Baker. Ian Hutcheon very kindly provided us with an updated version of the SOLMNEQ program. To all these people, we extend our appreciation. Finally, we commend the editors Vijai Shukla and Paul Baker for their efforts on our behalf.

REFERENCES

Baker, P. A., and Burns, S. J., 1985, Occurrence and formation of dolomite in organic-rich continental margin sediments; American Association of Petroleum Geologists Bulletin, v. 69, p. 1917–1930.
———, and Kastner, M., 1981, Constraints of the formation of sedimentary dolomite: Science, v. 213, p. 214–216.
Busenberg, E., and Plummer, L. N., 1985, Kinetic and thermodynamic factors controlling the distribution of SO_4 and Na in calcites and selected aragonites: Geochimica et Cosmochimica Acta, v. 49, p. 713–725.
Carpenter, A. B., 1980, The chemistry of dolomite formation 1: The stability of dolomite, *in* Zenger, D. H., Dunham, J. B. and Ethington, R. L., eds., Concepts and Models of Dolomitization: Society of Economic Paleontologists and Mineralogists Special Publication 28, p. 111–1212.
Froese, E., 1967, A note on strontium magnesium carbonate: Canadian Mineralogist, v. 9, p. 65–70.
Garrels, R. M., and Thompson, M. E., 1962, A chemical model for sea water at 25°C and one atmosphere total pressure: American Journal of Science, v. 260, p. 57–66.
Goldsmith, J. R., and Graf, D. L., 1958, Structural and compositional variations in some natural dolomites: Journal of Geology, v. 66, p. 173–186.
Hardie, L. A., 1987, Dolomitization: A critical review: Journal of Sedimentary Petrology, v. 57, p. 166–183.
Helgeson, H. C., Delaney, J. M., Nesbitt, H. W., and Bird, D. K., 1978, Summary and critique of the thermodynamic properties of rock-forming minerals: American Journal of Science, v. 278–A, 229 p.
Kastner, M., 1983, Origin of dolomite and its spatial and chronological distribution–A new insight (abst.): American Association of Petroleum Geologists Bulletin, v. 67, p. 2156.
———, 1984, Control of dolomite formation: Nature, v. 311, p. 410–411.
Kharaka, Y. K., and Barnes, I., 1973, SOLMNEQ: Solution-mineral equilibrium computations: U. S. Department of Commerce, NTIS Report PB 215-899, Springfield, Virginia, 81 p.
Land, L. S., 1980, The isotopic and trace element geochemistry of dolomite: The state of the art, *in* Zenger, D. H., Dunham, J. B., and Ethington, R. L., eds., Concepts and Models of Dolomitization: Society of Economic Paleontologists and Mineralogists Special Publication 28, p. 87–110.
Lippman, F., 1973, Sedimentary Carbonate Minerals: Springer-Verlag, New York, 228 p.
Lumsden, D. N., and Chimahusky, J. S., 1980, Relationship between dolomite nonstoichiometry and carbonate facies parameters, *in* Zenger, D. H., Dunham, J. B., and Ethington, R. L., eds., Concepts and Models of Dolomitization: Society of Economic Paleontologists and Mineralogists Special Publication 28, p. 123–137.
Morrow, D. W., and Ricketts, B. D., 1986, Chemical controls on the precipitation of mineral analogues of dolomite: The sulfate enigma: Geology, v. 14, p. 408–410.
Oomori, T., Kaneshima, K., Taira, T., and Kitano, Y., 1983, Synthetic studies of protodolomite from brine waters: Geochemical Journal, v. 17, p. 147–152.

PATTERSON, R. J., 1972, Hydrology and carbonate diagenesis of a coastal sabkha in the Persian Gulf: Unpublished Ph.D. Dissertation, Princeton University, Princeton, New Jersey, 498 p.

ROBIE, R. A., HEMINGWAY, B. S., AND FISHER, J. R., 1979, Thermodynamic properties of minerals and related substances at 298.15K and 1 bar pressure and at higher temperatures: U.S. Geological Survey Bulletin 1452, 456 p.

ROSENBERG, P. E., AND HOLLAND, H. D., 1964, Calcite-dolomite-magnesite stability relations in solutions at elevated temperatures: Science, v. 145, p. 700–701.

STOESSELL, R. K., 1987, Chemistry and environments of dolomitization—A reappraisal: Discussion: Earth-Science Reviews v. 24, p. 211–212.

APPENDIX.—STOICHIOMETRY AND CRYSTALLINITY OF DOLOMITE PRECIPITATES

Table and Run No.	Mole% $CaCO_3$[1]	Crystallinity Index
Table 4—Run 1	51.5	0.50
Table 4—Run 2	51.5	0.36
Table 4—Run 4	51.5	0.44–0.50
Table 5—Run 1	56.0	amount too little
Table 5—Run 2	50.0	0.31
Table 5—Run 3	51.0	amount too little
Table 5—Run 4	51.0	amount too little
Table 6—Run 1	54.5	amount too little
Table 6—Run 2	51.5	0.25
Table 6—Run 3	51.0	0.35
Table 6—Run 4	51.0	0.50
Table 6—Run 8	56.0	0.30
Table 6—Run 9	50.0	0.36
Table 6—Run 10	50.0	0.50
Table 6—Run 11	50.0	amount too little
Table 8—Run 1	51.5	amount too little
Table 8—Run 2	50.0	0.34
Table 8—Run 3	50.0	0.36
Table 8—Run 4	50.0	0.35
Table 8—Run 5	50.0	0.36
Table 8—Run 6	50.0	0.40

1. The mole percent of $CaCO_3$ in these dolomite precipitates was calculated according to the formula of Lumsden and Chimahusky (1980) based on the position of the $\mathbf{d}_{(104)}$ X-ray peak spacing (Goldsmith and Graf, 1958). Dolomite crystallinity was calculated, as outlined in Patterson (1972), as the ratio of the sum of the peak heights of the superstructural reflections (10.1), (01.5) and (02.1) to the prominent non-superstructural reflection (11.3).

SECTION II
ORGANOGENIC DOLOMITES

SECTION II: INTRODUCTION
ORGANOGENIC DOLOMITES

The two papers in this section describe dolomite in organic-rich Miocene formations in California.

Burns and others examine the origin of dolomite nodules in siliceous mudstones. The dolomite is interpreted to have formed without precursor biogenic calcite. Instead, carbonate precipitating from solution was subsequently replaced by dolomite.

Compton evaluates the covariance of dolomite, pyrite, organic matter, and various trace metals. Fine-grained dolomite is post-pyrite in origin and may even have recrystallized at depth into coarser grained dolomite.

THE FACTORS CONTROLLING THE FORMATION AND CHEMISTRY OF DOLOMITE IN ORGANIC-RICH SEDIMENTS: MIOCENE DRAKES BAY FORMATION, CALIFORNIA

STEPHEN J. BURNS[1] AND PAUL A. BAKER
Department of Geology, Duke University, Durham, North Carolina 27706
AND
WILLIAM J. SHOWERS
Department of Marine Earth and Atmospheric Sciences, North Carolina State University, Raleigh, North Carolina 27695

ABSTRACT: The Drakes Bay Formation is an upper Miocene sequence of siliceous mudstones containing many small dolomite nodules. The nodules probably formed without a precursor biogenic calcite supplying Ca or HCO_3^- for dolomitization. Dolomite formation preferentially took place in sediment layers slightly richer in organic C than the surrounding sediments. More extensive sulfate reduction in these layers raised the porewater HCO_3^- concentration and caused carbonate precipitation. The initial carbonate may have been dolomite or calcite, which was later converted to dolomite.

Carbon and oxygen stable isotope ratios vary systematically and clearly illustrate changes in the isotopic composition of dissolved CO_2 that occurred with depth. The isotopic analyses show that dolomite formation did not begin until the pore waters were free of dissolved sulfate.

The Ca contents of the dolomites decrease, and both the Mg and Fe contents increase, with depth of formation. Manganese contents correlate with Fe contents. Sodium contents of the dolomites are relatively high, probably reflecting their poor ordering and nonstoichiometry. Strontium contents of the dolomites are typical of those from hemipelagic sediments with moderate sedimentation rates.

INTRODUCTION

The Drakes Bay Formation is an upper Miocene sequence of hemipelagic siliceous mudstones that contains numerous small dolomite nodules. Dolomites in the Drakes Bay Formation are similar to well-studied dolomites from the Miocene Monterey Formation and from modern continental margins. Unlike many of these dolomites, however, dolomites found at Drakes Bay probably formed without a significant calcite precursor and thus are truly authigenic in nature. As a result, they may be particularly useful in studying the questions that remain about the formation of dolomite in marine hemipelagic sediments, including: (1) the exact role of sulfate reduction in dolomite formation; (2) the depths at which these dolomites form; (3) the controls on where dolomite is likely to form; and (4) the controls on the major and minor-element chemistry of these dolomites.

The environments of formation of dolomites from the Monterey Formation, from modern continental margins, and from the Drakes Bay Formation have several things in common. All were formed in continental-margin sediments rich in organic carbon, and their carbon-isotopic composition is strongly influenced by organic-matter diagenesis. Collectively, they have been termed "organic" (Spotts and Silverman, 1966) or "organogenic" (Compton, 1986) dolomites. Many aspects of their origin and geochemistry are well understood, particularly the diagenetic nature of the dolomite (Bramlette, 1946; Pisciotto, 1981) and the relationship between their highly varying carbon-isotopic ratios and various zones of organic-carbon diagenesis (Spotts and Silverman, 1966; Friedman and Murata, 1979; Irwin, 1980; Pisciotto and Mahoney, 1981; Kelts and McKenzie, 1982; and others).

Oxygen isotope values of organic dolomites have frequently been used to determine their temperature of formation and are often coupled with an assumed geothermal gradient and initial temperature and an assumed porewater $\delta^{18}O$ value to calculate depths of formation (Pisciotto, 1978, 1981; Pisciotto and Mahoney, 1981; Kelts and McKenzie, 1982, 1984; Mertz, 1984; Garrison and Graham, 1984; Kablanow and others, 1984; Kushnir and Kastner, 1984; Hennessy and Knauth, 1985). Calculated depths typically range from 100 to 1,000 m. The oxygen isotopes, however, are probably reset as the dolomite recrystallizes and becomes more ordered during deeper burial (Gaines, 1977; Katz and Matthews, 1977; Land, 1980). Calculations of the depths of dolomite formation on the basis of a diffusion-limited seawater source of Mg^{2+} constrain dolomite formation to much shallower burial depths, particularly for settings where a significant amount of dolomite is formed (Baker and Burns, 1985; Compton and Siever, 1986; Burns and Baker, 1987).

Relatively little work has been done on the minor- and trace-element compositions of organic dolomites. The Fe contents of these dolomites appear to be most closely related to the amount of terrigenous sediment deposited in a particular location (Kushnir and Kastner, 1984; Burns and Baker, 1987) and to the zone of organic-matter diagenesis in which the dolomite forms (Murata and others, 1972; Pisciotto, 1981). Manganese contents usually correlate with Fe and are thought to be controlled by the same mechanisms (Burns and Baker, 1987). The Sr contents of organic dolomites may be a function of the stoichiometry of the dolomitization reaction, which in turn is a function of the amount of dolomite that forms at a particular location (Baker and Burns, 1985).

In this report we present the results of a study of the sedimentology and geochemistry of the Drakes Bay Formation as they pertain to the origin of organic dolomites. These data are used to refine our understanding of the controls on dolomite formation and dolomite composition in organic-rich sediments.

[1]Present address: Department of Marine Geology and Geophysics, Rosenstiel School of Marine and Atmospheric Science, 4600 Rickenbacker Causeway, Miami, Florida 33149.

Sedimentology and Geochemistry of Dolostones, SEPM Special Publication No. 43

STRATIGRAPHY

The first detailed study of the upper Miocene and Pliocene sedimentary rocks of the Point Reyes Peninsula was by Galloway (1977), who named the sequence the Drakes Bay Formation. The rocks are a transgressive series of sandstones, siltstones, and siliceous mudstones lying unconformably on the Monterey Formation. They occupy a gently folded basin between Point Reyes on the west and the San Andreas Fault on the east (Galloway, 1977; Clark and others, 1984).

The study area for the present paper is the type section of the Drakes Bay Formation, located along Drakes Beach just northeast of Point Reyes (Fig. 1). At this location, the lower few meters of the Drakes Bay Formation are composed of a distinctive glauconitic arkosic sandstone. At the base of the sandstone is a conglomerate that contains clasts of the underlying Monterey Formation. This lower greensand is overlain by 20 to 30 m of thinly bedded, light brown siliceous mudstone. The mudstone grades upward into more thickly bedded, light tan to light grey siltstones and mudstones. Several thin (30–50 cm) arkosic sandstone beds are interbedded with the siltstones in the middle and upper parts of the formation (Fig. 2). Both the siliceous mudstone unit and the siltstones above it contain numerous layers of dolomite nodules. The top of the Drakes Bay has been removed by erosion.

In more recent work, Clark (1981) and Clark and others (1984) correlated the Drakes Bay Formation to rocks in the Santa Cruz Mountains and divided the formation into three units—the Santa Margarita Sandstone, the Santa Cruz Mudstone, and the Purisima Formation. Applying this stratigraphy to the study area, the glauconitic sandstone at the base of the Drakes Bay Formation corresponds to the Santa Margarita Sandstone, the siliceous mudstones overlying it correspond to the Santa Cruz Mudstone, and the interbedded siltstones and mudstones correspond to the Purisima Formation (Fig. 2). In and near the study area, the Drakes Bay Formation is largely composed of rocks that correlate with the Purisima Formation. The portion of the Drakes Bay Formation, which Clark and others (1984) correlated with the Santa Cruz Mudstone, thickens considerably to the southeast of the study area, becoming as much as 1,600 to 2,000 m thick near Bolinas (Clark and others, 1984). Because the rocks in the measured section in this study are quite homogeneous, and for the sake of simplicity, we will refer to them as the Drakes Bay Formation.

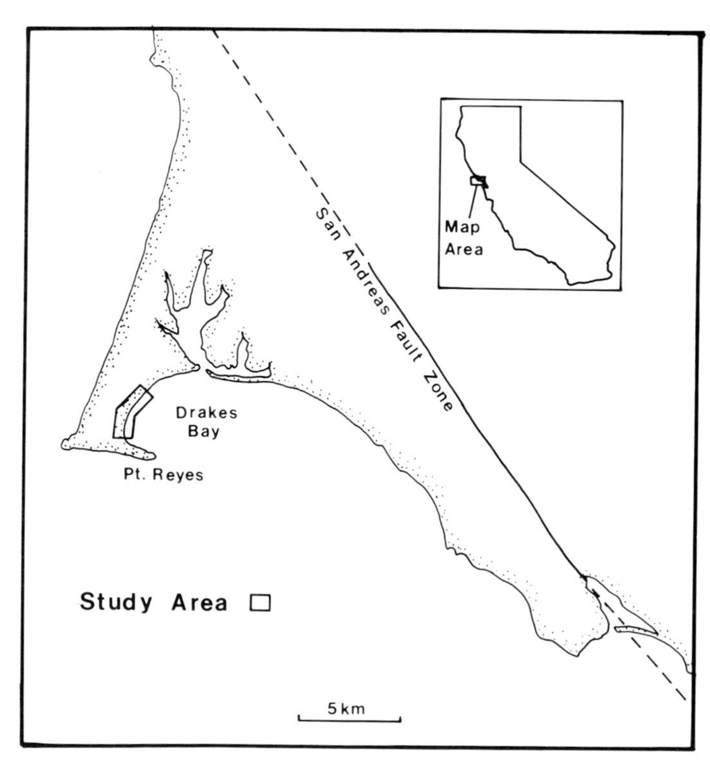

FIG. 1.—Location of study area.

The age of the Drakes Bay Formation is not firmly established. No fossils have been found in the glauconitic sandstone at the base of the formation. A K-Ar date of 9.3 ± 0.5 Ma was determined for glauconite from the sandstone (Galloway, 1977). This age matches fairly closely the age of the top of the correlative Santa Margarita Sandstone from the Santa Cruz area, which is approximately 9 Ma (Kleinpell, 1938), although the age of the Santa Margarita Sandstone may vary considerably (e.g., Barron, 1986a). Clark and others (1984) reported ages for the Santa Cruz Mudstone on the basis of arenaceous benthic foraminifera and diatoms collected from near Bolinas. The foraminifera belong to the *Bolivina obliqua* Zone, the lower portion of which is around 6.5 Ma (Kleinpell, 1980). The diatom flora was diagnostic of subzone a of the *Nitzschia reinholdii* Zone, the age of which overlaps the age of the lower portion of the *Bolivina obliqua* Zone (Barron, 1986b).

Within the study area, microfossil preservation is generally poor. Diatoms belonging to subzone b of the *Nitzschia reinholdii* Zone were found in sample DB-42 (J. A. Barron, pers. commun., 1986) from 127 m above the base of the measured section (Fig. 2). Clark and others (1984) also reported that a sample from within the study area, approximately 240 m above the base of the measured section, has a flora identified as belonging to subzone b of the *N. reinholdii* Zone. The latter sample may be somewhat younger than DB-42, 5.5–5.2 Ma versus 5.7–5.3 Ma for DB-42 (J. A. Barron, pers. commun., 1986). The radiolaria fauna could be confidently identified from only one sample, DB-56, from 197 m above the base of the section. This sample belongs to the *Stichocorys peregrina* Zone (R. E. Casey, pers. commun., 1985), which is approximately 5.5 to 6 Ma (Weaver and others, 1981). Combining all of the above data, a best estimate of the age of the measured section is approximately 6–6.5 Ma at the bottom and approximately 5 Ma, latest Miocene or earliest Pliocene, at the top (Fig. 2).

METHODS

A single section of the Drakes Bay Formation, at the type locality, was measured, described, and sampled. All layers of dolomite nodules were sampled, and samples of the surrounding lithologies were taken at 3- to 5-m intervals. Six

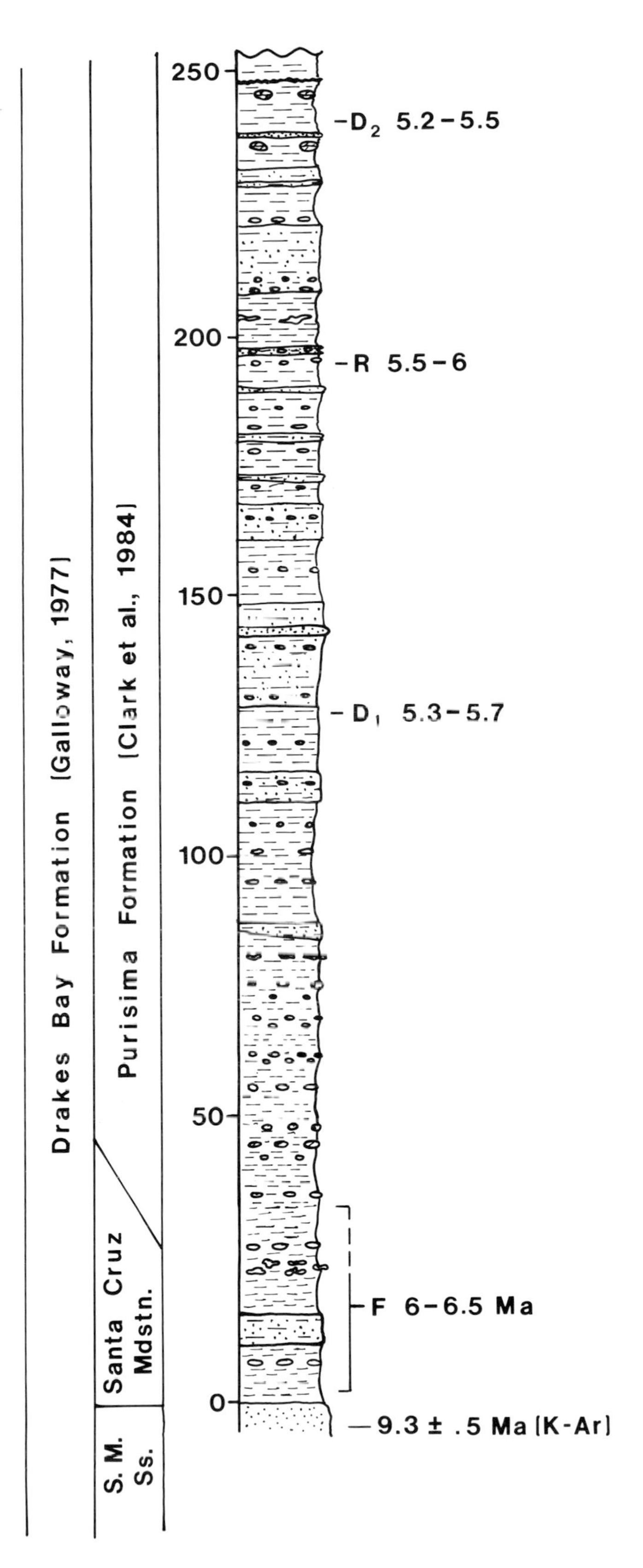

of the dolomite nodules were sampled several times from their exterior to their interior. For a total of 39 dolomite samples and 43 sediment samples, the weathered portion of each sample was removed and the sample was powdered. An oriented dry-powder mount of each sample was analyzed by X-ray diffraction (XRD) on a Phillips X-ray diffraction unit.

The dolomite samples were analyzed for the following: Ca, Mg, Fe, Mn, Sr, Na, and carbon and oxygen stable isotopes. For elemental analyses, the powdered sample was washed in distilled deionized water, dissolved in 1 M HCl for 15 min, and then filtered through 0.45-μm membrane filters. The solution was analyzed by standard flame atomic absorption spectrophotometry techniques on a Perkin-Elmer Model 5000 spectrophotometer. For isotopic analyses, a small portion of powdered washed sample was roasted in a vacuum at 320°C for 1 hr. The sample was dissolved overnight in phosphoric acid at 50°C (McCrea, 1950). Analyses of the separated CO_2 were done on a Finnegan MAT-251 stable isotope ratio mass spectrometer at North Carolina State University. All results are presented relative to the PDB standard, and no phosphoric acid fractionation factor for the difference between calcite and dolomite was applied to the isotope values. Analyses of replicate samples yielded values within 2 to 3% of each other for elemental values and estimated error for the isotope ratios is ±0.07‰ for $\delta^{13}C$ and ±0.120‰ for $\delta^{18}O$.

Whole rock elemental analyses were conducted on 17 dolomite and 15 noncarbonate sediment samples. The samples were washed in distilled water, fused with a lithium borate flux, and dissolved in 1% HNO_3. Samples were analyzed for Si, Al, Mn, Fe, Mg, Ca, Na, K, Ti, P, Sr, and Ba by standard techniques on a Perkin-Elmer Model 6000 inductively coupled plasma (ICP) spectrophotometer. The Si analyses were checked by the standard molybdate complex colorometric method (e.g., Gieskes, 1974). Not all of the fused sample could be recovered for subsequent dissolution. Thus, the results are reported as weight ratios to aluminum.

Total N, C, and S analyses were carried out on 22 dolomite and 20 sediment samples. All samples were run in triplicate. Dolomite samples were first leached in 1 M HCl for 15 min to remove the carbonate. One set of dolomite samples was analyzed without rinsing, and two sets were rinsed multiple times in a 0.5 M NaCl solution and finally rinsed in distilled water to remove any soluble-sulfate sulfur prior to analysis. One set of the noncarbonate sediment samples was run unwashed and two sets were treated the

←

FIG. 2.—Generalized stratigraphy of the studied section. F marks location of microfossil locality with foraminifera belonging to subzone a of the *Nitzschia reinholdii* Zone (Clark and others, 1984). D_1 = location of sample DB-42, containing diatoms belonging to the lower part of subzone b of the *N. reinholdii* Zone (J. A. Barron, pers. commun., 1986). R = location of sample DB-56 with radiolaria belonging to the *Stichocorys peregrina* Zone (R. E. Casey, pers. commun., 1985). D_2 = location of a microfossil locality with diatoms belonging to upper part of subzone b of the *Nitzschia reinholdii* Zone (Clark and others, 1984). K-Ar date from Galloway (1977).

same as the dolomite samples. The analyses were done on a Carlo Erba NA-1500 Nitrogen-Carbon-Sulfur Analyzer by combustion in an oxygen-enriched atmosphere at 1,200°C. Replicate samples run with and without vanadium pentoxide yielded the same results.

RESULTS AND DISCUSSION

Sedimentology.—

The rocks of the type section of the Drakes Bay Formation are quite uniform in mineralogy, chemical composition, organic-matter content, and texture. X-ray diffraction analyses show that the mudstones and siltstones surrounding the dolomite nodules are primarily composed of detrital quartz and feldspars, opal CT (a product of biogenic opal diagenesis), and clay minerals, usually chlorite and montmorillonite. The rocks contain essentially no calcite. When analyzed by combustion, both unwashed and acid-washed samples contained the same amount of C, indicating no contribution to the total C of the rocks from any carbonate minerals. Also, XRD analyses revealed no calcite. The SiO_2/Al_2O_3 weight ratio of the mudstones is fairly constant, with almost all values falling between 5 and 7 (Table 1). The FeO/Al_2O_3, Na_2O/Al_2O_3, and K_2O/Al_2O_3 ratios are also relatively constant (Table 1). The organic C values almost all fall between 0.5 and 1.0%. Texturally, the rocks are all nonlaminated, fine-grained mudstones and siltstones.

Depth and location of dolomite formation.—

Why did dolomite form in this environment, and what controls when and where it formed? Dolomite in the Monterey Formation (Bramlette, 1946) and in similar sediments at DSDP Site 479 (Kelts and McKenzie, 1982) may have preferentially formed in layers that had higher calcareous microfossil contents. Not all organic dolomites, however, had a calcite precursor (Pisciotto, 1981; Hennessey and Knauth, 1985). For dolomites from the Drakes Bay Formation, there is virtually no carbonate in the rocks surrounding the dolomite nodules, and there is also no evidence of precursor carbonate in the dolomite nodules themselves. Thin sections of the dolomite nodules only rarely contain any recrystallized foraminifera or other carbonate shelled organisms, although these could have been destroyed by dolomite formation. On the other hand, because the rocks above and below the dolomite nodules contain no calcite, the dolomite layers would have to have been very calcite rich, probably approximately 50% calcite, if they were to supply the precursor carbonate for all of the dolomite. The sediments adjacent to the dolomite nodules in the same stratigraphic level, however, are similar in lithology to the siltstones and mudstones above and below the nodules.

Dolomite nodules appear to have formed in sediment layers or lenses that were richer in organic C than the surrounding sediments. The results of the N-C-S analyses are presented in Figures 3 and 4 and Table 2. The percent or-

TABLE 1.—WHOLE ROCK ANALYSES

Sample#	m above base	SiO_2/Al_2O_3	FeO/Al_2O_3	MnO/Al_2O_3	Na_2O/Al_2O_3	K_2O/Al_2O_3
DB1	−1	9.07	2.70	0.44	0.17	0.68
DB5	12	5.92	0.26	0	0.13	0.16
DRK6	17	7.09	0.14	0	0.18	0.19
DRK12	33	4.75	0.26	0	0.14	0.19
DRK15	42	5.83	0.31	0	0.15	0.12
DRK23	68	5.65	0.20	0	0.12	0.13
DRK28	89	5.44	0.21	0	0.19	0.18
DRK33	111	4.65	0.35	0.001	0.11	0.13
DRK37	123	4.92	0.32	0	0.13	0.15
DRK41	140	6.22	0.21	0	0.19	0.21
DRK44	145	7.29	0.18	0	0.28	0.23
DRK46	156	5.15	0.17	0	0.18	0.14
DRK48	169	4.75	0.34	0.001	0.13	0.15
DRK51	182	5.02	0.34	0.001	0.15	0.17
DB54	192	5.79	0.23	0	0.15	0.14
DRK54	205	6.95	0.31	0	0.13	0.16
DRK57	231	5.54	0.26	0	0.15	0.15
DOLOMITES						
DB4	4	4.21	0.31	0.006	0.15	0.16
DB8	21	5.28	0.41	0.138	0.17	0.16
DB10a	25	5.34	0.95	0.092	0.17	0.14
DB10b	25	3.81	0.60	0.053	0.14	0.12
DB10c	25	3.98	0.60	0.056	0.15	0.13
DB23	69	5.46	0.51	0.033	0.17	0.14
DB27	81	5.69	0.42	0.035	0.17	0.14
DB30	91	5.05	0.48	0.022	0.15	0.13
DB37	110	5.18	0.54	0.047	0.20	0.16
DB43a	130	4.93	1.12	0.121	0.15	0.15
DB43c	130	3.72	1.09	0.092	0.10	0.11
DB49	162	6.09	0.38	0.011	0.15	0.14
DB53	185	5.14	0.45	0.019	0.15	0.13
DB63	227	5.60	0.71	0.027	0.16	0.12
DB67	243	5.01	0.59	0.030	0.13	0.11

All results in oxide ratios to aluminum. For multi-sampled nodules, samples labelled a, b, . . . are from the outside to the inside of the nodule.

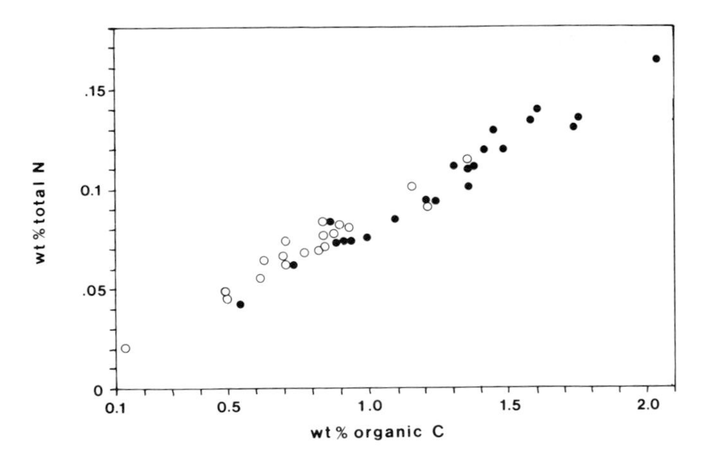

FIG. 3.—Weight percent organic C vs. weight percent total N for sediments and dolomites. Open circles are nondolomite sediment samples, solid circles are dolomites. Percentages for the dolomites are reported on a carbonate-free basis.

ganic C of the dolomites and the non-acid and non-water soluble percent S (which is believed to be representative of S from pyrite), reported on a carbonate-free basis, both tend to be higher in the dolomite nodules than in the surrounding rocks.

There are several possible explanations for this observation. The most plausible one is that there was a slightly higher original organic C content in the layers where the dolomite formed. Higher organic C levels would have caused more rapid and extensive sulfate reduction in those layers (Berner, 1978; Jorgensen, 1979), raising the dissolved sulfide concentration of the pore waters. High-sulfide concentrations coupled with dissolved Fe^{2+} in the pore waters would have lead to more extensive precipitation of Fe sulfides. The extensive sulfate reduction would also have greatly increased the porewater bicarbonate concentration. The combined effects of removing sulfate and adding bicarbonate would promote dolomitization (Baker and Burns, 1985). If these two processes occurred most rapidly in particular layerse, dolomite formation might be localized in those layers.

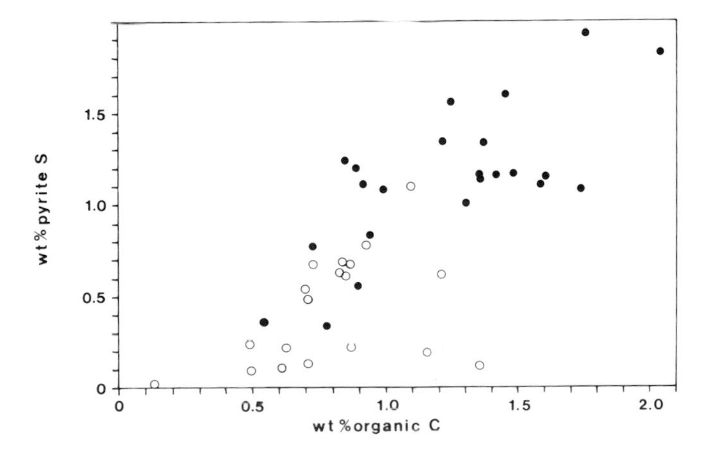

FIG. 4.—Weight percent organic C vs. weight percent pyrite S for sediments and dolomites. Open circles are nondolomite sediment samples, solid squarres are dolomites. Percentages for the dolomites are reported on a carbonate-free basis.

TABLE 2.—NITROGEN-CARBON-SULFER ANALYSES, IN WEIGHT PERCENTS

Sample#	Meters	Percent N	Percent C	Percent S	C/N	C/S
DB-5	12	0.0952	1.1873	0.2778	12.5	4.3
DB-6	17	0.0623	0.7036	0.1328	11.3	5.3
DRK-12	33	0.0721	0.8977	0.921	12.5	1.0
DB-15	42	0.0833	0.972	0.9796	11.7	1.0
DRK-20	55	0.0921	0.935	0.8039	10.2	1.2
DRK-23	68	0.0838	0.9878	0.3313	11.8	3.0
DRK-26	79	0.0683	0.8373	0.5196	12.3	1.6
DRK-28	89	0.0485	0.5306	0.3443	10.9	1.5
DRK-30	97	0.0577	0.6187	0.7355	10.7	0.8
DRK-33	111	0.0687	0.7803	0.9202	11.4	0.8
DRK-37	123	0.0769	0.7839	1.0339	10.2	0.8
DRK-41	140	0.0552	0.5653	0.1518	10.2	3.7
DRK-44	145	0.0194	0.1236	0.0177	6.4	7.0
DRK-46	156	0.0566	0.5949	0.1633	10.5	3.6
DRK-48	169	0.0878	0.8797	0.9803	10.0	0.9
DRK-51	182	0.0731	0.6757	0.6568	9.2	1.0
DRK-54	192	0.0614	0.7052	0.5133	11.5	1.4
DRK-57	231	0.0858	0.9975	0.9504	11.6	1.0
DB-62	223	0.0751	0.8615	0.9251	11.5	0.9
DB-68	245	0.1148	1.3172	0.1284	11.5	10.3
Dolomites						
DB-4	6	0.1133	1.4271	1.0696	12.6	1.3
DB-8	21	0.0594	0.6877	0.5346	11.6	1.3
DB-10a	25	0.1324	1.6651	1.1848	12.6	1.4
DB-10c	25	0.1301	1.4991	1.1493	11.5	1.3
DB-14	40	0.1368	1.1163	0.8316	8.2	1.3
DB-17	49	0.1127	1.365	1.4375	12.1	0.9
DB-23	69	0.0839	1.1305	1.2315	13.5	0.9
DB-27	81	0.0774	0.9763	0.9894	12.6	1.0
DB-30	91	0.0803	1.0111	1.2525	12.6	0.8
DB-32	96	0.0815	1.0385	1.5121	12.7	0.7
DB-35a	103	0.1147	1.429	1.2651	12.5	1.1
DB-35c	103	0.1635	2.0089	1.9175	12.3	1.0
DB 37	110	0.0935	1.1758	0.8371	12.6	1.4
DB 39	118	0.0575	0.666	0.796	11.6	0.8
DB-43a	130	0.1284	1.4496	1.8224	11.3	0.8
DB-43c	130	0.1387	1.8199	1.9641	13.1	0.9
DB-46	147	0.075	0.8628	1.2617	11.5	0.7
DB-49	160	0.1064	1.2655	1.5598	11.9	0.8
DB-51b	174	0.1353	1.6238	1.2313	12.0	1.3
DB-51d	174	0.1063	1.2269	1.515	11.5	0.8
DB-53	193	0.1209	1.4802	1.4157	12.2	1.0
DB-58	209	0.1009	1.2086	1.6424	12.0	0.7
DB-63	227	0.0829	0.8619	1.3619	10.4	0.6
DB-67	243	0.116	1.3451	1.2269	11.6	1.1

A second possibility is that the dolomites might have formed not where there is more organic C, but where the organic C is more reactive. The rate of sulfate reduction in sediments is a function of the reactivity of the organic C in the system in addition to the amount of organic C present (Goldhaber and Kaplan, 1975; Jorgensen, 1979; Westrich and Berner, 1984). Thus, the dolomites might form in layers where more rapid sulfate reduction is due to the presence of more reactive organic matter. This hypothesis is not supported by the C/N ratio data.

One possible measure of the reactivity of organic C could be its C/N ratio. Past studies of the C/N ratios of organic matter in coastal and marine sediments have shown that the greater the contribution of terrestrially derived carbon, the higher the C/N ratio (Nissenbaum and Kaplan, 1972; Tissot and others, 1974; Stuermer and others, 1978). Since marine organic C is more protein rich, and therefore more reactive than terrestrially derived organic matter (Goldhaber and Kaplan, 1975; Lyons and Gaudette, 1979), different C/N ratios may indicate different reactivities of the organic fraction. Typical values of C/N for continental-margin sediments range from 8 to 14 (Emery, 1960; Nissenbaum and Kaplan, 1972; Krom and Berner, 1981). Diagenesis of organic matter can change its C/N ratio. Nitrogen-bearing organic compounds are more reactive in many systems, and the C/N ratio of sedimentary organic matter frequently increases as diagenesis progresses (Sigl and others, 1978; Krom and Berner, 1981). Exceptions to that diagenetic trend are sediments with very low organic C/clay ratios, where, during diagenesis, significant N can be bound to clays in the form of ammonium, decreasing the C/N ratio (Muller, 1977).

In the Drakes Bay Formation, the C/N ratios of the dolomite nodules and of the surrounding sedimentary rocks are remarkably constant (Fig. 3). The average value of the ratio is 11, which falls about in the middle of the range found for continental-margin sediments. The constant C/N ratios are interpreted to mean two things. First, there was a relatively constant ratio of marine/terrestrial carbon input during deposition of the sediments. Second, diagenesis did not greatly alter the original signal. The fact that both dolomite and nondolomite samples fall on the same line (Fig. 3) implies that the dolomites did not form in layers where there was a different type of organic carbon.

Finally, it is possible there was initially a more constant percent organic C throughout the sediments, but early dolomite formation tightly cemented particular layers, preventing post-dolomite formation organic C or pyrite S loss in the nodules. Organic C could be lost through either methanogenesis or thermal cracking of the organic matter after dolomite formation. Both organic C and pyrite S could be preferentially lost from the more poorly indurated mudstones during weathering. There is some suggestion that the first mechanism operated for the dolomites that formed the most deeply. A plot of percent organic C vs. $\delta^{18}O$ (Fig. 5) shows that the dolomites with the lowest $\delta^{18}O$ values tend to contain the lowest amounts of organic C. For the majority of dolomites, however, there is no correlation between depth of formation, as interpreted from $\delta^{18}O$ values (discussed later) and their percent organic C.

The relatively constant C/N ratios of the insoluble fraction of the dolomites and of the rocks argue against the effects of surficial weathering (Fig. 3). It seems likely that more extensive weathering of the surrounding rocks would have altered their C/N ratios. The C/S ratios of the dolomites and sediments are also roughly the same through most of the section (Fig. 4). Again, it seems likely that differential weathering should have changed these ratios.

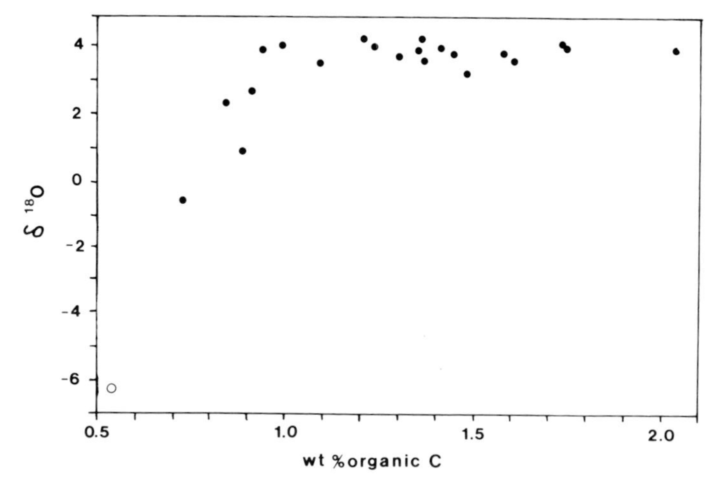

FIG. 5.—Weight percent organic C vs. $\delta^{18}O$ of dolomites in ‰ PDB.

For the Drakes Bay sediments, with little or no precursor calcite supplying Ca for dolomitization, it is likely that the primary stoichiometry of the dolomitization reaction was

$$Ca^{2+} + Mg^{2+} + 4HCO_3^- = CaMg(CO_3)_2 + 2H_2CO_3 \quad (1)$$

with the Ca and Mg both coming ultimately from sea water and the bicarbonate being produced during sulfate reduction. Dolomitization would remove both Ca and Mg from the pore waters, but because there is more than five times as much Mg in sea water as Ca, it is expected that the Ca would have been more rapidly depleted from the pore waters than Mg.

The implications of a seawater source of Ca are that dolomite formation must have taken place at shallow depths in the sediment column and that the rate and extent of dolomite formation were probably limited by the availability of Ca. The maximum depth of dolomite formation, based on a diffusion-limited source of Ca, may be calculated as follows.

Assuming that

(1) there is 1 weight percent dolomite in the sediments,
(2) the decompacted sedimentation rate is 200 m per million years,
(3) the average sediment grain density is 2.5 g/cm^3,
(4) the diffusion coefficient for Ca^{2+} in the sediment pore waters, D_s, was 4×10^{-6} cm^2/s (calculated from data in Li and Gregory, 1974), and
(5) the original porosity was 0.80,

then the flux of Ca^{2+} into the sediments, J, must have been:

$$\begin{aligned} J &= \text{(sedimentation rate)} \\ &\quad \text{(average mass Ca/volume of sediment)} \\ &= (6.4 \times 10^{-10}\ \text{cm/s}) \\ &\quad (2.7 \times 10^{-5}\ \text{moles Ca/cm}^3\ \text{of sediment}) \\ &= 17.3 \times 10^{-15}\ \text{moles} \\ &\quad \text{Ca/cm}^2\ \text{of sea floor/s}. \end{aligned}$$

From Fick's First Law for sediments (Berner, 1980, p. 38),

$$J = -\phi D_s (dC/dx) \quad (2)$$

where C is the concentration of dissolved Ca and x is the subbottom depth. Solving for the diffusion gradient yields

$$\begin{aligned} (dC/dx) &= -5.4 \times 10^{-9}\ \text{moles Ca/cm}^3\text{/cm} \\ &= -5.4 \times 10^{-4}\ \text{moles/L/m}. \end{aligned}$$

Using the seawater concentrations of 10.3×10^{-3} moles/L as the initial [Ca^{2+}], then [Ca^{2+}] = 0 at a depth of 19 m. In other words, if Ca was supplied by diffusion from overlying sea water, dolomite formation would have been limited to the upper 19 m of the sediment column.

Alternatively, the increased HCO_3^- concentrations resulting from sulfate reduction may first cause calcite or magnesian calcite to precipitate (e.g., Sholkovitz, 1973; Kelts and McKenzie, 1982, 1984; Kulm and others, 1984). This early formed calcite could then later be converted into diagenetic dolomite by a reaction such as that proposed by Baker and Burns (1985):

$$CaCO_3(s) + Mg^{2+} + 2HCO_3^- = CaMg(CO_3)_2 + H_2CO_3. \quad (3)$$

The Mg^{2+} for this reaction would probably have been supplied from the overlying sea water (Baker and Burns, 1985; Compton and Siever, 1986), and the maximum depth of dolomite formation would be limited by diffusion of Mg^{2+} rather than Ca^{2+}, into the sediments. Using the same assumptions as in the calculation above, this would limit dolomite formation to the upper 100 m of the sediment column. With the data on hand, it is impossible to determine whether Ca^{2+} or Mg^{2+} diffusion was the depth limiting factor.

In summary, the locus of dolomite formation was probably controlled by the amount of organic carbon in the sediments. The oxidation of organic matter via bacterially mediated sulfate reduction, by increasing the HCO_3^- concentration and removing dissolved sulfate from the sediment pore waters, initiated dolomitization. The depth of dolomite formation was limited by either the diffusion of Ca^{2+} or of Mg^{2+} into the sediments from sea water. This depth was probably less than 100 m subbottom.

Controls on the C and O isotopic compositions.—

The oxygen- and carbon-isotopic data for the dolomites are shown in Figure 6. The majority of the oxygen isotope ratios falls between +3.5 and +4.5‰ PDB. The carbon isotope ratios range from −14.6 to +12.8‰ PDB. Applying the method of Pisciotto (1978), the $\delta^{18}O$ values were used to calculate temperatures of formation for the dolomites. The calculated values range from approximately 5 to 10°C for the heavier values, to approximately 30 to 35°C for the two lightest values. These calculations assume a porewater $\delta^{18}O$ value of 0‰ (SMOW). The low calculated temperatures of formation for the majority of the dolomites imply dolomite formation at very shallow burial depths. Assuming that the temperatures of the bottom waters at the site of deposition were similar to the bottom water temperatures in present offshore basins of California, from about 4 to 8°C (Sholkovitz and Geiskes, 1971; Douglas, 1981), and that the near-surface geothermal gradient was about 60°C/km (not an excessive value for active margin basins, see Erickson, 1973; Pisciotto and Mahoney, 1981; Kelts and McKenzie, 1982), then the majority of the dolomites formed within a few tens of meters of the sediment/water interface. The deepest dolomites formed at approximately 300 to 400 m subbottom.

The carbon isotope ratio of the dolomites reflects the carbon isotope ratio of the dissolved CO_2 of the zone of organic-matter diagenesis in which they formed. Dolomites formed in or just below the zone of bacterial sulfate reduction should have negative $\delta^{13}C$ values, whereas those formed in the zone of methanogenesis should have positive $\delta^{13}C$ values (Claypool and Kaplan, 1974; Irwin, 1980; Pisciotto and Mahoney, 1981; Kelts and McKenzie, 1982).

Although this observation appears to be straightforward, it is not well understood what determines the zone of organic-matter diagenesis in which a dolomite is likely to form. Several workers have previously noted correlation between the sedimentation rate and the C isotope ratios of organic

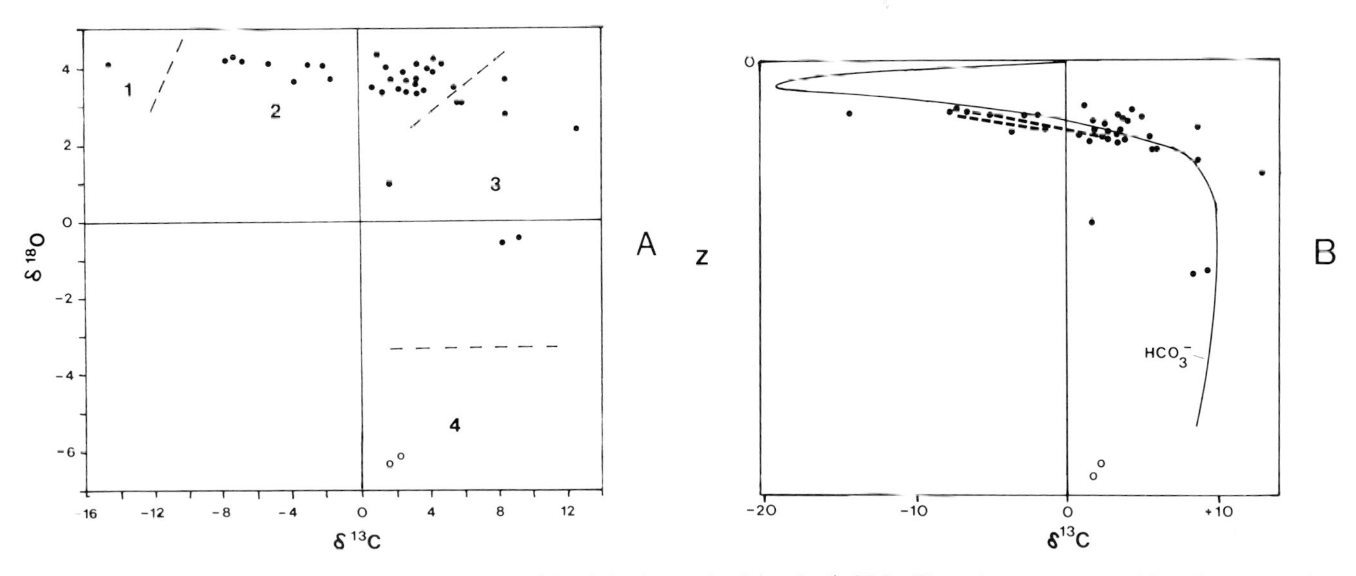

FIG. 6.—(A) Oxygen vs. carbon stable isotope ratios of the dolomites and calcites in ‰ PDB. The values are separated into four groups based on major-element content. The group 1 dolomite is very Ca rich and Fe poor. Group 2 dolomites are slightly Ca rich, becoming slightly Mg rich as $\delta^{18}O$ decreases, and with moderate Fe contents. Group 3 dolomites are Ca poor and Mg or Fe rich. Group 4 are Fe and Mg rich calcites. (B) Carbon isotope values vs. depth, as interpreted from oxygen isotope values, and a generalized $\delta^{13}C$ vs. depth curve for porewater bicarbonate. The carbonate isotope values fall only on the portion of the curve produced after complete sulfate reduction.

dolomites (Pisciotto and Mahoney, 1981; Kelts and McKenzie, 1982; Kulm and others, 1984; Kablanow and others, 1984; Burns and Baker, 1987, among others). Dolomites from areas with high-sedimentation rates often have positive values of $\delta^{13}C$, and those from areas with lower sedimentation rates often have negative values of $\delta^{13}C$. This observation is most likely a result of the relation between sedimentation rates and the depth at which sulfate is depleted in the sediment pore waters. Generally, the higher the sedimentation rate, the shallower the depth at which sulfate is completely reduced and the shallower the depth at which methanogenesis begins (Berner, 1978).

Because the zone of sulfate reduction is shallower than the zone of methanogenesis, it might be expected that there would be a relation between the oxygen and carbon isotopes of organic dolomites. The picture is complicated, however, because the depth ranges of the various zones of organic-matter oxidation vary as a result of differences in several factors, including: the sedimentation rate, the percent C_{org}, the reactivity of the organic carbon, and the rate of bioturbation (Berner, 1980, and references therein). Nonetheless, several studies of individual nodules have revealed systematic co-variations in $\delta^{13}C$ and $\delta^{18}O$ between the interior of a nodule, formed at shallow depths, and the exterior of a nodule, formed at greater depths (Irwin, 1980; Kelts and McKenzie, 1982; Kushnir and Kastner, 1984; Hennessey and Knauth, 1985).

The Drakes Bay Formation dolomites also show a systematic relation between their carbon and oxygen isotope ratios. Figure 6 shows that as the $\delta^{18}O$ values decrease, the $\delta^{13}C$ values shift from negative to positive and with greater depth begin to swing back toward negative values. The shape of this curve closely matches portions of the commonly observed $\delta^{13}C$ of dissolved CO_2 vs. depth curve for sediments where sulfate reduction and methanogenesis are occurring (Claypool and Threlkeld, 1983). The fact that the values for dolomites from throughout the section should fit a single curve of this type indicates that the depth of sulfate depletion remained fairly constant during deposition of the entire sequence. The $\delta^{13}C$ values for the Drakes Bay dolomites are thus a simple function of the depth at which they formed.

It is important to note that the $\delta^{13}C$ vs. $\delta^{18}O$ curve for the dolomites (Fig. 6) coincides with only the lower portion of the $\delta^{13}C_{CO_2}$ vs. depth curve of Claypool and Kaplan (1974). With one exception, all the dolomites fall near the part of the curve formed after sulfate reduction was complete. The values do not seem to fall on the upper portion of the curve where, *during* sulfate reduction, the dissolved CO_2 becomes isotopically lighter. Rather, they fall on the part of the curve where, after methanogenesis begins, the dissolved CO_2 becomes heavier. In other words, the dolomites form only when the pore waters are free of dissolved sulfate. This observation can also be made for other studies where it is possible to see a relation between $\delta^{18}O$ and $\delta^{13}C$ (see Hennessey and Knauth, 1985, Fig. 7; Mertz, 1984, Fig. 8; Kushnir and Kastner, 1984; Fig. 8), and is consistent with experimental evidence regarding the inhibiting effect of dissolved sulfate on dolomite formation (Baker and Kastner, 1981).

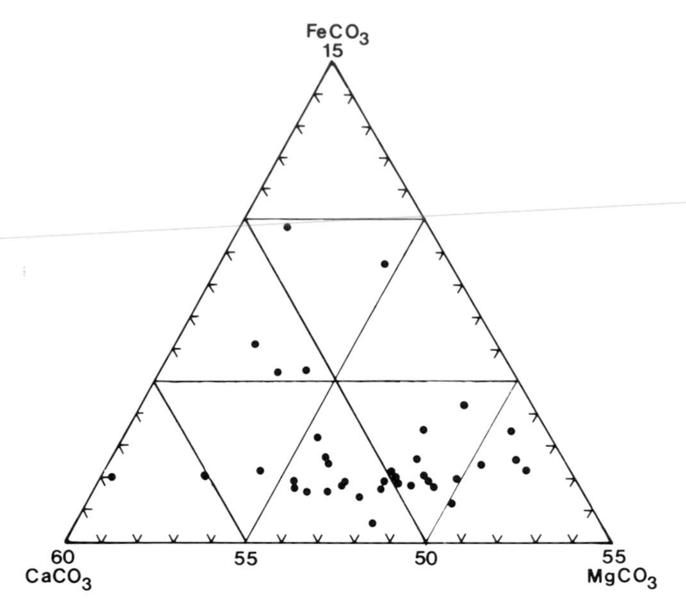

FIG. 7.—Triangular plot of Ca-Mg-Fe contents, in mole percents, of the dolomites.

Controls on the major-element chemistry: Ca, Mg, and Fe.—

Many studies of organic-type dolomites have shown that they are typically poorly to moderately ordered and usually Ca and Fe rich as compared to stoichiometric dolomite. Calcium contents of organic dolomites usually fall within the range of 51 to 55 mole percent Ca, and Fe values are commonly 2–3 mole percent with values as high as 15 mole percent (Murata and others, 1972; Pisciotto, 1978, 1981; Kelts and McKenzie, 1982; Shimmield and Price, 1984; Kushnir and Kastner, 1984; Compton, 1986; Burns and Baker, 1987). Magnesium contents are correspondingly low. The calcium-rich nature of these dolomites is probably a result of their initial formation as protodolomites and incomplete recrystallization to stoichiometric dolomites (Compton, 1986; Burns and Baker, 1987). The Fe content of many organic dolomites is primarily a function of the availability and content of Fe in the surrounding sediments (Kushnir and Kastner, 1984; Burns and Baker, 1987), and secondarily of the zone of organic-matter diagenesis in which the dolomite forms (Murata and others, 1972; Pisciotto, 1981). The Fe content is often correlated with $\delta^{13}C$, with dolomites that have negative $\delta^{13}C$ values having lower Fe contents than dolomites with positive $\delta^{13}C$ values (Murata and others, 1972; Pisciotto, 1981). The correlation can primarily be ascribed to the fact that dolomites with negative $\delta^{13}C$ values often form in areas where there is a low-sedimentation rate and therefore a relative lack of a terrigenous-sediment source of Fe. On the other hand, dolomites with positive $\delta^{13}C$ values often form in areas with high-sedimentation rates because of significant detrital terrigenous sedimentation (Pisciotto, 1981; Burns and Baker, 1987). Also, dolomites with negative $\delta^{13}C$ values form in or near the zone of sulfate reduction, where available Fe can be precipitated in pyrite rather than dolomite (Pisciotto, 1978, 1981).

The Ca, Mg, and Fe contents of the Drakes Bay Formation dolomites are shown in Figure 8 and Table 3. Oddly, rather than being Ca rich, many of the dolomites are Ca poor and Mg rich. Most of the dolomites do have high Fe contents (>1 mole percent), as expected for the high terrigenous-sediment content of the section. As indicated in Figure 6, the Fe contents of the dolomites tend to increase with the presumed depth of formation.

The depleted Ca contents and the variation in the Ca, Mg, and Fe contents may best be explained by the hypothesis that the dolomites formed without a calcite precursor and that the source of Ca and Mg was sea water. Since dolomitization removes both Mg^{2+} and Ca^{2+} from pore waters and since there is less Ca^{2+} than Mg^{2+} in sea water, the Ca^{2+} concentration of the pore water should decrease with increasing burial depth at a faster rate than the Mg^{2+} concentration. Dolomites formed at greater depths in the sediments should be depleted in Ca relative to those formed at shallow depths. Deeper formed dolomites should also be Fe rich. This relation between depth of formation, as interpreted from $\delta^{18}O$ values, and major-element chemistry is generally what is observed for the Drakes Bay Formation. Plots of mole percent Ca and Mg vs. $\delta^{18}O$ (Figs. 8, 9) reveal that dolomites with lighter $\delta^{18}O$ values generally have less Ca and more Mg than isotopically heavier dolomites. Relatively light dolomites also tend to have more Fe (Fig. 6). The two dolomites which fall off the trend were formed most deeply and have very high-iron contents (>8 mole percent).

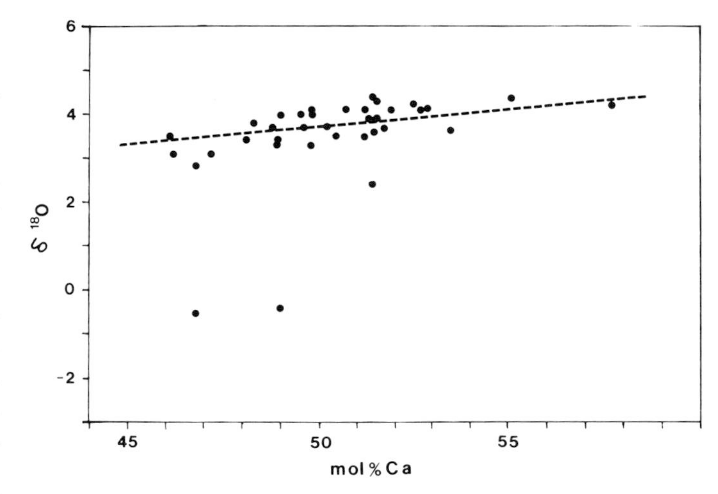

FIG. 8.—Mole percent Ca vs. $\delta^{18}O$ of the dolomites in ‰ PDB. The values, which fall off the trend of decreasing Ca with decreasing $\delta^{18}O$, are very Fe-rich.

For most organic dolomites, in contrast to the Drakes Bay samples, there was likely a calcite precursor to the dolomite that supplied the majority of the Ca for dolomite formation (Baker and Burns, 1985). Magnesium, on the other

TABLE 3.—ELEMENTAL ANALYSES OF CARBONATE FRACTION OF NODULES

Sample	Meters	Mol percent Ca	Mol percent Mg	Mol percent Fe	ppm Mn	ppm Sr	ppm Na	$\delta^{18}O$	$\delta^{13}C$	Percent DOL
DB-4	6	51.2	48.4	0.6	1091	335	511	4.1	−14.6	76
DB-7	20	83.2	12.4	4.4	12150	119	691	−6.1	2.15	
DB-8	21	91.3	2.9	5.8	11400	124		−6.3	1.53	
DB-10a	25	51.7	45.2	6.1	2530	437	1820	3.7	−1.65	69
DB-10b	25	55.1	42.8	2.1	2020	480	1400	4.3	−7.36	77
DB-10c	25	57.7	40.2	2.1	2370	512	1620	4.2	−7.76	75
DB-12	27	48.1	49.9	2.0	1830	345	686	3.4	2.56	74
DB-13a	32	48.9	49.2	1.9	3250	460	989	3.3	3.2	68
DB-13b	32	48.8	49.4	1.8	2870	455	817	3.7	2.5	75
DB-13c	32	50.2	47.9	1.9	1836	437	681	3.7	1.7	74
DB-13d	32	50.4	47.9	1.7	2110	445	707	3.5	2.2	73
DB-14	40	46.8	49.0	4.1	2790	475	575	2.8	8.5	66
DB-17	49	49.6	50.2	1.2	1180	380	557	3.7	3.2	79
DB-18	54	51.2	47.4	1.4	1430	390	697	3.5	0.73	76
DB-23	69	53.5	44.2	2.3	1820	446	667	3.6	−3.84	72
DB-27	81	49.5	48.5	1.9	2490	359	758	4	1.49	65
DB-28a	86	47.2	50.4	2.4	1730	308	502	3.1	5.63	66
DB-28b	86	46.2	51.2	2.6	1770	314	534	3.1	5.82	61
DB-28c	86	46.1	51.5	2.3	1600	319	552	3.5	5.34	58
DB-30	91	52.7	45.3	1.9	1270	491	595	4.1	−2.96	79
DB-32	96	49.8	48.3	1.8	2100	316	725	4	1.5	72
DB-35a	103	49.8	47.9	2.2	2300	303	960	3.3	1.29	73
DB-35b	103	51.9	46.5	1.6	1220	312	682	4.1	−5.23	85
DB-35c	103	52.5	46.0	1.5	1060	305	602	4.2	−6.72	89
DB-37	110	51.4	45.8	2.7	2120	401	1840	3.6	3.12	73
DB-39	118	46.8	44.4	8.7	4390	146	422	−0.53	8.25	60
DB-43a	130	51.5	43.2	5.3	2800	307	702	3.9	4.01	75
DB-43b	130	50.7	44.0	5.3	3690	336	852	4.1	4.79	78
DB-49	160	51.5	46.8	2.5	1070	482	562	4.3	4.14	86
DB-51a	174	48.9	48.5	2.6	1630	346	1050	3.4	3.62	66
DB-51b	174	51.3	46.8	1.9	1630	345	831	3.9	2.4	82
DB-51c	174	49.8	48.1	2.1	1760	331	704	4.1	3.24	80
DB-51d	174	49.0	49.0	2.0	1680	271	644	4	3.78	76
DB-53	185	51.4	46.8	1.8	1050	389	549	4.4	0.95	82
DB-55	193	49.0	41.1	9.8	5460	144	401	−0.43	9.16	72
DB-58	209	52.8	45.4	1.7	1380	454	642	4.1	−2.12	80
DB-63	227	51.4	45.1	3.3	1320	473	855	2.4	12.8	64
DB-67	243	48.3	48.2	3.5	1590	454	457	3.8	8.47	78

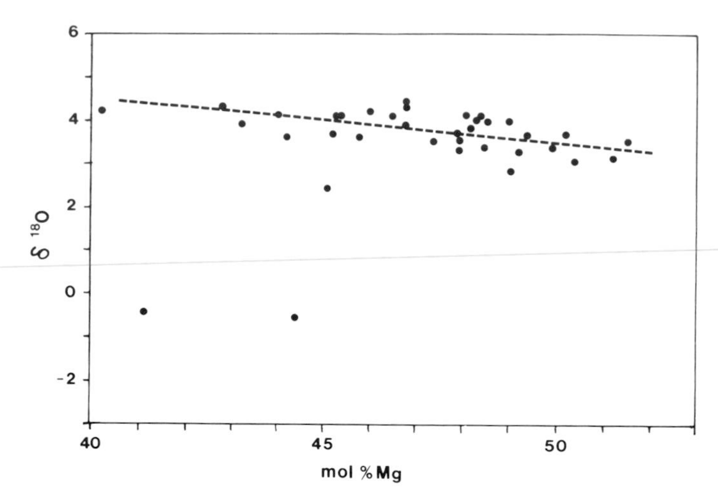

FIG. 9.—Mole percent Mg vs. $\delta^{18}O$ of the dolomites in ‰ PDB. The values, which fall off the trend of increasing Mg with decreasing $\delta^{18}O$, are very Fe-rich.

hand, was supplied by diffusion from overlying sea water. In this case, dolomitization caused Mg to be depleted in the pore waters and deeper formed dolomites were correspondingly depleted in Mg and richer in Fe. Because there was a readily available source of Ca, the dolomites remain Ca rich throughout the sediment column.

Controls on the trace-element chemistry: Mn, Na, and Sr.—

The Mn contents of dolomites from the Drakes Bay Formation were found to correlate roughly with their Fe contents (Fig. 10). This was also observed in dolomites from the Monterey Formation (Burns and Baker, 1987). Shortly after burial, the pore waters of these sediments became anoxic and, successively, Mn and Fe oxide coatings on detrital terrigenous sediment grains were microbially reduced and put into solution. The amounts of Mn and Fe released to the pore waters is probably a function of the amount of detrital minerals in the sediments (Burns and Baker, 1987). A comparison of the Mn content of whole rock analyses of the dolomites with whole rock analyses of the surrounding sediments (Table 1) shows that the dolomites are greatly enriched in Mn relative to the enclosing shales, implying extensive or complete reduction of available Mn oxides and efficient co-precipitation of the Mn as a trace metal in dolomite rather than in another mineral phase.

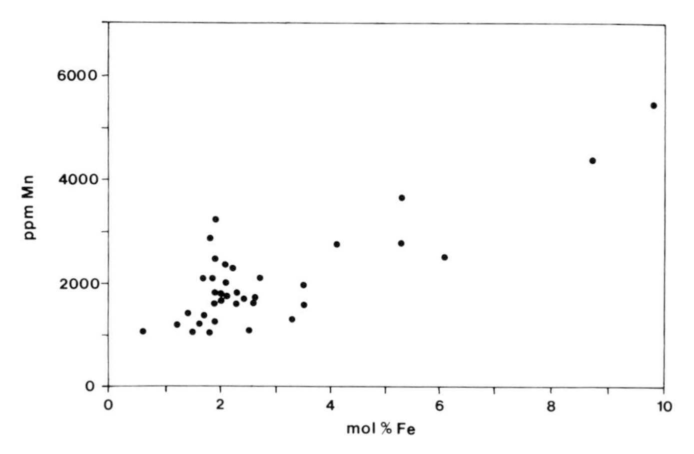

FIG. 10.—Mole percent Fe vs. ppm Mn of the dolomites.

The Na contents of the Drakes Bay dolomites (Table 3) are comparable to those of more recent Holocene dolomites, many of which are believed to have been formed from hypersaline solutions (Land and Hoops, 1973). The dolomites from the Drakes Bay Formation, however, almost certainly formed from pore waters of normal marine salinity. Some of the measured Na could have been leached from detrital minerals incorporated in the nodules. There is no relation, however, between the detrital mineral content of a nodule and the anlayzed sodium content of the dolomite in the nodule. Also, the Na contents of the Drakes Bay dolomites are comparable to dolomites from detritus-poor sections of the Monterey (unpubl. data).

The relatively high Na contents of the Drakes Bay dolomites are probably a function of their poor ordering and nonstoichiometry. Busenburg and Plummer (1986) concluded that the Na content of calcites is probably dependent on the number of lattice dislocations in the calcite in addition to the Na content of the precipitating solution. Presumably, these sites can accommodate Na atoms more easily than undisturbed lattice sites. The same hypothesis may be true of dolomite. Reeder (1981) has shown that poorly ordered Holocene dolomites have a higher number of growth defects than do more ordered ancient dolomites.

The Sr contents of the Drakes Bay dolomites all fall into the narrow range of 300 to 500 ppm Sr, with four exceptions (Table 3). This range is comparable to that observed in Monterey Formation dolomites from sections with intermediate sedimentation rates (Burns and Baker, 1987). The Sr/Ca contents of the dolomites are likely a function of changes in the porewater Sr^{2+}/Ca^{2+} ratio caused by the dolomitization process. Since the Sr^{2+}/Ca^{2+} ratio of sea water is higher than the Sr/Ca ratio of the precipitating dolomites, the porewater Sr^{2+}/Ca^{2+} ratio should increase as dolomitization proceeds. As a result, the dolomites have Sr contents above that predicted for a dolomite precipitating from a solution with a seawater Sr/Ca ratio (approximately 245 ppm, Baker and Burns, 1985).

CONCLUSIONS

The rocks of the type section of the Drakes Bay Formation are uniform siliceous mudstones and siltstones containing numerous layers of dolomite nodules. The type and amount of organic C deposited with the sediments and its rate of oxidation were also reasonably uniform through time. For this reason, the environment of diagenesis of the Drakes Bay Formation was probably a steady-state system during deposition of the sediments. As a result, depth-related changes in the porewater chemistry of the system are particularly well reflected in the authegenic dolomites, which formed within the pore waters.

The Drakes Bay sediments probably contained almost no biogenic calicite that could act as a precursor to dolomite formation. Bacterially mediated sulfate reduction, by removing sulfate from the pore waters and increasing the HCO_3^- concentration, may have caused dolomite to precipitate directly from the pore fluids. Alternatively, calcite may have been the first carbonate mineral to form. The calcite could later have been converted to dolomite. In the first case, the depth of dolomite formation would have been limited by diffusion of Ca^{2+} into the sediments to depths shallower than 19 m subbottom. In the second case, the depth of dolomite formation would have been limited by Mg^{2+} diffusion to depths shallower than 100 m subbottom.

The layers of dolomite nodules appear to be located in layers that are slightly richer in organic C than the overlying and underlying rocks. These layers could have been the location of somewhat more rapid and extensive sulfate reduction. The carbon and oxygen isotope ratios of the dolomites vary systematically. The $\delta^{18}O$ values are a function of temperature (burial depth). The $\delta^{13}C$ values mirror the changes in the $\delta^{13}C$ of the dissolved CO_2, caused mainly by microbial sulfate reduction and methanogenesis. Because the depth at which the shift from sulfate reduction to methanogenesis occurred was fairly constant throughout deposition of the sediments, the $\delta^{13}C$ of the dolomites is mainly a function of the depth at which they formed. Significantly, the relation between $\delta^{13}C$ and the depth of dolomite formation (as inferred from $\delta^{18}O$) indicates that dolomite did not begin to form until sulfate was completely reduced.

The major elemental chemistry of the dolomites also shows a variation with the depth of formation. Calcium contents decrease and Mg and Fe contents increase with depth. These changes are probably caused by increasing isolation of the site of dolomite formation from sea water, the likely source of Ca and Mg. The Fe contents of the dolomites are relatively high, as expected for dolomites forming in anoxic pore waters in sediments rich in terrigenous detrital minerals. Manganese contents correlate with Fe contents and are probably controlled by similar mechanisms. The Na contents of the dolomites are also fairly high. This may be due to the fact that the dolomites are poorly ordered and nonstoichiometric in composition, and therefore have abundant lattice imperfections which can accommodate Na. Finally, the Sr contents of the dolomites are higher than would be the case for dolomites formed from a solution with a seawater Sr^{2+}/Ca^{2+} ratio, because dolomitization raises the porewater Sr^{2+}/Ca^{2+} ratio.

ACKNOWLEDGMENTS

This work was supported by Texaco Inc. and by National Science Foundation Grant NSF-EAR-8516528 to Baker. We thank Mr. Steve Bourroughs for laboratory assistance gathering the stable isotope data, and Mr. Tony Rathburn for assistance with the C-S-N analyses. We especially thank Dr. Caroline Isaacs for a most helpful review.

REFERENCES

Baker, P. A. and Burns, S. J., 1985, The occurrence and formation of dolomite in organic-rich continental margin sediments: American Association of Petroleum Geologists Bulletin, v. 69, p. 1917–1930.

———, and Kastner, M., 1981, Constraints on the formation of sedimentary dolomites: Science, v. 213, p. 351–415.

Barron, J. A., 1986a, Updated diatom biostratigraphy for the Monterey Formation of California, *in* Casey, R. E., and Barron, J. A., eds., Siliceous Microfossil Studies of the Monterey Formation and Modern Analogs: Pacific Section, Society of Economic Paleontologists and Mineralogists Publication, v. 45, p. 105–119.

———, 1986b, Paleoceanographic and tectonic controls on deposition of the Monterey Formation and related siliceous rocks in California: Palaeogeography, Palaeoclimatology, Palaeoecology, v. 53, p. 27–45.

Berner, R. A., 1978, Sulfate reduction and the rate of deposition of marine sediments: Earth and Planetary Science Letters, v. 37, p. 492–498.

———, 1980, Early Diagenesis, A Theoretical Approach: Princeton University Press, Princeton, New Jersey, 241 p.

Bramlette, M. N., 1946, The Monterey Formation of California and the origin of its siliceous rocks: U.S. Geological Survey Professional Paper 212, 57 p.

Burns, S. J., and Baker, P. A., 1987, A geochemical study of dolomite in the Monterey Formation, California: Journal of Sedimentary Petrology, v. 57, p. 128–139.

Busenburg, E., and Plummer L. N., 1986, Kinetic and thermodynamic factors controlling the distibution of SO_4^{2-} and Na^+ in calcites and selected aragonites: Geochimica et Cosmochimica Acta, v. 49, p. 713–726.

Clark, J. C., 1981, Stratigraphy, paleontology and geology of the central Santa Cruz Mountains, California Coast Ranges: U.S. Geological Survey Professional Paper 1168, 51 p.

———, Brabb, E. E., Greene, G., and Ross, D. C., 1984, Geology of Point Reyes Peninsula and implications for San Gregorio Fault history, *in* Crouch, J. K., and Bachman, S. B., eds., Tectonics and Sedimentation along the California Margin: Pacific Section, Society of Economic Paleotologists and Mineralogists Publication, v. 38, p. 67–86.

Claypool, G. E., and Kaplan, I. R., 1974, The origin and distribution of methane in marine sediments, *in* Kaplan, I. R., ed., Natural Gases in Marine Sediments: Plenum Press, New York, p. 99–140.

———, and Threlkeld, C. N., 1983, Anoxic diagenesis and methane generation in sediments of the Blake Outer Ridge, DSDP Site 533, Leg 76, *in* Initial Reports of the Deep Sea Drilling Project, v. 76, p. 391–402.

Compton, J. S., 1986, Early diagenesis and dolomitization in the Monterey Formation, California: Unpublished Ph.D. Dissertation, Harvard University, Cambridge, Massachusetts, 174 p.

———, and Seiver, R., 1986, Diffusion and mass balance of Mg during early dolomite formation, Monterey Formation: Geochimica et Cosmochimica Acta, v. 50, p. 125–136.

Douglas, R. G., 1981, Paleoecology of continental margin basins: A modern case history from the borderland of southern California, *in* Douglas, R. G., Colburn, I. P., and Gorsline, D. S., eds., Depositional Systems of Active Continental Margin Basins: Short Course Notes, Pacific Section, Society of Economic Paleontologists and Mineralogists Publication, p. 121–156.

Emery, K. O., 1960, The Sea off Southern California: John Wiley and Sons, New York, 366 p.

Erickson, A., 1973, Initial report on downhole temperature and shipboard thermal conductivity measurements, *in* Initial Reports of the Deep Sea Drilling Project, v. 19, p. 643–656.

Friedman, I., and Murata, K. J., 1979, Origin of dolomite in Miocene Monterey Shale and related formations in the Temblor Range, California: Geochimica et Cosmochimica Acta, v. 43, p. 1357–1365.

Gaines, A., 1977, Protodolomite redefined: Journal of Sedimentary Petrology, v. 47, p. 543–546.

Galloway, A. J., 1977, Geology of the Point Reyes Peninsula, Marin County, California: California Division of Mines and Geology Bulletin 202, 72 p.

Garrison, R. E., and Graham, S. A., 1984, Early diagenetic dolomites and the origin of dolomite-bearing breccias, lower Monterey Formation, Arroyo Seco, Monterey County, California, *in* Garrison, R. E., Kastner, M., and Zenger, D. H., eds., Dolomites of the Monterey Formation and Other Organic-Rich Units: Pacific Section, Society of Economic Paleontologists and Mineralogists Publication, v. 41, p. 87–102.

Gieskes, J. M., 1974, Interstitial water studies, Leg 25, Deep Sea Drilling Project, *in* Initial Reports of the Deep Sea Drilling Project, v. 25, p. 362–394.

GOLDHABER, M. B., AND, KAPLAN, I. R., 1975, Controls and consequences of sulfate reduction in recent marine sediments: Soil Science, v. 119, p. 42–55.

HENNESSEY, J., AND KNAUTH, L. P., 1985, Isotopic variation in dolomite concretions from the Monterey Formation, California: Journal of Sedimentary Petrology, v. 55, p. 120–130.

IRWIN, H., 1980, Early diagenetic carbonate precipitation and pore fluid migration in the Kemmeridge Clay of Dorset, England: Sedimentology, v. 27, p. 577–591.

JORGENSEN, B. B., 1979, A comparison of methods for the quantification of bacterial sulfate reduction in coastal marine sediments. 2. Calculations from mathematical models: Journal of Geomicrobiology, v. 1, p. 29–51.

KABLONOW, R. I. SURDAM, R. C., AND PREZBINDOWSKI, D., 1984, Origin of dolomites in the Monterey Formation: Pismo and Huasna Basins, California, *in* Garrison, R. E., Kastner, M., and Zenger, D. H., eds., Dolomites of the Monterey Formation and Other Organic-Rich Units: Pacific Section, Society of Economic Paleontologists and Mineralogists Publication, v. 41, p. 119–140.

KATZ, A., AND MATTHEWS, A, 1977, The dolomitization of $CaCO_3$: An experimental study at 252–295°C: Geochimica et Cosmochimica Acta, v. 41, p. 297–308.

KELTS, K., AND MCKENZIE, J., 1982, Diagenetic dolomite formation in Quaternary anoxic diatomaceous muds of Deep Sea Drilling Project Leg 64, Gulf of California, *in* Initial Reports of the Deep Sea Drilling Project, v. 64, p. 553–570.

———, AND ———, 1984, A comparison of anoxic dolomites from deep sea sediments: Quarternary Gulf of California and Messinian Tripoli Formation of Sicily, *in* Garrison, R. E., Kastner, M., and Zenger, D. H., eds., Dolomites of the Monterey Formation and Other Organic-Rich Units: Pacific Section, Society of Economic Paleontologists and Mineralogists Publication, v. 41, p. 19–28.

KLEINPELL, R. M., 1938, Miocene Stratigraphy of California: American Association of Petroleum Geologists, Tulsa, Oklahoma, 450 p.

———, 1980, The Miocene Stratigraphy of California Revisited: American Association of Petroleum Geologists, Tulsa, Oklahoma, 349 p.

KROM, M. D., AND BERNER, R. A., 1981, The diagenesis of phosphorous in a nearshore marine sediment: Geochimica et Cosmochimica Acta, v. 45, p. 207–216.

KULM, L. D., SEUSS, E., AND THORNBURG, T. M., 1984, Dolomites in organic-rich muds of the Peru forearc basin: Analog to the Monterey Formation, *in* Garrison, R. E., Kastner, M., and Zenger, D. H., eds., Dolomites of the Monterey Formation and Other Organic-Rich Units: Pacific Section, Society of Economic Paleontologists and Mineralogists Publication, v. 41, p. 29–47.

KUSHNIR, J., AND KASTNER, M., 1984, Two forms of dolomite occurrences in the Monterey Formation: Concretions and layers—A comparative mineralogical, geochemical and isotopic study, *in* Garrison, R. E., Kastner, M., and Zenger, D. H., eds., Dolomites of the Monterey Formation and Other Organic-Rich Units: Pacific Section, Society of Economic Paleontologists and Mineralogists Publication, v. 41, p. 171–183.

LAND, L. S., 1980, The isotopic and trace element geochemistry of dolomite: The state of the art, *in* Zenger, D. H., Dunham, J. B., and Ethington, D. L., eds., Concepts and Models of Dolomitization: Society of Economic Paleontologists and Mineralogists Special Publication 28, p. 87–110.

———, AND HOOPS, G. K., 1973, Sodium in carbonate sediments and rocks: A possible index to the salinity of diagenetic solutions: Journal of Sedimentary Petrology, v. 43, p. 614–617.

LI, Y.-H., AND GREGORY, S., 1974, Diffusion of ions in sea water and deep-sea sediments: Geochimica et Cosmochimica Acta, v. 38, p. 703–714.

LYONS, W. B., AND GAUDETTE, H. E., 1979, Sulfate reduction and the nature of organic matter in estuarine sediments: Organic Geochemistry, v. 1, p. 151–155.

MCCREA, J. M., 1950, Isotopic chemistry of carbonates and a paleotemperature scale: Journal of Chemical Physics, v. 18, p. 849–857.

MERTZ, K. A., JR., 1984, Diagenetic aspects, Sandholdt Member, Miocene Monterey Formation, Santa Lucia Mountains, California: Implications for depositional and burial environments, *in* Garrison, R. E., Kastner, M., and Zenger, D. H., eds., Dolomites of the Monterey Formation and Other Organic-Rich Units: Pacific Section, Society of Economic Paleontologists and Mineralogists Publication, v. 41, p. 49–74.

MULLER, P. J., 1977, C/N ratios in Pacific deep sea sediments: Effect of inorganic ammonium and organic nitrogen compounds sorbed by clays: Geochimica et Cosmochimica Acta, v. 41, p. 756–776.

MURATA, K. J., FRIEDMAN, I., AND CREMER, M., 1972, Geochemistry of diagenetic dolomites in Miocene marine formations of California and Oregon: U.S. Geological Survey Professional Paper 724-C, 12 p.

NISSENBAUM, A., AND KAPLAN, I. R., 1972, Chemical and isotopic evidence for the *in situ* origin of marine humic substances: Limnology and Oceanography, v. 17, p. 570–582.

PISCIOTTO, K., 1978, Basinal sedimentary and diagenetic aspects of the Monterey shale, California: Unpublished Ph.D. Dissertation, University of California, Santa Cruz, 450 p.

———, 1981, Review of secondary carbonates in the Monterey Formation, California, *in* Garrison, R. E., and Douglas, R. G., eds., The Monterey Formation and related siliceous rocks of California: Pacific Section, Society of Economic Paleontologists and Mineralogists Publication, p. 273–283.

———, AND MAHONEY, J. J., 1981, Isotopic survey of diagenetic carbonates, Deep Sea Drilling Project Leg 63, *in* Initial Reports of the Deep Sea Drilling Project, v. 63, p. 595–609.

REEDER, R. J., 1981, Electron optical investigations of sedimentary dolomites: Contributions to Mineralogy and Petrology, v. 76, p. 148–157.

SHIMMIELD, G. B., AND PRICE, N. B., 1984, Recent dolomite formation in hemipelagic sediments off Baja California, Mexico, *in* Garrison, R.E., Kastner, M., and Zenger, D. H., eds., Dolomites of the Monterey Formation and Other Organic-Rich Units: Pacific Section, Society of Economic Paleontologists and Mineralogists Publication, v. 41, p. 5–18.

SHOLKOVITZ, E. R., 1973, Interstitial water chemistry of the Santa Barbara Basin sediments: Geochimica et Cosmochimica Acta, v. 37, p. 2043–2073.

———, AND GIESKES, J. M., 1971, A physical-chemical study of the flushing of the Santa Barbara Basin: Limnology and Oceanography, v. 16, p. 479–489.

SIGL, W., CHAMLEY, H., FABRICIUS, F., D'ARGOUD, G. G., AND MULLER, J., 1978, Sedimentology and environmental conditions of sapropels, *in* Initial Reports of the Deep Sea Drilling Project, v. 42, pt. 1, p. 445–465.

SPOTTS, J. H., AND SILVERMAN, S. R., 1966, Organic dolomite from Point Fermin, California: American Mineralogist, v. 51, p. 1144–1155.

STUERMER, D. H., PETERS, K. E., AND KAPLAN, I. R., 1978, Source indicators of humic substances and proto-kerogen. Stable isotope ratios, elemental compositions and electron spin resonance spectra: Geochimica et Cosmochimica Acta, v. 42, p. 989–997.

TISSOT, B., DURAND B. M., EPISTALIE, J., AND COMBAZ, A., 1974, The influence of nature and diagenesis of organic matter in the formation of petroleum: American Association of Petroleum Geologists Bulletin, v. 58, p. 499–506.

WEAVER, F. M., CASEY, R. E., AND PEREZ, A. M., 1981, Stratigraphic and paleoceanographic significance of early Pliocene to middle Miocene radiolarian assemblages from northern to Baja California, *in* Garrison, R. E., and Douglas, R. G., eds., The Monterey Formation and Related Siliceous Rocks of California: Pacific Section, Society of Economic Paleontologists and Mineralogists Publication, p. 71–86.

WESTRICH, J. T., AND BERNER, R. A., 1984, The role of sedimentary organic matter in bacterial sulfate reduction: The G model tested: Limnology and Oceanography, v. 29, p. 236–249.

SEDIMENT COMPOSITION AND PRECIPITATION OF DOLOMITE AND PYRITE IN THE NEOGENE MONTEREY AND SISQUOC FORMATIONS, SANTA MARIA BASIN AREA, CALIFORNIA

JOHN S. COMPTON

Department of Marine Science, University of South Florida, St. Petersburg, Florida 33701

ABSTRACT: A 1.2-km-thick section of the Miocene Monterey and overlying Pliocene Sisquoc formations in the Santa Maria basin area of California contains highly variable amounts of biogenic silica, detrital clay and silt, organic matter, carbonate, pyrite, and francolite. Organic-matter diagenesis resulted in the early precipitation of dolomite, pyrite, and francolite, and the concentration of trace metals. Dolostone horizons occur 1 to 10 m apart and consist of 50 to 95 weight percent pore-filling dolomite. The dolomite is low in Fe and Mn and contains an average of 0.8 to 5.3 mole percent excess Ca. Dolomite composition is related to texture in some samples and suggests several different episodes of dolomitization. There is a positive correlation between organic matter, pyrite, and the trace metals V, Cr, Ni, Cu, and Zn. Pyrite formation probably occurred below the sediment/seawater interface (noneuxinic basin) in the microbial-sulfate reduction zone, and was limited by Fe in sediment having a high organic matter-to-clay ratio and by reduced sulfur in sediment having a low organic matter-to-clay ratio.

INTRODUCTION

The organic-rich Miocene Monterey Formation is well known for its rhythmic bedding and as a source and reservoir rock of petroleum hydrocarbons. The establishment of coastal upwelling resulted in the deposition of a diatomaceous sediment containing highly variable amounts of detrital clay and silt, biogenic calcite, and organic matter. The initial sediment composition reflected the tectonic setting of the basin and the influence of climatic oscillations on oceanic circulation and surface water productivity.

The Monterey and Sisquoc formations were deposited in Neogene basins that developed in response to a major tectonic reorganization of the California margin from a convergent to a transform margin (Blake and others, 1978). The local zones of extension and compression of wrench tectonism resulted in rapid uplift and subsidence, creating deep, elongate, and subparallel localized pull-apart basins (Crowell, 1974; Hall, 1977). The topography consisted of deep basins separated by isolated bank top sills, some of which restricted deep-water circulation. Most basins were approximately 0.5 to 1.5 km deep (Ingle, 1981). Basins immediately adjacent to the continental borderland acted as sediment traps, leaving basins farther offshore starved of terrigenous sediment (Gorsline and Emery, 1959). The present geothermal gradient in the southern California borderland is 48 to 100°C/km (McCulloh and Beyer, 1979; Shipboard Scientific Party, 1981) and was probably significantly higher during deposition of the Monterey Formation (Heasler and Surdam, 1983).

Independent of the wrench fault tectonism responsible for Neogene basin formation along the California margin were large-scale oscillations in climate. A global climatic shift from nonglacial to glacial during the mid-Miocene (13–16 Ma) is indicated from faunal and stable isotope data (Ingle, 1981). During deposition of the Monterey Formation, when basins were at approximately their present latitude, the climate oscillated between cold glacial and cool interglacial. The increased vigor in atmospheric and oceanic circulation associated with cold, glacial climatic periods intensified upwelling. Intensified coastal upwelling increased the supply of relatively nutrient-rich water to the surface, allowing greater productivity in the surface waters and higher biogenic sedimentation rates.

Sedimentation rates ranged from 50 to 600 m/Ma for the Monterey Formation. Tectonic uplift during the Pliocene resulted in the overlying Sisquoc Formation having sedimentation rates as much as an order of magnitude higher than the Monterey Formation (Isaacs, 1983). Sedimentation rates varied widely in response to areal and temporal variations in surface productivity and differences in depositional environment, such as basinal vs. slope, and proximity to the continental borderland (Graham, 1976; Ingle, 1981; Pisciotto, 1981; Kelts and McKenzie, 1982; Isaacs, 1983; Mertz, 1984).

Diagenesis.

Extensive diagenesis of the Monterey Formation was driven in large part by the chemical instability of two of its major sedimentary components: opaline silica and organic matter. The siliceous component of the sediment undergoes a series of phase transformations with increasing burial depth. X-ray amorphous, hydrous, biogenic silica (opal A of Jones and Signet, 1971), present mostly as diatom frustules, recrystallizes by a solution/precipitation mechanism to a disordered cristobalite-tridymite phase opal CT (Wise and others, 1972; Murata and Larson, 1975; Murata and others, 1977). The recrystallization of opal A to opal CT with increasing burial depth depends on many factors, including temperature (geothermal gradient) and sediment composition, with recrystallization temperature increasing with the clay content of the sediment (Kastner and others, 1977; Isaacs, 1980). Upon deeper burial, there is an increase in the cristobalite relative to tridymite ordering in the opal CT (Mizutani, 1977), and eventually the opal CT recrystallizes to quartz by a solution/precipitation mechanism (Murata and Larson, 1975).

In organic-rich sediments, the bacterial degradation of organic matter can significantly alter the interstitial water chemistry and result in the early precipitation of diagenetic carbonate, pyrite, and francolite (carbonate fluorapatite, commonly referred to as apatite; Fig. 1). These minerals are termed organogenic because their precipitation is a direct result of organic-matter diagenesis. The degradation of organic matter takes place within a succession of microbial-depth zones: sulfate reduction, methanogenesis, and fermentation, followed by thermal degradation (Claypool and

Sedimentology and Geochemistry of Dolostones, SEPM Special Publication No. 43
 ISBN 0-918985-77-3

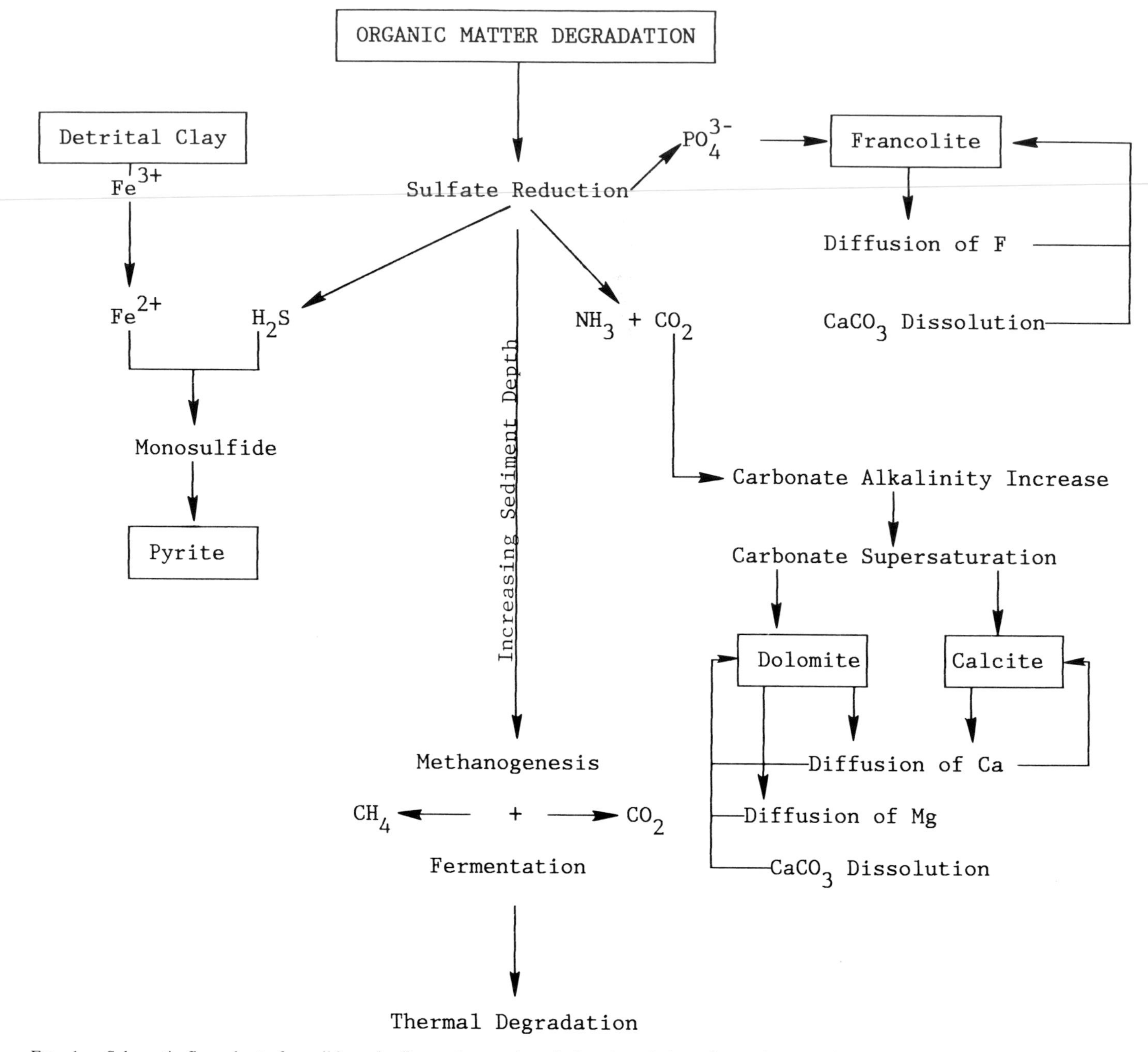

FIG. 1.—Schematic flow chart of possible early diagenetic reactions during degradation of organic matter in marine sediment.

Kaplan, 1974). In most organic-rich marine sediments, the diagenetic carbonate precipitated is dolomite, but other carbonate minerals, such as calcite, ankerite, or rhodochrosite, are possible depending on the availability of Mg, Ca, Fe, and Mn during diagenesis. Organogenic dolomite similar to that found in the Monterey Formation is known to occur in the Kimmeridge Clay of Dorset, England (Irwin, 1980), and in organic-rich, hemipelagic sediment of several Deep Sea Drilling Project (DSDP) sites (Baker and Burns, 1985). Organogenic calcite concretions have been reported from organic-rich Devonian shales (Dix and Mullins, 1987) and rhodochrosite from the Franciscan Complex, California (Hein and Koski, 1987). Organogenic dolomites are distinguished from other types of sedimentary dolomites by having a significant proportion of their carbon derived from the degradation of organic matter as well as the dissolution of biogenic calcite (Spotts and Silverman, 1966; Murata and others, 1969; Claypool and Kaplan, 1974; and others).

Sulfate reduction promotes dolomite precipitation by simultaneously removing the sulfate ion as a potential inhibitor to dolomite precipitation (Baker and Kastner, 1981) and greatly increasing the carbonate alkalinity. The large increase in the carbonate alkalinity results in highly supersaturated interstitial waters. For example, the interstitial waters from DSDP Site 467 (southern California borderland, Gieskes and others, 1981) and Site 479 (Gulf of California, Gieskes and others, 1982), where dolomite is believed to be precipitating currently at depth in the sediment, are calculated to be as much as 10^3 times supersaturated with respect to dolomite (Compton, 1988). The necessary Ca and Mg ions are supplied by diffusion from the overlying sea water and constrain significant dolomite precipitation to the uppermost sediment (Baker and Burns, 1985; Compton and Siever, 1986). The presence of a carbonate precursor enhances dolomitization by providing an additional source of Ca and carbonate ions, since the diffusion of Ca from the overlying sea water is limited by the significantly lower seawater concentration of Ca compared to Mg.

The details of pyrite formation in marine sediments are complex, but the major aspects are fairly well understood. The H_2S produced by the oxidation of organic matter by sulfate-reducing bacteria reacts with reduced iron to form monosulfides that can then react with elemental sulfur to form pyrite (Berner, 1970, 1984). A probable source of reactive Fe for pyrite formation is the reduction of Fe oxide coatings present on terrigenous detrital minerals (Carroll, 1958; Sholkovitz, 1973) with the necessary reducing conditions sustained by continual organic-matter oxidation.

The precipitation of francolite is promoted by the high reactive-phosphorus concentration in sulfate-reducing interstitial waters of organic-rich sediments (e.g., Sholkovitz, 1973; Gieskes and others, 1981). Like dolomite, francolite requires a source of Ca and carbonate ions; the importance of precursor biogenic calcite is suggested by replacement of foraminiferal tests by francolite (e.g., Ames, 1959). Similar to the diffusion of Mg in response to dolomitization, diffusion of fluoride from the overlying sea water in response to francolite precipitation is suggested by fluoride concentration gradients observed in Peru continental-margin phosphatic muds (Froelich and others, 1983). The Mg ion has been shown experimentally to inhibit strongly francolite precipitation (Martens and Harriss, 1970; Nathan and Lucas, 1976). Kastner and others (1984) suggest that an amorphous Ca phosphate may precipitate during early diagenesis in interstitial waters having a high-phosphorus concentration and a relatively high Mg/Ca ratio, and convert into crystalline francolite during later diagenesis at deeper burial, where the interstitial waters have a lower Mg/Ca ratio brought about by dolomite precipitation.

The present paper examines the compositional variation of a single, well-exposed section of the Neogene in the Santa Maria basin area and its relation to the precipitation of organogenic dolomite, pyrite, and francolite.

METHODS

Fresh, relatively unaltered samples were collected from actively exposed beach cliffs located in the Point Pedernales area of south-central coastal California (Fig. 2). The composition of the bulk sediment was determined by combining X-ray fluorescence elemental analyses, X-ray diffraction, and total carbon, carbonate carbon, and water measurements. Samples were bitumen-extracted with chloroform acetone, oven-dried at 110°C, and ground to 200 mesh. Organic-carbon content was determined by subtracting the carbonate carbon from the total carbon content. The weight percent organic matter (kerogen) was calculated by multiplying the weight percent organic carbon by 1.4. A computer program at Chevron Oil Field Research Company was used to calculate the relative amount (weight percent) of the minerals present by a linear programming technique solved by the Simplex Method. The computer program is set up as a series of linear equations, representing both minerals and physical properties of the sample, that are solved by optimizing a specified parameter.

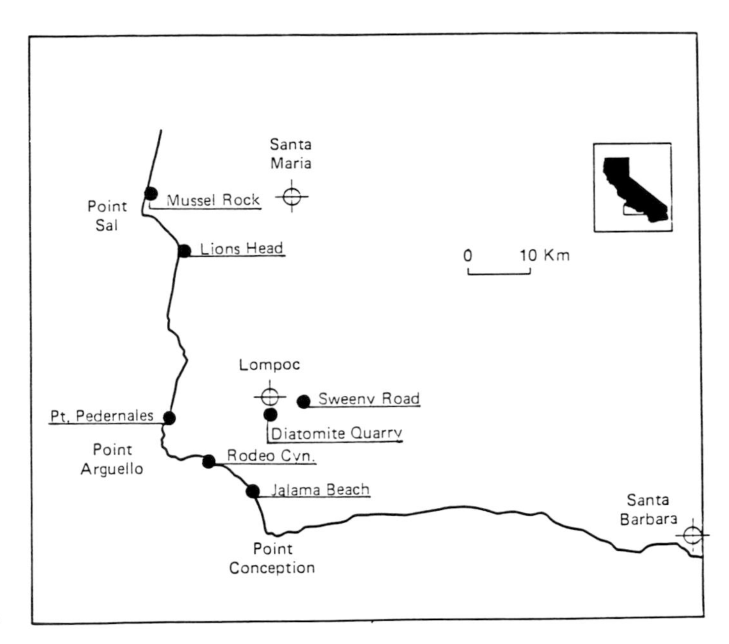

FIG. 2.—Location map of study area, south-central coastal California.

Dolomite composition was determined by electron microprobe analysis of polished dolostone thin sections. Individual dolomite grains were analyzed for Ca, Mg, Fe, Mn, Al, and S using a Cameca MBX microprobe operated at 8 KV with an incident beam of 5 to 40 μm in diameter for 20 to 30 sec. Analyses with greater than 0.2 weight percent Al were discarded because of the possible contribution of Ca or Mg from detrital minerals. The S analysis was used to subtract out the Fe contributed by any associated pyrite. A sample of near-ideal dolomite from Binnenthal, Switzerland (BINN STD), was used as a standard to insure reproducible results.

STRATIGRAPHY

The general stratigraphy of the Point Pedernales area, originally measured and described by Grivetti (1982), is shown in Figure 3. The member names are taken from Isaacs (1983). General aspects of the stratigraphy of the Santa Maria basin area were described by Woodring and Bramlette (1950) and Dibblee (1950). The detailed stratigraphy and dolostone occurrence in the Santa Maria basin area were described by Compton and Siever (1984).

The base of the section is in depositional contact with the underlying Tranquillon Volcanics. The lower calcareous

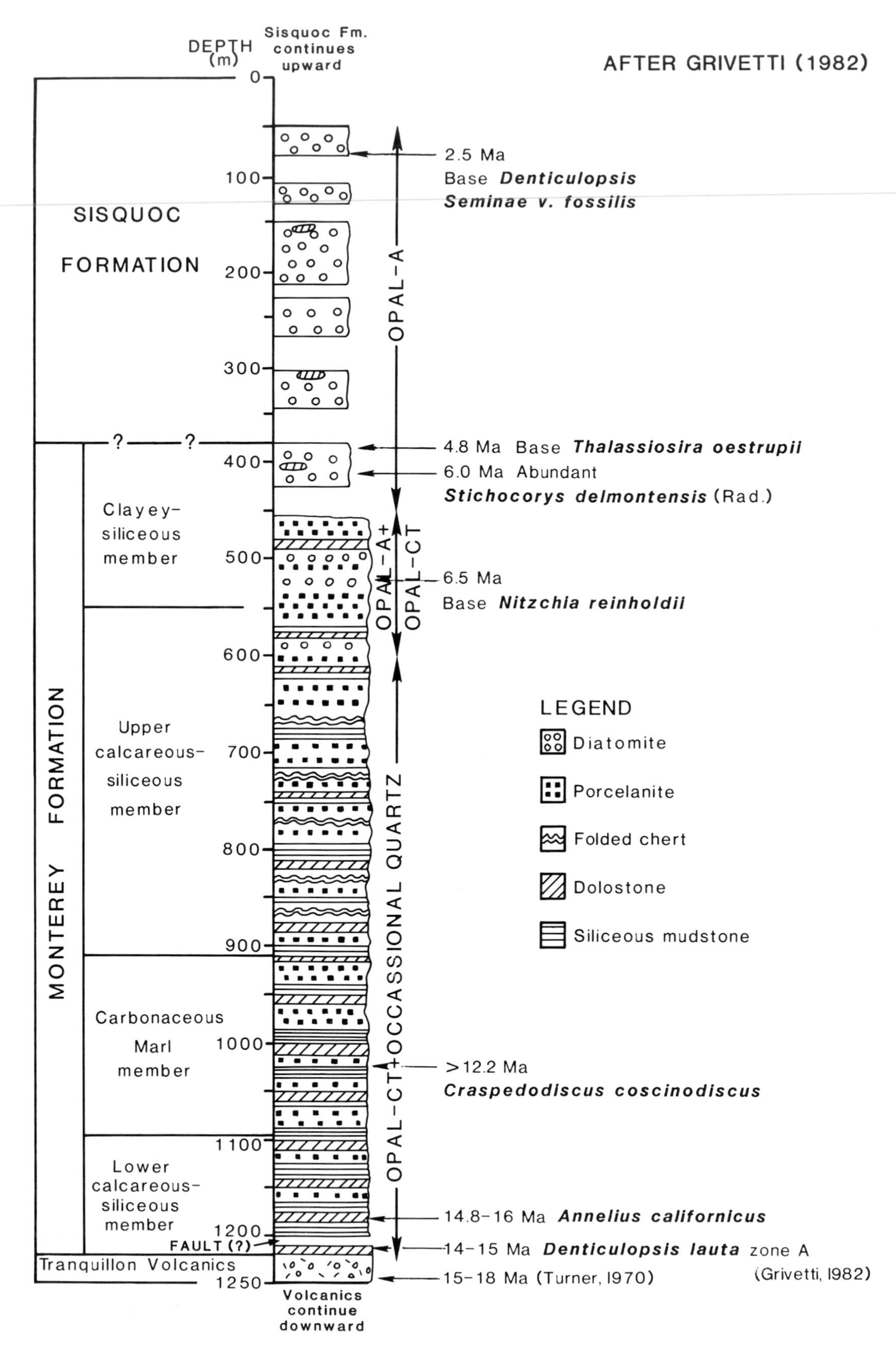

FIG. 3.—Generalized stratigraphic column for the Point Pedernales area, modified after Grivetti (1982). Fossil dates from the Chevron Oil Field Research Company Biostratigraphy Group; member names after Isaacs (1983).

siliceous and carbonaceous-marl members consist predominantly of interbedded porcelanite and siliceous mudstone and dolostone. The carbonaceous-marl member is distinguished by its abundant phosphatic (francolite) laminations and nodules. The upper calcareous siliceous member contains intensely folded, quartz chert beds as well as interbedded porcelanite and siliceous mudstone, and dolostone. The clayey siliceous member consists of diatomite and porcelanite and considerably less carbonate. There is a gradational transition into the overlying Sisquoc Formation marked by a general increase in the detrital-clay and silt content, a decrease in both organic-matter and carbonate content, coarser laminations, and more massive (bioturbated) intervals. The Point Pedernales section probably represents a hemipelagic basinal deposit, although a slope environment is suggested for the Sisquoc Formation by the occurrence of soft-sediment slump folds.

The predominant silica phase of the rocks in the section is shown in Figure 3. The uppermost 450 m of the section is coarsely laminated, clayey opal A diatomite. The transition from opal A diatomite to opal CT porcelanite or chert with increasing burial depth occurs over a stratigraphic interval about 150 m thick (450 m to 600 m in Fig. 3), where both opal A diatomite and opal CT porcelanite and chert occur interbedded on the scale of centimeters to meters. The large depth interval of the opal A to opal CT transition is consistent with the dependence of the silica phase transformation on sediment composition, as well as temperature, with clay inhibiting the transformation (Kastner and others, 1977; Isaacs, 1980). Most of the silica present in the chert, porcelanite, and siliceous mudstone below 600 m is opal CT. Minor amounts of diagenetic quartz occur throughout these rocks, but quartzose rocks are confined to occasional, intensely folded chert beds within the upper calcareous siliceous member. Apparently the Point Pedernales section was not buried deeply enough to precipitate pervasive quartz-grade rocks. Instead, some silica-rich beds, perhaps originally diatom oozes, were altered to quartz at relatively shallow burial depths.

One of the striking features of the Monterey Formation is the alternation of biogenic and clastic sedimentation on the scale of laminations and beds (Bramlette, 1946; Pisciotto and Garrison, 1981). Most of the Monterey Formation is finely laminated (0.1 to 0.5 mm), whereas the Sisquoc Formation is coarsely laminated (0.5 to 1.5 mm). The laminated sediment suggests that the oxygen content of the bottom waters during deposition was too low (suboxic to anoxic) to support a burrowing infauna (Calvert, 1964). Laminations are defined by differences in their silica, clay, or francolite content. Alternating silica- and detritis-rich lamination couplets appear to represent 10-yr, rather than annual, cycles in sedimentation, on the basis of constraints imposed by the average sedimentation rates determined for the section. There are intervals several meters thick of mottled to massive homogeneous sediment, particularly in the Sisquoc Formation, indicating fluctuations in the oxygen content of the bottom waters.

The approximate sedimentation rate is 130 m/Ma for the Sisquoc Formation and 88 m/Ma for the Monterey Formation, using the biostratigraphic age dates and present thicknesses of the Point Pedernales section (Fig. 3); however, this commonly used method for calculating sedimentation rates does not accurately reflect the surface sedimentation rate during deposition because of compaction. In addition, the Monterey Formation has experienced far more porosity reduction (compaction) than has the less deeply buried and less diagenetically altered overlying Sisquoc Formation. The decompacted sedimentation rates were calculated using the porosity-depth relation of the Point Pedernales section (Compton, in prep.) by expanding the thickness of the entire section to its original surface porosity of 86%. The decompacted sedimentation rates are approximately 340 m/Ma for the Sisquoc and 440 m/Ma for the Monterey. The large difference between the noncompacted and decompacted sedimentation rates for the section demonstrates the importance of taking compaction into account when determining sedimentation rates. Calculation of the decompacted sedimentation rate requires knowing the porosity-depth relation of the section.

The abundance and mode of dolomite occurrence are highly variable in the Point Pedernales area (Compton and Siever, 1984). The dolomite can occur disseminated throughout the sediment, but more commonly, it is concentrated in nodules and nodular to stratiform beds to form the rock type dolostone. Dolostone horizons occur more or less regularly, spaced 1 to 10 m apart, and constitute as much as 20 volume percent of the section. Dolostone is most abundant (12–20 volume percent) in the more organic-rich, calcareous, and phosphatic sediment of the lower three members of the Monterey Formation, and is far less abundant in the clayey siliceous member (1–4 volume percent) and Sisquoc Formation (<1–2 volume percent). Differential compaction of laminations suggests that most of the dolomite precipitated early, prior to significant compaction, at burial depths less than 100–200 m (Bramlette, 1946). Whale bones, bivalves, and foraminifera are commonly observed at the center of dolostone nodules, and siliceous lithologies typically contain little or no carbonate between dolostone beds.

SEDIMENT COMPOSITION

The Monterey and overlying Sisquoc Formation are extremely variable in composition (Table 1). The principal components of the sediment are opal A and opal CT silica, diagenetic and detrital quartz, detrital-clay and silt minerals, carbonate, organic matter, pyrite, and francolite. The originally deposited biogenic fraction consists of the remains or organisms and includes opal A silica, carbonate, and organic matter. Diatoms are the dominant siliceous organism, but radiolaria, silicoflagellates, and sponge spicules are observed in thin sections. Biogenic carbonate is composed of low-Mg calcite foraminiferal tests (mostly benthic foraminifers) and nannoplankton (coccolithophores).

The Monterey Formation is organic rich. Samples from the Point Pedernales area contain as much as 12.7 weight percent organic matter (Table 1), and samples from the Santa Barbara coastal area contain a mean of 8 weight percent organic matter (Isaacs, 1983). There is a general positive

TABLE 1.—BULK ROCK ANALYSIS: CALCULATED MINERALOGY (WT %)

Sample(1)	Depth(m)	Rock type	Quartz	Opal CT	Opal A	Dolomite	Calcite	Clay	Fldspr	Pyrite	OM(2)	Franc(3)	SUM
40701.24	0	diatomite	2.8	0.0	63.9	0.0	0.4	18.2	3.4	0.3	2.1	0.6	91.8
40512.8	55	diatomite	10.1	0.0	19.5	0.4	0.0	33.7	23.6	0.3	2.9	1.3	91.7
40472.6	110	diatomite	7.3	0.0	41.2	0.0	0.4	23.6	17.4	0.2	1.9	0.3	92.1
40512.3	125	diatomite	2.5	0.0	53.8	6.3	2.3	14.8	5.0	0.5	2.3	0.6	88.0
40472.58	160	diatomite	5.7	0.0	40.3	0.0	0.0	26.7	13.4	1.0	2.0	0.2	89.2
41744.4	167	dolostone	1.4	0.0	11.3	75.9	0.0	8.1	1.4	0.3	1.4	0.1	100.0
40472.56	230	diatomite	5.4	0.0	48.2	0.3	0.7	23.2	11.6	1.1	1.9	0.5	92.9
40472.54	260	diatomite	5.3	0.0	44.4	0.0	0.1	15.6	21.1	1.3	1.7	0.5	89.9
40472.51	380	diatomite	6.8	0.0	41.5	0.0	0.0	16.4	20.0	0.7	2.3	0.8	88.5
40472.5	420	diatomite	13.2	0.0	17.6	0.0	0.0	22.3	28.8	1.1	3.7	0.5	87.2
40701.18	450	diatomite	6.8	0.0	35.4	0.0	0.0	26.9	9.6	0.8	3.4	0.4	83.4
40472.48	480	chert	3.7	72.8	0.0	0.0	0.2	5.2	10.1	0.9	2.3	0.4	95.7
40701.15	505	diatomite	8.3	0.0	37.2	0.0	0.6	10.6	24.2	1.2	4.9	0.2	87.2
40701.14	525	diatomite	5.6	0.0	42.2	0.0	1.5	16.0	13.7	1.6	6.3	1.3	88.1
40472.37	630	porcelanite	5.8	49.8	0.0	0.0	0.3	15.1	13.6	2.4	6.2	0.6	93.7
40472.28	750	dolostone	2.1	3.2	9.5	58.2	6.1	10.6	3.6	0.8	3.2	2.1	99.4
41744.33Z	845	dolostone	1.5	0.0	0.4	95.4	0.0	1.4	0.5	0.2	0.4	0.3	100.0
411	860	chert	81.8	13.3	0.0	0.0	0.0	0.9	2.1	0.2	0.3	0.1	98.7
41744.33X	878	dolostone	0.0	0.0	1.9	96.3	0.0	0.6	0.8	0.1	0.0	0.3	100.0
41744.30	895	dolostone	0.0	0.0	1.4	94.5	0.0	1.0	0.5	0.4	1.2	0.3	99.1
41744.38M	898	dolostone	0.0	0.0	4.5	91.7	0.0	0.5	2.1	0.3	0.9	0.0	99.9
41744.38P	900	porc/dolo	7.6	27.6	0.0	53.2	0.0	2.1	0.3	0.1	1.4	0.0	92.2
41744.32	903	porcelanite	4.0	74.2	0.0	0.0	0.0	8.6	2.0	1.7	7.6	0.8	98.8
40472.21	905	porcelanite	9.2	29.1	0.0	0.0	0.0	15.6	20.3	2.9	12.3	0.9	90.4
41744.26	998	porcelanite	5.0	40.3	0.0	16.5	7.4	6.0	7.4	4.0	10.2	1.9	98.7
41744.39B	1000	dolostone	0.0	0.0	0.6	77.4	0.0	14.3	7.6	0.0	0.0	0.0	99.9
41744.27	1002	porcelanite	5.8	72.1	0.0	0.0	0.0	7.5	1.4	1.4	7.0	0.2	95.6
40472.14	1015	porcelanite	10.7	74.8	0.0	0.0	0.0	3.3	4.9	0.9	3.2	0.2	97.9
40701.3	1050	porc/dolo	4.6	21.4	0.0	33.7	2.2	8.1	11.1	1.7	6.9	6.6	96.1
40472.1	1065	porcelanite	3.2	53.1	0.0	19.6	2.2	6.6	5.4	1.4	6.0	2.5	99.9
41744.16	1087	dolostone	0.0	0.0	1.5	92.7	0.0	0.3	1.6	0.4	1.9	1.7	100.0
41744.15	1090	porc/dolo	4.0	31.8	0.0	23.2	12.9	9.3	2.7	2.0	7.4	6.2	99.4
40472.7	1110	dolostone	2.4	0.0	4.1	78.9	0.6	6.4	0.7	1.4	4.3	1.1	99.9
40472.4	1170	procelanite	3.7	71.7	0.0	0.0	0.0	6.9	9.2	1.8	5.7	0.4	99.5
41744.9	1185	dolostone	0.0	0.0	4.4	84.5	0.0	5.1	0.8	0.4	3.0	1.9	100.1
41744.8M	1193	porc/dolo	3.7	22.3	0.0	30.9	15.2	8.5	3.4	2.9	12.7	0.2	99.9
41744.8P	1195	porcelanite	3.7	76.2	0.0	0.9	1.3	0.2	7.1	1.9	8.9	0.1	100.3
41744.7	1216	dolostone	4.3	0.0	9.5	84.2	0.0	0.5	0.9	0.5	0.1	0.1	100.0
41744.5	1218	dolostone	1.8	0.0	2.1	73.9	0.0	10.5	4.7	1.4	2.7	2.8	99.9
40472.2	1220	dolostone	2.6	9.8	0.0	74.9	0.0	4.2	2.3	1.5	2.7	1.9	99.9

(1) All samples from the Point Pedernales area (Fig. 3). (2) Organic matter = 1.4 × organic carbon. (3) Francolite.

correlation between organic carbon and detrital minerals (weight percent Al; Fig. 4A), and a poor negative correlation between organic carbon and biogenic silica in the Monterey Formation (Fig. 4B). Similar correlations were observed by Isaacs (1983) in the Santa Barbara coastal area. No correlation was found between organic carbon and detrital minerals or biogenic silica for the overlying Sisquoc Formation, which has an overall higher detrital and lower organic-matter content than the Monterey Formation (Figs. 4A, B).

Terrigenous clay and silt constitute a significant component of many of the sediment samples in Table 1. There has been relatively little work done on the composition and mineralogy of the detrital-clay fraction in the Monterey Formation, but it appears to consist mostly of a mixed-layer illite/smectite (Isaacs, 1980; Kablanow and Surdam, 1983); the amount of authigenic clay has not been determined. The silt is composed of poorly sorted, angular to subangular feldspar, quartz, and mica grains. In the Santa Maria basin area, significant amounts of volcanogenic sediment are restricted to the lowermost 100 m of sediment, where the formation directly overlies the Tranquillon Volcanics; however, several altered volcanic-ash layers occur in places higher in the section.

Dolomite.—

Dolostone from the Monterey contains between 53 and 92 weight percent dolomite with sediment components such as opaline silica, detrital minerals, organic matter, pyrite, and francolite occurring between and included in the pore-filling dolomite grains (Table 1). The dolomite grains composing the sucrosic matrix are more or less homogeneous in size, shape, and clarity. Matrix grains generally range in size from 10 to 40 μm in diameter. No systematic change in the average matrix grain size is observed with increasing sediment depth; however, the uppermost Pliocene dolomites in the Sisquoc Formation, which occur as pore-filling nodules in opal A diatomite, have an extremely fine-grained matrix (<0.1 − 5 μm). The results of a preliminary X-ray diffraction study suggest that these younger, extremely fine-grained dolomites in the Sisquoc Formation are substantially disordered compared to the older, coarser grained dolomites lower in the section (Compton, 1986). Dolomite grains that partly or completely fill the void spaces of diatom frustules, foraminiferal tests, or fractures are usually clearer and larger than surrounding matrix dolomite, ranging in size from 10 to 160 μm.

Dolomites from the Point Pedernales area contain an av-

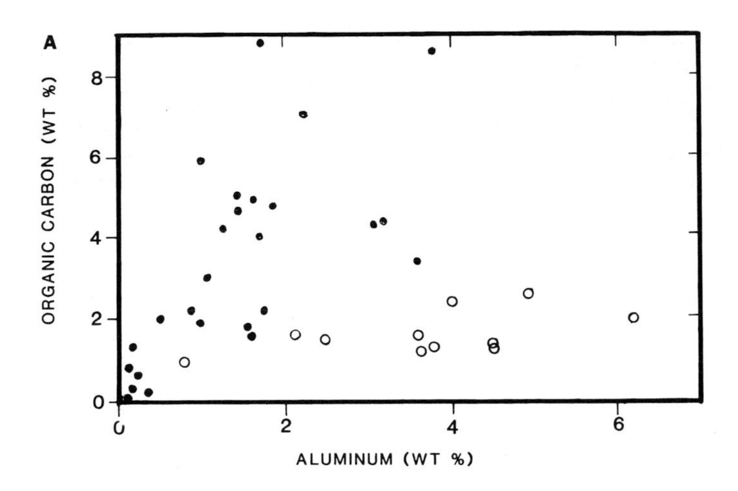

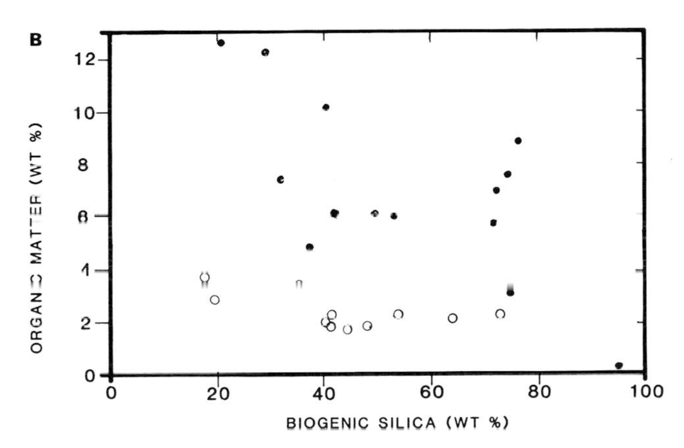

FIG. 4.—(A) Organic carbon (vs) aluminum (detrital minerals), Point Pedernales section. Solid spheres are samples from the Monterey Formation; empty spheres are samples from the Sisquoc Formation. (B) Organic matter (1.4 × organic carbon) (vs) biogenic silica, Point Pedernales section.

erage of 0.8 to 5.3 mole percent excess Ca and are low in both Fe and Mn (Table 2). The dolomite grains in a sample are either fairly homogeneous in composition regardless of grain size or equally heterogeneous for all grain sizes. No compositional zoning was observed from transects of individual probe data points or from fluorescence of the samples in the electron microprobe. There is no obvious trend in the mole percent excess Ca of the dolomites with stratigraphic depth. In several samples compositional variation is related to texture. For example, the matrix dolomite of sample 23J contains an average of 4.4 mole percent excess Ca, whereas the larger, clearer vein dolomite in the same sample (23JV) contains an average of only 0.9 mole percent excess Ca. In sample 42J, there are two texturally distinct populations of intergrown dolomite that have different average compositions of 0.8 and 5.3 mole percent excess Ca. The dolomite grains in sample 42J containing greater excess Ca are less euhedral and less clear.

Pyrite.—

Organic carbon correlates strongly with total sulfur (Fig. 5). Most of the sulfur is present as ubiquitous, fine-grained pyrite (FeS_2), which constitutes <1 to 4 weight percent of the sediment. Most of the analyzed samples lie near the line representing normal marine sediment (Fig. 5; Berner, 1982, 1984), suggesting that the depositional basin in the Point Pedernales area was not euxinic and, hence, pyrite probably precipitated below the sediment/seawater interface and not in the overlying water column (Leventhal, 1979; Raiswell and Berner, 1985). Therefore, whereas bottom waters were suboxic to anoxic, they were probably not sulfidic.

Samples with a high organic matter-to-clay ratio tend to fall above the pyrite (FeS_2) line in the plot of Fe vs. S (Fig. 6), suggesting pyrite formation was limited by available reduced Fe. The most likely source of Fe for pyrite precipitation was reduction of Fe oxide coatings present on detrital clay minerals. The amount of reactive Fe available for pyrite precipitation will be influenced by the precipitation of Fe carbonate minerals and the complexation of Fe with

TABLE 2.—CHEMICAL COMPOSITION OF DOLOMITE BY MICROPROBE ANALYSIS

Sample(1)	Depth(m)	No. of Analyses	Mol percent (2)				Excess Ca		
			Mg	Ca	Fe	Mn	(Mol percent)	(Min – Max)	esd
BINN STD	N.A.	43	49.48	50.38	0.04	0.09	0.38	(0.00–0.95)	0.23
Pt. Pedernales Area:									
102	1217	29	48.02	51.98	0.00	0.00	1.98	(0.40–4.06)	1.01
41744-5	1215	29	47.44	52.56	0.00	0.00	2.56	(0.00–4.43)	1.02
41744-7	1212	20	48.81	51.19	0.00	0.00	1.19	(0.00–2.26)	0.71
41744-9	1184	26	48.03	51.97	0.00	<0.05	1.97	(0.42–5.91)	1.52
41744-16	1087	26	48.04	51.96	<0.05	<0.05	1.96	(0.86–4.08)	1.07
41744-23J	996	15	45.65	54.35	0.00	0.00	4.35	(1.86–5.94)	1.16
41744-23JV	996	6	49.06	50.95	0.00	0.00	0.95	(0.30–1.51)	0.41
41744-35E	589	20	45.66	54.34	0.00	0.00	4.34	(2.86–5.70)	0.77
41744-38M	898	22	45.30	54.70	0.00	0.00	4.70	(0.87–6.59)	1.46
41744-38P	897	20	44.79	55.21	0.00	0.00	5.21	(0.61–6.91)	1.57
41744-39B1	998	17	47.00	53.00	0.00	0.00	3.00	(0.76–4.83)	1.01
41744-39B2	998	17	46.00	54.00	0.00	0.00	4.00	(0.37–6.02)	1.54
41744-42J1	765	11	49.22	50.78	0.00	0.00	0.78	(0.00–1.32)	0.36
41744-42J2	765	15	44.72	55.28	0.00	0.00	5.28	(3.11–6.25)	0.81
1250	307	35	48.49	51.51	<0.25	<0.05	1.51	(0.00–3.59)	1.08
PP-16-23	167	30	47.73	52.27	<0.10	0.00	2.27	(0.00–3.74)	0.95

(1) All samples from the Point Pedernales area (Fig. 3).
(2) Average mole percent (arithmetic mean of individual probe data points).

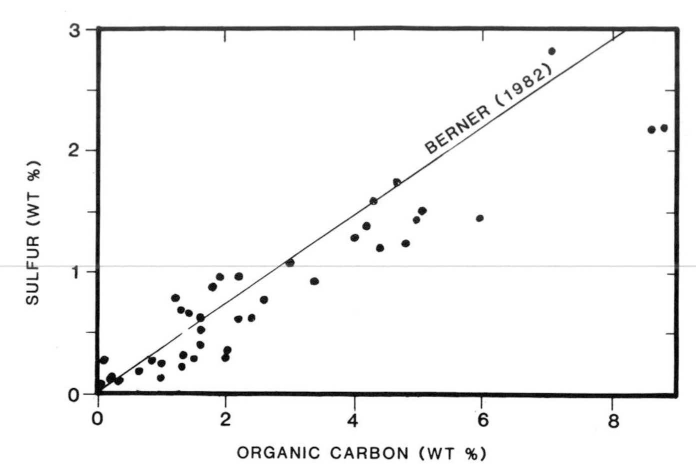

FIG. 5.—Sulfur (vs) organic carbon, Point Pedernales section. Line represents average for normal marine (noneuxinic) modern sediments (Berner, 1982).

organic matter. For example, Elderfield (1981) found that about 80 percent of the total Fe in the interstitial water of anoxic sediment from Narragansett Bay was organically bound. Detrital-rich samples having a low organic matter-to-clay ratio tend to fall below the pyrite (FeS_2) line in Figure 6, suggesting pyrite formation was limited by the amount of metabolizable organic matter available to sulfate-reducing bacteria.

Some of the organic-rich samples from Table 1 contain as much as 0.8 weight percent excess sulfur (total sulfur minus pyrite sulfur). The excess sulfur most likely occurs as organic-sulfur compounds (OSC). Assuming the excess sulfur is organically bound, the organic matter contains as much as 10 weight percent organic sulfur. Organic matter from the Monterey Formation has been reported to contain more than 9 weight percent organic sulfur (Orr, 1984). The trace-metal contents, discussed later, are too low for the amount of excess sulfur to be accounted for by metal sulfide minerals other than pyrite. Marine organisms generally

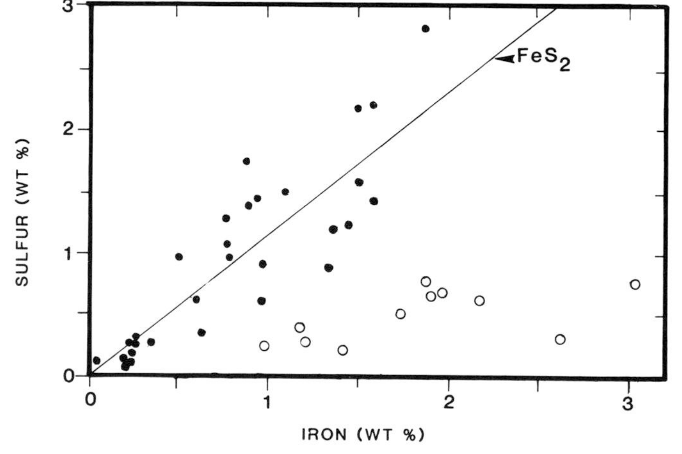

FIG. 6.—Sulfur (vs) iron, Point Pedernales section. Organic-rich samples above pyrite (FeS_2) line limited by available Fe, and detrital-rich samples below pyrite line limited by available reduced sulfur.

contain between 0.3 and 3.3 weight percent sulfur (Goldhaber and Kaplan, 1974). The high OSC content of the Monterey samples may be explained by the synthesis of additional OSC during the production of large amounts of H_2S by sulfate-reducing bacteria. In addition, the high-sulfur content of Monterey organic matter suggests that OSC may be preferentially retained during organic-matter diagenesis.

Francolite.—

Whitish francolite blebs, as well as brown phosphatic fish debris, are observed in minor amounts throughout the Point Pedernales section. The francolite content of the samples analyzed range from practically zero to over 6 weight percent (Table 1). There is no correlation between the phosphorus and organic-carbon contents of the samples; however, the phosphatic-rich facies is most commonly observed in the highly organic-rich carbonaceous-marl member. None of the samples in Table 1 represents the phosphatic-rich facies, which generally contains greater than 20 weight percent francolite. Cryptocrystalline francolite typically occurs as discrete white-to-tan particles that are concentrated to define laminations or nodules. The francolite in most places appears to have precipitated *in situ* with no evidence of sediment reworking or erosion.

Trace metals.—

A more or less positive correlation exists between organic matter and the trace metals V, Cr, Ni, Cu, and Zn. The correlation is best for Cu, Cr, and V and fairly poor for Ni and Zn (Fig. 7A–E). Many studies have found a strong correlation between trace metals and organic-carbon concentrations (e.g., Calvert and Price, 1970). The reported values (Fig. 7A–E) are within the range observed for most black shale sediments characterized as organic- and clay-rich (Wedepohl, 1978). The trace metals most likely occur either as metallo-organic complexes or as sulfides coprecipitated with associated pyrite. It is unlikely that the trace metals were derived simply by the concentration of the metals originally present in the organic tissue of organisms, although some organisms are known to have fairly high-metal contents (Nicholls and others, 1959). Instead, organic matter probably acts as a continual scavenger of metals throughout its depositional and diagenetic history (Vine and Tourtelot, 1970; Calvert and Price, 1970, 1983; Fischer and others, 1986). Elderfield (1981) found that most of the Fe, Cu, and Ni present in the interstitial waters of anoxic sediment was organically bound. In addition, many metallo-organic complexes (e.g., Ni and V) have a high stability and can survive the effects of burial diagenesis (Lewan and Maynard, 1982).

DISCUSSION AND SUMMARY

The Neogene sediments of the Point Pedernales area typify the organogenic mineral assemblage dolomite, pyrite, and francolite, as well as trace-metal enrichment, that can result from organic-matter diagenesis. The most important control on the precipitation of these minerals is clearly the deposition of an organic-rich sediment. Currently, organic-

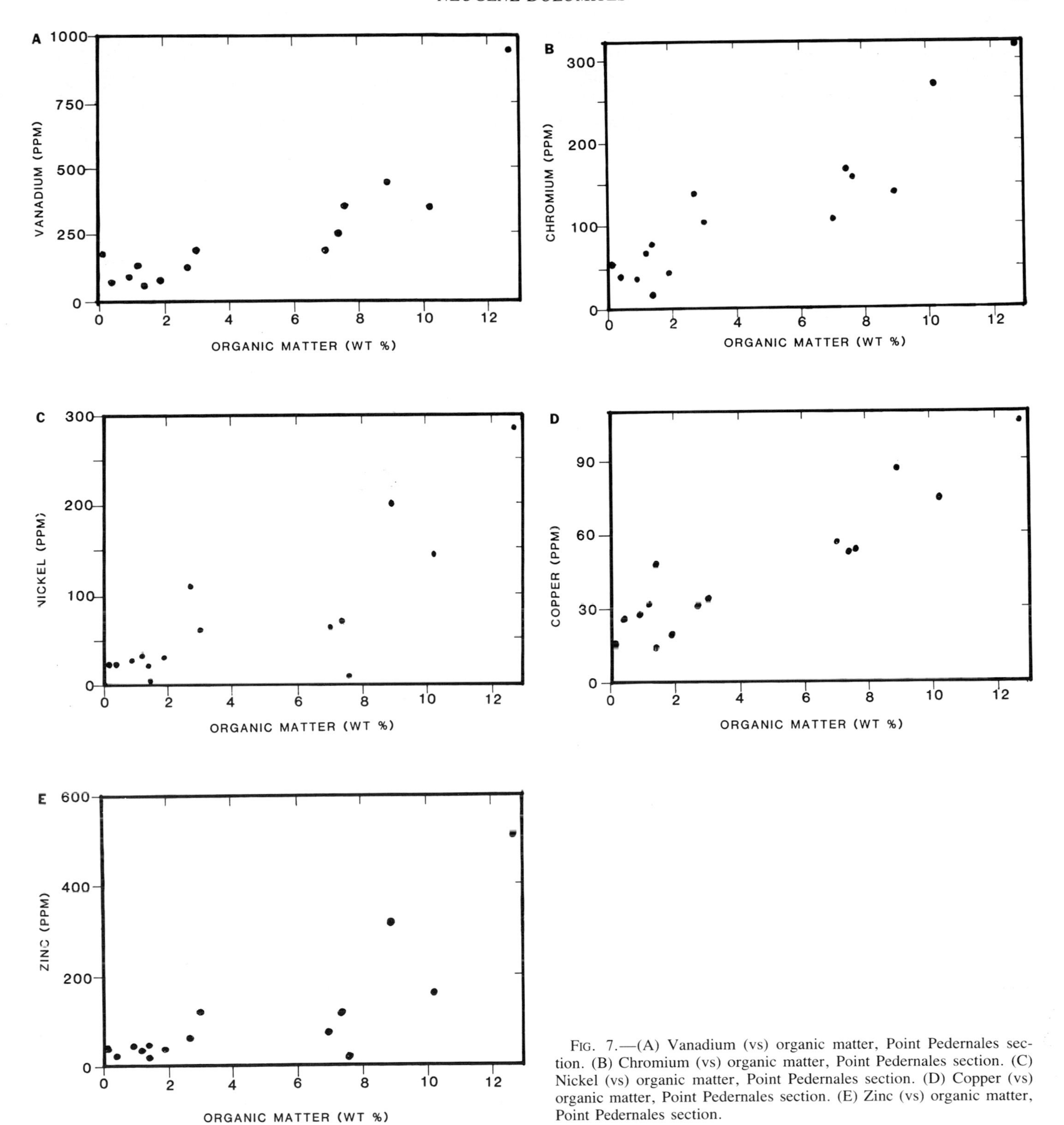

FIG. 7.—(A) Vanadium (vs) organic matter, Point Pedernales section. (B) Chromium (vs) organic matter, Point Pedernales section. (C) Nickel (vs) organic matter, Point Pedernales section. (D) Copper (vs) organic matter, Point Pedernales section. (E) Zinc (vs) organic matter, Point Pedernales section.

rich sediment is mostly confined to strong coastal-upwelling regions (e.g., Gulf of California, van Andel, 1964; offshore Peru, Veeh and others, 1973), and where basin topography restricts deep-water circulation (e.g., Cariaco trench, Shipboard Scientific Party, 1973; and Santa Barbara basin, Emery, 1960).

The compositional variation observed in the Monterey Formation, including organic-matter content, can be related in large part to climatic changes that affected surface water productivity and the vigor of atmospheric and oceanic circulation (Isaacs, 1983). The decrease in average global temperature during cold, glacial climatic periods produced a greater latitudinal temperature differential, which tended to increase atmospheric and oceanic circulation. Increased atmospheric circulation intensified coastal upwelling to promote a greater surface water productivity dominated by siliceous organisms. The increased demand for oxygen from the oxidation of organic debris raining down from high-productivity surface waters produced an expanded, stable oxygen-minimum layer (Bentor, 1980; Suess, 1981). Organic-matter preservation was enhanced where the oxygen-minimum layer intersected slope or basin depositional environments. The oxygen-minimum layer more likely intersected slope rather than basin depositional environments because of the shallow-water depth of the oxygen-minimum layer relative to the typically deep-water depth of these Neogene basins. The low organic-matter content of silica-rich beds can be explained by the rapid oxidation and recycling of the organic matter in the water column from the increase in oceanic circulation and greater deep-water renewal rates. Organic-rich sediment was restricted to slope environments that intersected the oxygen-minimum layer. These slope deposits usually contain more detrital clay than their basin counterparts because of their greater proximity to the continental landmass.

Conversely, during relatively warmer climatic periods, less intense upwelling decreased the input and dilution effect of biogenic silica, so that the calcite and detrital content of the sediment increased under slower rates of accumulation and a less extensive oxygen-minimum layer. Organic-matter preservation was enhanced in basinal environments by slower oceanic circulation and deep-water renewal rates, and diminished on slope environments from a less extensive oxygen-minimum layer. This generalized climatic scenario is a possible explanation for the observed positive correlation between detrital clay and organic matter and the negative correlation between biogenic silica and organic matter for Monterey sediments (Fig. 4A, B; Isaacs, 1983); as suggested by Isaacs (1983), the adsorption of organic matter by clay minerals may also enhance organic-matter preservation. The organic-matter content of the Sisquoc Formation was apparently diluted by a large increase in the detrital-accumulation rate, resulting in a lack of correlation between the organic matter and biogenic silica or detrital-clay contents.

The organic-carbon accumulation rates for Holocene marine sediments increase dramatically with the bulk sediment accumulation rate, because less time is available for oxidation of incoming organic matter at the sediment/seawater interface (Bralower and Thierstein, 1984). A high-sedimentation rate enhances organic-matter preservation, however, dilution by inorganic components or oxidation of settling organic matter in the water column can result in deposition of an organic-poor sediment. The influence of sedimentation rate on organic-matter preservation is less important where suboxic to anoxic water conditions exist in basins having restricted or sluggish deep-water circulation. For example, during the mid-Cretaceous, the association of low-surface productivity and low-sedimentation rates with high organic-carbon accumulation rates was most likely a result of much slower deep-water renewal rates (Bralower and Thierstein, 1984).

In addition to organic matter, the accumulation rates of calcite and detrital clay are important in the precipitation of organogenic dolomite, pyrite, and francolite. Detrital clay is an important source of iron in the precipitation of pyrite. The detrital-clay content depends on the proximity of the basin to the continental landmass and the amount of dilution from biogenic sedimentation. The importance of calcite as a source of Ca and carbonate ions in the precipitation, and perhaps nucleation, of dolomite and francolite is suggested by their common association with biogenic calcite. The calcite content depends on the relative surface productivity of calcareous to siliceous organisms, as well as the degree of calcite preservation en route from the surface waters to depth in the sediment.

Pyrite precipitates early in the sulfate reduction zone. Francolite, or an amorphous Ca phosphate precursor, probably also precipitates early in the sulfate reduction zone because of the high-phosphorous concentration found there and the greater availability of F by diffusion from the overlying sea water. If sulfate is a strong inhibitor to dolomitization, as indicated by experimental work (Baker and Kastner, 1981), then dolomite would be expected to precipitate below the sulfate reduction zone. The above sequence is supported by the entrapment of both pyrite and francolite within tightly cemented dolostone. The low Fe content of dolomites from the Point Pedernales area is probably due to the removal of reactive iron by the earlier precipitation of pyrite. Ferroan dolomites observed in other Monterey sediments may be explained by a reactive-iron concentration greater than that required for pyrite precipitation (high clay-to-organic matter ratio) or co-precipitation of dolomite and pyrite in the sulfate reduction zone. The association of compositional variations with texture in some of the dolomite samples suggests several different episodes of dolomitization or partial recrystallization. The importance of dolomite recrystallization is not clear, but it is possible that much of the deeper, coarser grained (10–40 μm) dolomites were initially extremely fine-grained (<0.1–5 μm), similar to the younger dolomites in the Sisquoc Formation.

Two factors that may have influenced the cyclical distribution of dolomite in the Point Pedernales section are the deposition of a carbonate precursor (Kelts and McKenzie, 1982) and the residence time of the dolomite-forming horizon in the uppermost sediment, where diffusion of seawater Mg is possible (Compton and Siever, 1986). Using the approximate sedimentation rates of the Point Pedernales section, the time span represented by the sediment thickness (1–10 m) between dolomite horizons is on the order of 0.01 to 0.1 Ma. Unfortunately, the time resolution of this section does not allow a more detailed analysis of the carbonate distribution; however, the approximate (within a factor of two) range in the frequency for dolomite intervals in the Point Pedernales section is consistent with the frequencies associated with variations in the geometry of the earth's orbit (Milankovitch theory), suggesting that the climatic oscillations produced from orbital variations may have influenced biogenic carbonate productivity and deposition. The observed irregular spacing of carbonate horizons is expected because of changes in the sedimentation rate, degree of carbonate preservation, and complex climatic patterns produced by the interference between the amplitude variations in the orbital parameters (Imbrie and Imbrie, 1980). The residence time near the sediment/seawater interface is important when a seawater source of Mg or Ca limits dolomite precipitation, since the diffusive flux decreases rapidly with increasing sediment depth. The irregular thickness and spacing of dolomite horizons are expected from variations in the sedimentation rate and the depth of dolomite nucleation.

ACKNOWLEDGMENTS

Financial support is gratefully acknowledged from the National Science Foundation (Grant #EAR-82-12261) and the Chevron Oil Field Research Company (COFRC). Special thanks go to A. B. Carpenter and W. E. Seixas at COFRC for their assistance; D. Lange assisted with the microprobe analyses. This work benefited from the comments and suggestions of P. A. Baker, M. Kastner, and R. Siever.

REFERENCES

AMES, L. L., 1959, The genesis of carbonate-apatite: Economic Geology, v. 54, p. 829–841.

BAKER, P., AND BURNS, S., 1985, Occurrence and formation of dolomite in organic-rich continental margin sediments: American Association of Petroleum Geologists Bulletin, v. 69, p. 1917–1930.

———, AND KASTNER, M., 1981, Constraints on the formation of sedimentary dolomite: Science, v. 213, p. 214–216.

BENTOR, Y. K., 1980, Phosphorites–The unsolved problems, *in* Bentor, Y. K., ed., Marine Phosphorites–Geochemistry, Occurrence, Genesis: Society of Economic Paleontologists and Mineralogists Special Publication 29, p. 3–18.

BERNER, R. A., 1970, Sedimentary pyrite formation: American Journal of Science, v. 268, p. 1–23.

———, 1982, Burial of organic carbon and pyrite sulfur in the modern ocean: Its geochemical and environmental significance: American Journal of Science, v. 282, p. 451–473.

———, 1984, Sedimentary pyrite formation: An update: Geochimica et Cosmochimica Acta, v. 48, p. 605–615.

BLAKE, M. C., CAMPBELL, R. H., DIBBLEE, T. W., JR., HOWELL, D. G., NILSEN, T. H., NORMARK, W. R., VEDDER, J. C., AND SILVER, E. A., 1978, Neogene basin formation in relation to plate-tectonic evolution of San Andreas fault system, California: American Association of Petroleum Geologists Bulletin 62, p. 344–372.

BRALOWER, T. J., AND THIERSTEIN, H. R., 1984, Low productivity and slow deep-water circulation in mid-Cretaceous oceans: Geology, v. 12, p. 614–618.

BRAMLETTE, M. N., 1946, The Monterey Formation of California and the origin of its siliceous rocks; U.S. Geological Survey Professional Paper 212, 57 p.

CALVERT, S. E., 1964, Factors affecting the distribution of laminated diatomaceous sediments in the Gulf of California, *in* Van Andel, T. J., and Shor, G. G., Jr., eds., Marine Geology of the Gulf of California, A Symposium: American Association of Petroleum Geologists Memoir 3, p. 311–330.

———, AND PRICE, N. B., 1970, Minor metal contents of Recent organic-rich sediments off South West Africa: Nature, v. 227, p. 593–595.

———, AND ———, 1983, Geochemistry of Namibian shelf sediments, *in* Coastal Upwelling-Its Sediment Record. Part A: Response of the Sedimentary Regime to Present Coastal Upwelling: NATO Conference Series IV, v. 10a, Plenum Press, New York, p. 337–375.

CARROLL, D., 1958, Role of clay minerals in the transportation of iron: Geochimica et Cosmochimica Acta, v. 14, p. 1–27.

CLAYPOOL, G., AND KAPLAN, I. R., 1974, The origin and distribution of methane in marine sediments, *in* Kaplan, I. R., ed., Natural Gases in Marine Sediments: Plenum Press, New York, p. 99–140.

COMPTON, J. S., 1986, Early diagenesis and dolomitization of the Monterey Formation, California: Unpublished Ph.D. Dissertation, Harvard University, Cambridge, Massachusetts, 174 p.

———, 1988, Degree of supersaturation and precipitation of organogenic dolomite: Geology, v. 16, p. 318–321.

———, AND SIEVER, R., 1984, Stratigraphy and dolostone occurrence in the Miocene Monterey Formation, Santa Maria basin area, California, *in* Garrison, R. E., Kastner, M., and Zenger, D. H., eds., Dolomites of the Monterey Formation and Other Organic-Rich Units, Volume 41: Pacific Section, Society of Economic Paleontologists and Mineralogists Special Publication, p. 141–154.

———, AND ———, 1986, Diffusion and mass balance of Mg during early dolomite formation, Monterey Formation: Geochimica et Cosmochimica Acta, v. 50, p. 125–135.

CROWELL, J. C., 1974, Origin of late Cenozoic basins in Southern California, *in* Dickinson, W. R., ed., Tectonics and Sedimentation: Society of Economic Paleontologists and Mineralogists Special Publication 22, p. 190–204.

DIBBLEE, T. W., JR., 1950, Geology of southwestern Santa Barbara County, California: California Division of Mines and Geology Bulletin 150, 95 p.

DIX, G. R., AND MULLINS, H. T., 1987, Shallow, subsurface growth and burial alteration of Middle Devonian calcite concretions: Journal of Sedimentary Petrology, v. 57, p. 140–152.

ELDERFIELD, H., 1981, Metal-organic associations in interstitial waters of Narragansett Bay sediments: American Journal of Science, v. 281, p. 1184–1196.

EMERY, K. O., 1960, The Sea Off Southern California, A Modern Habitat of Petroleum: John Wiley and Sons, New York, 366 p.

FISCHER, K., DYMOND, J., LYLE, M., SOUTAR, A., AND RAU, S., 1986, The benthic cycle of copper: Evidence from sediment trap experiments in the eastern tropical North Pacific Ocean: Geochimica et Cosmochimica Acta, v. 50, p. 1535–1543.

FROELICH, P. N., KIM, K. H., JAHNKE, R., BURNETT, W. C., SOUTAR, A., AND DEAKIN, M., 1983, Pore water fluoride in Peru continental margin sediments: Uptake from seawater: Geochimica et Cosmochimica Acta, v. 47, p. 1605–1612.

GIESKES, J. M., NEVSKY, B., AND CHAIN, A., 1981, Interstitial water studies, Leg 63, *in* Yeats, R., Haq, B. U., eds., Initial Reports of Deep Sea Drilling Project 63: U.S. Government Printing Office, Washington, D.C., p. 623–629.

———, ELDERFIELD, H., LAWRENCE, J. R., JOHNSON, J., MEYERS, B., AND CAMPBELL, A., 1982, Geochemistry of interstitial waters and sediments, Leg 64, Gulf of California, *in* Curray, J. R., Moore, D. G., and others, eds., Initial Reports of Deep Sea Drilling Project 64: U.S. Government Printing Office, Washington, D.C., p. 675–694.

GOLDHABER, M. B., AND KAPLAN, I. R., 1974, The sulfur cycle, *in* Goldberg, E. D., ed., The Sea, Volume 5: Wiley and Sons, New York, p. 569–655.

GORSLINE, D. S., AND EMERY, K. O., 1959, Turbidity-current deposits in San Pedro and Santa Monica Basins of southern California: Geological Society of America Bulletin, v. 70, p. 279–290.

GRAHAM, S. A., 1976, Tertiary sedimentary tectonics of the central Salinian block of California: Unpublished Ph.D. Dissertation, Stanford University, Stanford, California, 510 p.

GRIVETTI, M. C., 1982, Aspects of stratigraphy, diagenesis, and deformation in the Monterey Formation near Santa Maria-Lompoc, California: Unpublished M.S. Thesis, University of California, Santa Barbara, California, 155 p.

HALL, C. A., Jr., 1977, Origin and development of the Lompoc-Santa Maria pull-apart basin and its relation to the San Simeon-Hosgri strike-slip fault, Western California, *in* Silver, E. A., and Normark, W. R., eds., San Gregorio-Hosgri Fault Zone, California: California Division of Mines and Geology, Special Report 137, p. 25–31.

HEASLER, H. P., AND SURDAM, R. C., 1983, A thermally subsiding basin model for the maturation of hydrocarbons in Pismo basin, California, *in* Isaacs, C. M., and Garrison, R. E., eds., Petroleum Generation and Occurrence in the Miocene Monterey Formation, California: Pacific Section, Society of Economic Paleontologists and Mineralogists Special Publication, p. 69–74.

HEIN, J. R., AND KOSKI, R. A., 1987, Bacterially mediated diagenetic origin for chert-hosted manganese deposits in the Franciscan Complex, California Coast Ranges: Geology, v. 15, p. 722–726.

IMBRIE, J., AND IMBRIE, J. Z., 1980, Modelling the climatic response to orbital variations: Science, v. 207, p. 943–954.

INGLE, J. C., JR., 1981, Origin of Neogene diatomites around the North Pacific Rim, *in* Garrison, R. E., and Douglas, R. G., eds., The Monterey Formation and Related Siliceous Rocks of California: Pacific Section, Society of Paleontologists and Mineralogists Special Publication, p. 159–179.

IRWIN, H., 1980, Early diagenetic carbonate precipitation and pore fluid migration in the Kimmeridge Clay of Dorset, England: Sedimentology, v. 27, p. 577–591.

ISAACS, C. M., 1980, Diagenesis in the Monterey Formation examined laterally along the coast near Santa Barbara, California: Unpublished Ph.D. Dissertation, Stanford University, Stanford, California, 329 p.

———, 1983, Compositional variation and sequence in the Miocene Monterey Formation, Santa Barbara coastal area, California, *in* Larue, D. K., and Steel, R. J., eds., Cenozoic Marine Sedimentation, Pacific Margin, U.S.A., Volume 28: Pacific Section, Society of Economic Paleontologists and Mineralogists Special Publication, p. 117–132.

JONES, J. B., AND SEGNIT, E. R., 1971, The nature of opal. I. Nomenclature and constituent phases: Journal of the Geological Society of Australia, v. 68, p. 56–68.

KABLANOW, R. I., II, AND SURDAM, R. C., 1983, Diagenesis and hydrocarbon generation in the Monterey Formation, Huasna Basin, California, *in* Isaacs, C. M., and Garrison, R. E., eds., Petroleum Generation and Occurrence in the Miocene Monterey Formation, California: Pacific Section, Society of Economic Paleontologists and Mineralogists Special Publication, p. 53–68.

KASTNER, M., KEENE, J. B., AND GIESKES, J. M., 1977, Diagenesis of siliceous ooze–I. Chemical controls on the rate of opal-A to opal-CT transformation–An experimental study: Geochimica et Cosmochimica Acta, v. 41, p. 1041–1059.

———, MERTZ, K., HOLLANDER, D., AND GARRISON, R., 1984, The association of dolomitite-phosphorite-chert: Causes and possible diagenetic sequences, *in* Garrison, R. E., KASTNER, M., AND ZENGER, D. H., eds., Dolomites of The Monterey Formation and Other Organic-Rich Units, Volume 41: Pacific Section, Society of Economic Paleontologists and Mineralogists Special Publication, p. 75–86.

KELTS, K., AND MCKENZIE, J. A., 1982, Diagenetic dolomite formation in Quaternary anoxic diatomaceous muds of Deep Sea Drilling Project Leg 64, Gulf of California, *in* Curray, J. R., Moore, D. G., and others, eds., Initial Reports of Deep Sea Drilling Project 64: U.S. Government Printing Office, Washington, D.C., p. 553–569.

LEVENTHAL, J. S., 1979, The relationships between organic carbon and sulfide sulfur in recent and ancient marine and euxinic sediments: EOS, American Geophysical Union, v. 60, p. 282.

LEWAN, M. D., AND MAYNARD, J. B., 1982, Factors controlling enrichment of vanadium and nickel in the bitumen of organic sedimentary rocks: Geochimica et Cosmochimica Acta, v. 46, p. 2547–2560.

MARTENS, C. S., AND HARRISS, R. C., 1970, Inhibition of apatite precipitation in the marine environment by Mg-ions: Geochimica et Cosmochimica Acta, v. 34, p. 621–625.

MCCULLOH, T. H., AND BEYER, L. A., 1979, Geothermal gradients, *in* Geologic Studies of the Point Conception Deep Stratigraphic Test Well 1 OCS-Cal 78-164, Outer Continental Shelf, Southern California: U.S. Geological Survey Open-File Report 79-1218, p. 43–48.

MERTZ, K. A., JR., 1984, Diagenetic aspects, Sandholdt Member, Miocene Monterey Formation, Santa Lucia Mountains, California: Implications for depositional and burial environments, *in* Garrison, R. E., Kastner, M., and Zenger, D. H., eds., Dolomites of the Monterey Formation and Other Organic-Rich Units, Volume 41: Pacific Section, Society of Economic Paleontologists and Mineralogists Special Publication, p. 49–74.

MIZUTANI, S., 1977, Progressive ordering of cristobalitic silica in early stage of diagenesis: Contributions to Mineralogy and Petrology, v. 61, p. 129–140.

MURATA, K. J., FRIEDMAN, I., AND GLEASON, J. D., 1977, Oxygen isotope relations between diagenetic silica minerals in Monterey Shale, Temblor Range, California: American Journal of Science, v. 277, p. 259–272.

———, ———, AND MADSEN, B. M., 1969, Isotopic composition of diagenetic carbonates in Miocene marine formations of California and Oregon: U.S. Geological Survey Professional Paper 614-B, 24 p.

———, AND LARSON, R. R., 1975, Diagenesis of Miocene siliceous shale, Temblor Range, California: Journal of Research, U.S. Geological Survey, v. 3, p. 553–566.

NATHAN, Y., AND LUCAS, J., 1976, Experiences sur la precipitation discrete de l'apatite dans l'eau de mer: Implications dans la genesis des phosphorites: Chemical Geology, v. 18, p. 181–186.

NICHOLLS, G. D., CURL, H., AND BOWEN, V. T., 1959, Spectrographic analyses of marine plankton: Limnology and Oceanography, v. 4, p. 472.

ORR, W. L., 1984, Sulfur and sulfur isotope ratios in Monterey oils of the Santa Maria River basin and Santa Barbara Channel areas: Society of Economic Paleontologists and Mineralogists First Annual Midyear Meeting, San Jose, California, Abstracts, p. 62.

PISCIOTTO, K. A., 1981, Review of secondary carbonates in the Monterey Formation, California, *in* Garrison, R. E., and Douglas, R. G., eds., The Monterey Formation and Related Siliceous Rocks of California: Pacific Section, Society of Economic Paleontologists and Mineralogists Special Publication, p. 273–283.

———, AND GARRISON, R. E., 1981, Lithofacies and depositional environments of the Monterey Formation, California, *in* Garrison, R. E., and Douglas, R. G., eds., The Monterey Formation and Related Siliceous Rocks of California: Pacific Section, Society of Economic Paleontologists and Mineralogists Special Publication, p. 97–122.

RAISWELL, R., AND BERNER, R. A., 1985, Pyrite formation in euxinic and semi-euxinic sediments: American Journal of Science, v. 285, p. 710–724.

SHIPBOARD SCIENTIFIC PARTY, 1973, Site 147: Cariaco Basin, *in* Edgar, N. T., Saunders, J. B., and others, eds., Initial Reports of Deep Sea Drilling Project 15: U.S. Government Printing Office, Washington, D.C., p. 169–199.

SHIPBOARD SCIENTIFIC PARTY, 1981, Site report, *in* Yeats, R., Haq, B. U., and others, eds., Initial Reports of Deep Sea Drilling Project 63: U.S. Government Printing Office, Washington, D.C., 967 p.

SHOLKOVITZ, E., 1973, Interstitial water chemistry of the Santa Barbara Basin sediments: Geochimica et Cosmochimica Acta, v. 37, p. 2043–2073.

SPOTTS, J. H., AND SILVERMAN, S. R., 1966, Organic dolomite from Point Fermin, California: The American Mineralogist, v. 51, p. 1144–1155.

SUESS, E., 1981, Phosphate regeneration from sediments of the Peru continental margin by dissolution of fish debris: Geochimica et Cosmochimica Acta, v. 45, p. 577–588.

VAN ANDEL, T. H., 1964, Recent marine sediments of the Gulf of California, *in* Van Andel, T. J., and Shor, G. G., Jr., eds., Marine Geology of the Gulf of California, A Symposium: American Association of Petroleum Geologists Memoir 3, p. 216–310.

VEEH, H. H., BURNETT, W. C., AND SOUTAR, A., 1973, Contemporary phosphorite on the continental margin of Peru: Science, v. 181, p. 844–845.

VINE, J. D., AND TOURTELOT, E. B., 1970, Geochemistry of black shale deposits–A summary report: Economic Geology, v. 65, p. 253–272.

WEDEPHOL, K. H., ed., 1978, Handbook of Geochemistry: Springer-Verlag, New York, (unpaginated).

WISE, S. W., JR., BLIE, B. F., AND WEAVER, F. M., 1972, Chemically precipitated sedimentary cristobalite and the origin of chert: Eclogae Geologicae Helvetiae, v. 65, p. 157–163.

WOODRING, W. P., AND BRAMLETTE, M. N., 1950, Geology and paleontology of the Santa Maria district, California: U.S. Geological Survey Professional Paper 222, 185 p.

SECTION III
DOLOMITES IN MVT DEPOSITS

SECTION III: INTRODUCTION
DOLOMITES IN MVT DEPOSITS

The two papers in this section describe occurrence of epigenetic dolomite in Mississippi Valley-Type ore deposits of southeast Missouri.

Gregg describes coarse crystalline (low-Fe) and ferroan dolomites from Cambrian formations. These dolomites formed from warm basinal brines (coarse dolomite) and waters derived from clay transformations (ferroan dolomite).

Buelter and Guillmette continue the same theme in which dolomites formed from basinal brines and from mixing between these brines and meteoric waters.

ORIGINS OF DOLOMITE IN THE OFFSHORE FACIES OF THE BONNETERRE FORMATION (CAMBRIAN), SOUTHEAST MISSOURI

JAY M. GREGG[1]

Geological Research Laboratory, St. Joe Minerals Corporation, P.O. Box 500, Viburnum, Missouri 65566

ABSTRACT: The Bonneterre Formation (Cambrian), southeast Missouri, is characterized by dolomitized algal bioherms and associated shelf carbonates that were deposited around the Precambrian St. Francois Mountains, which were islands during Late Cambrian time. West of the dolomitized shelf carbonates, the offshore facies of the Bonneterre consists of a deeper water limestone and shale sequence composed of oolitic and skeletal wackestones and packstones interbedded with silty lime mudstones and green illitic shales. Individual limestone and shale beds range in thickness from less than 1 cm to several meters.

At the base of the offshore facies, immediately overlying the Lamotte Sandstone, is a regionally extensive basal dolomite that averages about 6 m thick. The basal dolomite contains coarse crystalline, nonplanar dolomite. This dolomite is relatively low in iron and nearly stoichiometric with regard to $CaCO_3$. Stable carbon and oxygen isotope values for the basal dolomite are similar to values obtained for epigenetic dolomite associated with nearby sulfide ore bodies.

The interbedded limestones and shales of the offshore facies contain abundant ferroan dolomite occurring as individual crystals and patches of crystals replacing limestones and floating in shale beds. This dolomite is commonly concentrated near solution seams, in argillaceous seams in limestones, in shale beds, and also may selectively replace allochems. The ferroan dolomite is commonly enriched in $CaCO_3$ and zoned with respect to iron. Stable carbon and oxygen isotope values are low, indicating elevated temperature and an organic source of carbon.

The basal dolomite was formed by warm basinal brines circulating through the Lamotte Sandstone aquifer. This water may have been genetically related to the fluids that produced nearby Mississippi Valley-type ore deposits. The presence of impermeable overlying shale beds had restricted these fluids (and the resultant dolomite) to the lower few meters of the offshore facies. The ferroan dolomite was formed after burial by Mg^{+2}- and Fe^{+2}-rich water evolved during the illitization of smectite in the interbedded shale.

INTRODUCTION

The importance of late diagenetic or epigenetic dolomitization has been a subject of increasing interest over the past several years (e.g., Freeman, 1972; Mattes and Mountjoy, 1980; Zenger, 1983; Machel and Mountjoy, 1986; Hardie, 1987; Gawthorpe, 1987). Lyle (1977, p. 430) stated, on the basis of petrographic evidence, that dolomite in the Bonneterre Formation (Cambrian), southeast Missouri, was "a late diagenetic product and probably formed in the deep subsurface." Gregg (1985) presented petrographic and stable isotope evidence that a thin, regionally extensive dolomite bed at the base of a limestone/shale sequence, composing the offshore facies of the Bonneterre Formation, formed as a result of warm basinal water circulating through the underlying Lamotte Sandstone. The "basal dolomite" was distinguished from ferroan dolomite in the overlying limestone and shale sequence (Gregg, 1985).

This paper compares and contrasts the petrographic and geochemical characteristics of both the basal and ferroan dolomites. It will be argued that both dolomites are late diagenetic products; however, an origin involving the diagenesis of the neighboring shale beds is argued for the ferroan dolomite, whereas a mechanism involving basinal water is applicable to the basal dolomite.

Geologic Setting

The Bonneterre Formation is part of a predominantly carbonate, Upper Cambrian sequence in southeast Missouri (Fig. 1). The Bonneterre overlies the quartz arenitic to arkosic Lamotte Sandstone or, in some cases, rests directly on Precambrian intrusives and volcanics of the Ozark Uplands. The Bonneterre is, in turn, overlain by the interbedded limestone, dolomite, and shale of the Davis Formation (Fig. 2). The Bonneterre Formation contains algal stromatolite bioherms (reefs) and associated grainstone, lagoonal mudstone, and planar stromatolites that had built up around the Precambrian St. Francois Mountains, which had existed as islands in a shallow epicontinental sea during Late Cambrian time (Figs. 1, 2). These rocks were deposited in shallow water (less than several meters) and were subjected to periods of subaerial exposure. The algal bioherms and associated facies have undergone pervasive dolomitization and host the classic Mississippi Valley-type (MVT) lead-zinc-copper sulfide ore bodies of southeast Missouri (including the Old Lead Belt, Fredericktown, Annapolis, Indian Creek, and Viburnum Trend subdistricts). The distribution of the carbonate facies and their relations to sulfide mineralization were first recognized by Ohle and Brown (1954) and are discussed by Gerdemann and Meyers (1972) and Gerdemann and Gregg (1986). The sedimentology of the Bonneterre Formation is discussed by Howe (1968) and Larsen (1977), and the geologic history, structure, and stratigraphy of the southeast Missouri region are described by Thacker and Anderson (1977). Snyder and Gerdemann (1968) treat the economic geology of the southeast Missouri lead-zinc district. The petrology of the mineralized facies of the Bonneterre is discussed by Lyle (1977).

Offshore facies.—

The "fore reef" of the Bonneterre is represented by a sequence of interbedded limestone and shale, referred to as the "offshore facies" (Gerdemann and Myers, 1972). The offshore facies is located to the west of the algal bioherm grainstone facies (Figs. 1, 2) and covers an area of more than 24,000 km^2 in southeast Missouri. A similar limestone and shale sequence exists to the east of the bioherm/grainstone/backreef facies on the eastern side of the St. Francois Mountains and extends into the Illinois Basin (see Ohle and

[1]Present address: Department of Geology and Geophysics, University of Missouri-Rolla, Missouri 65401.

Sedimentology and Geochemistry of Dolostones, SEPM Special Publication No. 43
 ISBN 0-918985-77-3

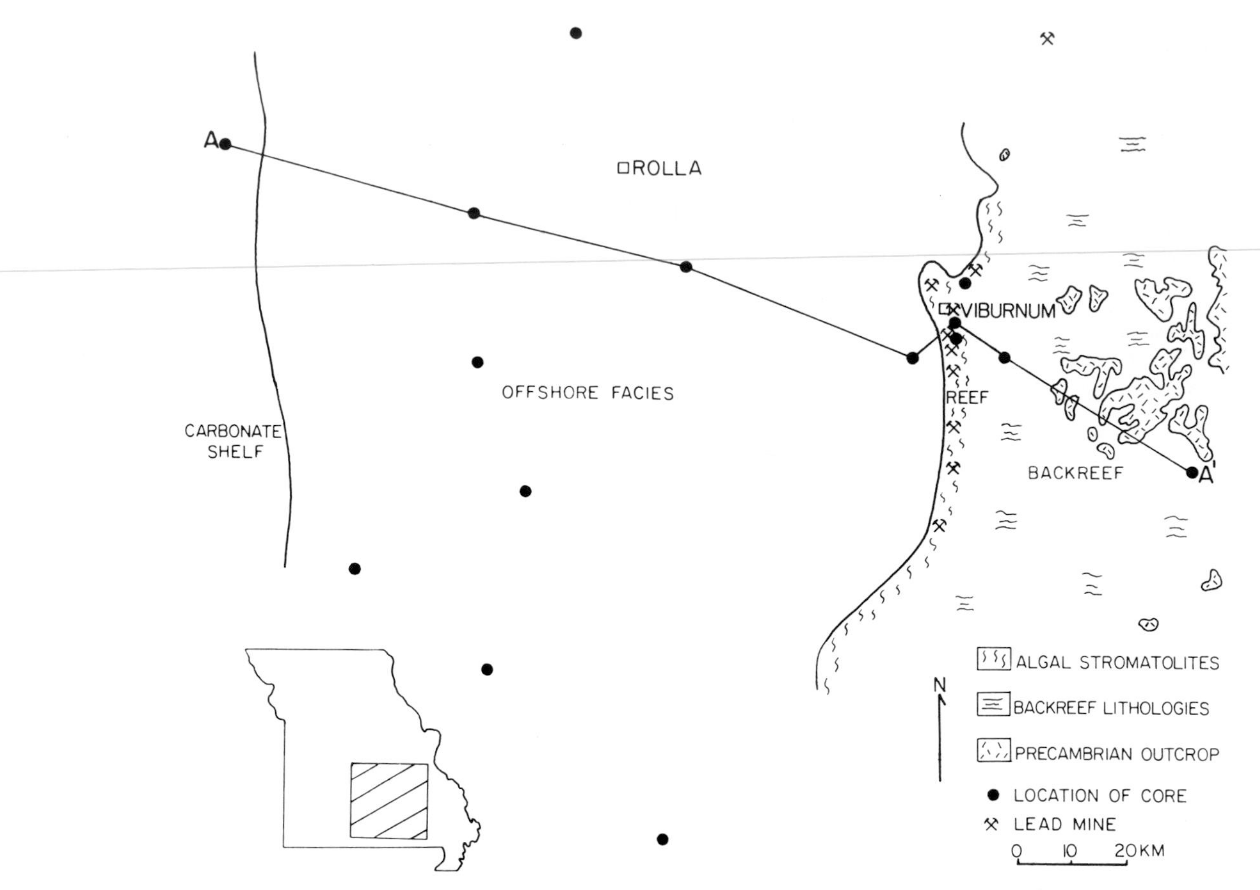

FIG. 1.—Location of study area, Precambrian outcrop, cores sampled for this study, and distribution of major facies of the Bonneterre Formation (from Gregg, 1985).

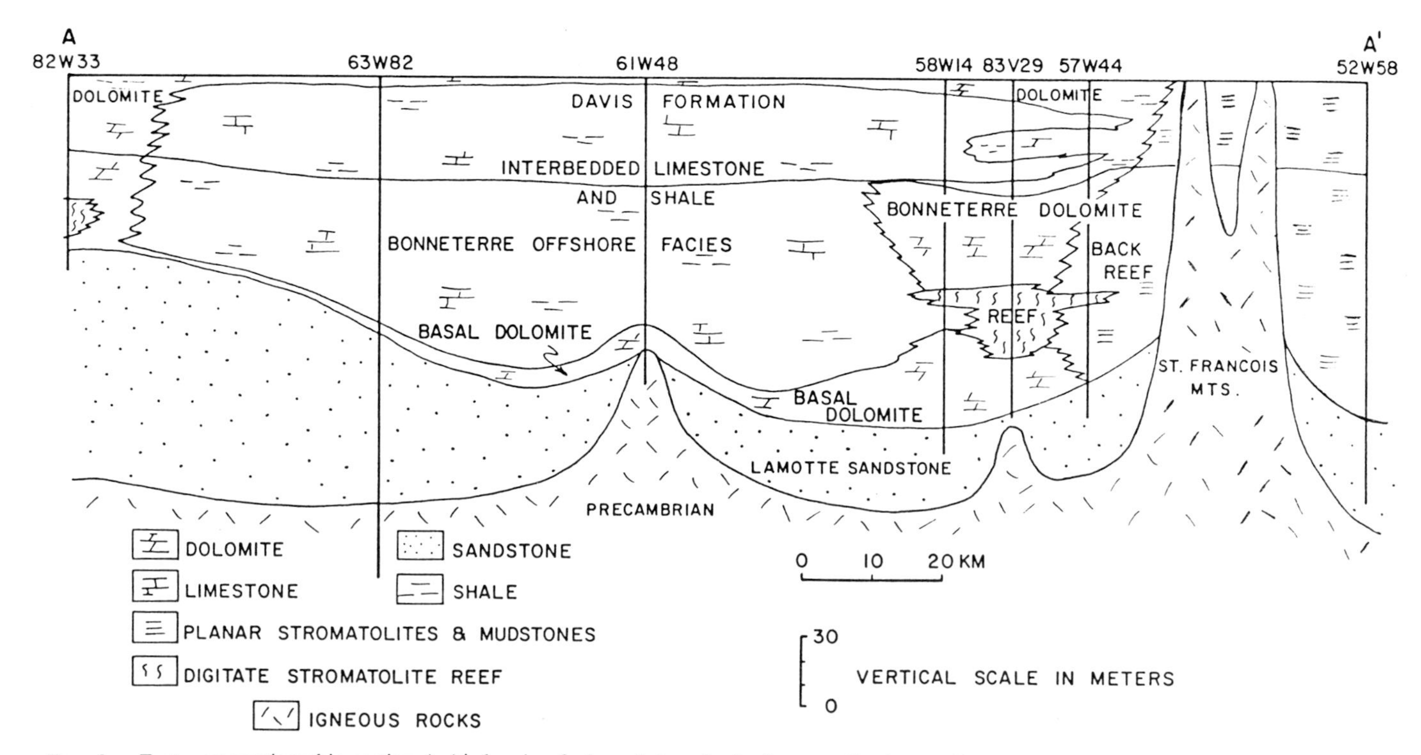

FIG. 2.—East-west stratigraphic section A-A′ showing facies relations in the Lamotte Sandstone, Bonneterre, and Davis formations (from Gregg, 1985).

Brown, 1954). Individual limestone and shale beds range in thickness from 1 cm or less to several meters. The facies pinches out into the dolomitized bioherm/grainstone facies to the east and dolomitized shelf carbonates to the west (Fig. 3).

The offshore facies of the Bonneterre Formation is composed mainly of glauconitic silty mudstone and fossiliferous wackestone and packstone interbedded with dark green silty shale. The sequence is sandy at the base, where it conformably overlies the Lamotte Sandstone. The offshore facies is overlain by the lithologically similar Davis Formation. This facies probably represents deposition under moderately deep-water conditions (below wave base but within the photic zone). Larson (1977) recognized two subfacies within the offshore facies: (1) a shaley lime mudstone subfacies deposited in a deeper water environment, and (2) a burrowed wackestone subfacies deposited in shallower water. In the upper part of the offshore facies sequence, a laterally continuous, tightly cemented, oolitic-grainstone bed is commonly encountered. This bed may represent an episode of shallow-water, higher energy conditions and is correlative with the thick grainstone beds overlying the stromatolitic bioherms to the east [the "upper Bonneterre facies" of Gerdemann and Myers (1972)]. Near the top of the Bonneterre is the Sullivan Siltstone Member that is composed of 1 to 2 m of argillaceous siltstone, which is commonly cemented by carbonate.

A regionally extensive dolomite bed occurs at the base of the offshore facies at the Lamotte/Bonneterre contact (Fig. 2). It averages about 6 m thick, becoming generally thinner toward the northwest (Fig. 4). The dolomitized bed was deposited under conditions similar to those of the overlying limestone beds and also contains thin shale beds. Trace sphalerite is commonly observed in the basal dolomite (Gregg, 1985).

Ferroan dolomite is commonly encountered in the limestone and shale beds overlying the basal dolomite. This dolomite is usually found as scattered individual crystals or as patches of crystals throughout the limestone and shale sequence.

METHODS

Thin sections were prepared from samples collected from 23 2.5-cm (1 in.) cores. Standard carbonate staining methods (Friedman, 1959) were employed for light microscopy. Cathodoluminescent properties of selected polished thin sections were examined using a Nuclide Luminoscope, model ELM-2, at 10–15 kv and 4–6 Ma, under a helium atmosphere.

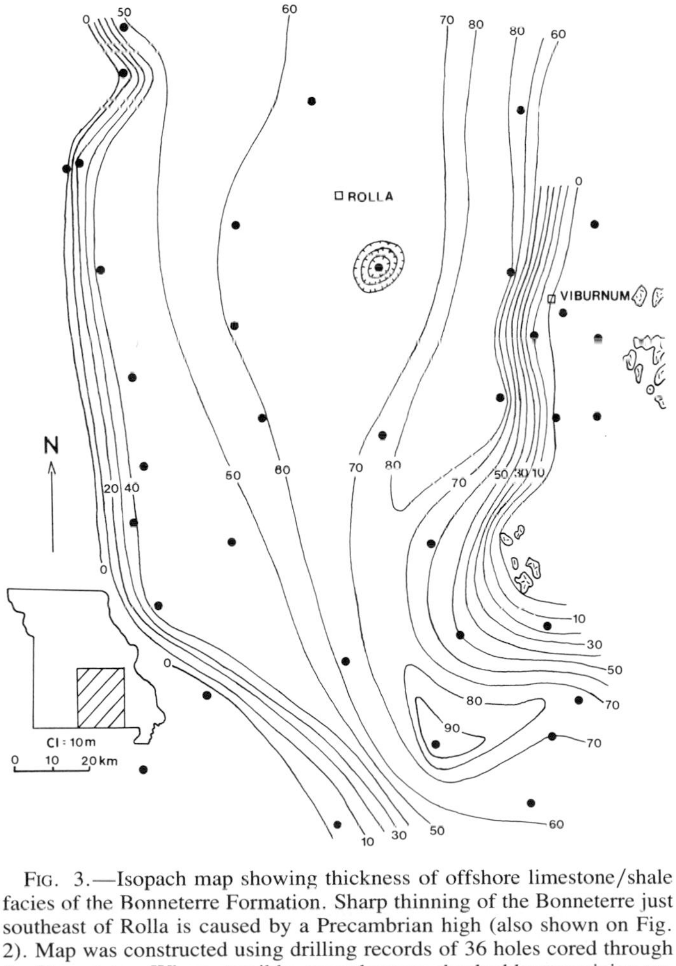

FIG. 3.—Isopach map showing thickness of offshore limestone/shale facies of the Bonneterre Formation. Sharp thinning of the Bonneterre just southeast of Rolla is caused by a Precambrian high (also shown on Fig. 2). Map was constructed using drilling records of 36 holes cored through the Bonneterre. Where possible, records were checked by examining existing core.

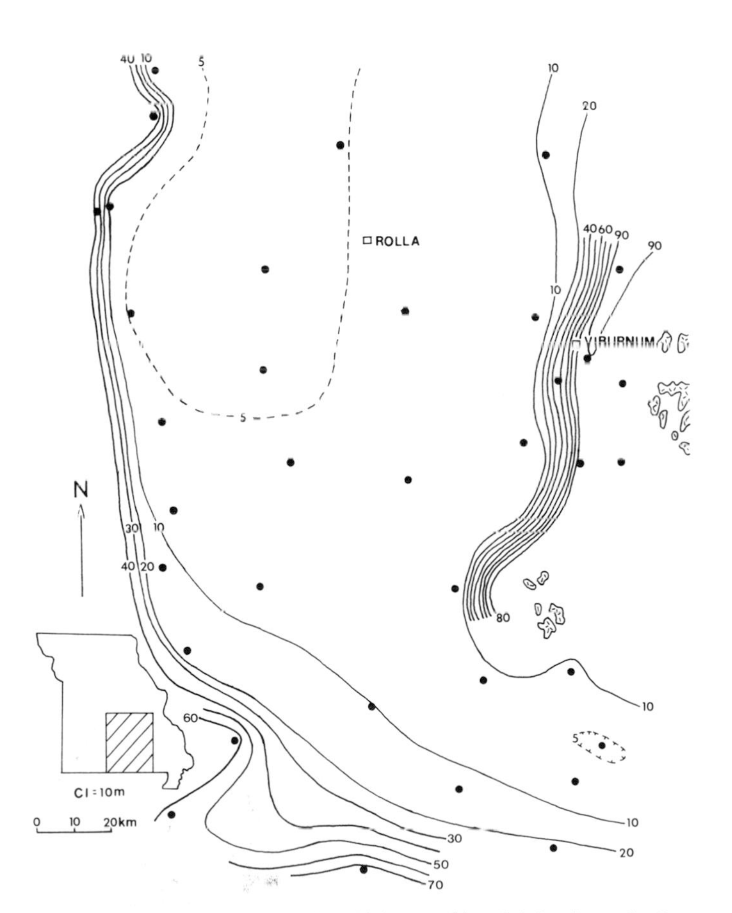

FIG. 4.—Isopach map showing thickness of basal dolomite at the Bonneterre/Lamotte contact. Dolomite is shown thickening upward into pervasively dolomitized units of the Viburnum Trend on the east and carbonate shelf facies on the west.

Polished thin sections were analyzed using a computer-automated JEOL 733 electron microprobe equipped with three crystal spectrometers, a backscattered electron detector, and energy-dispersive capabilities. The microprobe was operated at 15 kv and 30 na using maximum counting times of 80 sec and spot sizes of 10 μm. Spots were analyzed for Ca, Mg, Fe, and Mn employing Bence-Albee corrections and standards provided by the Smithsonian Institute (Jarosewich and Howard, 1983). Limits of determination for $FeCO_3$ and $MnCO_3$ were 0.09 and 0.06 mole percent, respectively (at a 0.997 confidence level).

Microsamples for isotope analysis were obtained from polished chips using a dental drill under a binocular microscope. Dolomites were reacted at 25°C with phosphoric acid, and isotopic enrichments were corrected for $\delta^{17}O$ contributions (Craig, 1957). No corrections for acid fractionation were applied to dolomite values.

Oriented clay mounts were prepared and examined with a computer-automated Phillips X-ray diffractometer using $Cu_{K\alpha}$ radiation. Standard clay mineral identification techniques were employed (Carrol, 1970).

RESULTS

Petrology

Basal dolomite.—

The basal dolomite is characterized by a uniform, coarse crystalline (0.2–0.5 mm), nonplanar texture, with undulatory extinction, typical of epigenetic dolomite (Fig. 5; the dolomite texture classification used here is that of Sibley and Gregg, 1987; see also Gregg and Sibley, 1984). More finely crystalline idiotopic dolomite occurs in the more argillaceous beds and along solution seams. The replacement dolomite has a relatively low-iron content and is not compositionally zoned. Void-filling dolomite cement in the basal dolomite displays a cathodoluminescent microstratigraphy, due to compositional zonation, similar to gangue dolomite associated with the nearby sulfide ores (Gregg, 1985; Voss and Hagni, 1985). The petrology of the basal dolomite, associated dolomite cements, and sulfide mineralization is more thoroughly discussed by Gregg (1985).

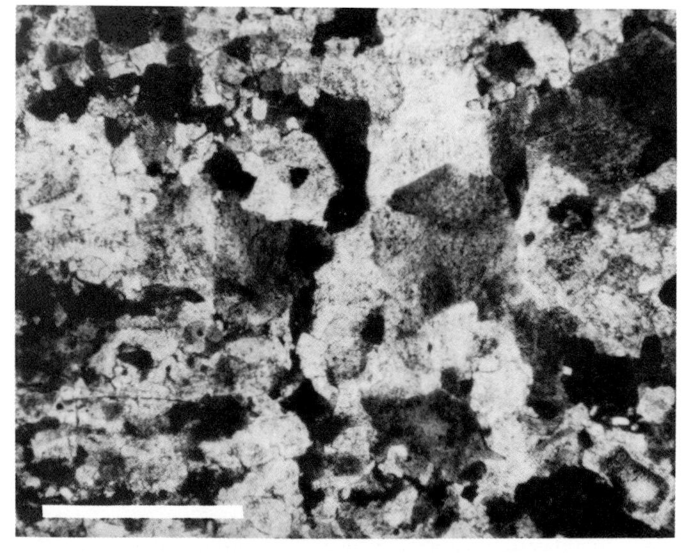

FIG. 5.—Photomicrograph of basal dolomite exhibiting the coarse crystalline nonplanar texture (Sibley and Gregg, 1987) characteristic of this unit. Cross-polarized light, scale = 1 mm.

Ferroan dolomite.—

Ferroan dolomite is found as scattered individual crystals and patches of crystals throughout the offshore facies. Difficulty was encountered in determining the total volume of dolomite because of problems with point counting the fine crystalline dolomite in argillaceous seams and shale. It is estimated, however, that the ferroan dolomite does not exceed 10% of the total volume of the offshore facies. Occasionally, limestone beds adjacent to the Sullivan Siltstone Member are completely replaced by ferroan dolomite, possibly due to increased permeability to dolomitizing fluids in the siltstone.

Ferroan dolomite occasionally occurs as void-filling saddle dolomite cement, commonly associated with late calcite cement. Such occurrences are rare, however, because there is very little porosity in the limestone of the offshore facies; primary porosity was usually occluded by early marine cements. The saddle dolomite is occasionally zoned with respect to iron but does not exhibit the same pattern of zonation observed in dolomite cements occuring in the basal dolomite. Small (<5 μm) two-phase fluid inclusions were observed in these cements. No analyses of these inclusions were made for this study.

Large (0.1–0.5 mm) ferroan dolomite rhombs are commonly observed selectively replacing allochems (Fig. 6A) or replacing micritic limestone, especially where associated with microfractures and stylolites (Fig. 6B). The dolomite rhombs have very strong growth zonation with respect to iron; some zones range to ankerite composition (Fig. 6C). Other large dolomite crystals (as large as 1 mm) do not exhibit a clear rhombic zonation but have a quiltlike pattern of iron zoning (Fig. 6D). Most of the large dolomite crystals have undulatory extinction in cross-polarized light. Occasional dedolomitization of ferroan dolomite also occurs in the offshore facies (see Lyle, 1977). Fine crystalline ferroan dolomite (<0.05 mm) is nearly always found associated with argillaceous seams in the limestone (Fig. 7A, B) and exists in the shale beds (Fig. 7C, D). These dolomite crystals are commonly zoned as discussed earlier (Fig. 6).

The ferroan dolomite is nonluminescent due to its high-iron content (see Pierson, 1981; Machel, 1985). By contrast, the hosting limestone exhibits a uniform bright orange luminescence, reflecting its low-iron content. (Mean values for electron microprobe analysis of selected limestone samples: $FeCO_3$ = 0.57 mole percent, $MnCO_3$ = 0.10 mole percent, n = 43.)

Petrology of shale beds.—

The carbonates of the offshore facies are interbedded with dark green shale. Shale composes approximately 35% of the total thickness of the offshore facies, with carbonate beds composing most of the remaining 65%, as determined by measuring lithologic thickness in cores 62W82 and 61W48

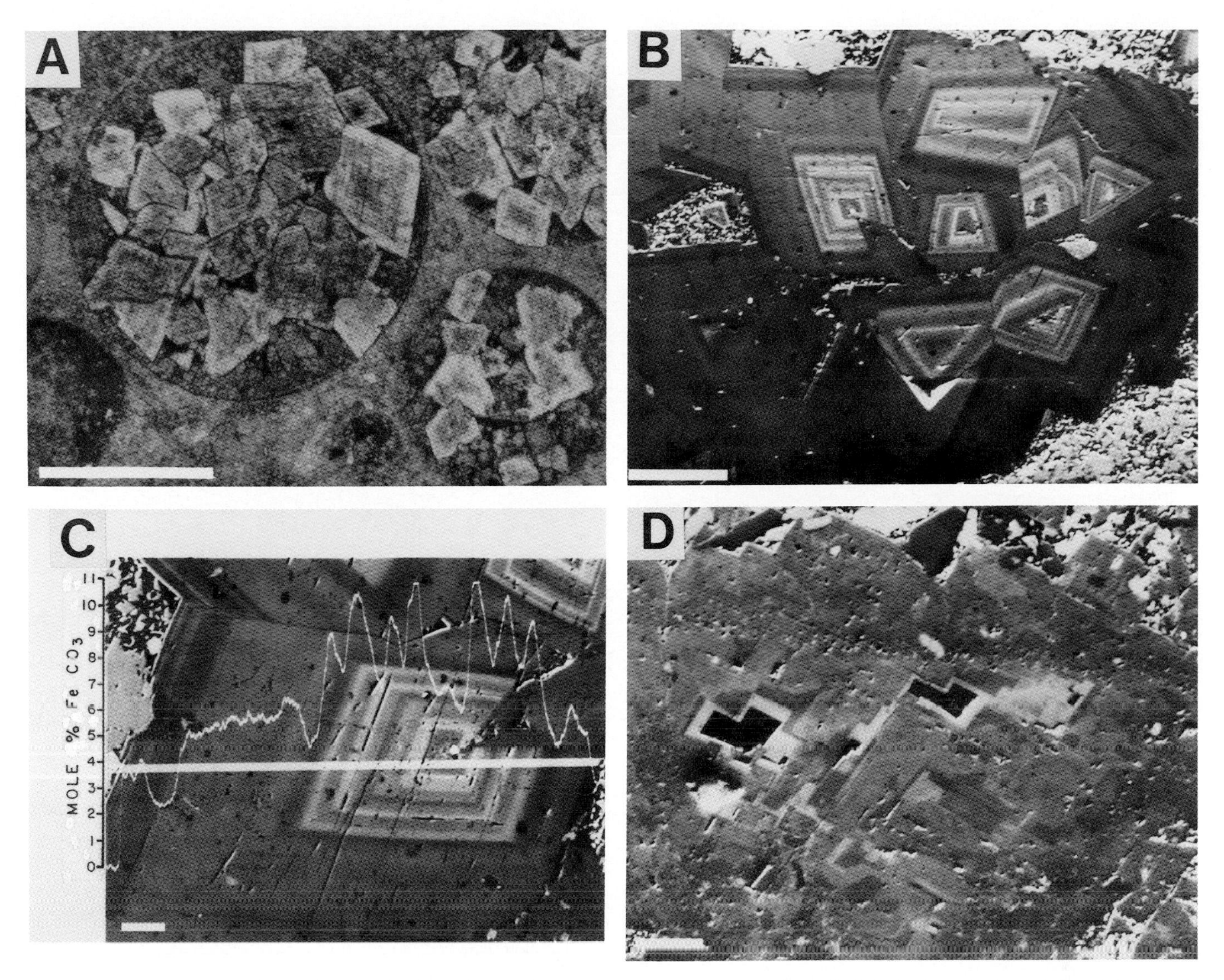

FIG. 6.—Coarse, Fe-zoned dolomite rhombs from offshore facies of the Bonneterre Formation. (A) Selectively replacing calcite ooids. Zonation is shown by potassium ferricyanide stain (dark areas). Plain polarized light, scale = 0.5 mm. (B) Fe-zoned (bright bands) dolomite rhombs associated with a solution seam in calcite micrite. SEM backscattered electron image, scale = 0.1 mm. (C) Closeup of one of the dolomite rhombs in (B) showing an electron microprobe traverse and $FeCO_3$ content ranging from 2.2 to 11 mole percent. Scale = 0.02 mm. (D) Large dolomite crystal replacing silty calcite mudstone. Note quiltlike Fe zonation (bright areas are high in iron). SEM backscattered electron image, scale = 0.05 mm.

(Fig. 2). Commonly, thin argillaceous seams are observed in thin section within the carbonate beds, although these are usually not distinguishable as individual shale beds in hand specimen or core. The shales are fissile, breaking into "poker chips" when in core, and are composed mainly of clay minerals and secondarily of silt-size detrital silicates. Occasionally, the shale contains significant dolomite (Fig. 7D) or more rarely, calcite. Pyrite was occasionally observed in the shale beds and in the neighboring limestone.

Clay mineral analyses were performed on three shale beds in the Bonneterre, one shale bed in the lower part of the Davis Formation, and one suspected volcanic-ash bed near the base of the sequence. Illite is the chief clay mineral in the offshore facies. No expandable clay component was observed. A trace of chlorite was observed in the Davis sample, and minor kaolinite was encountered in the Davis sample and in the basal dolomite.

Major- and Minor-Element Geochemistry

Ninety electron microprobe analyses for calcium, magnesium, iron, and manganese were made on samples from the basal dolomite, and 188 analyses were performed on samples of dolomite from the offshore facies of the Bonneterre Formation. Individual values are given in Appendix I.

Basal dolomite.—

The basal dolomite is slightly calcium enriched with $CaCO_3$ typically distributed about a mean of 50.8 mole percent (Fig.

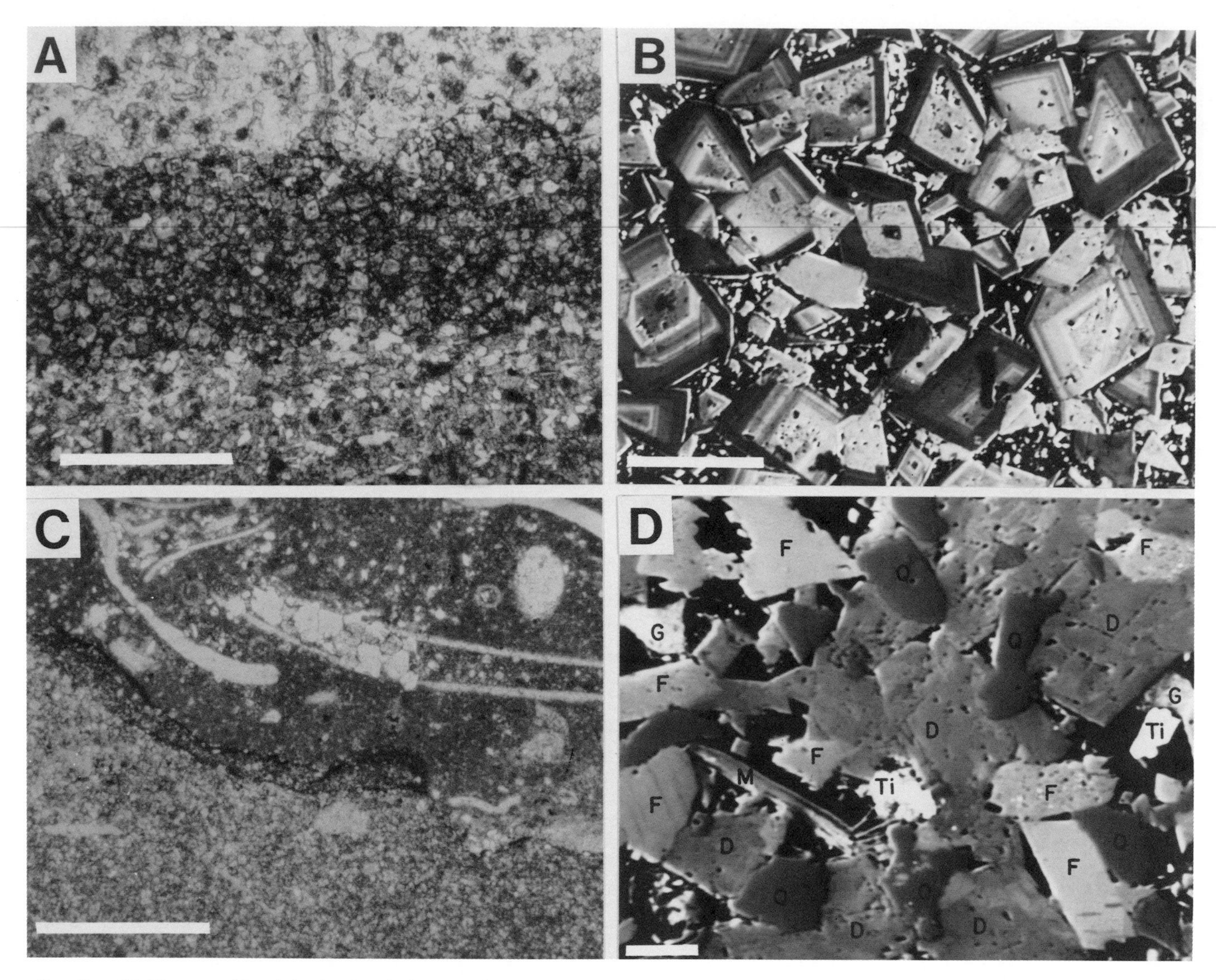

FIG. 7.—(A) Fine crystalline dolomite rhombs in an argillaceous seam (horizontal dark area, center) of a silty skeletal wackestone of the Bonneterre offshore facies. Plain polarized light, scale = 0.5 mm. (B) Closeup of dolomite rhombs in (A) showing Fe zonation. Clay minerals have been plucked from the inter-dolomite areas during polishing. SEM backscattered electron image, scale = 0.05 mm. (C) Contact between a skeletal wackestone (upper field) and a dolomitic shale bed (lower field) in the Bonneterre offshore facies. Plain polarized light, scale = 1 mm. (D) Closeup of dolomitic shale in (C) showing unzoned ferroan dolomite "D" and other minerals: quartz "Q", K feldspar "F", glauconite "G", a titanium mineral, probably rutile "Ti", and muscovite "M". Some K feldspar appears to be altered and partly replaced by glauconite (the bright specks). Note broken end of muscovite crystal (lower left), suggesting a detrital origin. Apparent pore space (black areas) was occupied by clay minerals that were plucked during sample preparation. SEM backscattered electron image, scale = 0.02 mm.

8). Scattered, anomalously high values give an appearance of bimodality to the population. Magnesium is somewhat depleted and bimodally distributed with geometric means at 43.5 and 47.3 mole percent $MgCo_3$ (Fig. 8).

The $FeCO_3$ content of the basal dolomite (excluding void-filling cements) averages 2.67 mole percent. Figure 9A shows the values plotted as a histogram. A cumulative frequency distribution of the values indicates that the data fit a log-normal distribution with a geometric mean at 2.3 mole percent $FeCO_3$ (Fig. 9B).

The $MnCO_3$ content of the basal dolomite averages 0.38 mole percent. The distribution of manganese is plotted as a histogram (Fig. 10A) and on a cumulative frequency diagram (Fig. 10B). The manganese curve appears to be more complex than that of iron and probably indicates a bimodal distribution geometric with means at 0.26 and 0.53 mole percent $MgCO_3$.

Ferroan dolomite.—

Calcium is enriched and bimodally distributed in the ferroan dolomite, with geometric means at 51.7 and 55.3 mole percent $CaCO_3$. Magnesium is depleted and also exhibits a bimodal distribution, with geometric means of 33.6 and 40.8 mole percent $MgCO_3$ (Fig. 8). Analyses having calcium

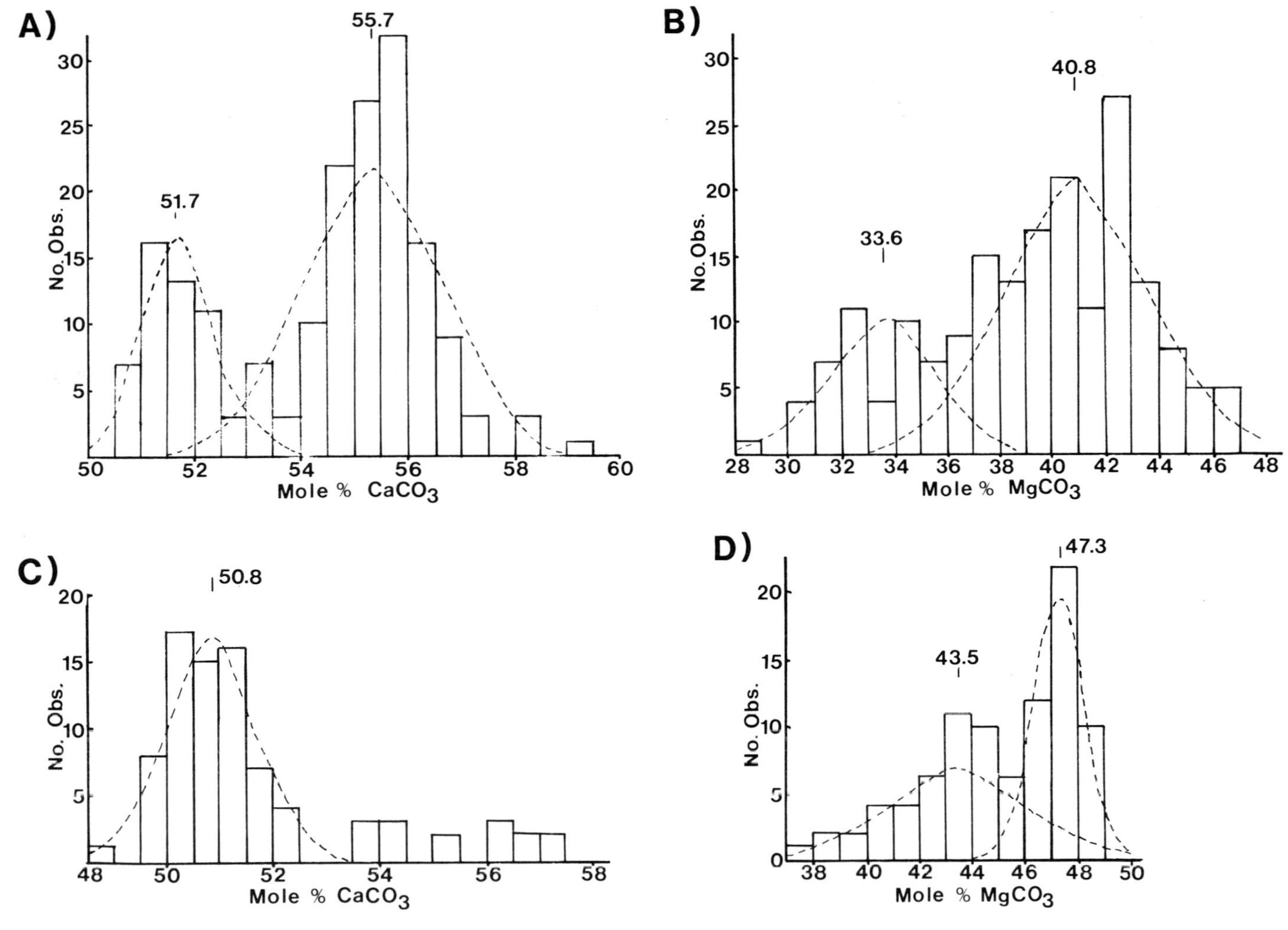

FIG. 8.—Histograms showing (A) $CaCO_3$ distribution in the ferroan dolomite, (B) $MgCO_3$ distribution in the ferroan dolomite, (C) $CaCO_3$ distribution in the basal dolomite, and (D) $MgCO_3$ distribution in the basal dolomite. Curves show theoretical normal distributions.

values that fall in the higher population tend to have magnesium values in the lower population, although this relation is not perfect.

The $FeCO_3$ content of the ferroan dolomite averages 6.25 mole percent. The iron is bimodally distributed with geometric means at 4.5 and 10.2 mole percent (Fig. 11A). A cumulative frequency diagram shows the population separated into two-component normal distributions (Fig. 11B). Values that fall within the higher iron component generally correspond to analyses made in the more iron-rich bands of the rhombic, zoned dolomite crystals (Figs. 6A,B,7B). Values falling into the lower iron component correspond to analyses made in the iron-poor bands of the same crystals as well as the iron-rich and iron-poor zones of the "quilt-like" dolomite crystals (Fig. 6D).

An average of 0.32 mole percent $MnCO_3$ occurs in the ferroan dolomite. A histogram (Fig. 12A) and cumulative frequency diagram (Fig. 12B) of the data show that manganese is lognormally distributed with a geometric mean at 0.28 mole percent $MnCO_3$.

Stable Isotope Geochemistry

Stable carbon and oxygen isotope values for the basal dolomite were presented by Gregg (1985) and will be discussed below. Stable isotope values for the ferroan dolomite are quite depleted, ranging from $\delta^{13}C = -0.24$ to -7.46 per mil (PDB) and $\delta^{18}O = -8.12$ to -17.67 per mil (PDB; Fig. 13; Table 1). Fractionation temperatures as high as 90°C were calculated for the oxygen isotopes using the dolomite/water fractionation equation of Matthews and Katz (1977). An assumed water composition of $\delta^{18}O = -7.86$ (SMOW) was obtained by using the calcite values of Gregg (1985) and assuming water in equilibrium with the calcite at a temperature of 15°C.

DISCUSSION

The ferroan dolomite of the offshore facies of the Bonneterre Formation is petrographically and geochemically distinct from the basal dolomite. The basal dolomite is restricted to the lowermost few meters of the Bonneterre,

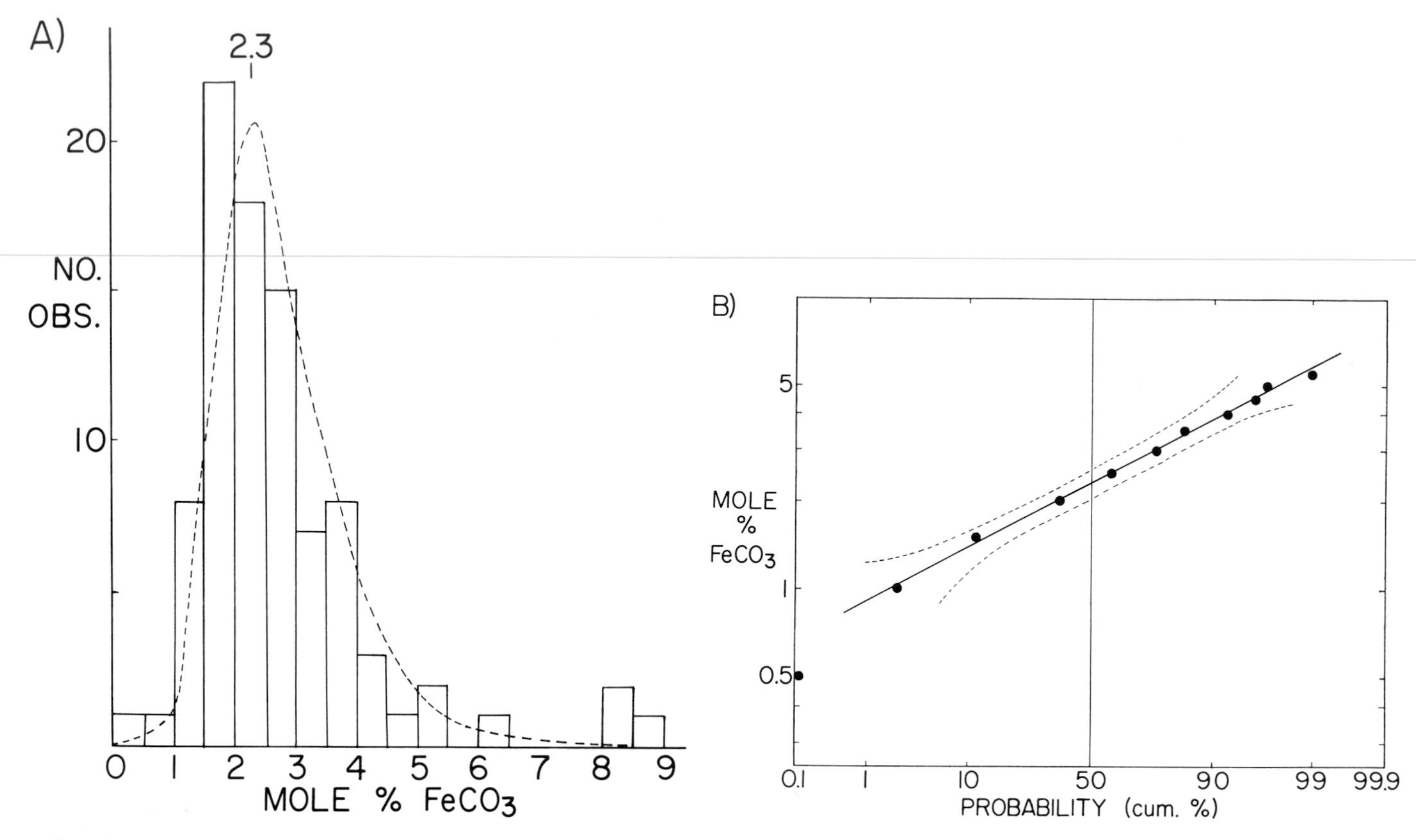

FIG. 9.—(A) Histogram showing mole percent $FeCO_3$ in the basal dolomite plotted against frequency of observations. Curve shows calculated lognormal distribution with a geometric mean at 2.3%. (B) Cumulative frequency distribution of mole percent $FeCO_3$ in the basal dolomite plotted on a logarithmic probability scale. Dashed curves give confidence limits taken at the 5% probability level around the line representing the distribution.

whereas the ferroan dolomite is scattered throughout the limestone/shale facies. In the basal dolomite, limestone is completely replaced, commonly by coarse crystalline nonplanar dolomite (Fig. 5). The ferroan dolomite occurs as scattered individual rhombic crystals or patches of crystals, selectively replacing allochems (Fig. 6A), occurring along solution seams (Fig. 6B), in argillaceous limestones, and in shale beds (Fig. 7). Ferroan dolomite rarely completely replaces a limestone bed.

The basal dolomite has near stoichiometric to slightly en-

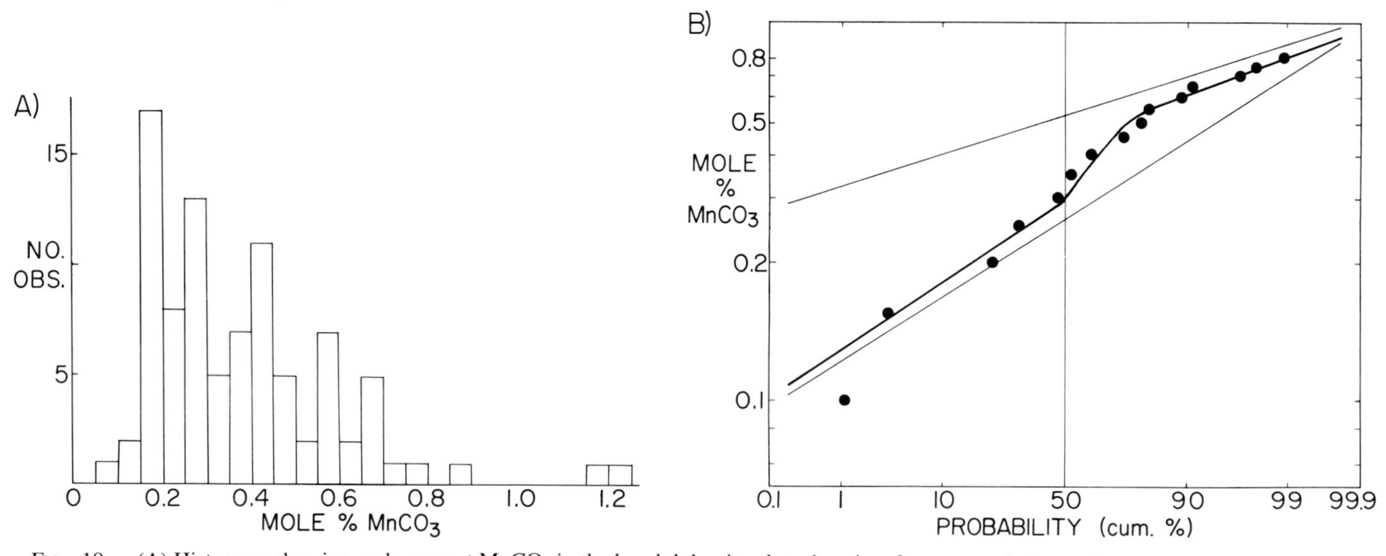

FIG. 10.—(A) Histogram showing mole percent $MnCO_3$ in the basal dolomite plotted against frequency of observations. (B) Cumulative frequency distribution of mole percent $MnCO_3$ in the basal dolomite plotted on a logarithmic probability scale. The two-component distribution curves were drawn by graphic separation of the total population curve (Sinclair, 1976).

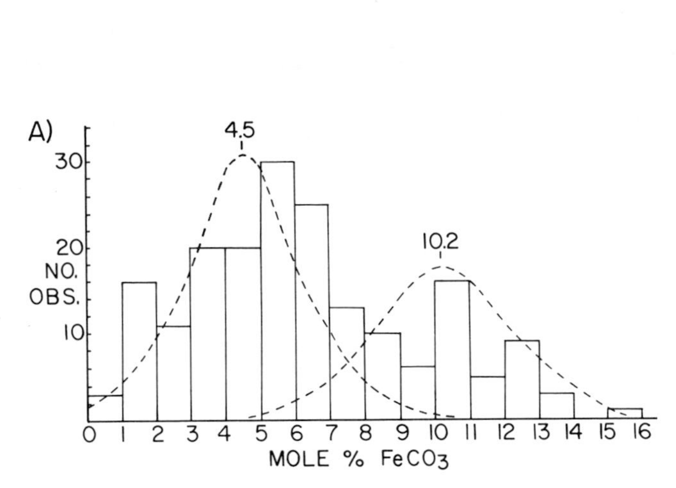

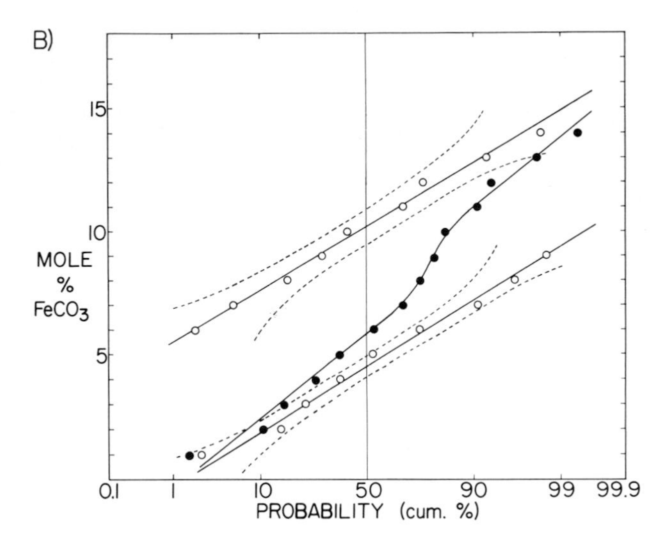

FIG. 11.—(A) Histogram showing mole percent $FeCO_3$ distribution in dolomite from the Bonneterre offshore facies plotted against frequency of observations. Broken curves show calculated normal distributions with means at 4.5 and 10.2 mole percent. (B) Cumulative frequency distribution of mole percent $FeCO_3$ in dolomite from the offshore facies plotted on an arithmatic probability scale. Solid circles and attendant curve (with inflection point near 75%) show cumulative percent of all data. Open circles show data separated into two-component normal distributions. The two-component curves were constructed by assigning individual analyses into one distribution or the other, based on petrographic observation, rather than a graphic separation of the total distribution. Confidence limits taken at the 5% probability level are shown around lines representing normal distributions.

riched values for calcium and is slightly depleted in magnesium (Fig. 8). Iron apparently substituted for the missing magnesium. The ferroan dolomite is more stoichiometrically enriched in calcium (i.e., >50 mole percent $CaCO_3$) and depleted in magnesium than the basal dolomite (Fig. 8). Stoichiometric enrichment in calcium is a characteristic common to other ferroan dolomites associated with shale (Irwin, 1980; Taylor and Sibley, 1986). There is a rough correlation of high-calcium with high-iron levels and low-magnesium levels in the ferroan dolomite. Iron and manganese in the basal dolomite, as well as manganese in the ferroan dolomite, have lognormal distributions (Figs. 9, 10,

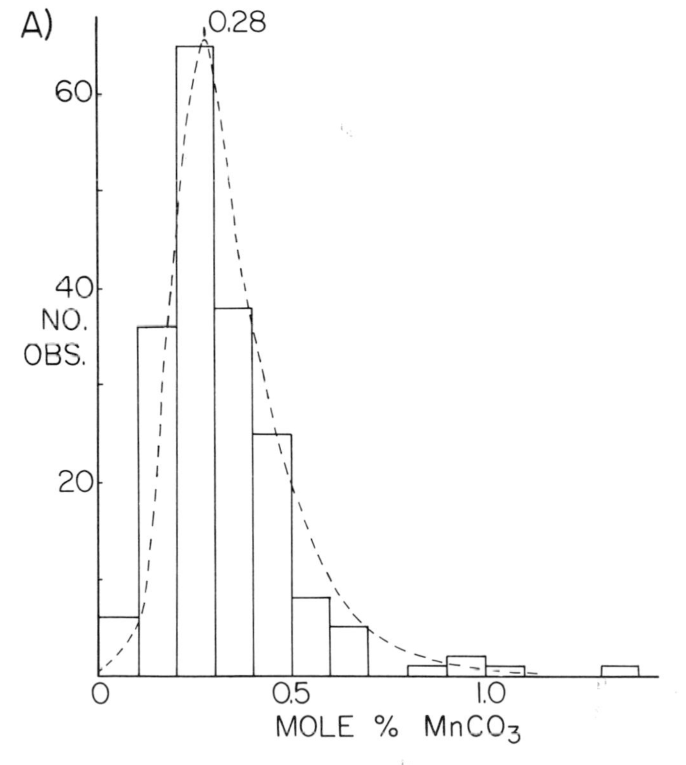

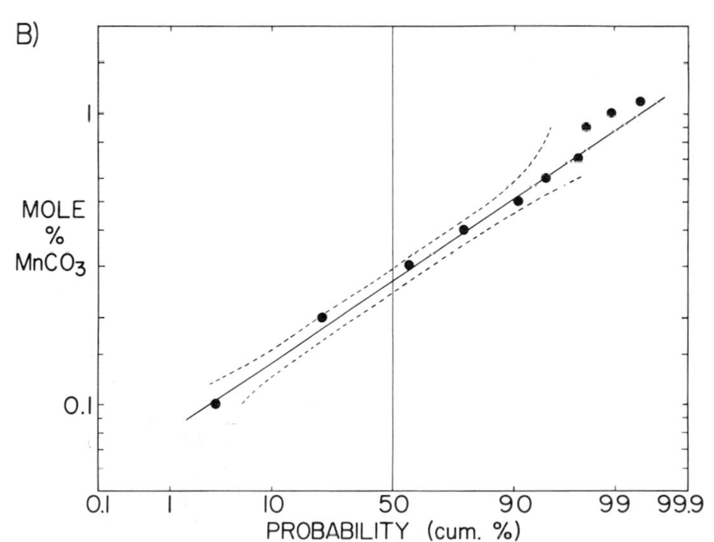

FIG. 12.—(A) Histogram showing mole percent $MnCO_3$ distribution in dolomite from the Bonneterre offshore facies plotted against frequency of observations. Broken curve shows calculated lognormal distribution with a mean at 0.28%. (B) Cumulative frequency distribution of mole percent $MnCO_3$ in the offshore facies plotted on a logarithmic probability scale.

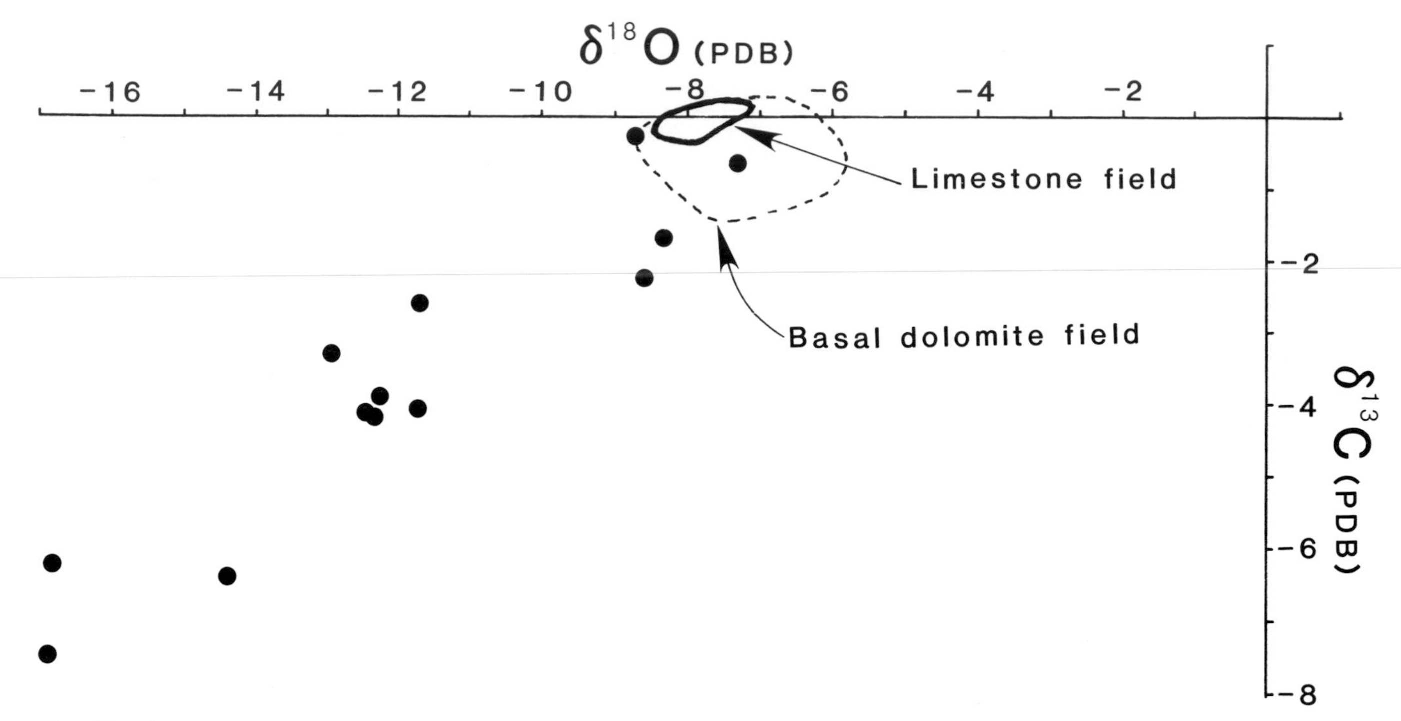

FIG. 13.—Stable oxygen and carbon isotope values for the ferroan dolomite in the offshore facies (closed circles). Fields representing isotope values for limestones in the offshore facies and values for the basal dolomite (Gregg, 1985) are also shown.

12), a characteristic of trace- and minor-element populations in nature (Shaw, 1961; Sinclair, 1976). Iron in the ferroan dolomite, on the other hand, is normally distributed (Fig. 11), a characteristic of major-element populations.

The marked differences in how major and minor elements are distributed in the basal dolomite and in the ferroan dolomite are consistent with the view that each was formed in a different diagenetic milieu.

Origin of the Basal Dolomite

Gregg (1985) argued that the basal dolomite was formed by warm, magnesium-rich water, probably related to the fluids that produced the nearby Mississippi Valley-type mineral districts. It was hypothesized that this water originated in the Arkoma Basin and moved north by means of a gravity flow drive (e.g., Garvin and Freeze, 1984). The water circulated through the underlying Lamotte Sandstone aquifer, dolomitizing the lowermost limestone beds of the offshore facies of the Bonneterre. Shale beds, acting as an aquiclude, prevented the basinal fluids from dolomitizing limestone stratigraphically higher in the section. The following evidence for this model was cited: (1) coarse crystalline nonplanar texture, characteristic of the basal dolomite, is typical of dolomite formed at elevated temperature (Gregg and Sibley, 1984); (2) cathodoluminescent microstratigraphy of dolomite cement in the basal dolomite closely matches that of gangue dolomite cement in the Viburnum Trend ore bodies; and (3) stable carbon and oxygen isotope compositions of the cements and replacement dolomite in the basal dolomite fall into the same field as those of epigenetic dolomite in the Viburnum Trend mineral district (Frank and Lohmann, 1986; Braunsdorf and Lohmann, 1983; confirmed by Gregg, unpubl. data). Further evidence for the origin of the basal dolomite comes from the major- and minor-element compositions obtained for this study, which indicate that the basal dolomite is chemically distinct from the overlying ferroan dolomite. In fact, the distribution of major and minor elements in the basal dolomite is similar to that of the massive dolomite associated with the ore bodies in the Viburnum Trend (Gregg, unpubl. data). Fluid inclusion data indicate that the mineralizing brines that formed the ores and associated dolomite (and therefore, presumably, the basal dolomite) were highly saline and at temperatures between 80° and 140°C (Roedder, 1977; Hagni, 1983).

TABLE 1.—CARBON AND OXYGEN ISOTOPE VALUES OF FERROAN DOLOMITES IN THE OFFSHORE FACIES OF THE BONNETERRE FORMATION

Sample	Description	$\delta^{18}O$ (PDB)	$\delta^{13}C$ (PDB)
SW, SW, Sec. 31, T37N, R10W, Pulaski Co., Missouri			
83-485	Large rhombs replacing ooids	−11.72	−2.60
83-485	Small rhombs in a shale seam	−12.96	−3.32
83-488	Patches of rhombs near a stylolite	−7.32	−0.66
83-490	Large anhedron along a microfracture	−8.35	−1.73
SE, SE, Sec. 36, T36N, R7W, Phelps Co., Missouri			
83-115	Large crystals in silty micrite	−8.68	−0.24
83-115	Large void-filling saddle crystals	−14.40	−6.41
83-116	Large rhombs replacing fossil	−12.54	−4.11
83-116	Small rhombs in argillaceous seam	−11.74	−4.06
83-120	Small rhombs in argillaceous micrite	−16.87	−7.46
83-120	Large crystals in silty micrite	−16.80	−6.22
83-123	Patches of large rhombs in micrite	−8.60	−2.26
83-124	Large rhombs replacing pellets	−12.34	−4.18
83-124	Patch of small rhombs in micrite	−12.31	−3.95

In a detailed petrographic study of the transition from the basal dolomite to the overlying limestone beds, Medary (1986) detected decreasing iron composition and gradual destructive recrystallization of zoned ferroan dolomite downward into the basal dolomite. This was interpreted as evidence that the ferroan dolomite predated the basal dolomite (Medary, 1986; Freemen and Medary, 1987).

Origin of the Ferroan Dolomite

Dolomitization by clay diagenesis.—

The possible role of clay in the origin of dolomite was discussed by Kahle (1965) and more recently was detailed by McHargue and Price (1982). Theoretically, water capable of producing dolomite can be evolved during the compaction of shale. This is a relatively early diagenetic process, occurring prior to the lithification of the sediment. Trapped sea water, organic matter, and adsorbed Mg^{+2} can be called on to provide magnesium for dolomitization by this model (McHargue and Price, 1982). With deeper burial, the conversion of smectite and mixed-layer illite/smectite (I/S) to illite can produce significant quantities of Mg^{+2} by the general equation:

$$\text{Smectite} + K^+ \rightarrow \text{Illite} + Ca^{+2} + Mg^{+2} + Fe^{+2} + SiO_2(aq) + Na^+ + \text{Water} \quad (1)$$

Ca^{+2} and $SiO_2(aq)$ are probably evolved early in this process with Mg^{+2} and Fe^{+2} evolved later (Boles and Franks, 1979). McHargue and Price (1982) believed that illitization of smectite was important in the formation of ferroan dolomite associated with mid continent Pennsylvanian shale. They also suggested, lacking a detrital feldspar or mica source for potassium, that organic matter may provide enough K^+ for the smectite-to-illite conversion. Smectite and mixed-layer I/S are common clay components of modern ocean sediments (Griffin and others, 1968; Grim, 1968; Weaver and Beck, 1971). Alteration of volcanic ash is thought to provide smectite to marine shale (Griffin and others, 1968; Grim, 1968). Mixed-layer I/S is provided by terrestrial sources (Środoń and Eberl, 1984). Shale beds in the Bonneterre Formation were derived, at least in part, from volcanic terrains (the St. Francois Mountains), and the region was volcanically active during the Late Cambrian (Snyder and Gerdemann, 1965). The fact that the shale beds are now composed almost entirely of illite (including a suspected volcanic-ash bed) is evidence that the smectite-to-illite conversion must have taken place in these rocks. Alteration of K feldspar and muscovite (Fig. 7D) in the Bonneterre shale beds provides a detrital source for potassium, not present in the shale studied by McHargue and Price (1982).

Dolomite associated with shale is commonly ferroan, and a number of mechanisms have been proposed to explain this observation. McHargue and Price (1982) believed that excess iron evolved during the smectite-to-illite conversion was incorporated into the growing dolomite crystals that they studied. Irwin (1980) noted the increasing iron content of dolomite across a carbonate bed in the Kimmeridge Clay of Dorset, England. Decreasing abundance of Mg^{+2} from overlying sea water and decreasing abundance of organic matter, coupled with increased availability of iron from reduction of detrital iron oxides as burial depth increased, were proposed to explain the increasing iron composition toward the edges of the bed. Dolomite precipitation at depths below the zone of bacterial sulfate reduction was suggested by Pye (1985) as the origin of compositionally zoned ferroan dolomite in the argillaceous Jet Rock Formation of northeast England. At such depths, H_2S is not available to react with the available Fe^{+2} which was then incorporated into the growing dolomite. The availability of reduced sulfur was called on by Taylor and Sibley (1986) to explain the distribution of ferroan dolomite at the contact of the Trenton Limestone and Utica Shale in the Michigan Basin. Here, low concentrations of organic matter limited the amount of H_2S produced by bacteria during early diagenesis, making Fe^{+2} available for dolomite (Taylor and Sibley, 1986). The ubiquity of glauconite in the offshore Bonneterre indicates that the sediments probably were not oxidized during early diagenesis (see Grim, 1968, p. 541–544). Therefore, dolomite formation below the zone of sulfate reduction (Pye, 1985) might best explain the availability of iron during dolomitization.

Most of the models proposed above (Irwin, 1980; Pye, 1985; Taylor and Sibley, 1986) involve early diagenetic processes during initial compaction and dewatering of argillaceous sediments. In the case of the Kimmeridge Clay and the Trenton Limestone, this is supported by relatively heavy carbon and oxygen isotope values (Irwin, 1980; Taylor and Sibley, 1986). The extremely depleted carbon and oxygen isotope values obtained for the ferroan dolomite in the Benneterre Formation (Fig. 13) indicate deeper burial and elevated temperatures. Pore water $^{18}O/^{16}O$ in ocean floor sediments may be lowered from 1 to 3 per mil during the alteration of volcanic ash to smectite (Lawrence and others, 1975). Since this process also results in lower Mg^{2+} and higher Ca^{2+} concentrations in pore water (Lawrence and others, 1975, fig. 1), it is unlikely that dolomitization occurred at this time. If the ^{18}O depleted pore water persisted into the deeper subsurface environment, then this process may partly explain the low $\delta^{18}O$ values of the ferroan dolomite. Thermal cracking of organic matter at depth seems the best explanation for the depleted carbon values, and fractionation of oxygen isotopes at temperatures as high as 90°C adequately explains the oxygen values of the ferroan dolomite in the Bonneterre (Fig. 13). Similar criteria were used by McHargue and Price (1982) and by Gawthorpe (1987) to explain negative carbon and oxygen values in shale-related dolomite that they studied.

Mechanism for dolomitization in the offshore facies.—

An explanation for the ferroan dolomite in the offshore facies of the Bonneterre Formation requires the following: (1) provide a source of magnesium and iron; (2) provide for the removal of calcium and other excess ions; (3) accommodate the depleted carbon and oxygen isotope values; (4) explain why dolomitization of this facies is incomplete; and (5) explain the iron zonation that occurs in some of the ferroan dolomite.

With increasing depth of burial and temperature, smectite

and smectite layers in mixed-layer I/S in the offshore facies began to convert to illite. The Mg^{2+} and Fe^{2+} evolved were incorporated into growing dolomite crystals in the limestone and shale. Some of the excess Ca^{2+} possibly precipitated as the late void-filling calcite cement that is occasionally observed in the offshore facies. Excess $SiO_2(aq)$, Na^+, Ca^{2+}, and water were expelled laterallly along the limestone beds.

The amount of ferroan dolomite in the offshore facies is limited by the amount of Mg^{2+} that could be produced during clay diagenesis. The total volume of ferroan dolomite is estimated to be 10% or less. A conservative estimate of the volume of clay in the offshore facies is about 28%, assuming that the shale beds and limestone beds contain 60 and 10% clay, respectively. Also assuming that for every mole of illite $[K_{1-1.5}Al_4(Si,Al)_8O_{20}(OH)_4]$ produced during the smectite- and mixed-layer I/S to-illite reaction, two moles of Mg^{2+} were evolved, and taking into account the difference in density between dolomite and illite, one volume of dolomite could be produced for every 2.4 volumes of illite. In the offshore facies, the ratio of dolomite to clay is estimated at 2.8:1, well within the range of available Mg^{2+}. Because the variables (i.e., volume and Mg content of original clay minerals and precise volume of dolomite) are not well known, this mass balance calculation should be regarded only as a rough etimate, but it does serve to illustrate the limiting factors of ferroan dolomite formation in the offshore facies of the Bonneterre.

The iron zonation in the ferroan dolomite crystals is difficult to explain. Possibly, pulses of connate water, moving laterally through the Bonneterre, precipitated the lower iron zones. This water may have had the effect of diluting the iron concentration in solution during dolomite precipitation or changing Eh conditions in such a way as to affect the activity of Fe^{2+} in solution. An episodic basin dewatering mechanism, as proposed by Cathels and Smith (1983), might explain such pulses of fluid. A major problem with this hypothesis is the difficulty of moving large amounts of water, presumably from the Arkoma basin to the south, through a rock unit of low permeability.

An alternate and possibly better explanation for iron zonation involves the kinetics of dolomite crystal growth. Self-patterning phenomena, such as iron zonation in the ferroan dolomite crystals, are commonly observed in geology (Merino, 1984). Under disequilibrium conditions, the smectite-to-illite conversion and the precipitation of dolomite can be considered irreversible thermodynamic reactions. In such cases, the kinetics of the reactions may be nonlienar (e.g., the concentrations of dissolved species in solution may oscillate over time). Compositional bands in crystals, such as those that occur in igneous or metamorphic rocks, are a result of this kind of behavior (Fisher and Lasaga, 1981). Oscillations in the molalities of Fe^{2+} and Mg^{2+} during diffusion of these species to the growing crystal surface wold result in iron zonation in the dolomite crystals.

CONCLUSIONS

The offshore facies of the Bonneterre Formation (Cambrian), southeast Missouri, consists of interbedded limestone and shale beds deposited in relatively deep water below wave base. This facies contains two dolomite types that can be distinguished petrographically and geochemically.

A basal dolomite occurs in the lowermost few meters of the offshore facies at the Bonneterre Formation/Lamotte Sandstone contact. It is characterized by coarse crystalline nonplanar texture typical of dolomite precipitated at elevated temperature. The basal dolomite is near stoichiometric in $CaCO_3$ content and has a relatively low $FeCO_3$ content. Iron and manganese have lognormal distributions, as is typical for minor-element populations. Carbon and oxygen isotopes in the basal dolomite resemble those in dolomite associated with the nearby sulfide ore bodies. The basal dolomite was formed when mineralizing basinal brines, moving through the underlying Lamotte Sandstone aquifer, dolomitized the lowermost limestone beds in the Bonneterre.

Ferroan dolomite occurs as individual crystals and patches of crystals throughout the interbedded limestone and shale of the offshore facies of the Bonneterre. The ferroan dolomite is enriched in $CaCO_3$ and commonly zoned with respect to $FeCO_3$. Iron has a normal distribution (typical of major-element populations) and is bimodal. Manganese has a lognormal distribution. Depleted carbon and oxygen isotope values are consistent with elevated temperature and an organic source of carbon during dolomitization. It is suggested that the ferroan dolomite was formed by Mg^{2+}- and Fe^{2+}-rich water evolved during illitization of smectite in the neighboring shales.

ACKNOWLEDGMENTS

I thank Robert W. Smith and Kevin L. Shelton for their advice on interpreting the stable isotope data and Richard D. Beane for his input on analysis of the minor-element data. Tom Freeman, William M. Murphy, and Bruce A. Ahler provided useful comments and criticisms during various stages of this study. Kevin L. Shelton analyzed samples of the ferroan dolomite for stable carbon and oxygen isotopes at the Yale University Stable Isotope Laboratory under the direction of Danny M. Rye. James B. Davis and James R. Palmer aided greatly in locating and collecting core samples for this study. Special thanks goes to David R. Brosnahan for contributing his expertise in scanning electron microscopy and microbeam analyses. Critical reviews by Duncan F. Sibley, Paul A. Baker, and Don L. Kissling greatly improved the paper. St. Joe Minerals Corporation granted permission to publish the paper.

REFERENCES

Boles, J. R., and Franks, S. G., 1979, Clay diagenesis in Wilcox sandstones of southwest Texas: Implications of smectite diagenesis on sandstone cementation: Journal of Sedimentary Petrology, v. 49, p. 55–70.

Braunsdorf, N. R., and Lohmann, K. C., 1983, Isotopic trends in gangue carbonates from the Viburnum Trend, Ozark lead mine, S.E. Missouri: Geological Society of America, Abstracts with Programs, v. 15, p. 532.

Catheles, L. M., and Smith, A. T., 1983, Thermal constraints on the formation of Mississippi Valley-type lead-zinc deposits and their implications for episodic basin dewatering and deposit genesis: Economic Geology, v. 78, p. 983–1002.

Carrol, Dorothy, 1970, Clay minerals: A guide to their X-ray identification: Geological Society of America Special Paper 126, 80 p.

CRAIG, HARMON, 1957, Isotopic standards for carbon and oxygen and correction factors for mass spectrometric analysis of carbon dioxide: Geochimica et Cosmochimica Acta, v. 12, 133–149.

FISHER, G. W., AND LASAGA, A. C., 1981, Irreversible thermodynamics in petrology, *in* Lasaga, A. C., and Kirkpatrick, R. J., eds., Kinetics of Geochemical Processes: Mineralogical Society of America, Reviews in Mineralogy, v. 8, p. 171–209.

FRANK, M. H., AND LOHMANN, K. C., 1986, Textural and chemical alteration of dolomite: Interaction of mineralizing fluids and host rock in a Mississippi Valley-type deposit, Bonneterre Formation, Viburnum Trend, *in* Hagni, R. D., ed., Process Mineralogy VI: The Metallurgical Society, Warrendale, Pennsylvania, p. 103–116.

FREEMAN, TOM, 1972, Sedimentology and dolomitization of Muschelkalk carbonates (Triassic), Iberian Range, Spain: American Association of Petroleum Geologists Bulletin, v. 56, p. 434–453.

———, AND MEDARY, TOM, 1987, Modification of precursor dolostone by a later dolomitizing event: The Bonneterre Formation (Cambrian), southern Missouri: SEPM Annual Midyear Meeting, Austin, Texas, Abstracts, p. 28.

FRIEDMAN, G. M., 1959, Identification of carbonate minerals by staining methods: Journal of Sedimentary Petrology, v. 29, p. 87–97.

GARVEN, GRANT, AND FREEZE, A. R., 1984, Theoretical analysis of the role of groundwater flow in the genesis of stratabound ore deposits. 1. Mathematical and numerical model. 2. Quantitative results: American Journal of Science, v. 284, p. 1085–1174.

GAWTHORPE, R. L., 1987, Burial dolomitization and porosity development in a mixed carbonate-clastic sequence: An example from the Bowland Basin, northern England: Sedimentology, v. 34, p. 533–558.

GERDEMANN, P. E., AND GREGG, J. M., 1986, Sedimentary facies in the Bonneterre Formation (Cambrian), southeast Missouri, and their relationship to ore distribution, *in* Sediment-Hosted Pb-Zn-Ba Deposits of the Midcontinent: Geological Society of America Annual Meeting, Field Trip No. 1, Missouri Department of Natural Resources, Division of Geology and Land Survey, p. 37–49.

———, AND MYERS, H. E., 1972, Relationships of carbonate facies patterns to ore distribution and to ore genesis in the southeast Missouri lead district: Economic Geology, v. 67, p. 426–433.

GREGG, J. M., 1985, Regional epigenetic dolomitization in the Bonneterre Dolomite (Cambrian), southeastern Missouri: Geology, v. 13, p. 503–506.

———, AND SIBLEY, D. F., 1984, Epigenetic dolomitization and the origin of xenotopic dolomite texture: Journal of Sedimentary Petrology, v. 54, p. 908–931.

GRIFFIN, J. J., WINDOM, HERBERT, AND GOLDBERG, E. D., 1968, The distribution of clay minerals in the world oceans: Deep-Sea Research, v. 15, p. 433–459

GRIM, R. E., 1968, Clay Mineralogy: McGraw-Hill, New York, 596 p.

HAGNI, R. D., 1983, Ore microscopy, paragenetic sequence, trace element content, and fluid inclusion studies of the copper-lead-zinc deposits of the southeast Missouri lead district, *in* Kisvarsanyi, G., Grant, S. K., Pratt, W. P., and Koenig, J. W., eds., International Conference on Mississippi Valley-Type Lead-Zinc Deposits, Proceedings, Rolla, Missouri, p. 243–256.

HARDIE, L. A., 1987, Dolomitization: A critical view of some current views: Journal of Sedimentary Petrology: v. 57, p. 166–183.

HOWE, W. B., 1968, Planar stromatolite and burrowed carbonate mud facies in Cambrian strata of the St. Francois Mountain area: Missouri Department of Natural Resources, Division of Research and Technical Information, Geological Survey Report of Investigations 41, 113 p.

IRWIN, HILARY, 1980, Early diagenetic carbonate precipitation and pore fluid migration in the Kimmeridge Clay of Dorset, England: Sedimentology, v. 27, p. 577–591.

JAROSEWICH, EUGENE, AND HOWARD, I. G., 1983, Carbonate reference samples for electron microprobe and scanning electron microscope analyses: Journal of Sedimentary Petrology, v. 53, p. 677–678.

KAHLE, C. F., 1965, Possible roles of clay minerals in the formation of dolomite: Journal of Sedimentary Petrology, v. 35, p. 448–453.

LARSON, K. G., 1977, Sedimentology of the Bonneterre Formation, southeast Missouri: Economic Geology, v. 72, p. 408–419.

LAWRENCE, J. R., GIESKES, J. M., AND BROECKER, W. S., 1975, Oxygen isotope and cation composition of DSDP pore waters and the alteration of layer basalts: Earth and Planetary Science Letters, v. 27, p. 1–10.

LYLE, J. R., 1977, Petrology and carbonate diagenesis of the Bonneterre Formation in the Viburnum Trend area, southeast Missouri: Economic Geology, v. 72, p. 420–434.

MACHEL, H.-G., 1985, Cathodoluminescence in calcite and dolomite and its chemical interpretation: Geoscience Canada, v. 12, p. 139–147.

———, AND MOUNTJOY, E. W., 1986, Chemistry and environments of dolomitization—A reappraisal: Earth Science Reviews, v. 23, p. 175–222.

MCHARGUE, T. R., AND PRICE, R. C., 1982, Dolomite from clay in argillaceous or shale-associated marine carbonates: Journal of Sedimentary Petrology, v. 52, p. 873–886.

MATTES, B. W., AND MOUNTJOY, E. W., 1980, Burial dolomitization of the Upper Devonian Miette buildup, Jasper National Park, Alberta, *in* Zenger, D. H., Dunham, J. B., and Ethington, R. L., eds., Concepts and Models of Dolomitization: Society of Economic Paleontologists and Mineralogists Special Publication 28, p. 259–297.

MATTHEWS, ALAN, AND KATZ, AMITAI, 1977, Oxygen isotope fractionation during the dolomitization of calcium carbonate: Geochimica et Cosmochimica Acta, v. 41, p. 1431–1438.

MEDARY, T. A., 1986, Replacement and recrystallization textures in Bonneterre dolostones: Unpublished M.S. Thesis, University of Missouri-Columbia, Missouri, 32 p.

MERINO, ENRIQUE, 1984, Survey of geochemical self patterning phenomena, *in* Nicolis, G., and Baras, F., eds., Chemical Instabilities: Applications in Chemistry, Engineering, Geology, and Materials Science: NATO Advancements in Science, Series C, v. 120, p. 305–328.

OHLE, E. L., AND BROWN, J. S., 1954, Geologic problems in the southeast Missouri lead district: Geological Society of America Bulletin, v. 65, p. 201–221 and p. 935–936.

PIERSON, B. J., 1981, The control of cathodoluminescence in dolomite by iron and manganese: Sedimentology, v. 28, p. 601–610.

PYE, K., 1985, Electron microscope analysis of zoned dolomite rhombs in the Jet Rock Formation (Lower Toarcian) of the Whitby area, U.K.: Geological Magazine, v. 122, no. 3, p. 279–286.

ROEDDER, EDWIN, 1977, Fluid inclusion studies of ore deposits in the Viburnum Trend, southeast Missouri: Economic Geology, v. 72, p. 474–479.

SHAW, D. M., 1961, Element distribution s in geochemistry: Geochimica et Cosmochimica Acta, v. 23, p. 116–134.

SIBLEY, D. F., AND GREGG, J. M., 1987, Classification of dolomite rock textures: Journal of Sedimentary Petrology, v. 57, p. 955–963.

SINCLAIR, A. J., 1976, Probability graphs in mineral exploration: The Association of Exploration Geochemists, Special Volume No. 4, 95 p.

SNYDER, F. G., AND GERDEMANN, P. E., 1965, Explosive igneous activity along an Illinois-Missouri-Kansas axis: American Journal of Science, v. 263, p. 465–493.

———, AND ———, 1968, Geology of the southeast Missouri lead district, *in* Ore Deposits of the United States, 1933–1967 (Graton-Sales Volume), v. 1: American Institute of Mining, Metallurgical, and Petroleum Engineers, New York, p. 326–358.

ŚRODOŃ, JAN, AND EBERL, D. D., 1984, Illite, *in* Bailey, S. W., ed., Micas: Mineralogical Society of America, Reviews in Mineralogy, v. 13, p. 495–544.

TAYLOR, T. R., AND SIBLEY, D. F., 1986, Petrographic and geochemical characteristics of dolomite types and the origin of ferroan dolomite in the Trenton Formation, Ordovician, Michigan Basin: Sedimentology, v. 33, p. 61–86.

THAKER, J. L., AND ANDERSON, K. H., 1977, The geologic setting of the southeast Missouri lead district–Regional geologic history, structure and stratigraphy: Economic Geology, v. 72, p. 339–348.

VOSS, R. L., AND HAGNI, R. D., 1985, The application of cathodoluminescence microscopy to the study of sparry dolomite from the Viburnum Trend, southeast Missouri, *in* Hausen, D. M., and Kopp, O. C., eds., Process Mineralogy V, Mineralogy–Applications to the Minerals Industry: American Institute of Mining, Metallurgical, and Petroleum Engineers, New York, p. 51–68.

WEAVER, C. E., AND BECK, K. C., 1971, Clay water diagenesis during burial: How mud becomes gneiss: Geological Society of America Special Paper 134, 96 p.

ZENGER, D. H., 1983, Burial dolomitization in the Lost Burrow Formation (Devonian), east-central California, and the significance of late diagenetic dolomitization: Geology, v. 11, p. 519–522.

Appendix I.—ELECTRON MICROPROBE DATA

Section	Description	Crystal	$CaCO_3$	$MgCO_3$	$FeCO_3$	$MnCO_3$
	BASAL DOLOMITE					
SE, SE, Sec. 36, T36N, R7W, Phelps Co., Missouri						
83-129	Coarse crystalline nonplanar	1	54.10	42.42	2.81	0.66
			56.24	41.42	2.17	0.16
		2	54.77	41.48	2.95	0.44
			53.84	42.49	3.22	0.45
83-130C	Fine crystalline planar in stylolite	1	51.82	43.59	4.02	0.57
			51.72	44.60	3.24	0.44
		2	51.84	43.62	3.97	0.57
			52.36	43.35	3.50	0.57
83-130D	Coarse crystalline nonplanar	1	52.28	43.92	3.50	0.30
			51.28	42.48	5.38	0.86
		2	52.49	44.44	2.78	0.29
			51.66	42.77	4.90	0.66
83-130E		1	51.23	43.71	4.36	0.69
			51.66	43.94	3.80	0.59
		2	51.59	43.96	3.83	0.61
			51.39	43.52	4.43	0.65
83-130F		1	51.25	45.16	3.10	0.49
			51.08	44.43	3.85	0.64
83-130G		1	51.02	44.65	3.60	0.73
			50.98	44.93	3.54	0.55
83-131A		1	50.96	47.20	1.66	0.18
			51.00	47.11	1.71	0.17
		2	51.23	46.96	1.65	0.15
			50.03	48.08	1.73	0.20
83-131B	Fine crystalline planar in shale seam	1	50.29	47.65	1.90	0.16
			51.09	46.89	1.86	0.15
		2	50.19	47.68	1.94	0.18
			50.84	47.07	1.92	0.16
83-134B	Coarse crystalline nonplanar	1	50.47	47.55	1.73	0.21
			49.93	48.17	1.63	0.23
		2	50.29	47.97	1.50	0.19
			49.98	47.65	2.08	0.29
83-134C	Fine crystalline planar	1	50.47	48.06	1.24	0.23
			50.01	48.53	1.24	0.22
		2	49.77	48.59	1.46	0.17
			50.31	47.97	1.56	0.15
SE, NW, Sec. 38, T34N, R2W, Dent Co., Missouri						
82-137	Coarse crystalline nonplanar	1	50.83	46.42	2.33	0.42
			50.34	46.90	0.26	0.26
		2	50.99	47.31	1.49	0.21
			50.34	47.47	1.83	0.35
		3	49.81	47.16	2.60	0.43
			50.85	46.15	2.51	0.45
		4	48.78	48.01	2.71	0.47
			49.69	47.27	2.63	0.44
SW, SW, Sec. 31, T37N, R10W, Pulaski Co., Missouri						
83-492A	Coarse crystalline nonplanar	1	50.15	47.04	2.37	0.43
			50.73	46.48	2.37	0.42
		2	50.70	46.52	2.34	0.44
			51.10	46.20	2.33	0.37
83-492B	Fine crystalline planar	1	50.97	46.54	2.09	0.40
			50.84	46.29	2.41	0.46
		2	50.51	47.15	2.04	0.30
			49.92	48.03	1.17	0.28
84-121D	Coarse crystalline planar near a stylolite	1	50.75	46.26	2.59	0.39
			50.07	48.08	1.69	0.16
		2	51.13	45.93	2.54	0.39
			50.56	46.76	2.40	0.28
84-121E	Fine crystalline planar	1	51.36	45.50	2.73	0.41
			51.28	45.70	2.66	0.35
		2	50.14	47.66	2.00	0.19
			51.29	45.19	3.15	0.36
85-657D	Coarse crystalline nonplanar	1	50.33	47.66	1.77	0.25
			50.88	48.99	0.00?	0.12
		2	50.07	47.86	1.84	0.23
			50.60	47.15	1.96	0.29
85-657E		1	49.92	47.87	1.96	0.25
			49.71	48.32	1.74	0.23
		2	50.39	47.83	1.64	0.14
			50.66	47.58	1.57	0.18

APPENDIX I.—*Continued*

Section	Description	Crystal	$CaCO_3$	$MgCO_3$	$FeCO_3$	$MnCO_3$
	BASAL DOLOMITE					
SE, NW, Sec. 9, T35N, R14W, Laclede Co., Missouri						
86-54A	Coarse crystalline planar replacing ooids	1	52.67	44.77	2.25	0.30
		2	52.39	44.97	2.37	0.26
		3	53.92	43.89	1.92	0.27
		4	51.88	40.83	6.09	1.20
		5	53.11	37.37	8.75	0.76
		6	52.17	41.36	5.30	1.17
86-54B	Coarse crystalline nonplanar replacing ooids	1	51.56	44.93	3.10	0.40
		2	51.63	45.04	2.94	0.39
		3	52.92	38.08	8.19	0.08
		4	52.88	38.18	8.26	0.69
SE, SW, Sec. 33, T36N, R3W, Crawford Co., Missouri						
86-52A	Fine crystalline planar in argillaceous bed	1	52.52	43.16	1.04	0.28
		2	57.03	39.64	3.02	0.31
		3	54.85	42.60	2.24	0.30
		4	56.43	40.31	2.72	0.54
		5	56.55	40.30	2.60	0.55
		6	54.26	44.88	0.71	0.15
86-52B	Coarse crystalline nonplanar	1	57.09	39.32	3.06	0.53
		2	55.48	42.81	1.43	0.27
		3	54.57	44.24	1.00	0.19
		4	56.28	41.40	2.14	0.18
		5	56.42	40.56	2.43	0.58
		6	55.12	43.15	1.46	0.27
	FERROAN DOLOMITE					
NW, NW, Sec. 33, T26N, R7W, Howell Co., Missouri						
85-537A	Coarse crystalline planar	1	52.22	43.20	4.45	0.13
			52.24	40.66	6.91	0.19
		2	51.34	44.03	4.48	0.51
			52.43	41.03	6.41	0.31
85-537B	Fine crystalline planar	1	51.95	43.83	4.08	0.14
			51.52	43.71	4.64	0.31
		2	51.81	44.93	3.15	0.11
			51.85	42.36	5.62	0.16
SW, Sec. 33, T40N, R8W, Maries Co., Missouri						
85-538A	Coarse crystalline planar zoned crystals, center to edge	1	55.21	36.55	7.98	0.26
			55.06	40.10	4.44	0.40
			54.19	37.12	8.34	0.35
			54.24	40.24	5.39	0.13
			54.40	43.50	1.97	0.13
85-538A		2	55.67	35.93	8.18	0.22
			55.54	39.70	4.42	0.33
			55.14	35.63	8.88	0.35
			54.09	39.48	6.26	0.17
			55.39	42.53	1.96	0.11
85-538B	Coarse crystalline planar zoned crystal, center to edge	1	55.72	34.32	9.63	0.33
			54.80	39.68	5.23	0.29
			54.90	36.93	7.89	0.28
85-539	Coarse crystalline planar zoned crystals, center to edge	1	55.93	31.59	11.55	0.93
			55.56	35.09	9.09	0.25
			55.71	39.49	4.59	0.21
		2	55.11	35.45	9.18	0.25
			56.10	38.64	5.04	0.21
			55.77	39.89	4.15	0.18
			55.62	30.21	13.69	0.48
85-540	Coarse crystalline planar zoned crystals, center to edge	1	56.02	38.04	5.72	0.22
			56.19	32.68	10.64	0.48
			56.45	34.49	8.78	0.28
			55.09	39.75	4.97	0.18
		2	55.63	31.17	12.79	0.42
			55.13	34.88	8.51	0.48
			55.65	32.84	11.09	0.39
			56.26	37.17	6.29	0.28
			55.52	35.63	8.56	0.28
NE, NE, Sec. 25, T32N, R10W, Texas Co., Missouri						
85-541A	Fine crystalline planar zoned crystals in argillaceous bed	1	58.88	39.87	1.23	0.02
			55.25	37.04	7.41	0.29
		2	55.70	35.97	8.10	0.22
		3	59.09	39.80	1.06	0.05
85-541B	Coarse crystalline planar zoned crystal, center to edge	1	54.74	38.80	6.26	0.20

APPENDIX I.—*Continued*

Section	Description	Crystal	$CaCO_3$	$MgCO_3$	$FeCO_3$	$MnCO_3$
	FERROAN DOLOMITE					
NE, NE, Sec. 25, T32N, R10W, Texas Co., Missouri (Continued)						
			55.07	38.76	5.92	0.24
			55.35	40.64	3.89	0.11
			55.77	37.26	6.76	0.21
			54.95	43.09	1.81	0.14
85-542A	Fine crystalline planar zoned crystals in argillaceous bed	1	54.80	34.22	10.56	0.42
		2	54.78	40.14	4.84	0.24
		3	56.49	32.04	10.95	0.52
85-542B	Coarse crystalline planar zoned crystals in silty micrite	1	55.78	37.51	6.38	0.32
			56.78	32.74	10.04	0.44
		2	56.62	38.10	5.00	0.28
			55.78	32.16	11.46	0.60
85-542C	cement, outer edge		55.56	41.02	3.17	0.24
			54.91	42.71	2.12	0.25
			55.16	42.66	1.98	0.19
	cement, centers		53.02	35.60	10.43	0.95
			54.99	31.21	13.11	0.68
			55.15	30.09	13.73	1.03
SE, SE, Sec. 36, T36N, R7W, Phelps Co., Missouri						
83-115A	Coarse crystalline planar unzoned crystals	1	50.93	38.30	10.24	0.52
			51.42	43.00	5.25	0.33
83-115B		1	51.58	37.92	9.83	0.66
			51.34	42.89	5.44	0.33
83-115C	Fine crystalline planar unzoned crystals	1	51.40	42.46	5.61	0.53
		2	50.89	42.84	5.87	0.40
		3	50.96	42.78	5.79	0.46
		4	51.23	42.59	5.81	0.37
83-115D	cement, center to edge	1	51.57	46.70	1.52	0.21
			52.11	42.14	5.39	0.35
			52.54	34.24	12.54	0.67
83-116A	Coarse crystalline planar unzoned crystals	1	51.20	42.86	5.51	0.43
			51.27	42.51	5.81	0.40
			54.68	43.07	2.02	0.22
		2	51.75	43.19	4.67	0.39
			52.16	39.90	7.47	0.46
83-116B		1	54.21	45.14	0.55	0.10
			51.15	42.20	6.24	0.41
			52.29	46.36	1.18	0.16
			51.55	42.10	5.91	0.44
83-116C		1	53.03	46.02	0.78	0.16
			52.09	40.57	6.90	0.43
			54.17	44.93	0.75	0.15
			51.86	41.44	6.30	0.40
83-116D	small crystals near stylolite	1	51.66	42.95	5.02	0.36
		2	52.01	41.31	6.30	0.38
83-120A	Coarse crystalline planar, faintly zoned crystals center to edge	1	51.38	42.57	5.65	0.40
			50.95	45.54	3.15	0.36
			52.64	36.43	10.42	0.50
			51.82	37.03	10.59	0.56
			51.09	44.08	4.53	0.29
			51.14	43.67	4.81	0.38
			51.04	44.76	3.82	0.37
83-120B		2	51.24	42.34	6.02	0.40
			51.41	42.86	5.39	0.34
			50.36	46.46	2.90	0.26
			50.19	46.63	2.89	0.28
83-120C	Fine crystalline planar, small unzoned crystals	1	50.83	42.51	6.26	0.40
			50.90	45.90	2.92	0.28
		2	50.78	45.94	3.01	0.26
		3	51.26	45.56	2.93	0.25
		4	51.50	44.81	3.40	0.28
83-123A	Coarse crystalline planar zoned crystal, center to edge	1	55.62	37.13	6.99	0.26
			55.25	40.29	4.27	0.18
			55.29	40.57	3.91	0.22
			55.24	43.00	1.69	0.07
83-123C	Coarse crystalline planar zoned crystals center	1	52.53	40.99	6.21	0.26
	middle		55.76	34.28	9.68	0.28
	edge		55.23	37.16	7.34	0.26
	center	2	58.58	36.19	4.94	0.29
	midle		55.35	34.44	9.90	0.30
	edge		55.15	36.98	7.61	0.26
83-123D	Coarse crystalline planar zoned crystal, center to edge	1	52.77	40.57	6.40	0.26
			52.17	41.61	6.01	0.21
			55.78	33.90	10.00	0.32
			56.01	36.38	7.42	0.19

Appendix I.—*Continued*

Section	Description		Crystal	$CaCO_3$	$MgCO_3$	$FeCO_3$	$MnCO_3$
	FERROAN DOLOMITE						
SE, SE, Sec. 36, T36N, R7W, Phelps Co., Missouri (Continued)							
				55.05	34.85	9.82	0.30
				54.99	38.57	6.11	0.33
				54.98	36.74	8.03	0.25
				55.12	42.40	2.19	0.28
83-124A	Coarse crystalline planar unzoned crystals	center	1	57.76	36.00	6.08	0.15
		edge		56.12	37.29	6.38	0.12
		center	2	55.67	37.93	6.17	0.23
		edge		55.80	39.65	4.38	0.17
83-124B	Fine crystalline planar zoned crystals in argillaceous bed		1	57.23	39.89	2.80	0.08
			2	57.79	40.89	1.30	0.03
			3	58.78	38.41	2.57	0.06
SW, SW, Sec. 31, T37N, R10W, Pulaski Co., Missouri							
83-485A	Small planar, unzoned crystal near stylolite		1	55.66	37.65	6.47	0.22
				54.73	40.04	5.06	0.17
83-485B	Coarse crystalline planar zoned crystal, center to edge		1	55.61	36.67	7.45	0.28
				55.50	39.38	4.71	0.41
				55.02	37.37	7.27	0.33
				54.99	43.02	1.88	0.12
83-485C	Fine crystalline planar unzoned crystals in shale		1	53.25	43.10	3.53	0.11
			2	53.74	42.15	3.98	0.13
			3	53.49	42.81	3.60	0.10
83-487A	Fine crystalline planar zoned crystals	center	1	55.95	33.49	10.15	0.41
		edge		55.67	40.31	3.74	0.27
		center	2	56.50	32.38	10.68	0.43
		edge		54.58	39.94	5.21	0.26
		center	3	56.43	31.03	12.20	0.33
		edge		55.45	38.28	5.99	0.28
83-487B	Coarse crystalline planar zoned crystals	center	1	56.57	33.00	10.04	0.38
		edge		55.82	40.90	3.02	0.26
		center	2	56.52	32.67	10.38	0.43
		edge		54.50	41.05	4.23	0.21
83-488A	Coarse crystalline planar zoned crystal, center to edge		1	56.61	31.51	11.48	0.39
				56.33	34.21	8.77	0.69
				56.66	32.17	10.63	0.55
				55.73	38.26	5.67	0.34
				55.45	38.02	6.27	0.25
				55.50	40.82	3.39	0.29
83-488B	Fine crystalline planar zoned crystals	center	1	58.88	34.19	6.75	0.18
		edge		56.73	30.52	12.26	0.49
		center	2	57.32	34.95	7.41	0.31
		edge		56.08	32.50	11.04	0.39
83-489A	Coarse crystalline planar zoned crystal, center to edge		1	57.97	31.72	12.73	0.58
				54.71	38.66	6.39	0.24
				55.88	31.50	12.23	0.37
				32.19	40.10	7.42	0.29
				55.09	41.60	3.00	0.31
				54.90	41.89	2.83	0.37
83-489B	Coarse crystalline planar zoned crystals near stylolite random analysis		1	55.47	41.26	3.02	0.25
			2	54.79	39.51	5.49	0.21
			3	54.42	39.36	5.99	0.22
			4	54.49	41.42	3.90	0.18
			5	54.85	42.45	2.58	0.11
	Fine crystalline planar in stylolite		1	51.84	40.58	7.28	0.29
			2	54.36	32.44	12.83	0.36
				53.63	40.30	5.87	0.19
83-489C	Fine crystalline planar in shale bed		1	53.37	39.17	2.22	0.23
			2	51.35	41.78	6.66	0.20
83-490A	cement crystal adjacent to a calcite-filled fracture, center to edge		1	54.78	43.34	1.68	0.20
				53.37	37.32	8.83	0.48
				54.48	28.89	15.33	1.30
		center	2	56.49	32.03	10.91	0.57
		edge		56.19	38.14	5.33	0.33
		center	3	57.00	30.37	12.14	0.49
		edge		56.69	37.55	5.47	0.29
	Coarse crystalline nonplanar		4	56.24	39.08	4.38	0.30
83-490B	Coarse crystalline planar zoned crystal, center to edge		1	54.01	42.38	3.14	0.46
				56.03	40.60	3.13	0.24
				53.46	44.79	1.53	0.22
	large zoned cement crystal adjacent to calcite-filled fracture, center to edge		1	55.30	40.60	3.80	0.30
				55.45	40.22	4.01	0.32
				55.63	42.28	1.85	0.24
				55.59	42.94	1.26	0.21
				56.23	42.47	1.11	0.18
				53.63	33.36	12.17	0.84

GEOCHEMISTRY OF EPIGENETIC DOLOMITE ASSOCIATED WITH LEAD-ZINC MINERALIZATION OF THE VIBURNUM TREND, SOUTHEAST MISSOURI: A RECONNAISSANCE STUDY

DONALD P. BUELTER[1] AND RENALD N. GUILLEMETTE

Department of Geology, Southern Illinois University, Carbondale, Illinois 62901

ABSTRACT: A major concern with Viburnum Trend lead-zinc deposits is the nature of the mineralizing fluids. The chemical compositions of secondary recrystallized and sparry dolomites in the Bonneterre and Davis formations were determined to help define the nature of the coexisting mineralizing fluids in the Viburnum Trend and surrounding areas.

Chemical compositions of the recrystallized host basal dolomite in the lowermost part of the Bonneterre Formation show a general south to north decrease of iron and manganese and an associated increase of strontium. This is consistent with a southern source for the mineralizing fluids. A similar trend of decreasing iron and manganese and increasing strontium upsection in the Viburnum Trend and the back reef indicates that the fluids moved from the underlying Lamotte Sandstone into the Bonneterre Formation.

Low-iron and manganese concentrations in the backreef sparry dolomite, as well as relatively constant strontium values in the entire area, suggest the mixing of basinal brines with meteoric waters.

A proposed sequence of dolomitization and ore-forming events can be summarized as: (1) dolomitization of the backreef unit by depositional marine or diagenetic waters, (2) updip movement of mineralizing brines from a southern source, possibly the Ouachita-Arkoma Basin, causing epigenetic dolomitization and ore deposition in the Viburnum Trend, and (3) changing of mineralizing conditions due to the mixing in the back reef of principal basinal brines with dilute meteoric waters from an eastern source.

INTRODUCTION

A major concern with the Viburnum Trend Mississippi Valley-type lead-zinc district is the source of the mineralizing fluids. Gregg (1985) concluded that mineralizing fluids migrating through the Lamotte Sandstone interacted with the lower Bonneterre Formation, forming an epigenetic basal dolomite. Other recent studies suggest that the formation of many of the Mississippi Valley-type deposits in the central and eastern United States is related to the expulsion of fluids during the Appalachian-Ouachita orogeny (Leach and others, 1984; Kaiser and Ohmoto, 1985; Oliver, 1986).

Other possible regional epigenetic dolomitization events have been discussed in the literature (Mattes and Mountjoy, 1980; Zenger, 1983). The significance of this type of dolomitization in the geologic record may be underestimated. Burial environments may be more favorable for dolomitization than many surficial environments. As temperature increases with depth, dolomite reaction rates increase rapidly (Mattes and Mountjoy, 1980). In addition, magnesium ions can be supplied by basinal waters that react with limestones (Zenger, 1983). For dolomitization to occur in the burial environment, the mass transfer constraints of providing water to the limestone host and driving enough water through the limestone to provide the magnesium mass for the conversion must be overcome (Hardie, 1987). In southeast Missouri, the Lamotte Sandstone provides access of basinal waters to the limestone, and gravity drive models provide a mechanism to move large amounts of water from basins to neighboring platforms (Gregg, 1985). Studies of the trace- and minor-element chemistry of epigenetic dolomites, using the Berthelot-Nernst distribution law, may lead to a better understanding of the movement and nature of sedimentary brines in the epigenetic environment. The Berthelot-Nernst distribution law states that as a mineral grows, trace components will partition between two phases in a characteristic manner, as long as equilibrium is maintained (McIntire, 1963). This study examines the areal and vertical distributions of iron, manganese, strontium, zinc, lead, and sodium in the Bonneterre and Davis formations and attempts to relate the distributions to the source of the mineralizing fluids of the Viburnum Trend lead-zinc deposits.

METHODS

Samples used in this study represent a spatial and/or vertical distribution throughout the region (Fig. 1). Samples were collected from the core stored at the Missouri Department of Natural Resources, Division of the Geological Survey in Rolla, Missouri, from St. Joe Minerals Corporation production cores and mines, and from outcrop locations.

Iron, manganese, zinc, and lead were analyzed using a Jarrell-Ash Atomcomp 975 inductively coupled plasma spectrograph. Strontium and sodium analyses were performed on a Perkin-Elmer Model 603 atomic absorption spectrophotometer; strontium was analyzed in the flame absorption mode, whereas sodium was analyzed in the flame emission mode. In order to minimize matrix and interference effects due to high background levels of calcium and magnesium, chemically ultrapure calcium and magnesium carbonates were used in the preparation of standard and blank solutions, matching the concentrations of those two elements to those found in the unknown solutions. Reagent grade hydrochloric acid was used in this study. The same bottle of acid was used in all carbonate digestions and in the preparation of all standard and blank solutions. The effects of parts per billion metal contaminants in the acid were eliminated by zeroing both analytical instruments on blank solutions containing the same concentrations of acid and chemically ultrapure carbonates as the unknowns and standards.

All samples selected for host rock analysis were examined by stereoscopic microscope to determine that they were free of any visible mineralization. The samples were then crushed with a Diamonite™ mortar and pestle and dried overnight at 135°C. Approximately 2 g of material were weighed to 0.1 mg and digested in the 1.25M hydrochloric

[1]Present address: Department of Chemistry and Geochemistry, Colorado School of Mines, Golden, Colorado 80401.

Sedimentology and Geochemistry of Dolostones, SEPM Special Publication No. 43

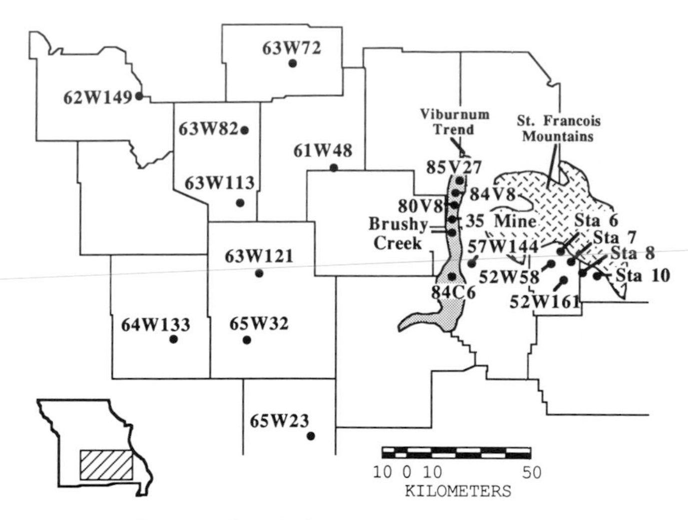

FIG. 1.—Sample locations within study area.

acid. After digestion, the samples were filtered through a Whatman No. 42 filter paper to remove insoluble residue.

Sparry dolomite crystals were removed from the samples with a hardened dental probe. Each sample was thoroughly rinsed in triple-deionized distilled water to remove any fine adhering carbonate fragments. The grains were then examined under a stereoscopic microscope and hand-picked to ensure that no host rock was attached to any of the grains. The grains were then dried at 135°C and weighed before digestion in hydrochloric acid.

GEOLOGIC SETTING

The southeast Missouri lead district is located in the central Mississippi Valley area, which is part of the stable interior of the United States. The deposits are located on the northwest margin of the Ozark Uplands. The Ozark uplift is the dominant feature in Missouri and had a major influence on marine deposition in the area.

Late Cambrian strata in southeast Missouri consist of clastic and carbonate rocks that represent essentially continuous deposition with numerous small hiatuses. The Lamotte Sandstone lies unconformably over Precambrian rocks and was probably the main aquifer for the mineralizing fluids (Fig. 2). Sulfide mineralization is almost totally restricted to the Bonneterre Formation and appears to be unit controlled (Gerdemann and Myers, 1972). The Bonneterre consists of four major units: the basal dolomite (Gregg, 1985), reef complex, back reef, and offshore shelf (after Lyle, 1977). The basal dolomite is gradational with the overlying offshore shelf unit and conformably overlies the Lamotte Sandstone (Gregg, 1985). The reef complex is the major ore producer in the region. The backreef unit consists of burrowed muds and planar stromatolites (Howe, 1968). The offshore shelf consists of interbedded shales and limestones, which grade into dolomite near and above the reef complex.

PETROGRAPHY

Two major dolomitization events occurred throughout the Bonneterre Formation. The first event was the massive dolomitization of the Bonneterre Formation, which later hosted the sulfide mineralization and will be referred to here as the host dolomite. Secondary dolomite, in the form of the sparry dolomite cement, was abundantly deposited during the latter stages of mineralization. The two dolomitization events provide information on both the early and late stages of the porewater chemistry.

The basal dolomite samples are medium to dark gray in color. Green shale and glauconite are abundant in the samples. Interparticle pores are partially filled by sparry do-

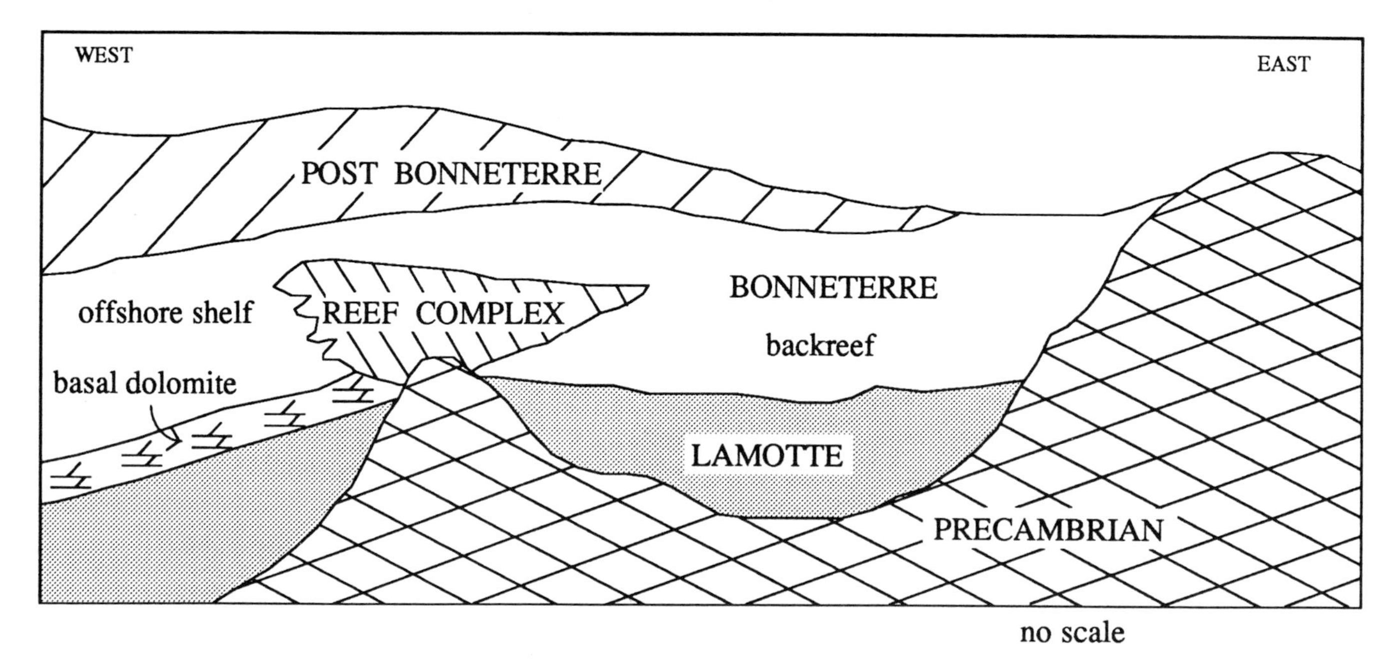

FIG. 2.—Schematic cross section showing the relation of the Lamotte Sandstone with the underlying Precambrian and the unit relations of the Bonneterre Formation in southeast Missouri (after Gerdemann and Myers, 1972) with units of Lyle (1977) and basal dolomite of Gregg (1985).

TABLE 1.—BACKREEF HOST DOLOMITE ANALYSES

Drill Hole	Depth[a]	Fm[b]	Mn	Fe	Zn	Sr	Pb
57W144	98 m	D	359	4194	7	>100	[c]bdl
	91 m	D	274	900	bdl	>100	bdl
	83 m	B	465	2021	2	19	bdl
	77 m	B	1194	6187	2	6	bdl
	64 m	B	398	1297	15	50	bdl
	61 m	B	345	1111	17	>100	bdl
	49 m	B	289	877	8	57	bdl
	39 m	B	566	1275	6	50	bdl
52W58	200 m	D	177	729	bdl	69	bdl
	174 m	B	364	1877	13	26	bdl
	150 m	B	329	1643	13	18	bdl
	130 m	B	1027	5281	14	60	bdl
52W161	165 m	D	373	1323	15	30	bdl
	141 m	B	245	817	12	24	bdl
	131 m	B	482	2523	6	15	bdl
	101 m	B	496	2603	12	15	1
	97 m	B	657	3642	4	10	bdl

(Concentrations in ppm)

a. All drill hole depths referenced to mean sea level.
b. Formations: B = Bonneterre, D = Davis.
c. Below detection limit.

TABLE 3.—REEF COMPLEX HOST DOLOMITE ANALYSES

Drill Hole	Depth[a]	Fm[b]	Mn	Fe	Zn	Sr	Pb
85V27	191 m	D	755	3793	bdl	>100	bdl
	183 m	B	355	2521	13	17	bdl
	175 m	B	1197	5356	bdl	24	bdl
	158 m	B	870	5237	6	12	bdl
	146 m	B	1014	4647	13	12	bdl
	136 m	B	1761	5768	13	12	5
	125 m	B	2807	13880	12	20	21
	118 m	B	4673	13803	12	12	3
84V8	172 m	D	741	13636	6	13	bdl
	151 m	D	365	1018	bdl	28	bdl
	136 m	B	644	2516	1	14	bdl
	123 m	B	1186	4919	bdl	14	bdl
	107 m	B	1108	3484	7	7	bdl
	94 m	B	2624	7349	5	13	1
	77 m	B	3248	13250	4	8	bdl
	65 m	B	356	1582	bdl	40	bdl
80V8	170 m	D	398	1590	6	>100	bdl
	159 m	B	544	1875	bdl	>100	bdl
	145 m	B	861	3546	1	>100	5
	138 m	B	2466	9490	8	16	55
	104 m	B	2366	9631	3	10	2
84C6	110 m	D	366	6305	bdl	8	bdl
	60 m	D	1322	6666	bdl	19	bdl
	28 m	B	2043	11409	bdl	10	bdl
	20 m	B	2233	10047	6	9	1
	2 m	B	3664	10479	12	9	2
	−7 m	B	1250	5679	7	9	bdl
	−30 m	B	2759	16231	4	4	bdl

(Concentrations in ppm)

a. All drill hole depths referenced to mean sea level.
b. Formations: B = Bonneterre, D = Davis.

lomite, which ranges in size from 100 to 500 μm. The backreef rocks are light gray in color; green clays are present in some samples. Vuggy porosity has developed along stromatolite planes, and the vugs are lined with white to pink sparry dolomites 0.5 to 2 mm in size. Samples from the mineralized areas are altered and are variable in color. Sparry dolomites ranging from 0.5 to 2 mm line vugs; calcite spar occurs in fractures.

Powder X-ray diffraction analysis was performed on one sample from each of the basal dolomite, backreef, and reef complex units. All three samples contain only dolomite, quartz, and potassium feldspar. The majority of the feldspar is found near stylolites or fractures in the samples. Backscatter imaging of polished sections using an ETEC Autoscan scanning electron microscope revealed the presence of many small iron sulfide inclusions. These sulfide inclusions are found throughout the host rock samples in an apparently random pattern.

RESULTS

Host Dolomite

Back reef.—

Strong regional differences exist for the trace- and minor-element concentrations in the host dolomite. The samples in the backreef unit differ greatly from samples in the reef complex or basal dolomite units (Tables 1–4). The concentrations of zinc and strontium are generally highest in the backreef unit, whereas concentrations of iron and manganese are much lower. Distinct vertical zonation patterns are observed in only one of the three backreef drill holes, 52W161. Iron and manganese both decrease upsection within the Bonneterre Formation in this drill hole, whereas strontium increases. In drill hole 52W58, the highest iron and manganese concentrations are found in the lowermost sample, but no other pattern can be distinguished. No vertical zonation patterns exist in drill hole 57W144.

Basal dolomite.—

Trace- and minor-element trends are evident in the basal dolomite (Fig. 3a–d). Iron and manganese concentrations decrease to the northwest, with iron generally being present in higher concentrations in this unit than in the others. Zinc concentrations are lowest in the north and northwest, with the highest values being found in the center of the basin. The most striking feature of the chemical data is the strong negative correlations of strontium with iron (correlation

TABLE 2.—BACKREEF OUTCROP SAMPLE ANALYSES

Sample Site	Mn	Fe	Zn	Sr	Pb	Na
Station 6	399	4292	14	>100	bdl[b]	—
Station 7	367 (*521*)[a]	550 (*1622*)	6 (*1*)	>100 (*54*)	bdl (*bdl*)	— (*140*)
Station 8	100	798	6	24	bdl	—
Station 10	303 (*553*)	707 (*2625*)	12 (*3*)	57 (*67*)	1 (*1*)	— (*178*)

(Concentrations in ppm)

a. Host dolomite analyses in plain type; sparry dolomite analyses italicized in parentheses.
b. Below detection limit.

TABLE 4.—BASAL DOLOMITE HOST DOLOMITE ANALYSES

Drill Hole	Mn	Fe	Zn	Sr	Pb
65W23	2316	14696	4	8	bdl
65W32	1586	11705	2	19	bdl
64W133	1300	14577	4	15	bdl
63W121	1553	13626	11	14	bdl
63W113	1296	12955	2	6	bdl
63W82	462	6385	bdl	82	bdl
62W149	337	2979	1	>100	bdl
61W48	1799	10171	2	35	bdl
63W72	1842	15478	bdl	12	bdl

(Concentrations in ppm)

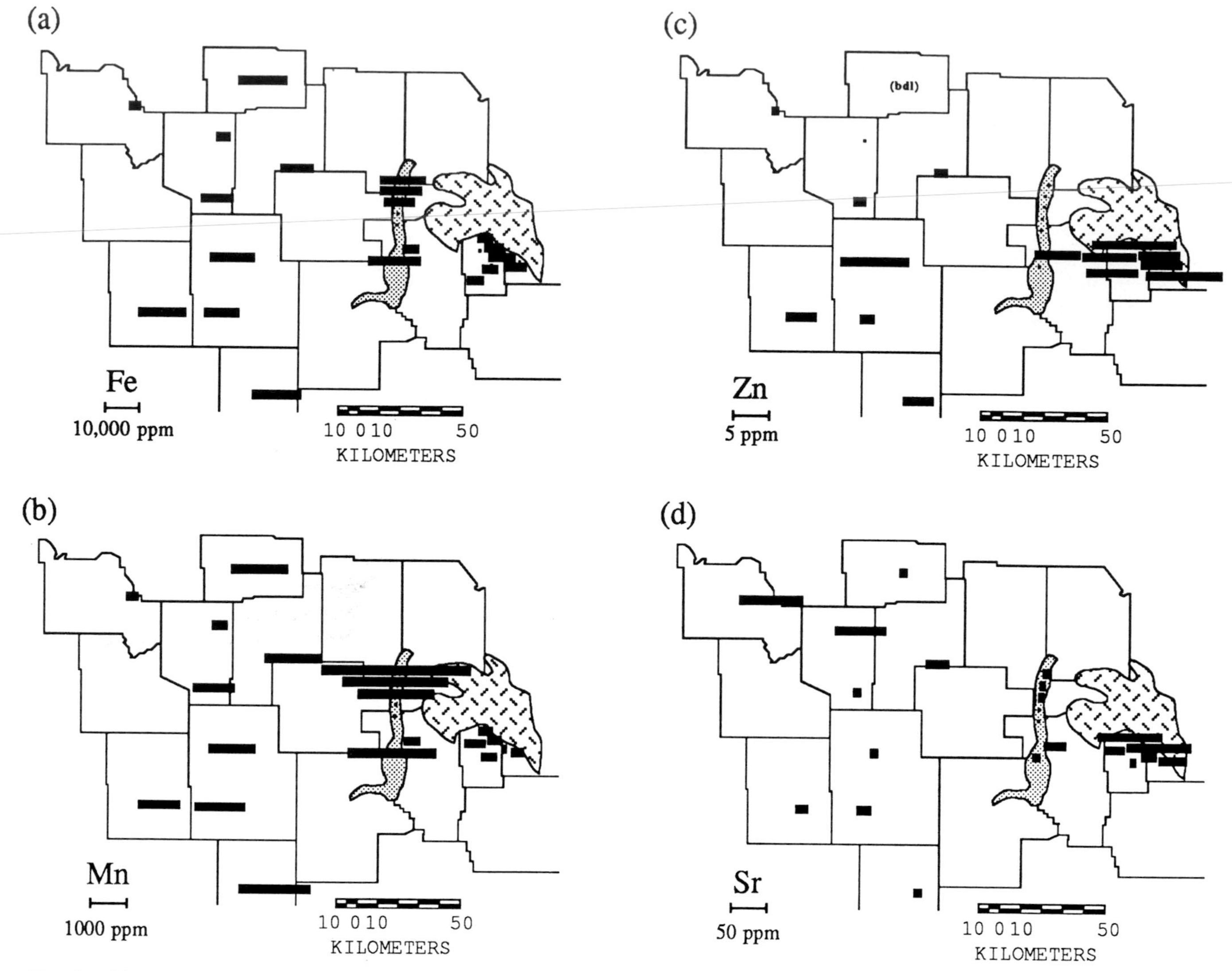

FIG. 3.—Trace- and minor-element distribution in host dolomite of (a) iron, (b) manganese, (c) zinc, and (d) strontium concentrations. Note the inverse relation between strontium and iron or manganese.

coefficient = −0.96) and manganese (−0.84) in the basal dolomite. Zinc shows a slight (−0.40) negative correlation with strontium.

Reef complex.—

Vertical patterns are discernible in the reef complex (Fig. 4; Table 3). Iron and manganese generally tend to decrease upsection in each of the sampled drill cores, whereas strontium may in some cases increase. A slight negative correlation of strontium with iron (−0.53) and manganese (−0.45) does exist. Iron and manganese values are low in the only backreef rock (84C6 at −7 m) analyzed in the section.

Sparry Dolomite

In addition to the host dolomite samples, sparry dolomite cements were also analyzed. Since these dolomites are zoned, the values obtained represent averages of the zones in each particular sample. Based on the evidence provided by paragenetic sequences, these cements formed at a time when major changes were occurring in the ore-forming process (e.g., Hagni, 1983). These changes are marked by a shift from octahedral to cubic galena habits as well as by the deposition of vaesite. This mineralogic evidence suggests a change in ore fluid chemistry (Hagni, 1983). Analyses of these sparry dolomite cements could therefore provide information concerning the chemistry of the coexisting ore-forming brines at that stage in their evolution.

The chemical data on the sparry dolomite are somewhat ambiguous even though certain trends appear to be present (Table 5; Fig. 5a, b). The iron and manganese concentrations exhibit a similar areal distribution pattern, with lower values in the backreef unit and higher, fairly constant concentrations elsewhere. Iron concentrations in the sparry cements are higher than those found in the host dolomite, whereas the manganese concentrations are similar. Strontium concentrations are generally higher than those of the

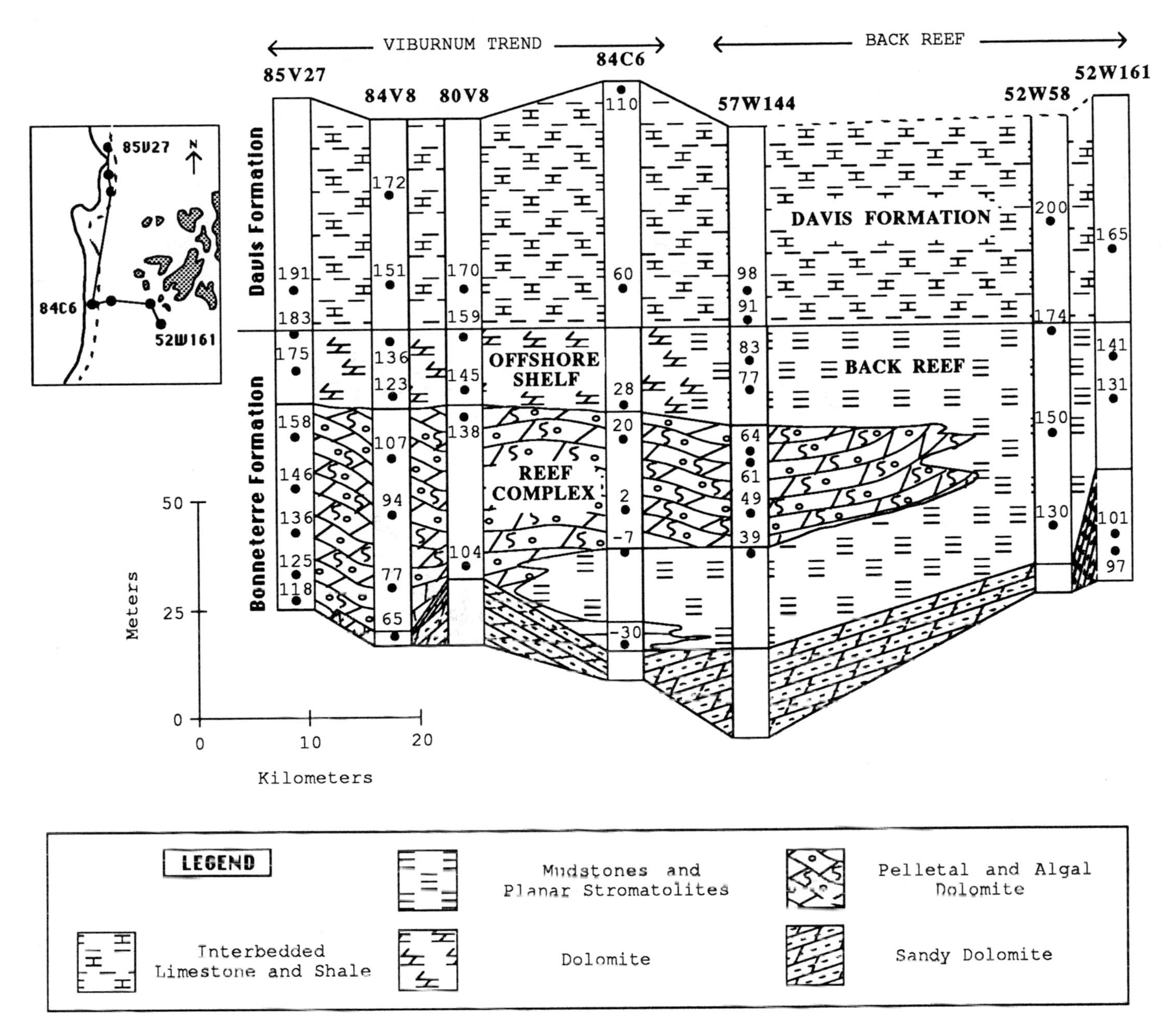

FIG. 4.—Unit distribution and sample locations in sampled drill cores. Common datum is the Davis-Bonneterre contact. Depths are referenced to mean sea level. See Tables 1 and 3 for individual analyses.

host dolomite and constant throughout the study area. Sodium concentrations, most likely reflecting sodium in fluid inclusions rather than in the carbonate lattice, are similar for the entire area, with the exception of the lower values found in the reef complex unit. Lead concentrations are higher in the reef complex samples; this may be caused by minute galena inclusions in the sparry dolomites.

Comparisons with Other Studies

Chemical analyses of Bonneterre Formation rocks in southeast Missouri have been made in other studies. Mosier and Motooka (1983) found lower iron concentrations in the back reef and higher iron and manganese concentrations in the lower Bonneterre Formation. Viets and others (1983) showed that iron and manganese concentrations in the Bonneterre Formation decrease with distance from the Lamotte-Bonneterre contact. Recent work by Gregg and Shelton (1986) support the iron and strontium trends observed for the basal dolomite in this reconnaissance study. They also found an interesting reverse trend of decreasing iron and increasing strontium from north to south adjacent to the Viburnum Trend in the lowermost basal dolomite. This may correlate with the higher iron and manganese and lower strontium concentrations observed in this study for the two northeasterly sample sites in the basal dolomite.

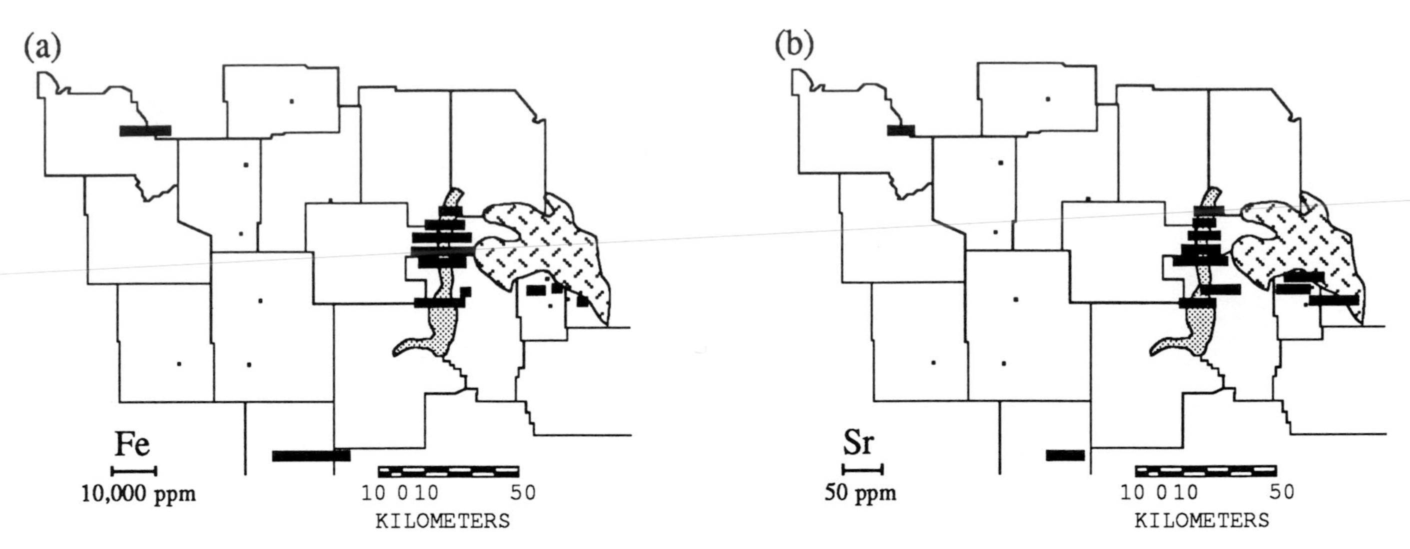

FIG. 5.—(a) Iron and (b) strontium concentration distributions in the sparry dolomites.

DISCUSSION

Host Dolomite

Limestone has been replaced by dolomite at the base of the Bonneterre Formation in the offshore shelf unit. The xenotopic texture of the basal dolomite suggests an epigenetic origin resulting from the interaction of the Bonneterre with warm fluids moving through the Lamotte Sandstone (Gregg, 1985). Disseminated pyrite is found throughout the basal dolomite; Gregg (1985) also found disseminated sphalerite. The presence of pyrite and sphalerite is consistent with the interpretation that the basal dolomite was affected by the first stages of the mineralizing fluid.

The mineralizing brines had a considerable chemical effect on the basal dolomite and, to a certain extent, on the reef complex dolomites. The trends observed in the basal dolomite may be the result of the conversion of limestone to dolomite. Because the dolomite/calcite distribution coefficients for iron and manganese are believed to be greater than one (Veizer, 1983), the dolomitization of limestone will decrease the concentrations of these two elements in the coexisting solution. Conversely, the concentration of strontium in solution will increase, since its dolomite/calcite distribution coefficient is less than one (Veizer, 1983). This will lead to the formation of concentration gradients for these elements in solution along a fluid flow path. Iron and manganese (and possibly zinc) concentrations will decrease in the downflow direction, whereas strontium will increase (Veizer, 1983; Machel, this volume). These solution gradients will be reflected and preserved in the chemistry of the dolomites formed at various locations along the flow path.

TABLE 5.—SPARRY DOLOMITE ANALYSES

Drill Hole	Depth[a]	Mn	Fe	Zn	Sr	Pb	Na
BASAL DOLOMITE							
65W23		1573	21549	3	50	bdl	169
62W149		1722	14120	3	34	bdl	181
REEF COMPLEX							
85V27		1353	6033	5	45	8	130
84V8		1513	10593	3	33	bdl	86
80V8		1570	16340	3	44	bdl	39
35Mine		1854	17933	5	56	2	176
Brushy Creek Mine		1028	12596	22	77	32	180
84C6[b]	60 m	879	13170	3	56	2	58
	20 m	6864	13904	2	48	bdl	59
	2 m	2833	16653	3	45	5	166
BACK REEF							
57W144		365	2430	5	52	bdl	182
52W58		586	4983	11	48	6	172

(Concentrations in ppm)

a. All drill hole depths referenced to mean sea level.

b. 60 m sample from Davis Formation; 20 m and 2 m samples from Bonneterre Formation.

The northwest portion of the study area contains much higher concentrations of strontium. The inverse relationship between strontium and iron or manganese suggests that the fluid became enriched in strontium northward and upsection within the Bonneterre Formation. These results indicate a southern source for the major mineralizing fluids, which flowed upward from the Lamotte Sandstone into the Bonneterre Formation. Although less obvious, the zinc distribution trend may also be consistent with this interpretation.

The movement of basinal waters from the Ouachita-Arkoma Basin to the Viburnum Trend could be explained by a gravity drive hydrodynamic model (Garven and Freeze, 1984). In the gravity drive mechanism, fluids flow from recharge areas in elevated regions to discharge areas in lower elevation regions. A gentle topographic slope will allow regional flow systems to develop over distances of several hundred kilometers. Because of forced convective heat transport, temperatures would be depressed in recharge areas and elevated in discharge areas.

Recent data obtained by Gregg and Shelton (1986) show that immediately west of the Viburnum Trend iron concentrations in the lowermost part of the basal dolomite decrease from north to south, whereas strontium concentrations increase. This information, coupled with the high-iron and manganese and low-strontium concentrations that this

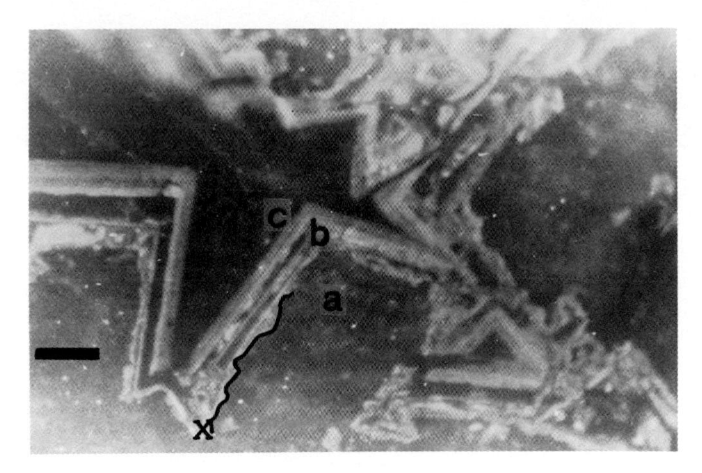

FIG. 6.—Cathodoluminescent microstratigraphy displayed by a sparry dolomite cement. "X" marks the dissolution boundary between zones "a" and "b." Bar 100 μm.

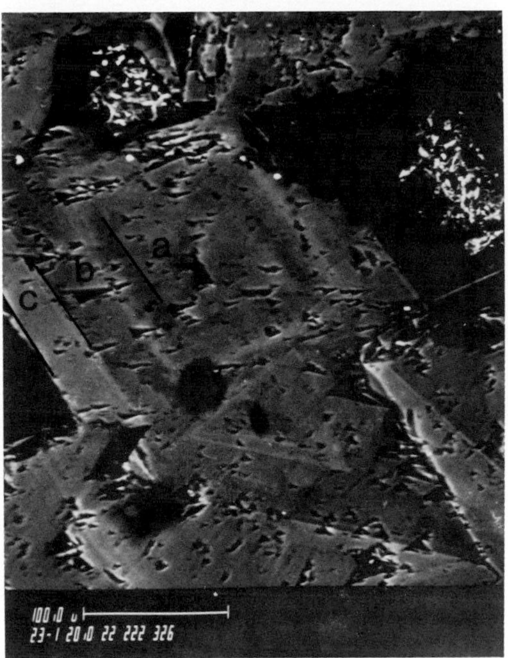

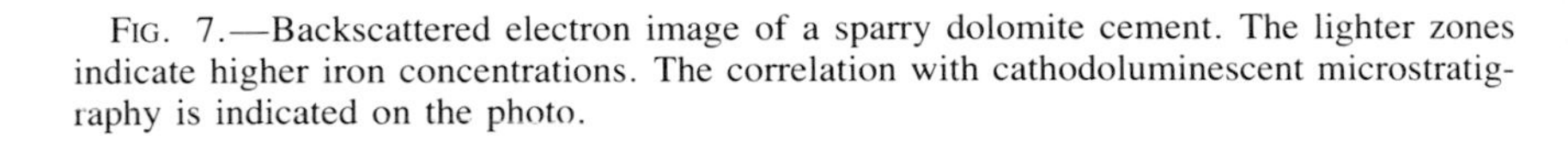

FIG. 7.—Backscattered electron image of a sparry dolomite cement. The lighter zones indicate higher iron concentrations. The correlation with cathodoluminescent microstratigraphy is indicated on the photo.

study found in the northeastern part of the basin, provides evidence of a potentially more complicated fluid flow system for the area. A secondary fluid originating from a source north of the Viburnum Trend would be suggested. The implications of such a fluid may have importance in the further exploration for new ore bodies in southeast Missouri.

The relatively high concentrations of zinc and strontium in the backreef unit, coupled with the low concentrations of iron and manganese, may be the result of diagenetic effects or dolomitization before the introduction of mineralizing fluids into the area. If epigenetic dolomitization had been the dominant factor in determining the chemistry of the backreef dolomites, a continuation of the distinct trends observed in the basal dolomite would be expected. Since the elemental concentrations are partly due to mineralogic change, the presence of only weaker trends in the back reef would be consistent with the interpretation that these rocks are neomorphosed pre-existing dolomites rather than epigenetically dolomitized limestones (Gregg and Sibley, 1984; Gregg, pers. commun., 1987).

Sparry Dolomite

Sparry dolomite cements were observed in all sample locations. The cathodoluminescent microstratigraphic correlation of these cements from the basin through the Viburnum Trend and into the back reef (Gregg, 1985; Voss and Hagni, 1985) indicates that all these cements formed from the same migrating solutions. The outer portions of the sparry cements are marked by distinct growth bands, suggesting a fluid composition that evolved with time (Fig. 6). Differences in minor- and trace-element chemistry between the host dolomite and sparry cements also reflect this change. Iron concentrations in the sparry dolomite are much higher than in the host dolomite, suggesting that the mineralizing fluid became more iron rich through time. This conclusion is also supported by the noncathodoluminescence of the outer zone in the sparry dolomite cements caused by increasing iron. Backscattered electron imaging of sparry dolomite crystals indicates that the zones present are due to discrete iron events. The crystals display a well-defined iron-rich outer zone as well as other iron-rich zones (Fig. 7).

Since the formation of cements in aquifers is mostly accomplished by direct precipitation from solution rather than replacement, the chemistry of these cements reflects the chemical composition and temporal variation of the bulk aquifer waters (Veizer, 1983). Strontium values in the sparry cements are generally higher and more consistent than in the host dolomite, suggesting that the mineralizing fluids had a higher strontium concentration when the cements formed. This may be due to the enhancement of strontium in the mineralizing fluids because of a lower strontium distribution coefficient for dolomite/solution than for calcite/solution. Lower iron and manganese concentrations in the back reef, as well as the observed mineralogical changes, support the possibility of a more dilute solution, possibly meteoric in origin, entering the backreef area from an easterly direction. Anomalous lead and zinc values may represent local concentrations of these metals, possibly due to the dissolution of nearby sulfides. The deficiency of sodium in the reef complex may indicate that fewer fluid inclusions were present in the reef complex dolomites. This might be due to the effects of recrystallization or to different growth rates. A larger sample population needs to be analyzed to determine if these are real trends for the area.

CONCLUSIONS

(1) Chemical data suggest that the principal mineralizing brines for the southeast Missouri deposits originated from the Ouachita-Arkoma Basin to the south, with a smaller

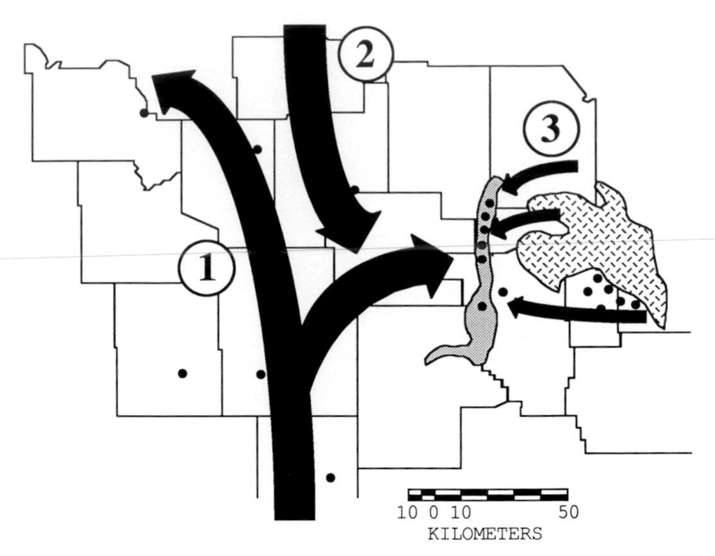

FIG. 8.—Interpretation of basinal-brine events for southeast Missouri. After early dolomitization of the backreef unit by depositional or diagenetic waters, basinal brines were introduced as mineralizing fluids, with a dominant source being the Ouachita-Arkoma Basin to the south (event 1). A smaller but still significant flow may have entered the area from the north (event 2). Event 3 marks the mixing of basinal waters with a secondary fluid having a source in the east.

flow possibly coming from a northerly source. The data also suggest that a mode dilute solution, possibly meteoric in origin, migrated westward through the backreef unit.

(2) Initial dolomitization of the back reef and possibly the reef complex was due to depositional and/or early diagenetic waters. The high-zinc and strontium concentrations, coupled with the low-iron and manganese values in the back reef, may have been fixed by this depositional or early diagenetic environment.

(3) As the basinal waters moved through the Lamotte Sandstone, interaction between the warm solutions and the Bonneterre Formation caused epigenetic dolomitization of the precursor limestones forming the basal dolomite (Fig. 8, event 1). The minor- and trace-element trends suggest the possibility of two separate fluids occurring in the basin burial environment. The dominant source of fluids would have been the Ouachita-Arkoma Basin to the south, with a smaller but potentially still significant flow entering the basin environment from the area to the north of the Viburnum Trend (Fig. 8, event 2).

(4) During the latter stages of the paragenetic sequence, a major period of growth of sparry dolomite cements occurred. Variations of fluid chemistry reflected by areal variation in sparry cement composition may be due to mixing of basinal water with a meteoric water having a source in the backreef area (Fig. 8, event 3).

(5) Since this reconnaissance study was completed, Gregg and Shelton (1986) have further delineated the observed concentration trends. Additional core samples from north of the Viburnum Trend should be examined to test the hypothesis that a northern fluid source may also have been involved in the formation of the Viburnum Trend lead-zinc deposits. The number of presently available cores from that area, however, may be inadequate to do this.

ACKNOWLEDGMENTS

St. Joe Minerals Corporation provided partial funding for this project. Jay M. Gregg of St. Joe Minerals was very helpful during the course of this study. Gregg provided many of the samples used in this study and offered encouragement and discussion dealing with the project. Charles Frank and Gary Salmon, Southern Illinois University, Carbondale, were very helpful in the laboratory stages of the project. We also thank James Palmer of the Missouri Department of Natural Resources, Division of the Geological Survey, for providing access to core samples stored in Rolla. Jay M. Gregg, Otto C. Kopp, and Vijai Shukla critically reviewed the paper.

REFERENCES

GARVIN, G., AND FREEZE, R. A., 1984, Theoretical analysis of the groundwater flow in the genesis of stratabound ore desposits. 1. Mathematical and numerical model. 2. Quantitative results: American Journal of Science, v. 284, p. 1085–1174.

GERDEMANN, P. E., AND MYERS, H. E., 1972, Relationships of carbonate facies patterns to ore deposition and to ore genesis in southeast Missouri lead district: Economic Geology, v. 67, p. 426–433.

GREGG, J. M., 1985, Regional epigenetic dolomitization in the Bonneterre Dolomite (Cambrian), southeastern Missouri: Geology, v. 13, p. 503–506.

———, AND SHELTON, K. L., 1986, Minor and trace element distributions at the Bonneterre Dolomite/Lamotte Sandstone contact (Cambrian), SE Missouri: Evidence for basinal fluid pathways: Geological Society of America, Abstracts with Programs, v. 18, p. 621.

———, AND SIBLEY, D. F., 1984, Epigenetic dolomitization and the origin of xenotopic dolomite texture: Journal of Sedimentary Geology, v. 54, p. 908–931.

HAGNI, R. D., 1983, Ore microscopy, paragenetic sequence, trace element content, and fluid inclusion studies of the copper-lead-zinc deposits of the southeast Missouri lead district, *in* Kisvarsanyi, G., Grant, S. K., Pratt, W. P., and Koenig, J. W., eds., International Conference on Mississippi Valley-Type Lead-Zinc Deposits, Proceedings, Rolla, Missouri, p. 243–256.

HARDIE, L. A., 1987, Dolomitization: A critical view of some current views: Journal of Sedimentary Geology, v. 57, p. 166–183.

HOWE, W. B., 1968, Planar stromatolite and burrowed carbonate mud facies in Cambrian strata of the St. Francois Mountain area: Missouri Department of Natural Resources, Division of Research and Technical Information, Geological Survey Report of Investigations 41, 113 p.

KAISER, C. J., AND OHMOTO, H., 1985, A kinematic model for tectonic structures hosting North America Mississippi Valley-type mineralization: Geological Society of America, Abstracts with Programs, v. 17, p. 622.

LEACH, D. L., VIETS, J. G., AND ROWAN, L., 1984, Appalachian-Ouachita orogeny and Mississippi Valley-type lead-zinc deposits: Geological Society of America, Abstracts with Programs, v. 16, p. 572.

LYLE, J. R., 1977, Petrography and carbonate diagenesis of the Bonneterre Formation in the Viburnum trend area, southeast Missouri: Economic Geology, v. 72, p. 420–434.

MCINTIRE, W. L., 1963, Trace element partition coefficients–A review of theory and applications to geology: Geochimica et Cosmochimica Acta, v. 27, p. 1209–1264.

MATTES, B. W., AND MOUNTJOY, E. W., 1980, Burial dolomitization of the upper Devonian Miette buildup, Jasper National Park, Alberta, *in* Zenger, D. H., Dunham, J. B., and Ethington, R. L., eds., Concept and Models of Dolomitization: Society of Economic Paleontologists and Mineralogists Special Publication 28, p. 259–297.

MOSIER, E. L., AND MOTOOKA, J. M., 1983, Induction coupled plasma–Atomic emission spectrometry: Analysis of subsurface Cambrian carbonate rocks for major, minor and trace elements *in* Kisvarsanyi, G., Grant, S. K., Pratt, W. P., and Koenig, J. W., eds., International Conference on Mississippi Valley-Type Lead-Zinc Deposits, Proceedings, Rolla, Missouri, p. 155–165.

OLIVER, J., 1986, Fluids expelled tectonically from orogenic belts: Their

role in hydrocarbon migration and other geologic phenomena: Geology, v. 14, p. 99–102.
VEIZER, J., 1983, Chemical diagenesis of carbonates: Theory and application of trace element technique, *in* Arthur, M. A., Anderson, T. F., Kaplan, I. R., Veizer, J., and Land, L. S., Society of Economic Paleontologists and Mineralogists Short Course No. 10, p. 3-1 to 3-100.
VIETS, J. G., MOSIER, E. L., AND ERICKSON, M. S., 1983, Geochemical variations of major, minor, and trace elements in samples transecting the Viburnum Trend Pb-Zn district in southeast Missouri, *in* Kisvarsanyi, G., Grant, S. K., Pratt, W. P., and Koenig, J. W., eds., International Conference on Mississippi Valley-Type Lead-Zinc Deposits, Proceedings, Rolla, Missouri, p. 174–186.
VOSS, R. L., AND HAGNI, R. D., 1985, The application of cathodoluminescence microscopy to the study of sparry dolomite from the Viburnum Trend, southeast Missouri, *in* Hausen, D. M., ed., The Metallurgical Society of the American Institute of Mining, Metallurgical, and Petroleum Engineers, Paul F. Kerr Memorial Symposium, Proceedings, New York, New York, p. 51–68.
ZENGER, D. H., 1983, Burial dolomitization in the Lost Burro Formation (Devonian), east-central California, and the significance of late diagenetic dolomitization: Geology, v. 11, p. 519–522.

SECTION IV
ROCK-WATER INTERACTIONS DURING DOLOMITIZATION

SECTION IV: INTRODUCTION
ROCK-WATER INTERACTIONS DURING DOLOMITIZATION

The two papers in this section trace various chemical interactions during dolomitization, from the element and isotope composition of dolomites.

Banner and others describe interactions in Mississippian dolomites within the framework established by cathodoluminescence. The various dolomites record complex and multiple episodes of rock-water interactions, which can be regionally correlated. This paper is one of two parts of a larger study; the other part is presented by Cander and others in section V.

Machel presents a qualitative mathematical model in which Sr, Mn, and Mg trends can be used to deduce paleoflow directions of dolomitizing fluids. This paper should be compared and contrasted with that of Sass and Bein (section VII).

WATER-ROCK INTERACTION HISTORY OF REGIONALLY EXTENSIVE DOLOMITES OF THE BURLINGTON-KEOKUK FORMATION (MISSISSIPPIAN): ISOTOPIC EVIDENCE

JAY L. BANNER[1], G. N. HANSON, AND W. J. MEYERS
Department of Earth & Space Sciences, State University of New York, Stony Brook, New York 11794

ABSTRACT: Two sequences of pervasive dolomitization are preserved in the Mississippian Burlington-Keokuk Formation of Iowa, Illinois, and Missouri. Cathodoluminescent petrography reveals (1) an early, post-depositional, dolomite-forming episode (dolomite I), and (2) a later dolomite (dolomite II), which replaced the first generation. These texturally and temporally distinct dolomites are correlative over 100,000 km^2 of outcrop and subsurface (see Cander and others, this volume) and have distinguishing isotopic and trace-element characteristics. Calculation of the simultaneous isotopic variations that occur during water-rock interaction demonstrates important differences in the relative rates at which the O, C, Sr, and Nd isotopic compositions of diagenetic carbonates are altered. These quantitative models are used to place constraints on the water-rock interaction history of the Burlington-Keokuk dolomites.

Dolomite I samples have a range of $\delta^{18}O$ (−2.2 to 2.5‰ PDB), $\delta^{13}C$ (−0.9 to 4.0‰ PDB) and ϵ_{Nd} (342) values (−6.0 to −4.7), and initial $^{87}Sr/^{86}Sr$ ratios (0.70757 to 0.70808) that encompass estimated marine dolomite isotopic compositions. These samples also have 107 to 123 ppm Sr, slightly lower than that of modern marine dolomites. Dolomite I formed from predominantly seawater-derived constituents with a small but significant non-marine component. A mixed-marine meteoric-fluid model can quantitatively account for the variations in dolomite I isotope and trace-element compositions, but the origin of the non-marine component is not well constrained.

Compared to dolomite I, dolomite II samples have radiogenic initial $^{87}Sr/^{86}Sr$ ratios (0.70885 to 0.70942), lower $\delta^{18}O$ values (−6.6 to −0.2‰ PDB), depleted Sr concentrations (50 to 63 ppm), similar $\delta^{13}C$ values (−1.0 to 4.1‰ PDB) and similar ϵ_{Nd} (342) values (−6.5 to −5). The isotopic composition and concentration of Sr in dolomite II preclude a source within the Burlington-Keokuk Formation for the Sr in dolomite II. Dolomite II apparently formed as a result of the recrystallization of the less stoichiometric dolomite I by extraformational subsurface fluids that migrated to shallow burial depths. The results suggest that the recrystallization process effectively exchanged nearly all of the Sr from dolomite I.

Oxygen isotopes equilibrate between dolomite and fluid at relatively low extents of water-rock interaction, and as a result, the $\delta^{18}O$ values of dolomite II may reflect only the last stages of recrystallization. The results of model calculations also suggest that the $^{87}Sr/^{86}Sr$ ratios of dolomite II preserve an earlier and larger record of water-rock interaction, whereas their C and Nd isotopic signatures are inherited from dolomite I precursors. Late-stage, vug-filling carbonates appear to have formed from extraformational fluids that experienced minimal interaction with Burlington-Keokuk host rocks. The petrology and geochemistry of Burlington-Keokuk dolomites document multiple episodes of pervasive water-rock interaction that can be correlated on a regionally extensive scale.

INTRODUCTION

Studies of carbonate diagenesis have endeavored to determine how processes of water-rock interaction have controlled the textural and geochemical evolution of the diagenetic products. Applied to dolomitization, this approach is increasingly complex and important, owing in part to the paucity of dolomite forming in modern settings compared to the extensive dolomite sequences in the Paleozoic and Mesozoic, which are host to significant petroleum and base-metal deposits. Several recent reviews have detailed the controversy regarding models for dolomitization, and some have disputed the utility of geochemical techniques for addressing this controversy (Land, 1980, 1985; Morrow, 1982; Machel and Mountjoy, 1986; Hardie, 1987). The dolomites of the Mississippian Burlington-Keokuk Formation provide a spatially correlative, temporal and textural framework that is well suited to test geochemical approaches for unravelling the water-rock interaction history of this thick-bedded, regionally extensive dolomite sequence.

We present here C, O, and Sr isotope data and Sr concentration data for two major dolomite generations and minor vug-filling carbonates in the Burlington-Keokuk Formation. These data will be used with other geochemical (Nd isotopes, rare-earth element, Fe and Mn concentrations) and petrologic information on the dolomites from previous studies in order to place limits on the processes of dolomitization. Neodymium isotope and trace-element data are treated in detail in Banner (1986) and Prosky and Meyers (1985 and in prep.), and the petrograhic framework of dolomitization is constructed in Harris (1982), Cander and others (this volume), Banner and others (in prep.), and by J. L. Prosky (pers. commun., 1987).

Quantitative models are developed to determine the *simultaneous* variations in several isotopic and elemental parameters as a function of water-rock interaction. This approach reduces the limitations inherent in using a particular geochemical system, such as uncertainties in values for distribution coefficients and fractionation factors. This study employs the method developed in Banner (1986), that of an iterative calculation using mass balance in order to simulate isotopic and elemental exchange during recrystallization. The variation in the isotopic composition and concentration of elements such as O, C, Sr, and Nd in diagenetic fluids and minerals results in important differences in the rates at which the different isotopic systems are altered during the progressive recrystallization of carbonate sediments. The utility of this approach lies not so much in the absolute values of the calculated water:rock ratios as in the *relative* differences calculated between different isotopic systems and between different diagenetic models. These relative differences result in diagnostic alteration trends on isotope-isotope and isotope-element variation diagrams and are used to constrain models for the water-rock interaction history of the Burlington-Keokuk dolomites.

PETROGRAPHY AND GEOCHEMISTRY OF BURLINGTON-KEOKUK DOLOMITES

The Burlington-Keokuk Formation crops out in southeastern Iowa, western Illinois, and eastern, central, and

[1]Present addresses: Division of Geological and Planetary Sciences, California Institute of Technology, Pasadena, California 91125 (corresponding), and Basin Research Institute, Louisiana State University, Baton Rouge, Louisiana 70803.

Sedimentology and Geochemistry of Dolostones, SEPM Special Publication No. 43
 ISBN 0-918985-77-3

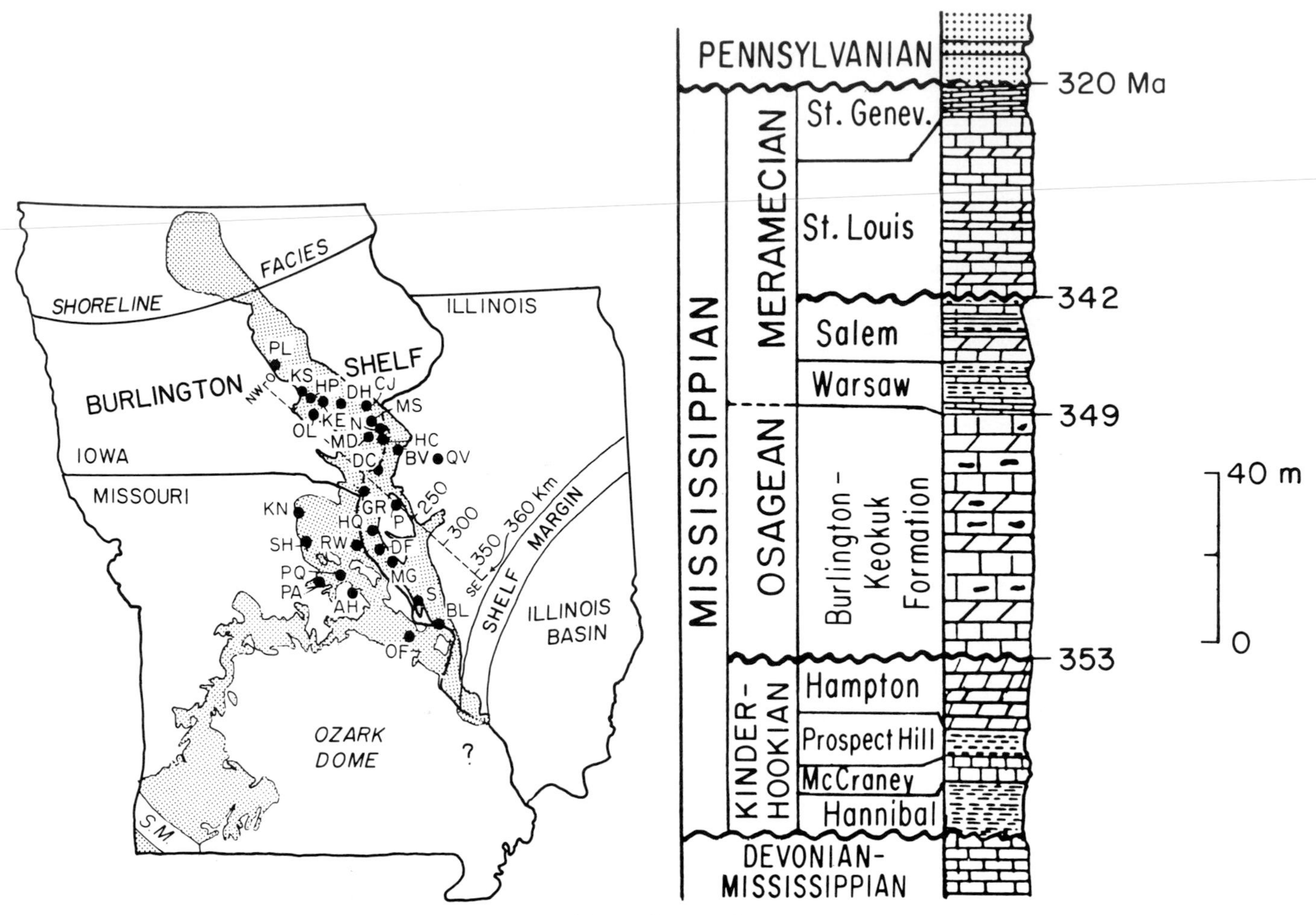

FIG. 1.—Paleogeographic setting of Burlington-Keokuk shelf during Osagean time and localities for samples analyzed in this study and listed in Table 1. Shaded area shows outcrop belt of Mississippian rocks. Paleogeography is taken from Lane and DeKeyser (1980) and Sixt (1983). Stratigraphic section for Iowa and western Illinois from Harris (1982). Time scale from Harland and others (1982).

southern Missouri as part of the Osage Series (Fig. 1). Carbonate deposition occurred on a broad, shallow, subtidal shelf on the southeast flank of the Transcontinental Arch, west of the Illinois Basin, and north and west of the Ozark Dome (Carlson, 1979; Lane and DeKeyser, 1980). Burial history curves indicate a maximum overburden of less than 0.5 km for Burlington-Keokuk strata at the end of Pennsylvanian time (O. A. Cox, unpubl. data). The Burlington-Keokuk Formation is comprised of medium- to thick-bedded, coarse-grained, crinoidal packstone and grainstone and interbedded wackestone and mudstone. A detailed diagenetic history for the Burlington-Keokuk Formation includes multiple episodes of dolomitization, calcite cementation, dedolomitization, chertification, and mechanical and chemical compaction. A paragenetic sequence can be correlated over an area of approximately 100,000 km^2 using calcite and dolomite zonal stratigraphies that are based on cathodoluminescent petrography (Harris, 1982; Cander and others, this volume; Kaufman and others, in prep.).

Two major generations of dolomite can be distinguished on the basis of cathodoluminescent characteristics (Fig. 2). Calcium-rich dolomite I (54.5–56.5 mole percent $CaCO_3$; Prosky and Meyers, 1985), the most common dolomite, consists of orange to light brown rhombs with concentric zoning. Dolomite I appears to have replaced nearly all lime mud as the earliest diagenetic phase of regional extent in the Burlington-Keokuk Formation. Evidence for the com-

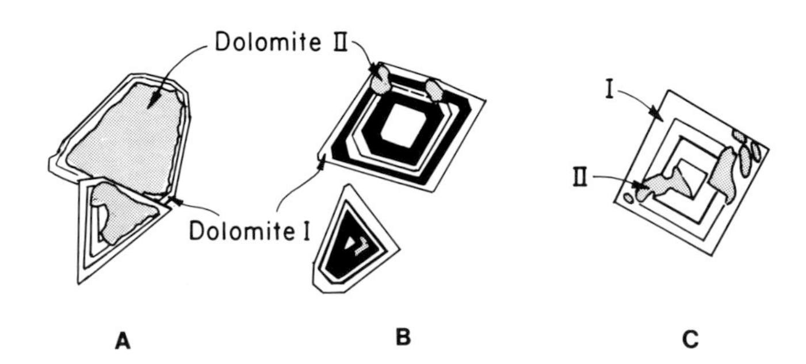

FIG. 2.—Sketches of different cathodoluminescent dolomite types and their morphologies. Pattern of concentric zoning in dolomite I is similar within some measured sections and varies between localities. Samples: (A) HP-4-GR; (B) HP-18-J; (C) MD-13. Replacement nature of dolomite II (shown in grey) is indicated by truncation of fine, concentric zones in dolomite I precursor and by the similar sizes of replaced and unreplaced dolomite I rhombs in the same samples. Most common replacement fabric is shown in (A). Larger rhombs are approximately 100 μm across. Dolomite petrography is detailed in Cander and others (this volume).

position and mineralogy of the precursor sediment to dolomite I has been obliterated by the dolomitization process. In addition to potential aragonite and Mg-calcite precursors, a syndepositional marine dolomite phase, calcium-rich and poorly ordered relative to dolomite I (e.g., Carballo and others, 1987), may have been recrystallized and overgrown by dolomite I.

Dolomite II occurs as a dull red, unzoned replacement of dolomite I and is more stoichiometric (51–52 mole percent $CaCO_3$). Dolomite II′ is less common and occurs as a very dull red to brown replacement of dolomites I and II. A progressive increase in Fe and Mn concentrations is observed through the paragenetic sequence dolomite I – dolomite II – dolomite II′. Iron and Mn concentrations are also correlated to luminescence variations within each dolomite generation (Prosky and Meyers, 1985). Dolomite I, II and II′ have similar $\epsilon_{Nd}(342)$ values and rare-earth element (REE) patterns (Banner, 1986). Two-phase fluid inclusions in dolomites I and II, characterized as primary, yield homogenization temperatures of 90–120°C and bulk salinities of as much as 20 weight percent (Smith, 1984).

In sections that have both dolomite I and II, there is a transition from strata dominated by dolomite II at the bottom to domination by dolomite I higher in the section. Cathodoluminescent petrography and regional stratigraphy have been used to constrain age ranges for the dolomites as follows: (1) dolomite I: post-Burlington-Keokuk deposition to pre-Pennsylvanian deposition (349–320 Ma), and (2) dolomite II: post-dolomite I to pre-Pennsylvanian or pre-Permian (Cander and others, this volume). Calcite spar, saddle dolomite, quartz, pyrite, marcasite, and sphalerite are found in solution vugs and fractures which postdate the major diagenetic episodes in the Burlington-Keokuk Formation.

METHODS

Nearly pure whole rock dolostones, comprised of predominantly one cathodoluminescent dolomite type, and physically separated dolomite (using heavy liquids and magnetic separation methods) from some less pure samples were chosen for analysis. Detailed petrographic descriptions and separation methods are given in Banner (1986). Samples were ground to less than 200 mesh. An approximately 0.2- to 1-mg split of each sample was roasted *in vacuo* at 380°C for 1 hr to remove volatile contaminants. Calcite samples were reacted with anhydrous H_3PO_4 at 50°C in an extraction line coupled to the inlet of a VG 602E ratio mass spectrometer. Dolomite separate and dolostone samples were reacted with anhydrous H_3PO_4 at 50°C in separate off-line vessels for 10 to 18 hr to enable complete digestion. Isotopic enrichments were measured relative to an intralaboratory standard reference gas, which was calibrated to PDB through daily analysis of NBS-20 calcite. All enrichments were corrected for ^{17}O contribution following the method of Craig (1957). No correction was applied for dolomite-phosphoric acid fractionation. Thirty-four analyses of NBS-20 calcite conducted during the course of this study indicate that precision at the one sigma level is ±0.13‰ for oxygen and ±0.09‰ for carbon. Replicate analyses on 14 unknown samples gave a mean deviation of ±0.11‰ for oxygen and ±0.05‰ for carbon. Standard and replicate data are given in Banner (1986).

Strontium and Nd isotope ratios were measured at Stony Brook using a NBS design surface emission mass spectrometer. Precision at the two sigma level for measured ratios was typically ±0.00004 for $^{87}Sr/^{86}Sr$ and ±0.00002 for $^{143}Nd/^{144}Nd$. Details of these methods are given in Banner (1986) and Banner and others (in prep). Fe, Mn, and Ca concentration data are from J. L. Prosky (pers. commun., 1987) and were measured using an ARL-EMX electron microprobe following procedures given in Reeder and Prosky (1986).

RESULTS

Table 1 presents analyses of dolomites I, II, and II′, including whole rock and mineral separate results for some samples. Analyses of vug-filling calcite and dolomite are also given. A comparison of $\delta^{18}O$ values for dolomite separate and whole rock analyses for six samples gives a mean deviation of ±0.57‰, which is greater than the analytical precision. The $\delta^{18}O$ values for all dolomite separates are greater than the corresponding whole rock values. This indicates that a component with a lower $\delta^{18}O$ value was removed during the separation process. The lower $\delta^{18}O$ value of the whole rock may be due to small amounts of late-stage calcite cements, which have $\delta^{18}O$ as low as −11.3‰ (Table 1). Five to 10% of such calcite included in a whole rock dolostone can account for the differences observed between whole rock and dolomite separate analyses. These percentages are higher than the amounts of modal calcite observed in the samples (0–5%). An additional effect may be a systematic removal of dolomite with low $\delta^{18}O$ during the separation procedures. There may be considerable intra-sample variation in dolomite $\delta^{18}O$ values similar to that observed for Ca, Mg, Mn, and Fe (Prosky and Meyers, 1985). $\delta^{13}C$ values for dolomite separate and whole rock analyses are essentially the same for five samples (mean deviation = ±0.15‰), whereas one sample (DH-8) shows a large difference of over 3‰.

The samples analyzed in this study probably all contain small amounts of calcite cements and solid and fluid inclusions with different isotopic compositions compared to the dolomite phase of interest. Based on petrographic criteria and mass balance calculations, the effects of these impurities are limited and will not change inferences or quantitative models based upon the analytical data to any significant degree.

Dolomites I, II, and II′ have the same range in $\delta^{13}C$ values, whereas $\delta^{18}O$ values vary widely between dolomite I and dolomites II and II′ (Fig. 3). For dolomite I, 21 out of 25 samples have $\delta^{13}C$ values between 2 and 4‰, whereas $\delta^{18}O$ values for all dolomite I samples are more evenly distributed over a 5‰ range. Oxygen and C isotope compositions show no distinct correlation for dolomite I samples. Similarly, dolomites II and II′ show a 6‰ span in $\delta^{18}O$ values, whereas most $\delta^{13}C$ values for these samples are between 2 and 4‰. The vug-filling carbonates are distinguished by their low $\delta^{18}C$ and $\delta^{18}O$ values.

The most definitive regional trend in dolomite stable isotope compositions is that in the $\delta^{18}O$ values for dolomite

TABLE 1.—OXYGEN, CARBON, AND STRONTIUM ISOTOPE COMPOSITIONS OF BURLINGTON-KEOKUK DOLOMITES AND ASSOCIATED PHASES

Sample No.	$\delta^{18}O$ PDB	$\delta^{13}C$ PDB	$^{87}Sr/^{86}Sr_i$	Sr, ppm
	Dolomite I			
PL-2	−0.95	1.99		
PL-9	−0.33	1.20		
PL-10	0.52	2.73		
PL-J-22	−0.07	2.93		
HP-18-J	1.72	2.41	0.70808	123.3
DH-4	−0.64	0.33		
DH-8-WR*	−0.40	0.15		
DH-8-DS	2.27	3.14	0.70777	108.0
MS-1	2.47	2.97		
MD-F	0.73	3.33	0.70796	110.2
BV-6.8-J-GRWR	0.29	2.85		
BV-6.8-J-GRDS	1.25	3.17		
BV-6.8-J-GFWR	−0.03	2.84		
BV-6.8-J-GFDS	0.98	2.52	0.70787	118.3
BV-11	0.34	3.35	0.70757	106.6
DC-2	−2.23	2.64		
DC-3	−0.36	3.14		
DC-13	−0.76	3.16		
MPA-17	−1.55	2.90		
IP-23	−0.47	3.74		
IQV-10	−1.25	2.16		
MSH-5a	−1.66	2.75		
IHQ-39	−0.37	3.73		
IDF-7	−0.70	3.64		
IDF-8	−0.48	3.54		
IDF-11	−0.31	4.03		
IDF-13	−1.80	2.61		
IDF-K	−1.40	3.54		
MPQ-11	−1.44	1.44		
MAH-14	−0.65	2.58		
IS-20	−2.19	−0.90		
IS-24	−2.20	3.12		
	Dolomite II			
KS-2	−2.24	2.06		
KS-5	−3.66	2.49		
KS-6	−2.81	1.36		
KS-8-WR	−3.37	1.04	0.70919	54.2
KS-8-DS	−2.49	1.54		
HP-3-J	−3.09	2.62	0.70942	53.7
HP-4-GRWR	−1.43	1.75		
HP-4-GRDS	−0.24	1.97		
HP-4-GFWR	−1.18	2.35		
HP-4-GFDS	−1.09	2.24	0.70885	62.9
HP-4.1-J	−2.41	2.25		
HP-10.1-J	−2.65	2.25	0.70905	53.5
KE-4	−4.38	2.73		
KE-5	−4.14	3.19		
KE-12	−4.14	2.71		
KE-16	−3.55	2.86		
N-12a	−4.47	2.62		
HC-3	−5.45	2.72		
HC-10	−4.66	3.47		
MKN-8	−4.56	3.41		
MKN-9	−3.52	3.61		
DC-18	−4.30	3.51	0.70931	49.9
GR-14	−4.14	3.79		
MPA-2	−6.43	4.11	0.70907	50.8
MPA-20	−5.03	4.04		
MRW-8	−5.50	3.64		
MRW-18	−5.68	3.55		
IMG-5	−3.94	−1.01		
MOF-16	−5.61	3.72		
IBL-O	−5.25	3.85		
IBL-2	−6.60	3.11		
	Dolomite II′			
HP-15-J	−2.31	2.47	0.70925	59.7
DC-7	−3.70	2.84		
DC-23	−4.09	3.58	0.70940	53.2
CJ-2	−3.66	0.63		
CJ-3	−2.00	−0.89		
	Vug Dolomite			
GRF-1	−4.96	−2.10	0.71024	91.3
GR5-J	−5.70	−0.56		

TABLE 1.—*Continued*

Sample No.	$\delta^{18}O$ PDB	$\delta^{13}C$ PDB	$^{87}Sr/^{86}Sr_i$	Sr, ppm
	Vug Calcite			
KE-5	−6.12	0.24		
DC-3	−5.34	1.75		
GR-6	−11.34	−0.14	0.70987	87.1
MPA-2	−9.81	1.14		

*WR = whole rock; DS = dolomite separate; whole rock analysis, where no designation given. Sample descriptions and localities given in Banner (1986) and Cander (1985) for I-samples and Kaufman (1985) for M-samples. These prefixes are omitted in Figure 1.

$$\text{For oxygen, } \delta^{18}O = \frac{(^{18}O/^{16}O)_{sample} - {}^{18}O/^{16}O)_{standard}}{(^{18}O/^{16}O)_{standard}} \times 10^3.$$

A similar expression can be written for $\delta^{13}C$. The PDB standard is used here. For Sr isotopes, $^{87}Sr/^{86}Sr_i$ = initial $^{87}Sr/^{86}Sr = (^{87}Sr/^{86}Sr)_{measured} - {}^{87}Rb/^{86}Sr(e^{\lambda T} - 1)$, where λ is the decay constant for the decay of ^{87}Rb ($\lambda = 1.42 \times 10^{-11}$ y^{-1}), calculated here for 342 Ma. NBS standard SRM 987 gives $^{87}Sr/^{86}Sr = 0.71034$.

II, which decrease from northwest to southeast from central Iowa to southwest Illinois (Fig. 4). These samples have a total range of nearly 6‰, and at any given locality the range is about 2‰ or less. Dolomite I shows a similar but less well-defined trend of decreasing $\delta^{18}O$ southeastward. $\delta^{13}C$ values for all dolomite types have an overall trend of slight depletion to the northwest, but the most striking feature of the dolomite $\delta^{13}C$ data is the restricted range of values for the principal dolomite types (Figs. 4, 5).

Dolomite I samples have initial $^{87}Sr/^{86}Sr$ ratios of 0.70757 to 0.70808, and Sr concentrations of 106.6 to 123.3 ppm (Table 1). Dolomites II and II′ have markedly higher initial $^{87}Sr/^{86}Sr$ ratios of 0.70885 to 0.70942 and lower Sr concentrations of 49.9 to 62.9 ppm. A summary of the isotope and trace-element geochemistry of the Burlington-Keokuk dolomites is given in Figure 5.

DISCUSSION

The geologic, petrographic, and geochemical characteristics of the Burlington-Keokuk Formation dolomites will be used toward determining the nature of the fluids and

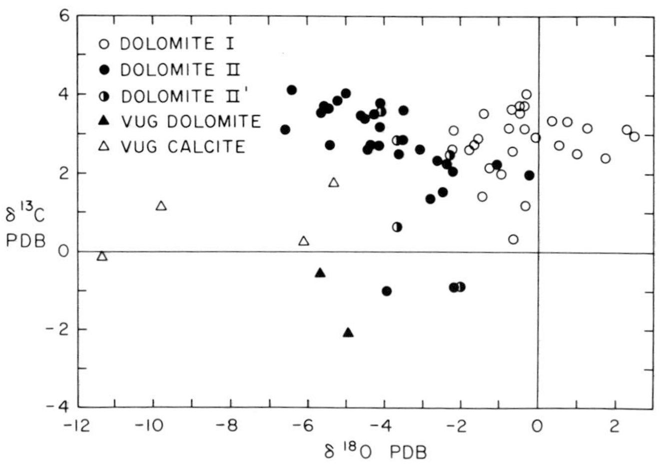

FIG. 3.—Carbon vs. oxygen isotope compositions for all Burlington-Keokuk phases analyzed in this study. In this and all succeeding figures, dolomite separate values are plotted for samples where available.

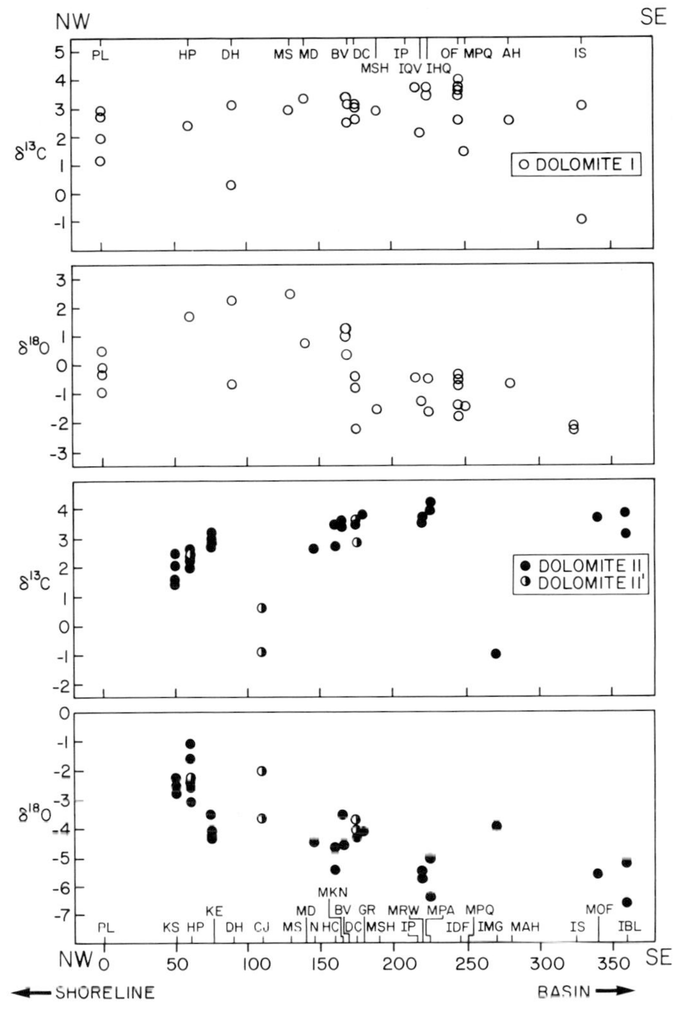

FIG. 4.—Regional variations in C and O isotopic compositions for Burlington-Keokuk dolomites. Sample localities are projected to plot along northwest-southeast transect shown in Figure 1.

processes that generated the dolomites. In order to evaluate the effects of diagenesis on the geochemistry of carbonate sediments, one must be able to estimate the composition of the sediments at the time of their formation in the marine depositional environment. We use the following estimated isotopic compositions for Osagean marine dolomite: $\delta^{18}O$ = 1.8–2.8‰ PDB, $\delta^{13}C$ = 4.0–4.6‰ PDB, $^{87}Sr/^{86}Sr$ = 0.7076, $\epsilon_{Nd}(342) = -7$ to -5, as given in Table 2 and Figure 5.

Calculation of Isotopic Variations During Water-Rock Interaction

The simultaneous variations in O, Sr, C, and Nd isotopic compositions of carbonates that occur during water-rock interaction are portrayed in Figure 6. The model curves were constructed using an iterative calculation procedure and represent changing rock compositions as a function of increasing molar water:rock ratio, $(W/R)_m$. In the model shown, the fluid flows through and recrystallizes the rock in increments. Isotopic exchange during each increment is calculated using mass balance relationships (Banner, 1986; see also Taylor, 1979, and Land, 1980). The results of the calculations demonstrate the different extents of water-rock interaction required to alter the different isotopic parameters (Fig. 6). It can be seen that these differences arise from the pronounced differences in the concentrations of O, Sr, C, and Nd within and between the fluid and solid phases. These differences are used to construct and evaluate models for the water-rock interaction history of the Burlington-Keokuk dolomites.

Petrogenesis of Dolomite I

Any model for the dolomitization of lime mud to produce dolomite I must account for: (1) the regional extent and early timing of dolomitization; (2) the replacement of lime mud and supply of Mg^{+2} to the site of dolomitization; (3) the calcium-rich, nonstoichiometric compositions; (4) the range of lower $\delta^{13}C$ and $\delta^{18}O$ values of dolomite I compared to the estimated marine dolomite value (EMD); (5) the range of moderately radiogenic Sr isotopic compositions encompassing the EMD and low Sr abundances relative to the EMD; (6) the high Fe and Mn concentrations relative to the EMD; and (7) the high temperatures and salinities of fluid inclusions.

If fluid inclusions have preserved a record of the fluids which crystallized the Burlington-Keokuk dolomites, then any model for the formation of the dolomites requires a fluid with high temperatures and salinities. Alternatively, the leakage of warm saline fluids along fractures in previously crystallized dolomite without recrystallizing the dolomite would produce secondary inclusions with high temperatures and salinities. The common observation of dolomite II replacement of only the inner portions of some dolomite I rhombs (and, less commonly, intra-rhomb pores) indicates that post-crystallization fluids have entered some dolomite I rhombs without leaving an apparent trace of their pathway. Evidence from experimental and natural systems suggests that fluid inclusions in calcite can exchange with post-crystallization fluids (Comings and Cercone, 1986; Goldstein, 1986). Fluid inclusions in the Burlington-Keokuk dolomites may reflect the passage of warm saline fluids through the Burlington-Keokuk Formation subsequent to dolomitization.

Normal sea water and hypersaline dolomitization.—

Models for the formation of ancient dolomites involving sea water or sea water modified by evaporation are based on modern occurrences and can account for the introduction of the large amounts of Mg^{+2} necessary to form dolomite (see review by Land, 1985). Since evaporation of sea water will only increase its $\delta^{18}O$ value and leave its $^{87}Sr/^{86}Sr$ ratio unchanged, these models cannot account for the depleted $\delta^{18}O$ and radiogenic $^{87}Sr/^{86}Sr$ values of some dolomite I samples relative to the EMD.

The migration of marine waters through older sediments has been proposed as a model for dolomitization in the Floridan aquifer (Kohout and others, 1977; Simms, 1984) and in the Enewetak Atoll (Saller, 1984a). Pre-Pennsylvanian sea water did not have the requisite Sr isotopic com-

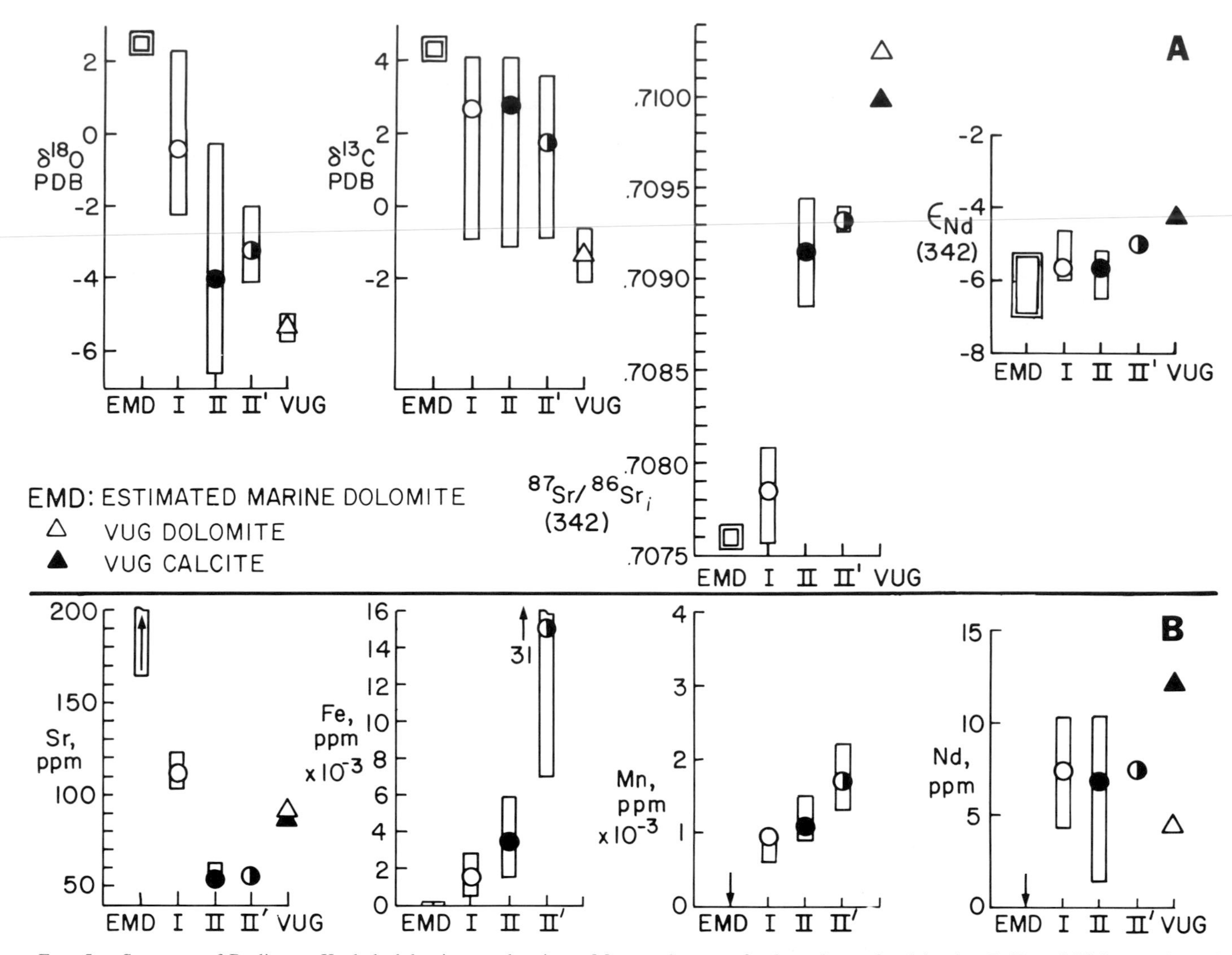

FIG. 5.—Summary of Burlington-Keokuk dolomite geochemistry. Mean and range of values shown for dolomites I, II, and II′ for (A) isotopic compositions of O, C, Sr, and Nd, and (B) Sr, Fe, Mn, and Nd abundances. Dolomite III is discussed in Cander and others (this volume). Estimated marine dolomite (EMD) values discussed in Table 2. Neodymium data are from Banner (1986). Iron and Mn data represent sample averages (J. L. Prosky, pers. commun., 1987).

positions to account for the range in Burlington-Keokuk dolomite I values through interaction with marine carbonate precursors with Osagean $^{87}Sr/^{86}Sr$ ratios (Burke and others, 1982; Popp and others, 1986). If pre-Pennsylvanian sea water, circulating through Burlington-Keokuk sediments, was modified by interaction with local clastics such as the Warsaw Shale or the thin argillaceous carbonate seams in Burlington-Keokuk strata, then the sea water could attain slightly higher and locally variable $^{87}Sr/^{86}Sr$ ratios, Fe contents, and $\delta^{18}C$ values and impart these signatures to dolomite I; however, this model cannot account for the dolomite I $\delta^{18}O$ values that range to 4‰ lower than the EMD.

Burial/basinal-brine dolomitization.—

Diagenetic carbonates formed from subsurface saline fluids often have distinctive geochemical signatures relative to contemporaneous marine carbonate. Isotopic data for dolomite I samples encompass the estimated marine dolomite values. The deviations in $\delta^{13}C$ values from the EMD for some samples suggest a relativley high water:rock ratio system during dolomite I crystallization (Fig. 6). As will be discussed in a later section, a basinal brine in such a system would be expected to impart lower $\delta^{18}O$ and higher $^{87}Sr/^{86}Sr$ values to the samples. A basinal-brine model for dolomitization is not supported by the geochemical data.

Marine meteoric mixing.—

In this model, dolomitization is facilitated by high rates of circulation and mixing of fluids in coastal seawater-freshwater interfaces (Badiozamani, 1973; Wigley and Plummer, 1976; Magaritz and others, 1980). In evaluating marine meteoric-mixing models for the generation of dolomite I, we consider both proximal coastal sources and more distal sources for the meteoric waters. The Transcon-

TABLE 2.—COMPARISON OF ESTIMATED MARINE DOLOMITE ISOTOPIC COMPOSITION (EMD) AND BURLINGTON-KEOKUK DOLOMITE I

Parameter	EMD*	Method of Estimation	Range of Values for Dolomite I
$\delta^{18}O$ (PDB)	1.8 to 2.8	1	−2.2 to 2.5
$\delta^{13}C$ (PDB)	4.0 to 4.6	2	−0.9 to 4.0
$^{87}Sr/^{86}Sr$	0.7076	3	0.70757 to 0.70808
ε_{Nd} (T)	−7 to −5	4	−6.0 to −4.6
Sr (ppm)	150 to >200	5	108 to 123
Fe (ppm)	16 to 57	5	500 to 3000
Mn (ppm)	19 to 22	5	600 to 1100

*Estimates for theoretical marine dolomite as follows:

(1) Estimated marine calcite $\delta^{18}O$ from concurrence of values for: (a) estimate from Osagean Lake Valley Formation (Meyers and Lohmann, 1985); (b) three heaviest nonluminescent brachiopod analyses (Kaufman, 1985); and (c) convergence of crinoid trends by locality (Chyi and others, 1985). To this value of −1.75 to −1.2‰, a dolomite-calcite fractionation factor of 3.6 to 4.0 was applied to bracket the value of 3.8 proposed by Land (1985).

(2) Estimated marine calcite $\delta^{13}C$ from concurrence of Burlington-Keokuk brachiopod and crinoid and Lake Valley data, similar to oxygen estimate above. No calcite-dolomite fractionation was applied.

(3) Estimated from lowest nonluminescent brachiopod analysis.

(4) No independent estimate from the Burlington-Keokuk or other Osagean formation available. Based on modeling of REE mobility (Banner, 1986), dolomite I samples which have seawater Sr isotopic compositions should also record seawater Nd isotopic compositions. A range was chosen to encompass most dolomite I $\varepsilon_{Nd}(342)$ values and the upper end of a range of values for older Mississippian conodonts that preserve marine $^{87}Sr/^{86}Sr$ ratios (Shaw and Wasserburg, 1985).

(5) Analyses of Enewetak dolomites (Saller, 1984b) used for all trace-element estimates. Calculations and observations of Baker and Burns (1985) for Deep Sea Drilling Project dolomites used for lower Sr limit and summary by Land (1980) used for upper Sr limit. The values adopted here are in general agreement with broader studies of secular isotopic variations (e.g., Veizer and Hoefs, 1976; Burke and others, 1982), and are considered the best estimates for the Burlington-Keokuk Formation.

tinental Arch could have provided regionally extensive recharge to the Burlington-Keokuk shelf following the pre-St. Louis or pre-Pennsylvanian regressions.

The depletions in $\delta^{18}O$ and $\delta^{13}C$ values of some Burlington-Keokuk dolomite I samples relative to the EMD may be the result of a freshwater component in the fluid that produced dolomite I. If meteoric waters with relatively low $\delta^{18}O$ and $\delta^{13}C$ values remain largely unmodified during seaward migration, then the mixing of fresh water with sea water will result in a series of curves in $\delta^{13}C$ vs. $\delta^{18}O$ space, which increasingly deviate from linearity as the difference between the total dissolved carbon (TDC) contents of the end-member fluids becomes larger (Fig. 7). Thus, in a mixing zone in which the TDC concentration of the recharging fresh water varied with time or position in the mixing zone, a field of isotopic compositions could be generated that fans out from the composition of sea water to lower $\delta^{13}C$ and $\delta^{18}O$ values.

Interaction between the meteoric water and Mississippian marine carbonates during migration to the zone of mixing would first drive the $\delta^{13}C$ and then the $\delta^{18}O$ value of the fresh water toward the marine values. In addition, dolomitization of lime mud in the Burlington-Keokuk Formation should reflect some buffering of the fluid by the precursor sediment $\delta^{13}C$ and $\delta^{18}O$ values. These water-rock interaction processes will produce an inverted L-shaped compositional trend in $\delta^{13}C$ vs. $\delta^{18}O$ space (Meyers and Lohmann, 1985; Banner, 1986) for each fluid composition. In this manner, the combined processes of fluid-fluid mixing and water-rock interaction that can occur during mixing zone dolomitization may produce a broad field of C and O isotopic compositions, rather than a distinct trend or composition.

Figure 8 illustrates that Burlington-Keokuk dolomite I samples have a similar range in $\delta^{13}C$ values and lower $\delta^{18}O$ values by about 0–3‰ relative to most Quaternary samples of proposed sea water, evaporite-modified sea water (Abu Dhabi), or mixed seawater-freshwater origin (Yucatan, Jamaica). They also have lower $\delta^{18}O$ values relative to dolomites of Meramecean age from Illinois, which are proposed to have formed in a marine meteoric-mixing zone (Choquette and Steinen, 1980).

Fresh waters that are recharged through soil horizons and fresh waters that have interacted with pre-Mississippian carbonates having $\delta^{13}C$ <4‰ (Lohmann, 1983) will have strongly to moderately depleted $\delta^{13}C$ compositions relative to Osagean marine carbonates. Interaction of these fluids with Osagean marine carbonates will produce diagenetic phases with $\delta^{13}C$ <4‰ only at relatively high water:rock ratios, as shown in Figure 6. Assuming that the precursor sediment to dolomite I was predominantly lime mud with a marine $\delta^{13}C$ signature of 4.0‰, the calculations suggest that the formation of most dolomite I samples was a relatively high water:rock ratio process (i.e., $(W/R)_m > 3000$). These results are consistent with calculations based upon Mg^{+2} requirements for dolomitization by seawater: freshwater mixtures ranging from 0–90% fresh water [$(W/R)_m$ = 1100–8900; Land, 1985].

Fresh waters that are recharged through young marine carbonates will derive their dissolved trace elements almost entirely from interaction with the carbonates, which would have $^{87}Sr/^{86}Sr$ values of contemporaneous sea water. The radiogenic initial Sr isotope compositions of most dolomite I samples relative to the seawater value preclude this type of mixing zone model from being applicable in the case of the Burlington-Keokuk Formation. In contrast to fresh waters with proximal sources, meteoric waters recharging from and interacting with older rocks can have radiogenic $^{87}Sr/^{86}Sr$ ratios relative to the contemporaneous seawater value (see Table 3 for river water and seawater analyses). If the lower Paleozoic rocks of the Transcontinental Arch provided radiogenic Sr to regionally recharged meteoric waters, then a significant source of meteoric water with radiogenic Sr may have been available during late Mississippian time. Because fresh waters have low Sr concentrations (<1 ppm) relative to sea water (8 ppm), mixtures of the two would have only slightly higher $^{87}Sr/^{86}Sr$ ratios compared to sea water and lower Sr concentrations than sea water (Banner, 1986). Interaction of mixtures of continentally derived fresh waters and Osagean sea water with marine carbonates at relatively large water:rock ratios could account for the range in Sr isotope compositions and concentrations in dolomite I samples.

The Fe and Mn concentrations of dolomite I rhombs are orders of magnitude higher than expected for marine dolomite (Fig. 5). The mineral fluid K_D values for Fe and Mn are greater than unity, and these elements have low concentrations in sea water and fresh water (Veizer, 1983), properties that are intermediate between Sr and Nd. By comparison with the models for Sr and Nd in Figure 6, it

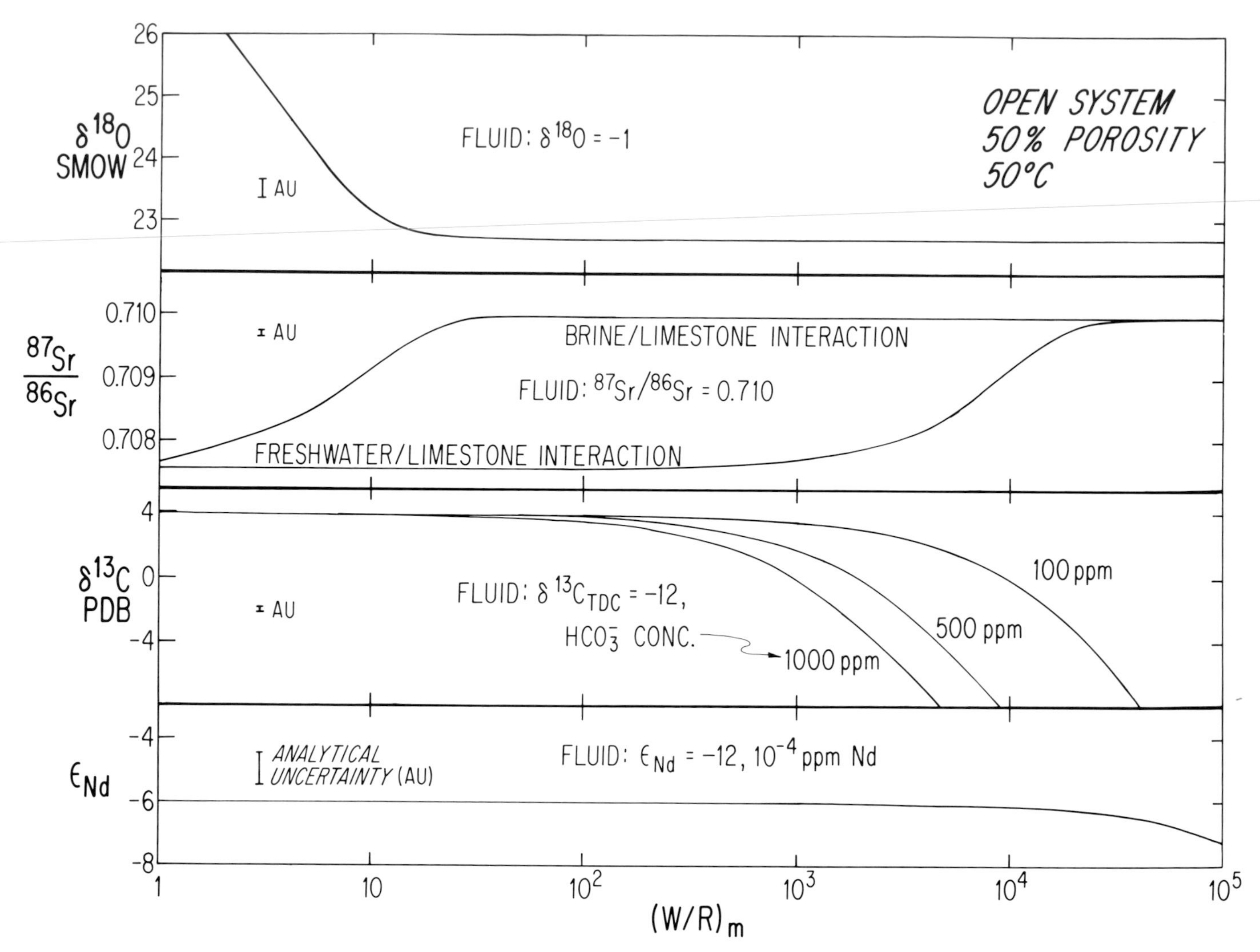

FIG. 6.—Simultaneous variations in the isotopic composition of O, Sr, C, and Nd during open-system recrystallization of a limestone as a function of increasing molar water:rock ratio, $(W/R)_m$ = (moles of water)/(moles of rock). In this model, a fluid is given composition flows through a given volume of rock in increments. Each increment of fluid fills the available porosity and recrystallizes calcite until isotopic equilibrium between the fluid and rock is attained. The resulting changes in the isotopic and trace-element composition of the rock are calculated using mass balance relations. Each new fluid increment displaces the previous one and the process is repeated. The curves illustrate the changes in the composition *of the rock* as a function of the total molar water:rock ratio (i.e., the sum of the increments). Initial isotopic compositions of limestone are: $\delta^{18}O$ = 28‰ SMOW, $^{87}Sr/^{86}Sr$ = 0.7076, $\delta^{13}C$ = 4.0‰ PDB, and ϵ_{Nd} = −6.0, 10 ppm Nd (ϵ_{Nd} as defined in DePaolo and Wasserburg, 1976). Fluid isotopic compositions as follows: $\delta^{18}O$ = −1.0‰ SMOW, $^{87}Sr/^{86}Sr$ = 0.710, $\delta^{13}C$ = −12‰ PDB, ϵ_{Nd} = −12. T = 50°C, porosity = 50%. Two cases illustrated for Sr isotopes: (1) the interaction between a diagenetic carbonate with 200 ppm Sr and a brine with 100 ppm Sr and 20,000 ppm Ca; and (2) the interaction between a marine carbonate with 1,345 ppm Sr and a fresh water with 0.5 ppm Sr and 20 ppm Ca. Calcite-water exchange distribution coefficient is $K_D(Sr/Ca) = (m_{Sr}/m_{Ca})_{calcite}/(m_{Sr}/m_{Ca})_{water} = 0.05$, where m_{Sr}, m_{Ca} are molar concentrations. The water-rock interaction pathways for carbon isotopes demonstrate the effects of dissolved bicarbonate concentrations (in ppm) on the rate at which the $\delta^{13}C$ value of the recrystallizing limestone changes. Note large differences in the water:rock ratios at which the O and C isotopic signatures significantly deviate from the original rock composition. Strontium-isotopic compositions respond at much more variable rates depending on the fluid and rock concentrations. For Nd isotopes, no significant changes from the original rock composition are obtained at water:rock ratios of as much as 5×10^4 using $K_D(Nd/Ca)$ = 100 (Palmer, 1985), and fluid concentrations of 0.0001 ppm Nd and 1,500 ppm Ca. For oxygen isotopes, we use the relationship $\Delta_{calcite\text{-}water} = 2.78 \times 10^6 T^{-2}$ (°K) − 2.89 (Friedman and O'Neil, 1977) in this study for calcite, and the relationship $\Delta_{dolomite\text{-}calcite}$ = 3.8, as discussed in Land (1985). Model calculations for dolomite recrystallization give very similar results to those shown here for calcite. Equations and procedures of the model are detailed in Banner (1986).

would be expected that relatively high water:rock ratios are needed to produce high Fe and high Mn dolomites in a mixing zone. Uncertainties in the effects of redox control on the supply of locally derived Fe and Mn from sulfides, oxides, hydroxides, and silicates in Burlington-Keokuk strata, however, obscure constraints on the water-rock interaction history of dolomite I that are based on dolomite Fe and Mn concentrations.

In summary, the petrology and geochemistry of Burlington-Keokuk dolomite I indicate that the dolomites formed from predominantly Late Mississippian marine waters shortly after deposition was complete. A minor but distinctly non-

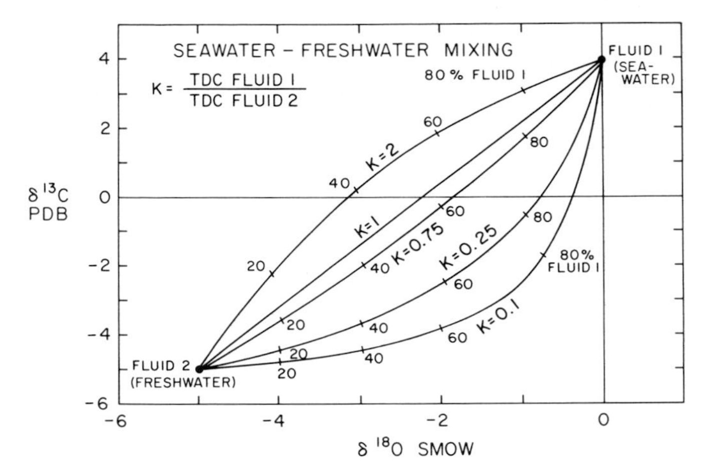

FIG. 7.—Mixing curves of fluid compositions for seawater-freshwater mixing in $\delta^{13}C$ vs. $\delta^{18}O$ space. Fresh water portrayed has typically low $\delta^{18}O$ and $\delta^{13}C$ values relative to sea water (Meyers and Lohmann, 1985). K is the concentration ratio of total dissolved carbon (TDC) as bicarbonate of fluid 1/fluid 2. Sea water has $\delta^{18}O$ = 0‰ (SMOW), $\delta^{13}C$ = 4‰ (PDB); fresh water has $\delta^{18}O$ = −5‰ and $\delta^{13}C$ = −5‰. Mixtures of fluids with equal TDC concentrations will define a straight line. For end members with different TDC concentrations, fluid mixtures describe a curve that will be concave up if the lighter $\delta^{13}C$ end member also has the higher TDC concentration. Method of curve calculation discussed in Banner (1986).

marine component in the water-rock interaction history of dolomite I is required by the isotopic and trace-element data. Although the source of this non-marine component is not clear, meteoric waters recharging through distal, older formations that mixed with sea water may have had the appropriate compositions for the genesis of dolomite I.

Petrogenesis of Dolomite II and II′

The geochemistry of dolomites II and II′ is summarized in Figure 5. From the close similarities in C, O, and Sr isotope compositions, Sr and major-element concentrations, and replacement textures between dolomites II and II′, it is evident that these two phases were formed by very similar processes. Any model for the generation of dolomites II and II′ must satisfy the constraints of the following observations:

(1) the replacement of dolomite I by dolomites II and II′, the closer approximation to stoichiometry of dolomites II and II′, and the higher Fe and Mn concentrations in dolomites II and II′ compared to those in dolomite I;

(2) the vertical (upsection) transition from dolomites II and II′ to dolomite I, indicating that the fluid that formed dolomites II and II′ migrated laterally and upward or simply upward into the Burlington-Keokuk Formation from older formations;

(3) the significant enrichment of radiogenic Sr in dolomites II and II′, and the relatively large range in $\delta^{18}O$ values over a narrow range of $^{87}Sr/^{86}Sr$ ratios and Sr concentrations. Mass balance calculations show that the relatively low $^{87}Sr/^{86}Sr$ ratios of shale horizons in the Burlington-Keokuk Formation during Mississippian time, combined with their composing ≤5% of the strata, make these shale horizons an insufficient source of radiogenic Sr to account for the $^{87}Sr/^{86}Sr$ ratios of dolomites II and II′ (Chyi and others, 1985). In addition, the juxtaposition of strata dolominated by dolomite I in between strata dominated by dolomite II and the Warsaw Shale makes this extensive clastic sequence an unlikely source of constituents for dolomites II and II′. Significant amounts of Sr must have been introduced from allochthonous sources.

(4) the low $\delta^{18}O$ values for dolomites II and II′, as much as 9‰ lower than the heaviest dolomite I sample, and the same mean and range of $\delta^{13}C$ values between 29 dolomite I samples and 28 dolomite II samples;

(5) similar ranges of Nd-isotopic compositions and REE abundances between dolomites I, II, and II′.

Two considerations should be borne in mind throughout the following discussion. First, one can determine from cathodoluminescence that dolomite II and II′ samples are all related by the process of dolomite I recrystallization. Therefore, there is a precise knowledge of the initial rock composition and porosity and an indication of the process involved for constructing quantitative models of water-rock interaction. Second, since the formation of dolomites II and

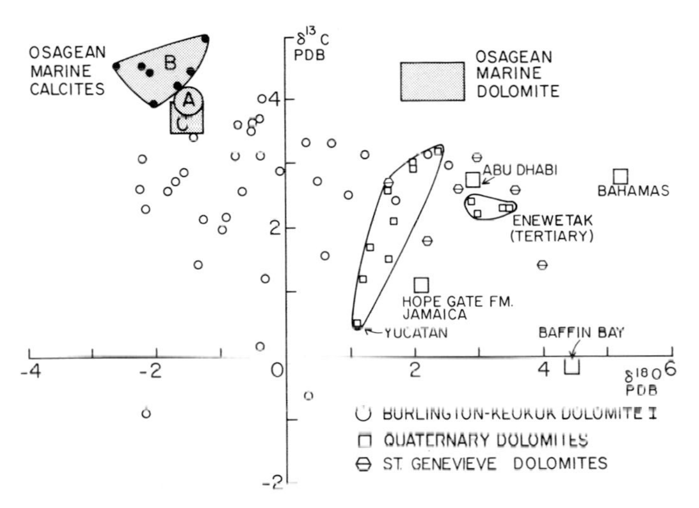

FIG. 8.—$\delta^{13}C$ vs. $\delta^{18}O$ for Burlington-Keokuk dolomite I and other ancient and Quaternary dolomites. Estimated Osagean marine calcite values are: A = Lake Valley Formation, New Mexico (Meyers and Lohmann, 1985); B = Burlington-Keokuk nonluminescent brachiopods, individual analyses shown (Kaufman, 1985); and C = Convergence of trends for Burlington-Keokuk crinoids (Chyi and others, 1985). Dolomite-calcite fractionation of 3.6 to 4.0‰ for oxygen used to estimate Osagean marine dolomite isotopic composition (see Table 2).

Mississippian dolomites from the St. Genevieve Formation (Meramecean) of Illinois are from Choquette and Steinen (1980). Mean values for Quarternary dolomite suites (large squares) are taken in part from compilation in Ward and Halley (1985) and include Baffin Bay, Texas (Behrens and Land, 1972); Little Bahama Bank (Mullins and others, 1985); Abu Dhabi, Persian Gulf (McKenzie, 1981); and the Hope Gate Formation, Jamaica (Land, 1973a). Unstippled fields enclose data for Pleistocene Yucatecan dolomites (Ward and Halley, 1985) and Tertiary to Recent dolomites from Eocene strata in Enewetak atoll (Saller, 1984a). Analyses of proposed mixing zone dolomites from Jamaica (Land, 1973b) and Israel (Magaritz and others, 1980) lie between $\delta^{13}C$ = −10 to −7 and $\delta^{18}O$ = −3 to −1 and are not shown here. Accounting for a 1‰ difference between modern and Mississippian carbonate, Burlington-Keokuk dolomite I samples have approximately 0–3‰ lower $\delta^{18}O$ values compared to most Quaternary dolomites.

TABLE 3.—Sr ISOTOPE COMPOSITIONS AND Rb, Sr, AND Ca CONCENTRATIONS (mg/l) OF SELECTED NATURAL WATERS

Sample/Locality	Rb	Sr	Ca	Sr/Ca-molar	$^{87}Sr/^{86}Sr$
1. Sea water	0.12	8.0	411	0.00885	0.7091
2a. North American Rivers	—	0.01–0.16	1.4–183	0.001–0.007	0.7077–0.71549
2b. Canadian Rivers	—	0.01–0.08	0.8–29	0.0005–0.006	0.7111
3. Illinois Basin:					
a. Mississippian	—	128–745	2,109–14,790	0.0041–0.060	0.7079–0.7104
b. Devonian	—	177–424	2,694–11,735	0.030–0.0164	0.7096–0.7101
c. Silurian	—	2–908	45–5,499	0.018–0.082	0.7091–0.7108
4. Miocene, Israel	0.168	57	1,400	0.019	0.7087
5. Canadian Shield	<1	1.7–2,060	19,000–64,000	0.001–0.024	0.711–0.740
6. Jurassic, Arkansas	—	2,930	44,200	0.0301	0.7101
7. Texas	0.15	97	1,320	0.033	—
8. Osagean-Meramecian, Kansas	0.8–0.9	40–46	1,440–1,670	0.012–0.013	0.7221–0.7230
9. Osagean, Missouri	<1	8.6	315	0.012	—

Sample Key

1. Present-day seawater from Drever (1982) and Burke and others (1982).
2a. North American rivers from Goldstein and Jacobsen (1987).
2b. Canadian rivers from Wadleigh and others (1985). Ranges for concentrations and weighted mean for isotopic composition.
3. Ranges for brines sampled from Missippian through Silurian strata, Illinois Basin (Hetherington and others, 1986 and A. M. Stueber, pers. commun., 1987).
4. Brine sampled from Mavqiim clastics, evaporites, and clastics, Upper Miocene, Israel coastal plain (Starinsky and others, 1983).
5. Range for brines sampled from Precambrian Shield (Frape and others, 1984; McNutt and others, 1984).
6. Brine sampled from Kerlin Field, Upper Jurassic Smackover Formation (Stueber and others, 1984, and Trout, 1974).
7. Brine sampled from High Island Field, Offshore Texas (Kharaka and others, 1985).
8. Ranges for brines sampled from Osagean to Meramecian carbonates, Hodgeman County, Kansas (Chaudhuri and others, 1987).
9. Ground water sampled from Burlington-Keokuk Formation, Saline County, Missouri (Carpenter and Miller, 1969).

II′ is a dolomite-to-dolomite recrystallization process, an external source of Mg^{+2} is not required. The nonstoichiometric composition of dolomite I was probably a driving mechanism in the formation of dolomite II, as calcium-rich dolomites are susceptible to recrystallization by a range of fluid compositions (Land, 1985).

Marine meteoric mixing.—

The high $^{87}Sr/^{86}Sr$ ratios of dolomites II and II′ are difficult to explain in terms of a mixing zone model. A fresh water with 1 ppm Sr and a $^{87}Sr/^{86}Sr$ ratio of 0.710 (Table 3) requires mixtures of more than 90 percent fresh water in order to exceed mixture values of 0.7088. The interaction of such a fluid with dolomite I would produce dolomites with $^{87}Sr/^{86}Sr$ ratios of less than 0.7089, the minimum dolomite II value. Fresh waters typically range to less than 0.1 ppm Sr. Using these low concentrations would lead to mixtures requiring greater than 99 percent fresh water to give $^{87}Sr/^{86}Sr$ ratios of 0.7089 in dolomites forming from the mixture.

Fresh water dolomitization.—

The end-member fresh water could produce the Sr isotopic compositions for dolomites II and II′ by recrystallization of dolomite I. Figure 9 graphically illustrates the results of model calculations simulating the recrystallization of dolomite I by various fluids, in the same manner as the results of calculations presented in Figure 6.

Owing to the small Sr concentrations in fresh waters, this recrystallization process would require approximately 12,000 moles of water for each mole of rock (Fig. 9A), or nearly 20,000 pore volumes of fresh water at 15 percent porosity to produce dolomites with the $^{87}Sr/^{86}Sr$ ratios and Sr concentrations of dolomites II and II′. Accounting for the distance over which an extraformational fluid had to have traveled to produce the regional distribution pattern for dolomites II and II′ (>360 km), and the amount of time during which these dolomites were probably generated (≤20 million years), a calculated water:rock ratio will determine a fluid flow velocity. The water:rock ratio for the freshwater, open-system model gives a flow velocity of >250 m/yr, which is more than an order of magnitude higher than flow rates in sedimentary aquifers, as determined by Back and Hanshaw (1970) and Bethke (1986). A Sr-Ca-rich brine, with the same m_{Sr}/m_{Ca} ratio as fresh water will exchange more Sr with a mineral during water-rock interaction than an equivalent amount of fresh water (Fig. 9B), thus requiring orders of magnitude less fluid (and geologically realistic flow rates) to achieve dolomite II and II′ compositions.

Normal, hypersaline, and sulphate-reduced seawater dolomitization.—

Normal seawater and sea water that has been modified by evaporation or sulphate reduction will all bear a marine $^{87}Sr/^{86}Sr$ ratio. The initial $^{87}Sr/^{86}Sr$ ratios determined for dolomites II and II′ have a maximum at 0.70942, which is higher than all estimated seawater compositions for Phanerozoic time (Burke and others, 1982). Most other chemical signatures—Fe, Mn, Sr contents, and C, O isotopes—also show significant depletion or enrichments from the estimated marine value. The failure of a seawater model to account for nearly all geochemical features of the later replacement dolomites II and II′ make most seawater models untenable.

Burial-compaction/basinal-brine dolomitization.—

Burial history curves for the Burlinton-Keokuk Formation limit maximum burial depths to less than 0.5 km (O. A. Cox, unpubl. data). If basinal brines were the dominant diagenetic agent in the formation of dolomite II, then they had to have migrated to these shallow burial depths from a deeper, allochthonous source. This is consistent with the conclusions that the intraformational shales and the Warsaw Shale were not important sources of Sr for dolomite II.

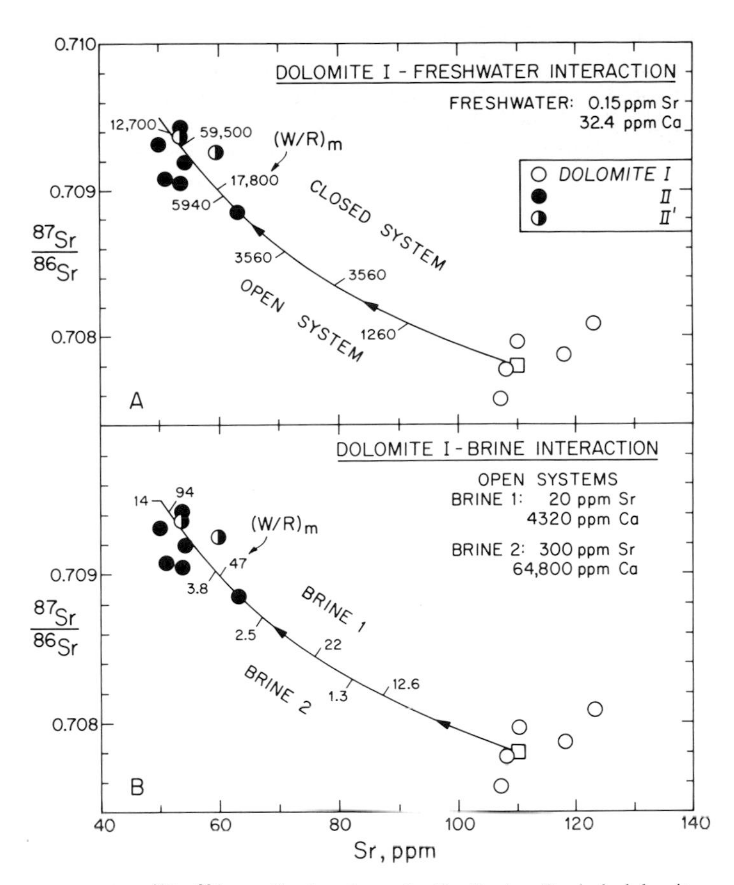

FIG. 9.—$^{87}Sr/^{86}Sr$ vs. Sr abundance for Burlington-Keokuk dolomites I, II, and II', shown as open, filled, and half-filled circles, respectively. Curves show calculated pathways of changing dolomite compositions during water-rock interaction between dolomite I and diagenetic fluids for dolomite-water $K_D(Sr/Ca) = 0.05$. Porosity = 15%. Fluid $^{87}Sr/^{86}Sr = 0.7095$ in all models. Curve shape and position is dependent on $K_D(Sr/Ca)$, fluid composition and starting rock composition. Method of model calculation presented in Figure 6. Molar water:rock ratios, $(W/R)_m$, are numbered on all curves. Square indicates starting parameters for dolomite I in all models; $^{87}Sr/^{86}Sr = 0.7078$; 110 ppm Sr. (A) Curves illustrate open- and closed-system water rock pathways for fresh water with 0.15 ppm Sr and 32 ppm Ca. (B) Water-rock curves for brine-dolomite I interaction. Brine 1 has 20 ppm Sr and 4,320 ppm Ca. Brine 2 has 300 ppm Sr and 64,800 ppm Ca. Note differences in water:rock ratios required to produce dolomite II and II' compositions between the different diagenetic fluid regimes and between open and closed systems.

Subsurface waters have a spectrum of major- and trace-element concentrations and isotopic compositions that often is distinct compared to those of sea water and fresh waters (Table 3; Hanor, 1979). Qualitatively, the marked increases in $^{87}Sr/^{86}Sr$ ratios and Fe concentrations, and the decreases in $\delta^{18}O$ values of dolomites II and II' relative to the EMD are expected for the recrystallization of carbonates by basinal fluids at elevated temperatures. The following section will examine the effects of subsurface fluids on the chemical and textural evolution of shelf carbonates and attempt to quantify these effects in the case of the Burlington-Keokuk Formation. Recognizing the wide range of salinities and sources of subsurface fluids, we use the term "brine" herein to connote a subsurface fluid that is enriched in certain cations relative to sea water (e.g., Sr, Ca) for comparative purposes in evaluating models.

Quantitative Models for Brine-Rock Interaction

Integrated studies of the geology, petrology, and geochemistry of carbonate-hosted lead-zinc ore deposits in the midcontinent region have proposed hypotheses for the origin of these Mississippi Valley-type (MVT) deposits involving the migration of warm saline brines from distant, deep basinal sources to carbonate shelves. Case studies have documented the large scale on which some fluid migration events appear to have occurred (Gregg, 1985; Leach and Rowan 1986), and various hydrologic models have been proposed (Jackson and Beales, 1967; Cathles and Smith, 1983; Garven, 1985; Bethke, 1986; Oliver, 1986). Calcite and dolomite cements associated with ore deposits (Kessen and others, 1981; Sverjensky, 1981; Gregg, 1985) and deep burial environments (Mattes and Mountjoy, 1980; Moore, 1985; Scholle and Halley, 1985) have had their origins attributed to the interaction of basinal brines with carbonate host rocks. These minerals may have preserved the isotopic signatures of the brines that crystallized them.

Strontium isotopes and Sr concentrations.—

Many gangue carbonates have significantly higher $^{87}Sr/^{86}Sr$ ratios relative to their host rocks, which have close to the estimated marine value (Fig. 10). Basinal brines are inferred to be the source of radiogenic Sr (Kesson and others, 1981; Chaudhuri and others, 1983). Late calcite and dolomite cements (vug-filling carbonates) in the Burlington-Keokuk Formation have similar high $^{87}Sr/^{86}Sr$ ratios (Fig. 10). In contrast to the results of other studies, however, Burlington-Keokuk Formation dolostone host rocks can have *both* radiogenic (dolomites II and II') and near-marine Sr isotopic compositions (dolomite I). As discussed earlier, the radiogenic Sr in the replacement dolomites was likely derived from sources external to the Burlington-Keokuk Formation. Consider that dolomite composes roughly one-third to one-half of the Burlington-Keokuk Formation over nearly the entire 100,000 km^2 study area, and that dolomites II

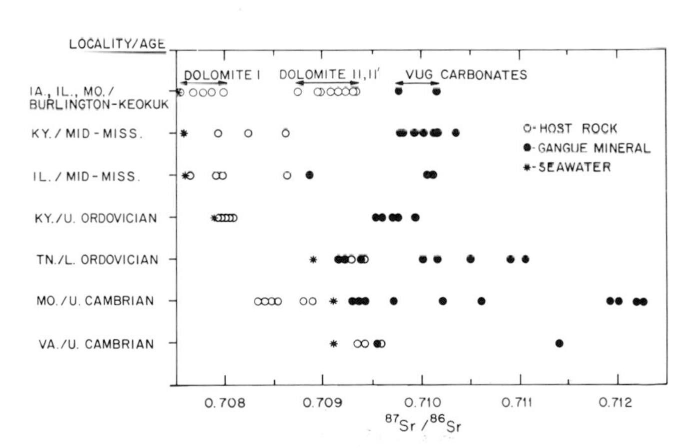

FIG. 10.—Comparison of $^{87}Sr/^{86}Sr$ ratios for gangue minerals associated with Mississippi Valley-type mineralization in host carbonate rocks of various age in the central United States and in Burlington-Keokuk carbonates. Data sources: Kessen and others (1981), Chaudhuri and others (1983), and this study. Burlington-Keokuk data adjusted to a value of $^{87}Sr/^{86}Sr = 0.71014$ for the NBS 987 Sr standard for comparative purposes. Sea water estimates for other studies are from Burke and others (1982).

and II′ compose one-third to one-half of all dolomite. If the radiogenic Sr in dolomites II and II′ was derived through the interaction of extraformational brines as they migrated through previously dolomitized (dolomite I) sediments, then the Burlington-Keokuk Formation may contain the most regionally extensive and correlative record of pervasive brine-rock interaction in carbonates yet reported.

Examination of the covariation of Sr abundances and Sr isotopic compositions for dolomites I and dolomites II and II′ shows relatively narrow and distinct ranges for the two parameters for the different dolomite generations. Pathways of water-rock interaction for various fluids interacting with dolomite I can be determined in order to model these two parameters in dolomite II. For a closed system, higher water:rock ratios are required to attain the same Sr isotopic compositions and Sr abundances relative to open-system calculations (Fig. 9A). As discussed earlier, the freshwater model gives approximately two to three orders of magnitude higher water:rock ratio values in both open and closed systems, compared to the results of the same calculations using various brine compositions (Fig. 9B).

The narrow range of Sr concentrations and isotopic compositions in dolomite II and II′ samples, with six of eight values within the range 49.9–54.2 ppm and $^{87}Sr/^{86}Sr$ = 0.7091–0.7094, for localities covering a distance of 160 km is indicative of an open system in which the diagenetic phases record the unbuffered m_{Sr}/m_{Ca} and $^{87}Sr/^{86}Sr$ ratios of the fluid. In such a system, the recrystallization has progressed to a stage in which the original Sr from dolomite I has essentially been completely removed from the system. If this advanced stage had not been reached, then the lower extents of fluid-rock interaction would have produced a wider range in Sr abundances and $^{87}Sr/^{86}Sr$ ratios, unless all samples from a 160-km regional extent have experienced the same amounts of water-rock interaction with similar fluids. The summary of data in Table 3 shows that present subsurface brines from both sedimentary sequences and granitic basement have high $^{87}Sr/^{86}Sr$ ratios relative to marine values. Brines extracted from Silurian through Devonian strata in the Illinois Basin have $^{87}Sr/^{86}Sr$ ratios of 0.7091 to 0.7108, similar to the range of values determined for dolomites II, II′, and the vug carbonates. In contrast, brines extracted from Keokuk and Warsaw strata in western Kansas have quite radiogenic and uniform isotopic compositions ($^{87}Sr/^{86}Sr$ = 0.7221–0.7230). It was suggested earlier that Sr isotopic compositions in dolomites II and II′ are the result of crystallization in an open system from an end-member fluid composition, making brines with isotopic compositions similar to those from the Illinois Basin more appropriate for the formation of dolomites II and II′.

In order to crystallize dolomite with 50 ppm Sr for a dolomite-fluid exchange $K_D(Sr/Ca)$ = 0.05, a fluid with a molar Sr/Ca ratio of less than 0.005 is required (Fig. 9). Nearly all of the subsurface fluids listed in Table 3, including some sampled from the Burlington-Keokuk Formation, have molar Sr/Ca ratios that are considerably higher than 0.005. To invoke such fluids in the formation of dolomites II and II′ would require a K_D of <0.01, which is considerably lower than any published values. Similar discrepancies in other diagenetic carbonate systems have been attributed to differences between experimental and natural systems (Bein and Land, 1983; Moore, 1985). Such comparisons involving the trace-element and isotopic compositions of present subsurface fluids are necessarily limited by the fact that the fluids available at the time of recrystallization of the dolomites may have had significantly different compositions.

In summary, Sr isotope and concentration data for Burlington-Keokuk dolomites II and II′ indicate an extraformational, older source of Sr for the dolomites. Dolomite I recrystallization was likely effected by brines that migrated to shallow burial depths on a scale of nearly 100,000 km^2, probably prior to Pennsylvanian time. Subsurface Keokuk and Warsaw dolomites from western Kansas have $^{87}Sr/^{87}Sr$ ratios of 0.7091–0.7095 (data from Chaudhuri and others, 1987; adjusted to SRM 987 = 0.71034). If the stratigraphic and isotopic similarities between the dolomites from Kansas and the dolomites from Iowa, Illinois, and Missouri are the result of related fluid migration events, then this would increase by a factor of two the large scale on which water-rock interaction appears to have occurred.

Oxygen isotopes.—

The depleted $\delta^{18}O$ values for dolomites II and II′ (−6.6 to −0.2‰) relative to dolomite I (−2.2 to 2.5‰) are likely due to a combination of the influences of: (1) elevated temperatures; (2) the $\delta^{18}O$ value of the replacement fluid; and

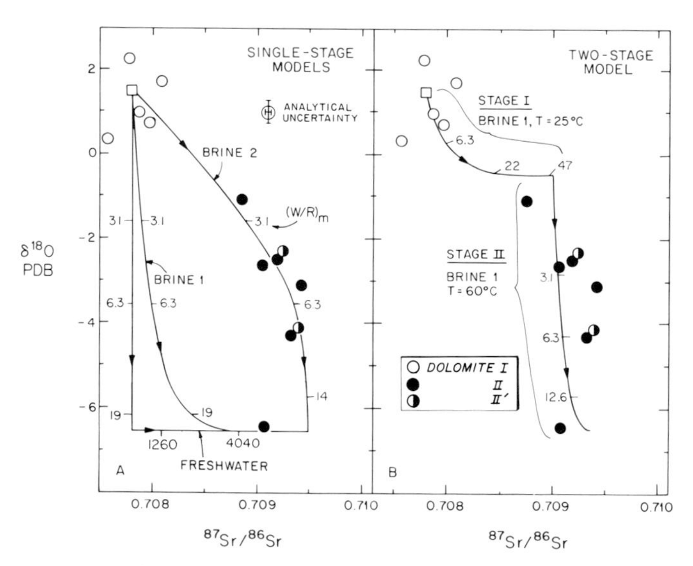

FIG. 11.—$^{87}Sr/^{86}Sr$ vs. $\delta^{18}O$ for Burlington-Keokuk dolomites I, II, and II′, illustrating narrow range of Sr isotopic compositions relative to oxygen-isotopic compositions for second generation dolomites. Calculated curves illustrate changing dolomite compositions during progressive water-rock interaction. (A) Single-stage water-rock interaction models shown are calculated for the recrystallization of dolomite I by three different fluids: fresh water, brine 1 and brine 2. (B) A two-stage model is shown for the recrystallization of dolomite I by brine 1. A single-stage model involving a saline brine and a multistage model involving increasing temperature and a dilute brine of constant composition can account for the dolomite II and II′ data. All models are for open systems. Fluid compositions, porosities, and K_D's as in Figure 9. All fluids have $\delta^{18}O$ = −2‰ SMOW. T = 60°C, except where noted otherwise.

(3) the extent of water-rock interaction. The northwest to southeast gradient of decreasing $\delta^{18}O$ values of dolomites II and II′ (Fig. 4) can be ascribed to changes in temperature, fluid composition, or extent of water-rock interaction on a regional scale.

Interpretation of the oxygen isotope data is least equivocal when examined in conjunction with other geochemical data. The covariation of $\delta^{18}O$ and $^{87}Sr/^{86}Sr$ for dolomites II and II′ can be modeled in the same manner as presented earlier. Figure 11 illustrates that $^{87}Sr/^{86}Sr$ ratios are relatively invariant compared to the range of $\delta^{18}O$ values for dolomites II and II′. Several possible evolutionary pathways involving fluid-rock interaction and fluid-fluid mixing can be calculated to give a reasonable fit to the data. These models are:

(1) Fluid-fluid mixtures of <80% sea water and >20% Sr-Ca-rich brine will produce mixing trends with a range of $\delta^{18}O$ values and a narrow range of $^{87}Sr/^{86}Sr$ ratios, similar in shape to the brine 2 curve in Figure 11A (see also Fig. 3–7 in Banner, 1986). Mixtures of brines with different oxygen isotope compositions and the same $^{87}Sr/^{86}Sr$ ratio will produce the same trend. Separate water-rock interaction curves could produce the dolomite II and II′ trend using the range of these fluid-fluid mixtures as end-member fluids and dolomite I as the initial rock composition. This model requires that the water-rock interaction process has proceeded to the same extent for each sample and that the mixtures have maintained their integrity for the duration of the process.

(2) Using a saline fluid with high Sr and Ca contents in the recrystallization calculation (brine 2, Fig. 11A) would produce a water-rock interaction curve that would change the $^{87}Sr/^{86}Sr$ ratio of the recrystallized product faster than it would change $\delta^{18}O$ during progressive water-rock interaction. This model produces a range of $\delta^{18}O$ and constant $^{87}Sr/^{86}Sr$ at low (<20 molar) water:rock ratios. Note that a minimum fluid salinity is required in order for this model to account for the dolomite data.

(3) Several stages of water-rock interaction involving a brine with moderate Sr and Ca contents (brine 1, Fig. 11B) can also explain the dolomite II and II′ trend. A water-rock pathway for such a brine recrystallizing dolomite I will be nearly L-shaped (brine 1 pathway in Fig. 11A). Extensive water-rock interaction will drive the replacement dolomites first to low $\delta^{18}O$ at low water:rock ratios (beginning of stage I in Fig. 11B) and then to higher and fairly uniform $^{87}Sr/^{86}Sr$ ratios at higher water:rock ratios (end of stage I in Fig. 11B). If there was a change in the temperature or the $\delta^{18}O$ value of the fluid during the last stages of recrystallization (stage II in Fig. 11B), then a range of $\delta^{18}O$ values could have been imposed on dolomites II and II′ through small extents (water:rock ratio <20) of water-rock interaction. Due to the low water:rock ratio, $^{87}Sr/^{86}Sr$ ratios in the recrystallized products will change only slightly during this stage. As a consequence, a new vertical limb of an L-curve (stage II) would be extended from the end of the horizontal limb of the previous L-curve (stage I). In this manner, the regional gradient in $\delta^{18}O$ for dolomites II and II′ could be accounted for by progressively increasing water-rock interaction from northwest to southeast during the last stages of recrystallization. In this multistage model, it can be seen that the $\delta^{18}O$ values of the dolomites record a relatively small and late segment in the water-rock interaction history of the dolomites, whereas dolomite Sr isotope signatures reflect an earlier and larger segment of the same history.

Carbon isotopes.—

In contrast to the oxygen isotope results for the Burlington-Keokuk dolomites, carbon-isotopic compositions of 28 samples of dolomites II and II′ have nearly the identical mean and range of values as observed for 29 samples of dolomite I (Fig. 5). For the few localities for which isotope analyses are available for samples of dolomite I, II, and II′, there are small but distinguishable differences between the average $\delta^{13}C$ values of each locality (Fig. 12). Dolomite I, II, and II′ samples from the same localities have similar ranges of values. Only the vug carbonates show significantly different $\delta^{13}C$ compositions. This suggests that dolomites II and II′ inherited their carbon-isotopic signatures from dolomite I and that the recrystallization process did not affect these values.

Quantitative modeling of the change of $\delta^{13}C$ values as a function of progressively increasing water:rock interaction during recrystallization supports these general observations. During the recrystallization of dolomite, oxygen in the dolomite will equilibrate with oxygen in the fluid at much lower water:rock ratios compared to the values required for carbon equilibration (Fig. 12). This produces an inverted L-shaped curve in $\delta^{13}C$ vs. $\delta^{18}O$ space. For a fluid that has total dissolved carbon (as bicarbonate) with $\delta^{13}C = -12‰$ and concentrations of 300 mg/l, a molar water:rock ratio value of approximately 3,000 is needed to establish a 2‰ change in the isotopic composition of the rock. Thus, because $\delta^{13}C$ values are essentially unchanged between dolomites I, II, and II′, the recrystallization process probably involved less than 3,000 moles of fluid for each mole of rock. The water-rock calculations for Sr contents and Sr isotopes predicted that replacement dolomitization via brine 1 required on the order of 50 to 100 moles of fluid for each mole of rock reacting. The independent constraints of the two isotopic systems are consistent and indicate geologically reasonable fluid flow velocities as discussed earlier.

An alternative explanation for the distribution of carbon isotopes in the Burlington-Keokuk dolomites is that the fluid that recrystallized dolomite I had dissolved bicarbonate with an isotopic composition similar to the $\delta^{13}C$ values of dolomite I (via interaction with older marine carbonates), and therefore any exchange that may have taken place during the formation of dolomite II would not be detectable. This hypothesis is less tenable than the model of retention of original dolomite I carbon in dolomite II because: (1) small but distinct differences in $\delta^{13}C$ are retained at several localities for dolomites I, II, and II′ (Fig. 12), as well as for crinoids (Chyi and others, 1985), and (2) as discussed in the final section, the low $\delta^{13}C$ and $\delta^{18}O$ values and high $^{87}Sr/^{86}Sr$ ratios of the vug carbonates indicate that extraformational fluids with light $\delta^{13}C$ migrated to the Burlington-Keokuk Formation without being buffered by carbon from older marine carbonates.

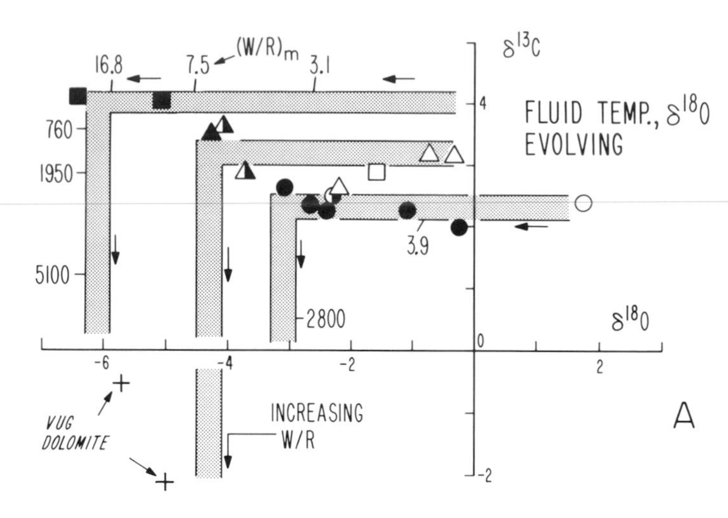

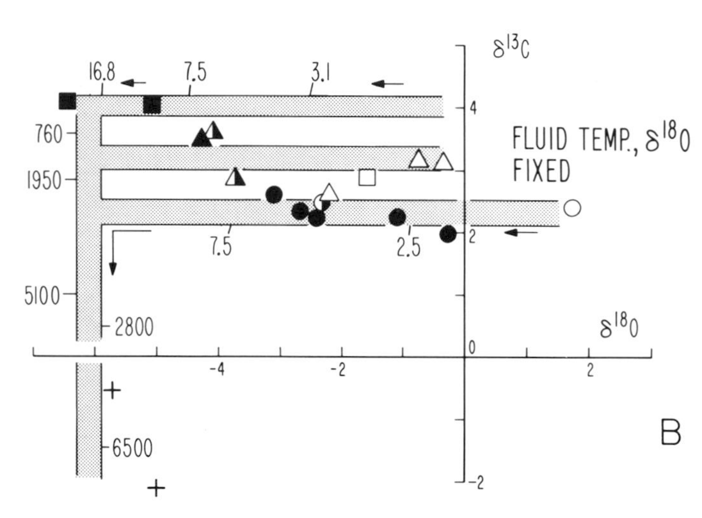

FIG. 12.—$\delta^{13}C$ vs. $\delta^{18}O$ (PDB) for localities where dolomites I, II, and II′ have been analyzed. Localities as follows: circles—HP; squares—MPA; triangles—DC. Dolomites I, II, and II′ are open, filled, and half-filled symbols respectively. Vug dolomites are from locality GR. Shaded trends are calculated pathways of water-rock interaction between brine and dolomite. Arrows indicate direction of increasing water-rock interaction. All models are for open systems. Three separate starting dolomite I compositions are used in both (A) and (B): (1) $\delta^{18}O = -0.5$, $\delta^{13}C = 4$; (2) $\delta^{18}O = -0.5$, $\delta^{13}C = 3.25$; (3) $\delta^{18}O = 1.75$, $\delta^{13}C = 2.3$ (all in ‰). The fluid has $\delta^{13}C = -12$ and 300 ppm HCO_3^- and the rock has 15% porosity in all models. (A) Water-rock pathways evolve to distinct dolomite $\delta^{18}O$ values. The curve extending from the dolomite I composition at $\delta^{18}O = -0.5$, $\delta^{13}C = 4$ (MPA locality) is a pathway calculated using a fluid $\delta^{18}O$ value of −1.6‰ (SMOW) and a temperature of 60°C. Differences in second generation dolomite $\delta^{18}O$ values between localities can be accounted for by either a progressive increase in fluid $\delta^{18}O$ values of 3‰ or a progressive 20°C decrease in temperature between the MPA and HP localities. (B) A single-fluid isotopic composition and temperature will produce similar isotopic variations in second generation dolomite. Differences between localites are a function of local variations in extents of water-rock interaction. Fluid $\delta^{18}O = -1.6$‰ (SMOW), T = 60°C at all localities. The similar $\delta^{13}C$ values among different dolomite types at individual localities suggest that dolomite I recrystallization occurred at moderate $(W/R)_m$ values of ≤3,000.

Rare-earth elements.—

The similarities in REE patterns and Nd isotopic compositions between dolomites I, II, and II′ are consistent with the invariance in $\delta^{13}C$ values between the dolomites, because even higher water:rock ratios would be required to change the REE distributions during the recrystallization of dolomite I compared to the results for carbon, as shown in Figure 6.

Final stages of brine-rock interaction.—

Although only limited analyses have been performed on the late-stage carbonates from solution vugs in Burlington-Keokuk strata, quantitative modeling of their isotope geochemistry suggests that they crystallized in a fluid-dominated environment with only a minor influence from the host rock. The vug carbonates have elevated $^{87}Sr/^{86}Sr$ ratios of 0.7099–0.7102, suggesting that they formed from similar but distinct fluids compared to those that formed dolomites II and II′. The vug dolomites have depleted $\delta^{18}O$ (−5.7 to −5.0‰) and $\delta^{13}C$ values (−2.1 to −0.6‰) relative to most dolomite II and II′ samples, and lie on a vertical limb of an inverted L-shaped brine-dolomite I interaction curve in $\delta^{13}C$ vs. $\delta^{18}O$ space (Fig. 12). The calculated alteration pathway transects the dolomite II and II′ data at low water:rock ratios and extends through the vug dolomites at high water:rock ratios.

Basinal hydrocarbon sources of light $\delta^{13}C$ have been proposed in many cases of MVT gangue and burial carbonate cements. As the calculations in Figure 12 demonstrate, significant depletions in carbon-isotopic signatures that are attributable to water-rock interaction imply relatively large water:rock ratios during crystallization. The solution vugs developed late in the diagenetic history of the Burlington-Keokuk Formation, after most cementation and recrystallization occurred and thus at a time when the rocks were least porous. Such high water:rock ratios do not require especially large volumes of fluid. Rather, they may reflect vanishingly small amounts of host rock taking part in the interaction. As observed for the C isotopic data, the REE patterns and ϵ_{Nd} value for the vug carbonates (Banner, 1986) are distinct relative to the overlapping values in dolomites I, II, and II′, consistent with higher water:rock ratios during vug carbonate formation.

Whereas the vug carbonates may preserve the C and REE signatures of an extraformational fluid, dolomites II and II′ that have replaced pre-existing sediments appear to record the C and REE signatures of the precursor dolomite I. If the majority of constituents for the vug calcites and dolomites is fluid-derived with little host rock influence, then these fluids would also have been capable of dolomitization. In contrast, the source of Mg for dolomites II and II′ could have been predominantly autochthonous.

CONCLUSIONS

(1) Two major episodes of regionally extensive (100,000 km^2) dolomitization can be correlated in the Burlington-Keokuk Formation of Iowa, Illinois, and Missouri. Dolomite I and its recrystallized product, dolomite II, have dis-

tinctive isotopic compositions and trace- and major-element concentrations.

(2) The results of model calculations that simulate isotopic exchange during water-rock interaction illustrate important differences in the relative rates at which different isotopic systems are altered during carbonate diagenesis. During dolomite recrystallization, dolomite $\delta^{18}O$ values are reset at relatively low water:rock ratios, whereas several orders of magnitude larger extents of water-rock interaction are required to alter dolomite $\delta^{13}C$ values. Strontium-isotopic compositions are changed at variable extents of water-rock interaction, depending on the Sr isotopic compositions and Sr and Ca concentrations of the dolomite and fluid.

(3) Burlington-Keokuk dolomite I samples have O, C, and Sr isotopic compositions that encompass estimated marine dolomite values and range to slightly lower $\delta^{18}O$ and $\delta^{13}C$ values and slightly higher $^{87}Sr/^{86}Sr$ ratios. These results can be explained by a marine meteoric-mixing model in which the diagenetic constituents that comprise dolomite I are predominantly marine derived.

(4) Dolomite II samples preserve a petrographic and geochemical record of the recrystallization of the nonstoichiometric dolomite I. This process imparted lower $\delta^{18}O$ values and Sr concentrations and higher $^{87}Sr/^{86}Sr$ ratios and Fe concentrations to the recrystallized dolomites. It appears that extraformational subsurface fluids migrated into Burlington-Keokuk strata at shallow burial depths during dolomite II formation. The results of quantitative modeling suggest that these fluids exchanged nearly all of the original Sr in the dolomites during recrystallization, and that the Sr signature of dolomite II samples reflects a large segment of their water-rock interaction history. In contrast, the $\delta^{18}O$ values of the dolomites may have been reset during the last stages of recrystallization, whereas their C and Nd isotopic signatures were probably inherited from dolomite I precursors. Dolomitization in the Burlington-Keokuk Formation occurred via water-rock interaction processes that can be correlated on a regionally extensive scale.

ACKNOWLEDGMENTS

The senior author is grateful to K. C. Lohmann, J. M. Budai, and D. Dettman for the use of and technical assistance with the mass spectrometer facilities at the University of Michigan, where carbon and oxygen isotope analyses were carried out. H. S. Cander and J. Kaufman carefully sampled the dolomites from Illinois and Missouri for isotopic analysis. We acknowledge comments by K. R. Cercone, R. J. Reeder, and D. A. Sverjensky, discussions with H. S. Cander, J. Kaufman, and J. Prosky, and reviews by E. Busenburg, P. I. Nabelek, and V. Shukla. This research was supported by grants from the Petroleum Research Fund of the American Chemical Society (PRF-14913AC2) and the Department of Energy (DE-AC02-83ER13112).

REFERENCES

Back, William, and Hanshaw, B. B., 1970, Comparison of chemical hydrogeology of the carbonate peninsulas of Florida and the Yucatan: Journal of Hydrology, v. 10, p. 330–368.

Badiozamani, K., 1973, The Dorag dolomitization model—application to the Middle Ordovician of Wisconsin: Journal of Sedimentary Petrology, v. 43, p. 965–984.

Baker, P. A., and Burns, S. J., 1985, Occurrence and formation of dolomite in organic-rich continental margin sediments: American Association of Petroleum Geologists Bulletin, v. 69, p. 1917–1930.

Banner, J. L., 1986, Petrologic and geochemical constraints on the origin of regionally extensive dolomites of the Mississippian Burlington-Keokuk Formation, Iowa, Illinois and Missouri: Unpublished Ph.D. Dissertation, State Univerity of New York, Stony Brook, New York, 368 p.

Behrens, E. W., and Land, L. S., 1972, Subtidal holocene dolomite, Baffin Bay, Texas: Journal of Sedimentary Petrology, v. 42, p. 155–161.

Bein, A., and Land, L. S., 1983, Carbonate sedimentation and diagenesis associated with Mg-Ca-chloride brines: The Permian San Andres Formation in the Texas Panhandle: Journal of Sedimentary Petrology, v. 53, p. 243–260.

Bethke, C. M., 1986, Hydrologic constraints on genesis of the Upper Mississippi Valley mineral district from Illinois Basin brines: Economic Geology, v. 81, p. 233–249.

Burke, W. H., Denison, R. E. Hetherington, E. A., Koepnick, R. B., Nelson, H. F., and Otto, J. B. 1982, Variation of seawater $^{87}Sr/^{86}Sr$ throughout Phanerozoic time: Geology, v. 10, p. 516–519.

Cander, H. S., 1985, Petrology and diagenesis of the Burlington-Keokuk Limestone, Illinois: Unpublished M.S. Thesis, State University of New York, Stony Brook, New York, 403 p.

Carballo, J. D., Land, L. S., and Miser, D. E., 1987, Holocene dolomitization of supratidal sediments by active tidal pumping, Sugarloaf Key, Florida: Journal of Sedimentary Petrology, v. 57, p. 153–165.

Carlson, M. P., 1979, The Nebraska-Iowa region, *in* Craig, L. C., and Varnes, K. L., eds., Paleotectonic Investigations of the Mississippian System: U.S. Geological Survey Professional Paper 1010-F, p. 107–114.

Carpenter, A. B., and Miller, J. C., 1969, Geochemistry of saline subsurface water, Saline County (Missouri): Chemical Geology, v. 4, p. 135–167.

Cathles, L. M., and Smith, A. T., 1983, Thermal constraints on the formation of Mississippi Valley-type lead-zinc deposits and their implications for episodic basin dewatering and deposit genesis: Economic Geology, v. 78, p. 983–1002.

Chaudhuri, S., Broedel, V., and Clauer, N., 1987, Strontium isotopic evolution of oil-field waters from carbonate reservoir rocks in Bindley Field, central Kansas, USA: Geochimica et Cosmochimica Acta, v. 51, p. 45–53.

———, Clauer, N., and Ramakrishnan, S., 1983, Strontium isotopic composition of gangue carbonate minerals in the lead-zinc sulfide deposits at the Brushy Creek Mine, Viburnum Trend, southeast Missouri, *in* Kisvarsanji, G., Grant, S. K., Pratt, W. P., and Koenig, J. W., eds., Proceedings, International Conference on Mississippi Valley-Type Lead-Zinc Deposits, University of Missouri, Rolla, Missouri, p. 140–144.

Choquette, P. W., and Steinen, R. P., 1980, Mississippian non-supratidal dolomite, Ste. Genevieve Limestone, Illinois Basin: Evidence for mixed-water dolomitization, *in* Zenger, D. H., Dunham, J. B., and Ethington, R. L., eds., Concepts and Models of Dolomitization: Society of Economic Paleontologists and Mineralogists Special Publication 28, p. 163–196.

Chyi, M. S., Hanson, G. N., and Meyers, W. J., 1985, Isotope geochemistry of crinoids from the Burlington-Keokuk Formation: implications for diagenesis: Geological Society of America, Abstracts with Programs, v. 17, p. 547.

Comings, B. D., and Cercone, K. R., 1986, Experimental contamination of fluid inclusions in calcite: Society of Economic Paleontologists and Mineralogists Annual Midyear Meeting, Raleigh, North Carolina, Abstracts, v. 3, p. 24.

Craig, H., 1957, Isotopic standards for carbon and oxygen and correction factors for mass-spectrometric analysis of carbon dioxide: Geochimica et Cosmochimica Acta, v. 12, p. 133–149.

DePaolo, D. J., and Wasserburg, G. J., 1976, Nd isotopic variations and petrogenetic models: Geophyical Research Letters, v. 3, p. 249–252.

DREVER, J. I., 1982, The Geochemistry of Natural Waters: Prentice Hall, Englewood Cliffs, New Jersey, 388 p.

FRAPE, S. K., FRITZ, P., AND MCNUTT, R. M., 1984, The role of water-rock interaction in the chemical evolution of groundwaters from the Canadian Shield: Geochimica et Cosmochimica Acta, v. 48, p. 1617–1627.

FRIEDMAN, IRVING, AND O'NEIL, J. R., 1977, Compilation of stable isotope fractionation factors of geochemical interest, *in* Fleischer, M., ed., Data of Geochemistry, 6th edition, U.S. Geological Survey Professional Paper 440-KK, 12 p.

GARVEN, GRANT, 1985, The role of regional fluid flow in the genesis of the Pine Point deposit, Western Canada Sedimentary Basin: Economic Geology, v. 80, p. 307–324.

GOLDSTEIN, R. M., 1986, Reequilibration of fluid inclusions in low temperature calcium carbonate cement: Geology, v. 14, p. 792–795.

GOLDSTEIN, S. J., AND JACOBSEN, S. B., 1987, The Nd and Sr isotopic systematics of riverwater dissolved material: Implication for the sources of Nd and Sr in seawater: Chemical Geology (Isotope Geoscience Section) v. 66, p. 245–272.

GREGG, J. M., 1985, Regional epigenetic dolomitization in the Bonneterre Dolomite (Cambrian), southeastern Missouri: Geology, v. 13, p. 503–506.

HANOR, J. S., 1979, The sedimentary genesis of hydrothermal fluids, *in* Barnes, H. L. ed., Geochemistry of Hydrothermal Ore Deposits, 2nd edition, Wiley and Sons, New York, p. 137–172.

HARDIE, L. A., 1987, Dolomitization: A critical view of some current views: Journal of Sedimentary Petrology, v. 57, p. 166–183.

HARLAND, W. D., COX, A. V., LLEWELLYN, P. G., PICKTON, C. A. G., SMITH, A. G., AND WALTERS, R., 1982, A Geologic Time Scale: Cambridge University Press, Cambridge, England, 131 p.

HARRIS, D. C., 1982, Carbonate cement stratigraphy and diagenesis of Burlington Limestones (Mississippian), southwestern Iowa and western Illinois: Unpublished M.S. Thesis, State University of New York, Stony Brook, New York, 297 p.

HETHERINGTON, E. A., STUEBER, A. M., AND PUSHKAR, PAUL, 1986, Strontium isotopic study of subsurface brines from Illinois Basin (Abs.): American Association of Petroleum Geologists Bulletin, v. 70, p. 600.

JACKSON, S. A., AND BEALES, F. W., 1967, An aspect of sedimentary basin evolution: The concentration of Mississippi Valley-type ores during the late stages of diagenesis: Bulletin of Canadian Petroleum Geologists, v. 15, p. 393–433.

KAUFMAN, J., 1985, Diagenesis of the Burlington-Keokuk Limestones (Miss.), Eastern Missouri: Unpublished M.S. Thesis, State University of New York, Story Brook, New York, 326 p.

KESSON, K. M., WODDRUFF, M. S., AND GRANT, N. K., 1981, Gangue mineral $^{87}Sr/^{86}Sr$ ratios and the origin of Mississippi Valley-type mineralization: Economic Geology, v. 76, p. 913–920.

KHARAKA, Y. F., HULL, R. W., AND CAROTHERS, W. W., 1985, Water-rock interactions in sedimentary basins, *in* Gautier, D. L., ed., Relationship of Organic Matter and Mineral Diagenesis: Society of Economic Paleontologists and Mineralogists Short Course, No. 17, p. 79–174.

KOHOUT, F. A., HENRY, H. R., AND BANKS, J. E., 1977, Hydrogeology related to geothermal conditions of the Floridan Plateau, *in* Smith, D. L., and Griffin, G. M., eds., The geothermal natural of the Floridan Plateau: Florida Bureau of Geology, Special Publication 21, p. 1–41.

LAND, L. S., 1973a, Holocene meteoric dolomitization of Pleistocene limestones, North Jamaica: Sedimentology, v. 20, p. 411–424.

———, 1973b, Contemporaneous dolomitization of Middle Pleistocene reefs by meteoric water, North Jamaica: Bulletin of Marine Science, v. 23, p. 64–92.

———, 1980, The isotopic and trace element geochemistry of dolomite: The state of the art, *in* Zenger, D. H., Dunham, J. B., and Ethington, R. L., eds., Concepts and Models of Dolomitization: Society of Economic Paleontologists and Mineralogists Special Publication 28, p. 87–110.

———, 1985, The origin of massive dolomite: Journal of Geological Education, v. 33, p. 112–125.

LANE, H. P., AND DEKEYSER, T. L., 1980, Paleogeography of the late Early Mississippian (Tournaisian 3) in the central and southwestern United States, *in* Fouch, T. D., and Magathan, E. R., eds., Paleozoic Paleography of West-Central United States: West-Central United States Paleogeography Symposium I: Rocky Mountain Section, Society of Economic Paleontologists and Mineralogists, Denver, Colorado, p. 149–162.

LEACH, D. L., AND ROWAN, E. L., 1986, Genetic link between Ouachita foldbelt tectonism and the Mississippi Valley type lead-zinc deposits of the Ozarks: Geology, v. 14, p. 931–935.

LOHMANN, K. C., 1983, Unravelling the diagenetic history of carbonate reservoirs: Integration of petrographic and geochemical techniques, *in* New Ideas and Methods for Exploration for Carbonate Reservoirs: Dallas Geological Society Short Course, Section 5, p. 1–41.

MACHEL, H.-G., AND MOUNTJOY, E. W., 1986, Chemistry and environments of dolomitization: a reappraisal: Earth Science Reviews, v. 23, p. 175–222.

MAGARITZ, M., GOLDENBERG, L., KAGRI, U., AND ARED, A., 1980, Dolomite formation in the seawater-freshwater interface: Nature, v. 287, p. 622–624.

MATTES, B. W., AND MOUNTJOY, E. W., 1980, Burial dolomitization of the Upper Devonian Miette buildup, Jasper National Park, Alberta, *in* Zenger, D. H., Dunham, J. B., and Ethington, R. L., eds., Concepts and Models of Dolomitization: Society of Economic Paleontologists and Mineralogists Special Publication 28, p. 259–320.

MCKENZIE, JUDITH, 1981, Holocene dolomitization of calcium carbonate sediments from the coastal sabkhas of Abu Dhabi, U.A.E.: A stable isotope study: Journal of Geology, v. 89, p. 185–198.

MCNUTT, R. H., FRAPE, S. K., AND FRITZ, P., 1984, Strontium isotopic composition of some brines from the Precambrian Shield of Canada: Isotope Geoscience, v. 2, p. 205–215.

MEYERS, W. J., AND LOHMANN, K. C., 1985, Isotope geochemistry of regionally extensive calcite cement zones and marine components in Mississippian limestones, New Mexico, *in* Schneiderman, N., and Harris, P. M., eds., Carbonate Cements: Society of Economic Paleontologists and Mineralogists Special Publication 36, p. 223–240.

MOORE, C. H., 1985, Upper Jurassic subsurface cements: A case history, *in* Schneiderman, N., and Harris, P. M., eds., Carbonate Cements: Society of Economic Paleontologists and Mineralogists Special Publication 36, p. 291–308.

MORROW, D. W., 1982, Diagenesis 2. Dolomite—Part 2: Dolomitization models and ancient dolostones: Geoscience Canada, v. 9, p. 95–106.

MULLINS, H. T., LAND, L. S., WISE, S. W., JR., SIEGEL, D. I., MASTERS, P. M., HINCHEY, E. J., AND PRICE, K. R., 1985, Authigenic dolomite in Bahamian slope sediment: Geology, v. 13, p. 292–295.

OLIVER, J., 1986, Fluids expelled tectonically from orogenic belts: Their role in hydrocarbon migration and other geologic phenomena: Geology, v. 14, p. 99–102.

PALMER, M. R., 1985, Rare earth elements in foraminifera tests: Earth and Planetary Science Letters, v. 73, p. 285–298.

POPP, B. N., PODOSEK, F. A., BRANNON, J. C., ANDERSON, T. F., AND PIER, J., 1986, $^{87}Sr/^{86}Sr$ ratios in Permo-Carboniferous seawater from the analyses of well-preserved brachiopod shells: Geochimica et Cosmochimica Acta, v. 50, p. 1321–1328.

PROSKY, J. L., AND MEYERS, W. J., 1985, Nonstoichiometry and trace element geochemistry of the Burlington-Keokuk dolomites: Society of Economic Paleontologists and Mineralogists, Annual Midyear Meeting, Golden, Colorado, Abstracts, v. 2, p. 73.

REEDER, R. J., AND PROSKY, J. L., 1986, Compositional sector zoning in dolomite: Journal of Sedimentary Petrology, v. 56, p. 237–247.

SALLER, A. H., 1984a, Petrologic and geochemical constraints on the origin of subsurface dolomite: An example of dolomitization by normal seawater, Enewetak Atoll: Geology, v. 12, p. 217–220.

———, 1984b, Diagenesis of Cenozoic Limestones on Enewetak Atoll: Unpublished Ph.D. Dissertation, Louisiana State University, Baton Rouge, Louisiana, 363 p.

SCHOLLE, P. A., AND HALLEY, R. B., 1985, Burial diagenesis: Out of sight, out of mind! *in* Schneidermann, N., and Harris, P. M., eds., Carbonate Cements: Society of Economic Paleontologist and Mineralogists Special Publication 36, p. 309–333.

SHAW, H. F., AND WASSERBURG, G. J., 1985, Sm-Nd in marine carbonates and phosphates: Implications for Nd isotopes in seawater and crustal ages: Geochimica et Cosmochimica Acta, v. 49, p. 503–518.

SIMMS, M., 1984, Dolomitizaton by groundwater-flow systems in carbonate platforms: Gulf Coast Association of Geological Societies, Transactions, v. 34, p. 411–420.

SIXT, S. C. S., 1983, Depositional environments, diagenesis and stratigraphy of the Gilmore City Formation (Mississippian) near Humboldt,

north-central Iowa: Unpublished M.S. Thesis, University of Iowa, Iowa City, Iowa, 164 p.
SMITH, F., 1984, A fluid inclusion study of the dolomite-calcite transition in the Burlington-Keokuk Limestones (Mid-Miss.), S.E. Iowa, W. Illinois: Unpublished M.S. Thesis, State University of New York, Stony Brook, New York, 201 p.
STARINSKY, A., BIELSKI, M., LAZAR, B., STEINITZ, G., AND RAAB, M., 1983, Strontium isotope evidence on the history of oilfield brines, Mediterranean Coastal Plain, Israel: Geochimica et Cosmochimica Acta, v. 47, p. 687–695.
STUEBER, A. M., PUSHKAR, P., AND HETHERINGTON, E. A., 1984, A strontium isotopic study of Smackover brines and associated solids, southern Arkansas: Geochimica et Cosmochimica Acta, v. 48, p. 1637–1649.
SVERJENSKY, D. A., 1981, The origin of a Mississippi Valley-type deposit in the Viburnum Trend, southeast Missouri: Economic Geology, v. 76, p. 1848–1872.
TAYLOR, H. P., 1979, Oxygen and hydrogen isotope relationships in hydrothermal mineral deposits, *in* Barnes, H. L., ed., Geochemistry of Hydrothermal Ore Deposits, 2nd edition, Wiley and Sons, New York, p. 236–277.
TROUT, M. L., 1974, Origin of bromide-rich brines in southern Arkansas: Unpublished M.A. Thesis, University of Missouri, Columbia, Missouri, 79 p.
VEIZER, J., 1983, Chemical diagenesis of carbonates: Theory and application of trace element technique: Society of Economic Paleontologists and Mineralogists Short Course No. 10, p. 3-1 to 3-100.
———, AND HOEFS, J., 1976, The nature of $^{18}O/^{16}O$ and $^{13}C/^{12}C$ secular trends in sedimentary carbonate rocks: Geochimica et Cosmochimica Acta, v. 40, p. 1387–1395.
WADLEIGH, M. A., VEIZER, J., AND BROOKS, C., 1985, Strontium and its isotopes in Canadian rivers: Fluxes and global implications: Geochimica et Cosmochimica Acta, v. 49, p. 1727–1736.
WARD, W. C., AND HALLEY, R. B., 1985, Dolomitization in a mixing zone of near-seawater composition, Late Pleistocene, northeastern Yucatan Peninsula: Journal of Sedimentary Petrology, v. 55, p. 407–420.
WIGLEY, T. M. L., AND PLUMMER, N. L., 1976, Mixing of carbonate waters: Geochimica et Cosmochimica Acta, v. 40, p. 989–995.

FLUID FLOW DIRECTION DURING DOLOMITE FORMATION AS DEDUCED FROM TRACE-ELEMENT TRENDS

HANS G. MACHEL

Department of Geology, University of Alberta, Edmonton, Alberta T6G 2E3

ABSTRACT: A qualitative mathematical model applying a variant of the Heterogeneous Distribution Law indicates that elements with distribution coefficients smaller than one (e.g., Sr) should increase downflow if (1) dolomite is formed exclusively or predominantly as a cement, and if (2) dolomite replaces calcium carbonate and the dolomitizing fluid has a molar Sr/Ca ratio that is equal to or lower than that of the calcium carbonate. If the dolomitizing fluid has a molar Sr/Ca ratio greater than that of the calcium carbonate precursor, Sr should decrease downflow. Trace elements with distribution coefficients greater than one (e.g., Mn) should decrease in the downflow direction for dolomite cementation. In the case of dolomitization of calcium carbonate, Mn trends could be sympathetic or antipathetic to those of Sr, depending on the composition of the fluid. Furthermore, trace-element trends may be pronounced, weak, or absent, depending on several interacting factors, such as fluid composition, flow rate, flow direction, water/rock ratio, degree of recrystallization, redox potential, and amount of clay or organic impurities.

Trace-element trends, such as those predicted by the model, occur in massive dolostones that range from the Cambrian to Eocene. Such trends may be more common than previously recognized, and they may be useful indicators of the direction of fluid flow during dolomitization.

INTRODUCTION

The origin of dolomites and dolostones (rocks with more than 75% dolomite by volume) has been the subject of intensive research for many years (Friedman and Sanders, 1967; Chilingar and others, 1979; Morrow, 1982a, 1982b). It is now clear that dolomites and dolostones form under a variety of geochemical, geological, and hydrological conditions. Two types of dolomite formation are common in diagenetic environments: (1) cementation, and (2) replacement of calcium carbonate (dolomitization). Dolomite cement precipitates in clastic and in carbonate sediments in a variety of depositional and diagenetic environments. More commonly, however, dolomite forms as a pervasive replacement of shallow-water carbonates. Massive, replacive dolostones are formed where long-lasting fluid flow facilitates extensive magnesium (and minor carbonate) input, i.e., in shallow subtidal and various subsurface environments (Land, 1985; Machel and Mountjoy, 1986, 1987a; Hardie, 1987). Extensive dolostones may also form as layered bodies via diffusion in hemipelagic sediments (Baker and Burns, 1985), or stacked due to repeated transgressions and regressions in coastal arid and semiarid environments such as sabkhas (Patterson and Kinsman, 1982). The subjects of this article are the first two types, dolomites as a cement and as a pervasive replacement of shallow-water carbonates. Hemipelagic and sabkha dolomites are not considered here because they are petrographically and geochemically distinct.

Pervasively dolomitized shallow-water carbonates are the most abundant dolostones in nature but genetically the least understood. This is because identical fabrics and geochemical characteristics may be present in dolostones that were formed in vastly different depositional and diagenetic environments (Sass and Katz, 1982; Machel and Mountjoy, 1986, 1987a). Standard petrography and even geochemical composition may not sufficiently characterize the environment of dolomitization. In these cases, determination of the direction (vertical and upward) or only the orientation (vertical, upward, or downward) of fluid flow during dolomitization might help to decide between various alternatives. For example, the predominant flow direction during shallow burial compaction is vertical and upward (Einsele, 1977), whereas regional topography-driven flow near the center of an uplifted basin is nearly horizontal (Garven and Freeze, 1984). Nonlinear and/or variable flow directions are realized in flow systems such as 'Kohout convection' or thermal convection cells in continental basins (Simms, 1984). Where such flow systems result in massive dolomitization, geochemical characteristics such as trace elements, rare earth elements, or isotopes in the dolostones plotted along traverses might indicate the direction or the orientation of fluid flow during dolomitization.

The objective of this paper is to present a model that permits deduction of the direction or orientation of fluid flow during dolomitization utilizing the most readily measurable quantity, i.e., the bulk trace-element composition of massive dolostones along one or several traverses. It is shown that trace-element trends indicating the direction of fluid flow during dolomitization may result under favorable circumstances. Trace-element trends are also predicted by models for recrystallization of metastable carbonates in meteoric aquifers (Morrow and Mayers, 1978; Brand and Veizer, 1980). The approach taken in this paper is different, however, because the diagenetic solutions are not meteoric, and the ionic contributions from dissolution of the calcium carbonate precursors are either negligible or have effects counteracting those in meteoric aquifers.

TRACE-ELEMENT MODEL AND FLOW DIRECTIONS

The model proposed in this section *qualitatively* describes the trace-element distribution in dolostones relative to the flow direction of the dolomitizing fluid(s). The model is qualitative because: (1) quantification of some of the variables of the model is possible only where a specific environment of dolomitization is considered, but the model is meant to be applicable for all natural types of dolomitization by fluid flow; and (2) the model contains at least 11 variables that are not known and/or cannot be estimated with reasonable accuracy for a given environment of dolomitization. These variables are (1) bulk fluid composition; (2) molar trace-element-to-calcium ratios of those trace elements under consideration; (3) fluid saturation states with

Sedimentology and Geochemistry of Dolostones, SEPM Special Publication No. 43

respect to dolomite and calcium carbonate; (4) temperature; (5) pressure at the time of dolomitization; (6) flow rate(s); (7) flow direction(s); (8) water/rock ratio; (9) composition and porosity of the affected sediment or rock; (10) trace-element distribution coefficients, which depend on bulk fluid composition, temperature, pressure, precipitation rate, and composition of the solid; and (11) degree of recrystallization. These variables interact and determine how much and how fast dolomite is formed (Machel and Mountjoy, 1986). They also determine the trace-element content of the dolomites (Land, 1980).

Trace-Element Distribution Coefficients and Distribution Laws

The trace-element distribution coefficients are of particular importance in this context because they determine not only the magnitude(s) of trace-element incorporation but also the direction of concentration changes, i.e., trace-element enrichment or depletion in the downflow direction. Unfortunately, few distribution coefficients for dolomite can be utilized at the present time. Distribution coefficients between calcite and dolomite have been suggested for 11 elements, but only for a temperature of about 650°C (Kretz, 1982). These coefficients are not applicable in the present context. Distribution coefficients between dolomite and water (here called λ) for diagenetic temperatures have been published only for strontium and sodium (λ_{Sr} = 0.025–0.060: Jacobson and Usdowski, 1976; Katz and Matthews, 1977; and λ_{Na} = 0.00002–0.00003: White, 1978). Strontium probably substitutes for Ca in the lattice (Land, 1980), which makes the application of trace-element distribution laws for this element meaningful. On the other hand, variable and commonly large proportions of Na are contained in liquid and solid inclusions (Bein and Land, 1983), which precludes a rigorous application of trace-element distribution laws for this element and an accurate determination of the Na/Ca ratio of the dolomitizing fluid. λ_{Na} between dolomite and solution also depends strongly on salinity (Sass and Katz, 1982), which further complicates the application of distribution laws to Na. For these reasons, Sr is taken to represent trace elements with a distribution coefficient smaller than one. Regarding trace elements with distribution coefficients greater than one, no reliable numerical values have been published for diagenetic temperatures. There is no doubt, however, that the coefficient of Mn, one of the most useful trace elements in calcite, is greater than one also for dolomite (Veizer, 1983). Therefore, Mn is taken to represent trace elements with distribution coefficients greater than one.

Distribution laws can be used to calculate trace-element concentrations where thermodynamic equilibrium between solid and solution is attained or at least approached. Dolomite formation probably takes place with bulk solution equilibrium, because precipitation rates of dolomites in natural environments are very low relative to flow rates (bulk solution disequilibrium, such as discussed by Pingitore, 1982, and Veizer, 1983, for recrystallization of calcium carbonate, is therefore not realized or is insignificant in most cases of natural dolomite formation). If distribution equilibrium is attained between a crystal surface and a solution, the trace-element (T) and main-element (M) concentrations on the crystal surface are proportional to their respective solution concentrations:

$$\left(\frac{^{m}T}{^{m}M}\right)_{\text{CRYSTAL SURFACE}} = \lambda\left(\frac{^{m}T}{^{m}M}\right)_{\text{SOLUTION}} \quad (1)$$

where ^{m}T = molar trace-element concentration, ^{m}M = molar main-element concentration, and λ = distribution coefficient. If, with progressing precipitation in a *closed* system with relatively *low* water/rock ratio, the trace-element and main-element concentrations on the crystal surface remain proportional to their respective solution concentrations, equation 1 yields through integration:

$$\log\left(\frac{^{m}T_i}{^{m}T_f}\right) = \lambda \cdot \log\left(\frac{^{m}M_i}{^{m}M_f}\right) \quad (2)$$

where i = initial and f = final molar concentrations in solution (equation 2 is equivalent to equation 7 in Gordon and others, 1959, p. 110). This equation is used to characterize the Heterogeneous Distribution Law (Doerner and Hoskins, 1925; Gordon and others, 1959). Continuing precipitation results in a radial concentration gradient in the crystal in which the trace-element ions are distributed throughout in a logarithmic manner. The law presumes a state of equilibrium between the solution and an infinitesimal crystal surface layer but not with the crystal as a whole. In other words, the crystal does not recrystallize during precipitation. Alternatively, if several crystals nucleate and grow in succession in such a way that only one crystal grows at a time, they would have different compositions mimicking the formerly radial concentration gradient.

The Heterogeneous Distribution Law is applicable for dolomitization in the above form with M = Ca for all elements that substitute for Ca in the lattice (Mg must be incorporated in the equations in the case of trace elements that also substitute for Mg). For λ = 1, the molar trace-element-to-calcium ratio ($^{m}T/^{m}$Ca) of the solid is equal to this ratio in the liquid (equation 1). For λ smaller than one, the $^{m}T/^{m}$Ca ratio in solid solution is smaller than that in the liquid, by an amount depending on the deviation of λ from unity. For example, if the initial Sr/Ca concentration ratio in water is u, the respective initial concentration in dolomite is p = (u × 0.025) to (u × 0.060), depending on which of the above values for λ is chosen. As precipitation proceeds, the Sr/Ca concentration ratio becomes greater and greater in both the water (equation 2) and dolomite (equation 1). If the final Sr/Ca concentration ratio in the water becomes v, the ratio in dolomite becomes q = (v × 0.025) to (v × 0.060). Conversely, the $^{m}T/^{m}$Ca ratio in solid solution is greater than that in the liquid for elements with λ greater than one (e.g., Mn), by an amount depending on the deviation of λ from unity. In these cases, the $^{m}T/^{m}$Ca ratios become smaller and smaller in both the liquid and the solid with progressive precipitation.

It must be emphasized that application of the Heterogeneous Distribution Law *sensu strictu* is not permissible for modeling extensive dolomitization, because extensive dolomitization requires open systems with high water/rock ratios (discussed later). The law does indicate, however, the directions of concentration changes for trace elements with distribution coefficients smaller and greater than one, respectively, even in an open system with large water/rock ratios, if the large water body is divided into small volumes that are considered one by one, which simulates small water/rock ratios with ionic changes in the individual water volumes due to mineral precipitation. This may be called an application of a 'variant' of the Heterogeneous Distribution Law.

Assumptions

Most natural dolostones consist of crystals too small to be isolated and/or analyzed individually in great quantities. Therefore, the model is designed in such a way as to permit utilization of bulk rock analyses. Nevertheless, large zoned crystals can also be used.

In its simplest form, which is described first, the following assumptions are made regarding the variables of the model:

(1) fluid composition is constant;
(2) the molar Sr/Ca ratio can have any value in the case of dolomite cementation; in the case of dolomitization, the molar Sr/Ca ratio in the fluid is lower than that in the calcium carbonate precursor;
(3) the fluid is supersaturated with respect to dolomite, which should result in dolomite formation;
(4) temperature is constant;
(5) pressure is constant;
(6) the flow rate is constant and dominates ion transport by diffusion;
(7) the flow direction is linear;
(8) the water/rock ratio is constant (which follows from assuming saturation state and flow rate to be constant), and this ratio is large;
(9) the to-be-dolomitized rock is compositionally and texturally homogeneous;
(10) the distribution coefficients for Sr and Mn are smaller and larger than one, respectively;
(11) the dolomite crystals must not be recrystallized, or recrystallization must have taken place isochemically (in a closed system).

Most of these assumptions are realistic for many natural situations; others apply to rather specific situations (discussed later). It should be emphasized, however, that large water/rock ratios (assumption 8) are *always* necessary for both extensive dolomite cementation and extensive calcium carbonate replacement because of the relatively low Mg concentration of natural waters. Depending on salinity, the saturation states with respect to dolomite and calcium carbonate, and the porosity of the rock, several tens to hundreds (hypersaline) to several thousands (brackish/dilute) of cubic meters of water are needed to dolomitize 1 m^3 of limestone (Land, 1985).

Within the framework of the above assumptions, numerical divisions in the model have been chosen somewhat arbitrarily, or no scale was given where any order of magnitude would do. Some numerical choices were necessary, and they were made as realistically as possible. The effects of varying these numerical choices and the effects of varying the assumptions are discussed later.

Dolomite Formation

The model (Fig. 1) is valid for dolomite cementation and for replacement of calcium carbonate within the framework of the above assumptions. In fulfillment of the water/rock ratio requirement, a large quantity of water is set to flow through a given, comparatively small, volume of limestone. The water is divided into equal volume units 1, 2, 3, ... *n* (illustrated by crossed diagonals, top left in Fig. 1), and the limestone is arbitrarily divided into equal volume units A, B, C, D, and E. As water volume 1 (*wv* 1) enters the limestone, it forms a specific volume of dolomite, represented by three black squares (row *wv* 1). This volume of dolomite is *not* to scale and depends mainly on the saturation state, flow rate, and temperature of the water in a natural situation. In rock volume A, *wv* 1 loses all its Mg in excess of the dolomite saturation state, that is, *wv* 1 changes its saturation state from supersaturated to saturated with respect to dolomite, thereby changing its composition (illustrated by omitting the crossed diagonals). This altered water is inert with respect to further dolomite formation. If other mineral reactions are disregarded, this water passes through the rest of the limestone without further compositional changes, and without forming any more dolomite.

Water volume 2 (*wv* 2) forms exactly the same volume of dolomite (three black squares), because the composition of the incoming water is constant. Wv 2 does not form all its dolomite in rock volume A, however, because A is already partially dolomitized. The first two-thirds of all new dolomite are fixed to be formed in A and the last third in B, illustrated by two additional black squares in A and one in B (Fig. 1). Wv 2 therefore advances the dolomite formation front in the downflow direction and continues dolomite formation in the already partially dolomitized limestone (this is in accord with observations from natural environments, where dolomitization could be related to a flow direction). After having passed through A and B, *wv* 2 too has lost all its excess Mg, and passes through the rest of the limestone without further compositional changes. All conditions remaining constant, every new water volume will have the same effect as the previous ones: advancing the dolomitization front, and at the same time completing the extent of dolomite formation in the upflow part of the aquifer. This process continues until flow stops, or until all calcium carbonate is dolomitized. In Fig. 1, the first two-thirds of dolomite of every new water volume are set to be formed along the flow path in the already partially dolomitized rock, and the last third at the new dolomitization front. This ratio can be varied, but changes do not affect the key implications (discussed later).

Mg Depletion

Decreases in the dolomite saturation state are best illustrated by monitoring the Mg depletion of the dolomitizing fluid through time and along the flow path (Fig. 1, center column). The unit rectangles of the water volumes are now

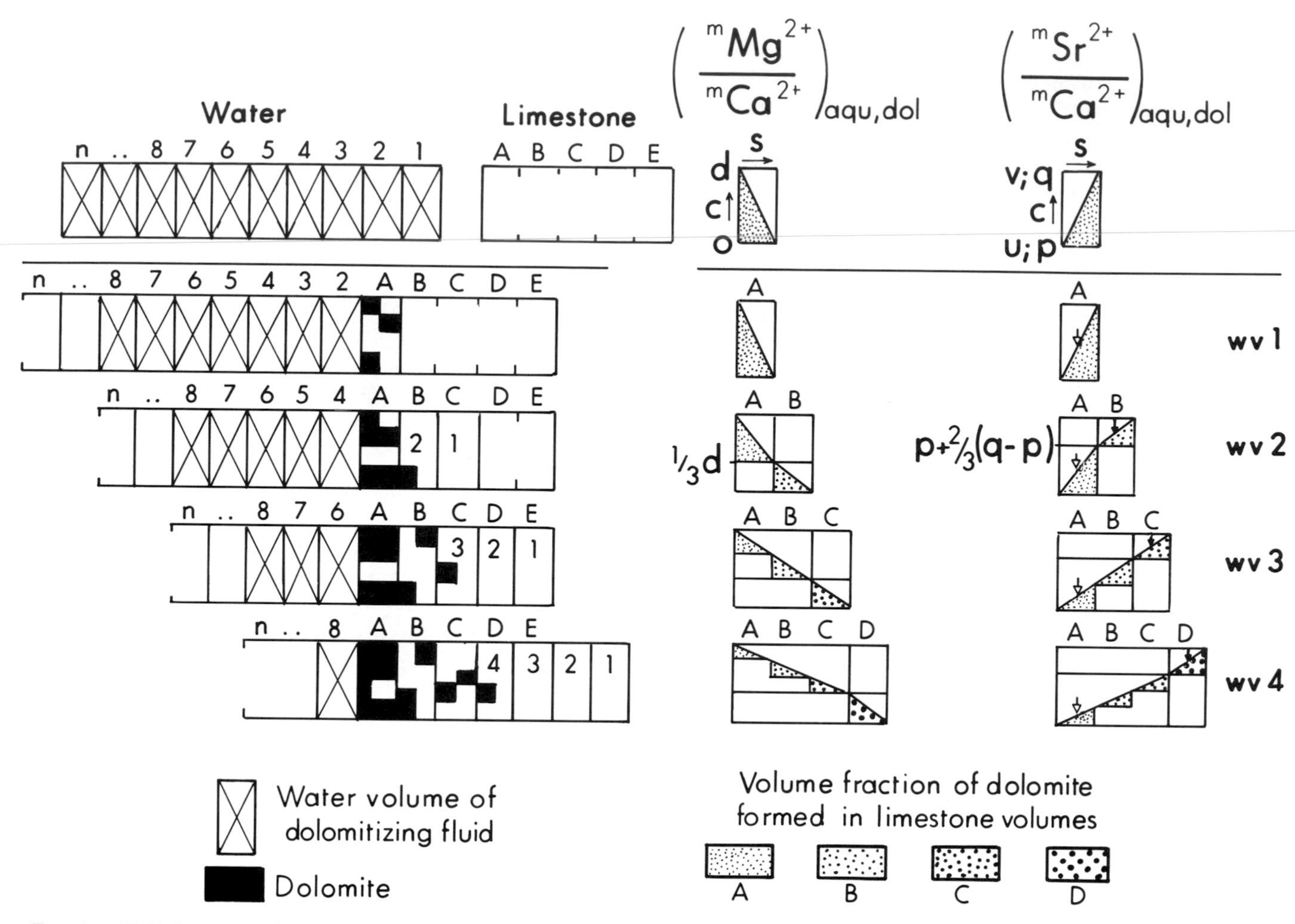

FIG. 1.—Fluid flow model for dolomitization. Water/rock ratio is large. Fluid (water) and rock are divided into equal, arbitrary unit volumes (1, 2, 3, ..., and A, B, C ..., respectively, left column). As successive water volumes enter the limestone, dolomitization (black areas) advances downflow and becomes more complete in the upflow part of the limestone. Molar Mg/Ca (center column) and Sr/Ca (right column) ratios are illustrated as *x-y* plots with concentration *c* along *y* and flow direction *s* along *x*. The Mg/Ca ratio of the fluid decreases from *d* to *o* along the flow path (*d* = supersaturation, *o* = saturation, with respect to dolomite). The Sr/Ca ratios increase from *u* (fluid) and *p* (solid) to *v* (fluid) and *q* (solid) at the same time. Triangular areas below diagonal concentration lines represent volume (mass) fractions of dolomite formed. See text for further explanation.

considered as *x-y* plots, with concentration *c* along *y*, and flow direction *s* along *x*.

In the case of dolomite cementation, the fluid composition will change from supersaturation to saturation with respect to dolomite. Thus, the saturation state with respect to calcium carbonate could be anywhere between undersaturation to supersaturation. During dolomitization, the fluid composition will change from supersaturation with respect to dolomite and undersaturation with respect to calcium carbonate to three-phase equilibrium calcite-dolomite-water (Machel and Mountjoy, 1987b). In either case, the $^mMg/^mCa$ ratio of the incoming water is *d* and drops along the diagonal line to *o* (first row, Fig. 1; *o* represents the saturation level with respect to dolomite after dolomitization). This designates the loss of all Mg in excess of the saturation state, not a drop in the $^mMg/^mCa$ ratio to c = zero. Analogous to the left column in Figure 1, *wv* 1 loses all its excess Mg in rock volume A (center column, row *wv* 1, Fig. 1). Wv 2 was set to form two-thirds of its dolomite in A and the remainder in B. This is represented by two diagonal concentration lines (center column, row *wv* 2, Fig. 1), where the $^mMg/^mCa$ ratio drops to one-third of its initial value *d* after passage through A, and then to the saturation level *o* after passage through B. The Mg depletions of further water volumes are plotted in an analogous manner.

Sr Increase

Figure 1 also illustrates the changes in $^mSr/^mCa$ changes of the water volumes and the newly formed dolomites along the flow path (right column). The unit rectangles of the water and rock volumes are again considered as *x-y* plots, with concentration along *y* and flow direction along *x*. The letters *u* and *v* designate arbitrary $^mSr/^mCa$ ratios in water, *p* and *q* designate the respective ratios in dolomite. The $^mSr/^mCa$ change of any water volume accompanying dolomite formation, i.e., during the drop in $^mMg/^mCa$ from *d* to *o*, is represented by an increase from the initial $^mSr/$

^{m}Ca of u to the final value of v (along the diagonal concentration line in the first row of Fig. 1). Correspondingly, ^{m}Sr/^{m}Ca in dolomite changes from p to q. In the following discussion, the term Sr concentration is used for the molar Sr/Ca ratios as well as for Sr content in dolomite by weight (i.e., ppm), because the Ca content in natural dolomites is very large and nearly constant compared to the much smaller and more variable contents of Sr.

The increase in ^{m}Sr/^{m}Ca results from partitioning according to application of the 'variant' of the Heterogeneous Distribution Law (as discussed previously). As *wv* 1 forms its three-volume units of dolomite, the Sr concentration of this dolomite increases from p to q (row *wv* 1, Fig. 2). The Sr increase occurs (a) either as multiple centrifugal zonation in a single hypothetical large crystal, if the fluid flow rate (velocity) is close to zero, or (b) many unzoned small crystals form, whereby the crystals that form first (upflow) have Sr concentrations of p and the crystals that form last (downflow) have Sr concentrations of q, which is the case if the fluid flow rate is extremely high. These cases represent end members of a spectrum of zonation patterns. Most dolostones contain intermediate patterns with many small crystals that are zoned, albeit not much. Thus, individual crystals along the fluid flow path represent only one or very few of the zones of the hypothetical large crystal. Hence, the intracrystalline Sr concentration ranges constitute only a very small part of the total increment from p to q. This permits utilization of whole-rock Sr data. If, for example, a sampling traverse in a homogeneous dolostone is 20 m long, one would sample a maximum of 2–5 g every 1 or 2 m. Each sample would contain numerous small, possibly zoned, dolomite crystals with slightly different mean compositions. A chemical analysis of the sample would give the average composition of the means of all crystals. This average composition accurately determines the relative position of the whole sample between p and q, if the size of the sample (commonly 1–2 cm^3, equivalent to 2–5 g) is small compared to the length of the flow path along which the water volume formed its dolomite (at least several decimeters to several meters). This is a reasonable assumption for most natural situations.

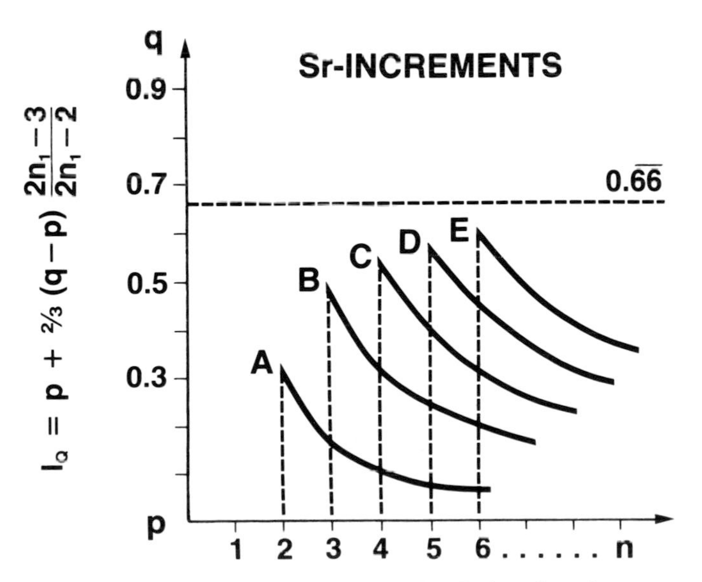

FIG. 2.—Dolomite Sr increments (equation 7) downflow. Increments in rock volumes A, B, C, D, and E plot along the respective curves. Stippled horizontal line at $0.\overline{66}$ is equivalent to $p + 2/3(q - p)$ in Figure 1.

Analogous to the Mg depletion, *wv* 2, and the dolomite formed from it, experience two-thirds of their total Sr increase in rock volume A, and the last third in rock volume B (along the Sr concentration lines in row *wv* 2, Fig. 1). Then *wv* 3 and the dolomite formed from it increase to two-thirds of their total Sr increase in A and B and to the total (v/p) in C (row *wv* 3, Fig. 1), and so on.

The average Sr concentration of all dolomite crystals formed can be expressed numerically. The mean Sr concentration of dolomite formed by *wv* 1 in rock volume A (hollow arrow, row *wv* 1 in Fig. 1) is $(p + q)/2$. All further increments in A (I_A = hollow arrows in rows *wv* 2, *wv* 3, and *wv* 4 in Fig. 2) can be expressed as an arithmetic series:

$$I_A = p + \frac{2}{3}(q - p)\frac{1}{2n_1 - 2} \tag{3}$$

where number of water volume $= n_1 = 2, 3, 4 \ldots n$. In rock volumes B, C, D... the Sr concentration of the first dolomite formed is always $p + (5/6)(q - p)$ (solid arrows, rows *wv* 2 to *wv* 4 in Fig. 1). All further increments can again be expressed as arithmetic series:

$$I_B = p + \frac{2}{3}(q - p)\frac{3}{2n_1 - 2}, \tag{4}$$

$$I_C = p + \frac{2}{3}(q - p)\frac{5}{2n_1 - 2}, \tag{5}$$

$$I_D = p + \frac{2}{3}(q - p)\frac{7}{2n_1 - 2}, \tag{6}$$

where number of water volume $= n_1 = 3, 4, 5 \ldots n$. Equations 3 to 6 can be generalized to:

$$I_Q = p + \frac{2}{3}(q - p)\frac{2n_1 - 3}{2n_1 - 2} \tag{7}$$

where Q = rock volume; number of water volume $= n_1 = 2, 3, 4 \ldots n$.

The quantity, which can be measured analytically, is the Sr concentration of dolomite after n water volumes. Calculation of this quantity must also take the volume (or mass) fractions of the successive increments into account. For example, *wv* 1 forms all (volume fraction = 1) its dolomite in rock volume A with Sr $= (p + q)/2$ (represented by the triangular area between the Sr concentration line and the baseline, row *wv* 1 in Fig. 1). Then *wv* 2 forms 2/3 (= volume fraction) of its dolomite in A, *wv* 3 forms 1/3 (= volume fraction) of its dolomite in A (represented by the smaller triangular areas between the Sr concentration line and the baseline, rows *wv* 2 and *wv* 3, Fig. 1), and so on. The measured (total) Sr concentration after n water volumes is the sum of all the partial Sr concentrations (hollow arrows, Fig. 1) multiplied by their respective volume fractions, divided by the total volume of dolomite formed up to that point:

$$\mathrm{Sr_A} = \left[\left(\frac{p+q}{2}\right)1 + \left(p + \frac{2}{3}(q-p)\frac{1}{2}\right)\frac{2}{3} + \left(p + \frac{2}{3}(q-p)\frac{1}{4}\right)\frac{1}{3} + \ldots\right]\frac{1}{1 + \frac{2}{3} + \frac{1}{3} + \ldots}, \quad (8)$$

or in generalized form:

$$\mathrm{Sr_A} = \left[\frac{p+q}{2} + \sum_{n_1=2}^{n} \cdot \left(p + \frac{2}{3}(q-p)\frac{1}{2n_1 - 2}\right)\right]\frac{1}{1 + \frac{2}{3}\sum_{n_1=2}^{n}\frac{1}{n_1 - 1}}. \quad (9)$$

The total Sr concentrations in rock volumes B, C, D ... can be obtained in an analogous way. They merely differ in the Sr concentration of the first dolomite formed $[p + (5/6)(q - p)$ instead of $(p + q)/2]$, and in other respects follow their respective incremental series, as exemplified by rock volume B:

$$\mathrm{Sr_B} = \left[p + \frac{5}{6}(q-p) + \sum_{n_1=3}^{n}\left(p + \frac{2}{3}(q-p)\frac{3}{2n_1 - 2}\right)\right] \cdot \frac{1}{1 + \frac{2}{3}\sum_{n_1=3}^{n}\frac{1}{n_1 - 1}}. \quad (10)$$

Equations 9, 10, and analogous equations for the other rock volumes, can also be written in integrated form instead of arithmetic series, in order to consider arbitrary partial water and rock volumes. This, of course, would be necessary in the case of quantification of the model in order to approach natural situations.

Implications

The implications of this model follow from two lines of evidence: (a) the results of equations 9 and 10, and analogous equations, specifically from their common incremental series (equation 7); and (b) the effects of subsequent modification of the original assumptions.

Results.—

The incremental series (equation 7) in rock volumes A, B, C, D, and E as functions of n plot along asymptotic curves that never intersect (Fig. 2). Therefore, the dolomite Sr concentration in rock volume A is always lower than in B, C, ..., the dolomite Sr concentration in rock volume B is always lower than in C, D, ..., and so on, regardless of the respective volume subdivisions, and regardless of how many water volumes passed through the rock (this even permits utilization of compositional data from incompletely dolomitized rocks). Furthermore, the above ratio of 2:1 (of dolomite forming in previously partially dolomitized limestone to that forming at the new dolomitization front) can also be varied at will. Different ratios would merely change the constants 2/3 and 5/6 in the equations, and therefore the positions of the Sr increment curves along the concentration axis, their distance to one another, and their slopes; but they would retain their general relative attitude and would not intersect. One main result of this model, therefore, is that, within the framework of the above assumptions, the *Sr concentration* of dolomite *increases downflow* from p to q, regardless of the choices of the parameters in question. This is also true for all other trace elements with a distribution coefficient smaller than one that largely or completely substitute for Ca in the dolomite lattice. Analogous application of the 'variant' of the Heterogeneous Distribution Law predicts a *decrease in the downflow direction* for trace elements with a distribution coefficient greater than one that largely or completely substitute for Ca in the dolomite lattice, such as Mn. Another result of the model is that the trace-element increases and decreases along the flow path are finite, because p and q are predetermined by the composition and saturation states of the fluid, and by the distribution coefficient(s).

Variations of assumptions.—

The assumptions made to establish the model, which are equivalent to 'conditions' in a natural environment, determine the orders of magnitude of the trace-element increases or decreases downflow, and their rates (which may be expressed as the differential of the incremental curves such as in Fig. 2). In other words, some assumptions/conditions result in 'excellent' trace-element trends extending over distances in the same order of magnitude as the flow systems (e.g., several tens of meters to several hundreds of kilometers), with a large trace-element concentration range $(q - p)$ between the upflow and downflow parts. In the other extreme, different assumptions/conditions result in the absence of trace-element trends, i.e., trends that are smaller than sample size. Intermediate cases are possible where trends are 'useful' or 'useless' extending over distances that are large to short (meters) relative to the size of the flow system, or undetectable (smaller than the sample traverse).

These possibilities are best discussed by making concrete specifications of the assumptions of the model. For example, for 'excellent' or 'useful' trace-element trends to develop, the assumptions/conditions should be as follows (using the numerals 1 to 11 for the same parameters):

(1′) fluid composition is constant;
(2′) the initial Sr/Ca ratio of the fluid can have any value in the case of dolomite cementation, and Sr/Ca of the dolomite will increase downflow; in the case of replacement of calcium carbonate, Sr in dolomite increases in the downflow direction if the initial Sr/Ca ratio of the fluid is lower than that of the calcium carbonate; if the initial Sr/Ca ratio of the fluid is greater than that of the calcium carbonate, Sr in dolomite decreases in the downflow direction (discussed later);
(3′) the fluid is only slightly supersaturated with respect to dolomite;

(4′) temperature is constant;
(5′) pressure is constant;
(6′) flow rate is constant and high relative to crystal nucleation and growth rates, and relative to ion diffusion rates;
(7′) flow direction is linear;
(8′) water/rock ratio is constant and large;
(9′) the to-be-dolomitized rock is homogeneous and contains few or no impurities (e.g., clay minerals, organic matter) that release or absorb trace elements during dolomite formation;
(10′) distribution coefficients greatly deviate from unity, implying that $(q - p)$ is large, which is typical for elements such as Sr and Mn;
(11′) dolomite is not recrystallized, or recrystallization took place isochemically with respect to the trace elements under consideration.

These assumptions are nearly identical to those originally made to establish the model, with the important additional specifications of a low degree of supersaturation with respect to dolomite, a relatively high flow rate, and a lithology poor in clay minerals and organic matter. These conditions assure low crystal nucleation and growth rates relative to the flow distance over which the crystals are formed, and a trace-element signal that is a function of only the carbonates. Such conditions may be typical for shallow-water reef and platform carbonates penetrated by compaction fluids, topography-driven flow, 'Kohout convection', or thermal convection of subsurface brines.

Assumption (2′) deserves special attention. The dependence of the sense of Sr/Ca change on the initial Sr/Ca ratio of the fluid results from the amounts of Ca released during replacement of calcium carbonate (Sass and Katz, 1982; Fig. 3). Where the Sr/Ca ratio of the fluid is greater than that of the calcium carbonate, the amounts of Ca released during dolomitization outweigh the downflow Sr/Ca increase due to trace element partitioning according to the Heterogeneous Distribution Law (assuming that replacement of calcium carbonate by dolomite takes place mole-per-mole, or nearly so). Regarding calculation of the Sr/Ca change in the downflow direction, this factor would result in Sr/Ca downflow depletions similar to the Mg/Ca depletions in Figure 1, inverting the curves in Figure 2 into those typical for elements with distribution coefficients greater than one. Furthermore, the above considerations do not apply only to the replacement of calcite and aragonite via sea water (as discussed by Sass and Katz, 1982). Any diagenetic fluid supersaturated with respect to dolomite could be plotted in a diagram such as Figure 3. For example, a dolomitizing fluid could have a Sr/Ca ratio lower than that of calcite (i.e., between curve C and the abscissa, e.g., D in Fig. 3), so that Sr/Ca may increase downflow also in the case of dolomitization of calcite. Analogous considerations for trace-element trends with distribution coefficients greater than one will result in inversions of their respective trends. Consequently, *it is possible that Sr and Mn vary sympathetically in the downflow direction in the case of calcium carbonate replacement.*

Individual variations of the other assumptions/conditions have variable effects on potential trace-element trends. For example, marked changes in fluid composition, saturation states, temperature, and/or flow rate during dolomitization may result in increases or decreases of the rate of dolomite formation, and therefore in unsteady trends with kinks and temporary reversals (if plotted vs. distance along the flow path). Variations in the concentrations of impurities may blur or obliterate trends along the flow path, an effect that could be pronounced for elements that have high affinities for clay minerals and organic matter (i.e., Mn), but weak affinities for elements such as Sr. Even the redox potential may influence trends where elements such as Mn or Fe are considered, which are significantly incorporated in dolomite only in their divalent forms. Alternatively, some factors have very little effect on potential trace-element trends. For example, pressure and distribution coefficients are not expected to vary greatly or to have significant effects during a particular dolomitization event. In any case, *not all conditions have to be perfect for 'useful' trace-element trends to develop, and trends with kinks and/or temporary reversals may also be 'useful'*. On the other hand, one or

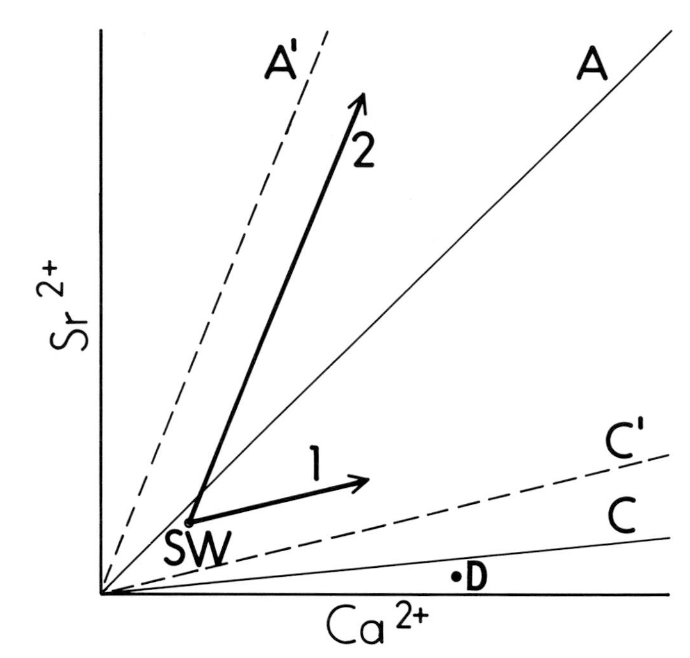

FIG. 3.—Molar Sr and Ca changes in solution during replacement of calcium carbonate by dolomite (modified after Sass and Katz, 1982). In this type of diagram, Sr/Ca ratios are depicted as lines in *x-y* space. The Sr/Ca ratio in solution approaches a value (C′ or A′), which is about twice that of the replaced carbonate (C or A, respectively), if dolomitization takes place as a mole-per-mole replacement (limiting values C′ and A′ are functions of the stoichiometry of the replacement reaction: see Sass and Katz, 1982; Machel and Mountjoy, 1986). If the Sr/Ca ratio of the initial fluid (SW) is higher than that of the replaced carbonate C and the limiting value C′, the Sr/Ca ratio of the fluid decreases (direction 1). Conversely, if the initial Sr/Ca ratio of the fluid is similar to, or lower than that of, the replaced carbonate A, the Sr/Ca ratio of the fluid increases (direction 2). Although this type of diagram was used to represent dolomitization of calcite (C) and aragonite (A) by sea water (SW) by Sass and Katz (1982), it can also be used to illustrate dolomitization by other diagenetic fluids. An increase in the molar Sr/Ca ratio would result for dolomitization of marine calcite, if the diagenetic fluid plots between line C and the abscissa (D).

more of the model conditions may vary during dolomitization, supersaturation with respect to dolomite may be very large, the flow direction may be tortuous, the rock may be inhomogeneous and/or contain variable amounts of impurities, or the dolomites are non-isochemically recrystallized. A combined occurrence of all these conditions may be unrealistic for most natural situations, but *'useless' or 'absent' trends may result even if only some of these unfavorable conditions are realized*. For example, a high degree of supersaturation relative to the flow rates would result in high crystal nucleation and growth rates and 'dump' all the dolomite over a very short flow distance. No measurable trends would result even in the case of elements such as Sr (whose distribution coefficient greatly deviates from unity). This may be typical for normal to hypersaline intertidal to supratidal flats (e.g., Bahamas: Shinn and others, 1965; Persian Gulf: Patterson and Kinsman, 1982), where the actual flow distance along which dolomitization takes place is only a few centimeters to about 2 m, respectively. In addition, such dolomites, and those of other hypersaline near-surface environments, are usually formed as protodolomites (*sensu* Graf and Goldsmith, 1956), which invariably recrystallize to more ordered dolomites acquiring new trace-element characteristics (Bein and Land, 1983; Land, 1985). In general, where dolomites are recrystallized, original trace-element trends may be altered or destroyed, and a trace-element signal indicative of the process of recrystallization may be generated.

NATURAL EXAMPLES

Trace-element trends such as those predicted by the model in the previous sections are not only hypothetical. They have been found in natural dolostones and reported from several locations ranging from Cambrian to Eocene and from a few tens of meters to about 200 km. Some of these examples are briefly discussed below, beginning with the most thoroughly studied location with respect to trace-element trends.

The Nisku Formation

The Upper Devonian Nisku Formation in the subsurface of central Alberta has been used as a test case for the applicability of the flow model discussed herein (Machel, 1984, 1985, 1986b). The study area bridges the shelf/slope/basin transition of a shallow Late Devonian sea. The shelf (bank) facies close to the shelf margin, a buildup fringing the shelf margin (bank edge reef), and several buildups on the slope toward the adjacent basin are pervasively dolomitized. Most dolostones consist of matrix-selective, medium- to coarse-grained dolomites. Samples of about 1–2 g consisting of several tens to hundreds of crystals were taken with a dental drill and each analyzed for 27 trace elements, by inductively coupled plasma. Several sample traverses were laid vertically over distances of 50 to 100 m (which is cross-formational because the strata are nearly horizontal), across the paleo-strike over distances of about 10 km, and along the structural dip (which is only about 0.5° and coincides with the paleo-strike) over a distance of about 70 km (note: all plots are available from the author upon request; the complete data set is tabulated in Machel, 1985).

Trace-element trends were found only in the vertical, cross-formational traverses, two of which are shown in Figures 4 and 5. Figure 4 is a traverse across the bank (well 8-1) to bank edge reef (well 3-20) to slope buildup (well 15-31) for Sr vs. depth, and Figure 5 is a traverse across the bank (well 8-1) to bank edge reef (well 2-12) to bank edge reef (well 3-20) for Mn vs. depth. Well 2-12 in Figure 5 has been substituted for well 15-31 in Figure 4, so that (1) all three facies domains are represented, and (2) it can be shown that trends in adjacent wells in one facies domain overlap (i.e., wells 2-12 and 3-20). In most cases, the cross-formational traverses in each figure are separated along the vertical axis due to the structural dip. Traverses at the top are from the updip wells; traverses at the bottom are from the downdip wells.

The following observations are important in this context. (1) In each well the concentration for Sr increases and for Mn decreases upward (a pronounced upward increase also occurs for Ba, and decreases occur for P, Fe, B, and V; 20 other elements do not show sufficiently clear trends). (2) Most data points are located on nearly linear regression lines (which crosscut the sedimentary cycle boundaries), except for the Mn data points of well 2-12, which lie on two regression lines with different slopes; the break between these two lines is not resolved because of a lack of data points (thus, they could be part of one continuous trend with a 'dog leg'). (3) The slopes of the Mn trends in the upper part of wells 2-12 and 8-1 are about the same, and the data points overlap. (4) The slopes of the Mn trends in

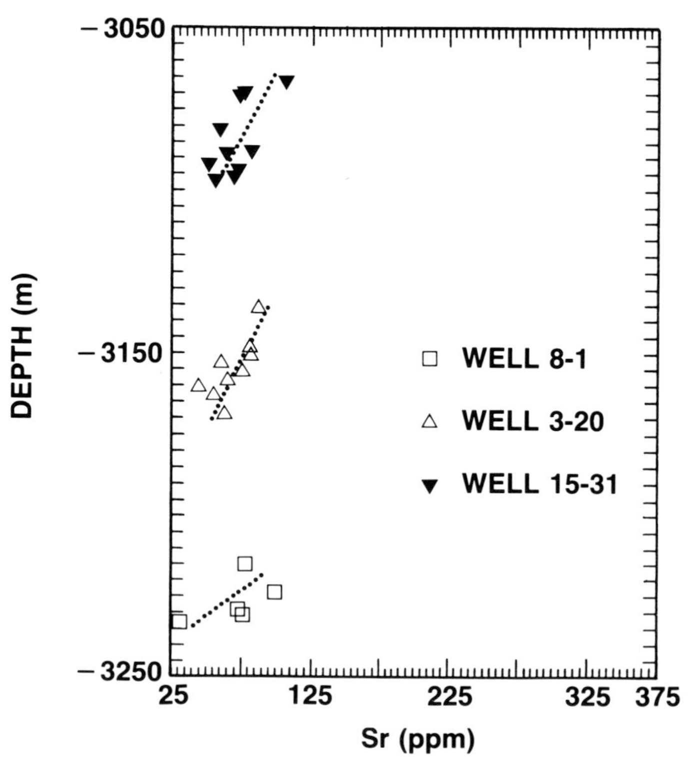

FIG. 4.—Sr vs. depth cross-plot of matrix dolomites from three Nisku wells in a traverse across the bank (well 8-1), bank edge reef (well 3-20), and reef on the slope (well 15-31). Sr increases upward in all three facies domains.

the lower part of wells 2-12 and 3-20 are about the same. (5) If the structural dip is removed by shifting the trends along the *y* axis until they are at a common depth, the trends overlap.

Applying the model, the upward increasing Sr and decreasing Mn concentrations (Figs. 4, 5) indicate that the predominant fluid flow direction during dolomitization was vertical. Calculation of the probable molar Sr/Ca ratio of the dolomitizing fluids suggests that it was lower than that of the precursor calcite (Machel, 1985, 1986a, fig. 21). Hence, it is likely that the sense of fluid flow was upward, and that the dolomitizing fluids entered and passed through the Nisku Formation predominantly from below. This interpretation is supported by all supplementary data. Dolomitization postdated limestone diagenesis (i.e., lithification and recrystallization mainly in marine phreatic and shallow burial environments: Machel, 1986a, 1986b) and took place after Nisku time, because the trends crosscut most cycle boundaries. The 'dog leg' in well 2-12 is probably due to permeability differences resulting in different fluid flow rates, because it coincides with the change from fairly tight marls below the reef to the more permeable reef facies. Furthermore, the dolomitizing fluids moved through all the different facies domains (bank, bank edge reef, reef on the slope) at the same time because the trends overlap once the structural dip is removed. Because the Nisku was being buried progressively to about 1,000 m between deposition (Late Devonian time) and Late Mississippian time, the dolomitizing fluids most probably were compaction fluids. It is difficult to conceive other hydrologic regimes during this time and burial interval, and mass balance calculations (in Anderson, 1985) suggest that the fluids in the underlying and adjacent strata contained enough Mg for dolomitization of the Nisku trend if they were funneled toward it.

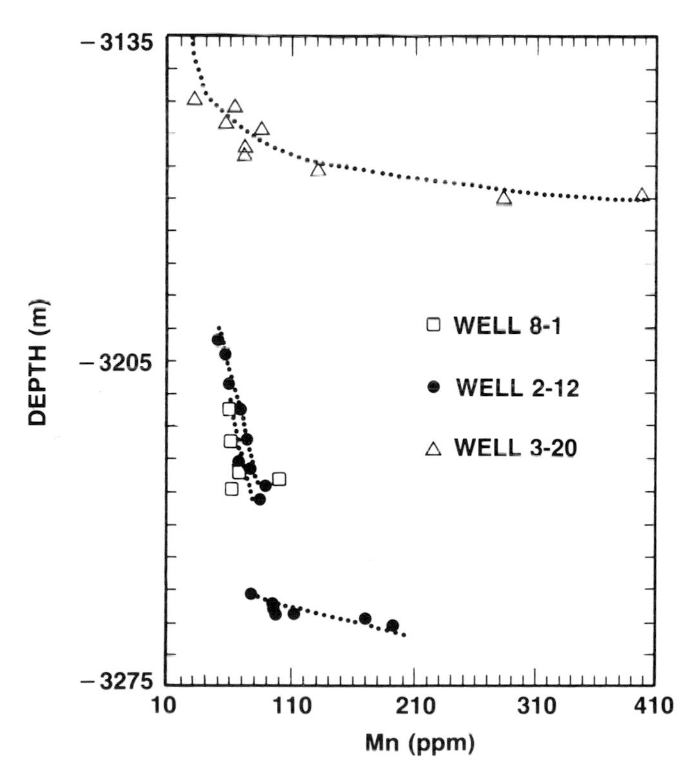

FIG. 5.—Mn vs. depth cross-plot of matrix dolomites from three Nisku wells in a traverse across the bank (well 8-1), bank edge reef (well 2-12), and bank edge reef along strike (well 3-20). Mn decreases in all three wells, which penetrated two facies domains. The trends for wells 2-12 and 3-20 overlap.

The fact that only seven elements display the patterns predicted by the model does not argue against it, because there are numerous possibilities for the absence of trends for the other elements. For example, (a) their distribution coefficients may not deviate much from unity, (b) the rocks contain impurities, and/or (c) some of the dolomites may be recrystallized, which may have resulted in concentration changes for all but those elements that do show trends (as predicted by the recrystallization model of Pingitore, 1982).

The Nisku dolostones are a test case for the trace-element model, because most conclusions regarding the dolomitizing fluids could be drawn even *without* the trace-element trends. In particular, the combined interpretation of sedimentologic, petrographic, absolute trace-element concentrations, and isotopic data suggests that dolomitization took place at depths of about 300–1,000 m by fluids which were sea water that was modified by rock-water interaction and possibly by injection of other fluids via fault systems (Anderson, 1985; Machel, 1985; Machel and Anderson, in prep.). The Nisku trends therefore confirm the model, in the sense that they support upward fluid flow in a situation where the combination of all other data already suggested this flow direction.

Other Examples

In the Cambrian Bonneterre Formation of Missouri, Buelter and Guillemette (1986), Gregg and Shelton (1986), and Gregg and others (this volume) documented trends for Sr, Na, Mn, and Fe for regionally extensive, replacive dolostones over a distance of 200 km. Several independent lines of evidence support the inferred direction of fluid flow during dolomitization (Gregg and Hagni, 1986), although the exact mode of fluid flow is still being debated.

In the dolostones of the Ordovician Allen Bay Formation of Canada, Sr increases upward almost linearly in several sections over a thickness of as much as about 700 m (Land and others, 1975), suggesting upward fluid flow during dolomitization. This, however, does not necessarily imply that the dolomitizing fluids were compaction fluids. In fact, Land and others (1975) favored meteoric waters. According to his more recent views, however, Land (1985) discards meteoric water as a potential agent for massive dolomitization. Furthermore, Mn does not show antithetic or sympathetic upward trends. Instead, Mn increases horizontally toward the adjacent basinal shale facies. Land and others (1975) attributed this pattern to Mn being preferentially "held in clays or organic matter," which is a plausible interpretation.

Land and others (1975) also reported systematic "into-the-basin" trends over a distance of about 5 km for Sr and Na in Miocene dolostones from Egypt, and a similar pattern was detected by Mueller (1975) in Cretaceous dolostones from central Texas over a distance of about 50 km. Furthermore, Veizer and Demovic (1974) and Rudolph (1978) reported systematic regional variations in the Sr contents of dolomitized platform sequences over a distance of about 100 km. No trends have been reported for Mn or other elements with distribution coefficients greater than one from these locations, but most rocks were not analyzed for these elements.

The examples reported in the literature suggest that Na may also be a useful trace element with respect to the model. Although much of the Na is not incorporated in lattice sites, the distribution coefficient of Na greatly deviates from unity, and the error introduced by Na from interstitial sites and fluid inclusions is apparently quite small in at least some of the cases. Also, the effects of Na in interstitial sites and fluid inclusions can be eliminated to a large degree by careful sample preparation (Bein and Land, 1983). On the other hand, Na may be more useful as an indicator for salinity or paleoenvironment because of the strong dependence of λ_{Na} on salinity (Veizer and Demovic, 1974; Sass and Katz, 1982).

CONCLUSIONS

The qualitative mathematical model applying a variant of the Heterogeneous Distribution Law presented in this paper predicts that systematic trace-element trends indicating fluid flow direction(s) can result during dolomitization. Unfortunately, *any* transition is possible between (1) a large upflow and downflow trace-element difference stretched over a large flow distance, to (2) a small upflow and downflow trace-element difference over a negligible flow distance, depending on the interplay of *all* involved parameters. In the case of dolomite cementation, trace elements with distribution coefficients smaller than one increase, and those with a distribution coefficient greater than one decrease, in the downflow direction. In the case of replacement of calcium carbonate, trace-element trends, if present, indicate at least the *orientation* of fluid flow, e.g., vertical or horizontal. The *direction* of fluid flow, e.g., upward or downward in the case of vertical trends, can be determined only where the trace-element-to-calcium ratio of the fluid relative to that of the precursor carbonate can be estimated with reasonable accuracy, or where other data are available (i.e., compaction curves, and so on).

Trace-element trends have been documented in several Phanerozoic dolostone sequences. The absence of such trends in other dolostone sequences may be real in a certain percentage of the cases, because dolomitization does not necessarily yield trace-element trends. On the other hand, many such trends may have gone undetected, and it is suggested that more systematic trace-element trends can be found that indicate the flow direction during dolomitization.

ACKNOWLEDGMENTS

Energetic and not always consenting opinions by P. A. Baker, R. Hesse, E. W. Mountjoy, D. W. Morrow, especially E. Sass, and J. Veizer, refined the thoughts presented in this paper. P. A. Baker, J. A. Mackenzie, N. E. Pingitore, and V. Shukla critically read earlier versions of the manuscript, and their numerous suggestions are greatly appreciated. This research was supported by Natural Sciences and Engineering Research Council Grant No. 55-47-939, McGill University, Natural Sciences and Engineering Research Council, Grant no. A2128 to E. W. Mountjoy, the Basin Research Institute at Louisiana State University, and a Central Research Fund grant from the University of Alberta.

REFERENCES

Anderson, J. H., 1985, Depositional facies and carbonate diagenesis of the downslope reefs in the Nisku Formation, Central Alberta, Canada: Unpublished Ph.D. Dissertation, University of Texas at Austin, Texas, 393 p.

Baker, P. A., and Burns, S. J., 1985, Occurrence and formation of dolomite in organic-rich continental margin sediments: American Association of Petroleum Geologists Bulletin, v. 69, p. 1917–1930.

Bein, A., and Land, L. S., 1983, Carbonate sedimentation and diagenesis associated with Mg-Ca-chloride brines: Journal of Sedimentary Petrology, v. 53, p. 243–260.

Brand, U., and Veizer, J., 1980, Chemical diagenesis of a multi-component carbonate system. I. Trace elements. Journal of Sedimentary Petrology, v. 50, p. 1219–1236.

Buelter, D. P., and Guillemette, R. N., 1986, Trace element distribution in epigenetic dolomite associated with lead-zinc mineralization of the Viburnum Trend, southeast Missouri (abst.), *in* Gregg, J. M., and Hagni, R. D. (convenors): Symposium on the Bonneterre Formation (Cambrian), Southeastern Missouri: Department of Geology and Geophysics, University of Missouri, Rolla, Missouri, p. 6.

Chilingar, G. V., Zenger, D. H., Bissell, H. J., and Wolf, K. H., 1979, Dolomites and dolomitization, *in* Larsen, G., and Chilingar, G. V., eds., Diagenesis in Sediments and Sedimentary Rocks: Developments in Sedimentology 25A, Elsevier, Amsterdam, p. 423–535.

Doerner, H. A., and Hoskins, W. M., 1925, Coprecipitation of radium and barium sulphates: Journal of the American Chemical Society, v. 47, p. 662–675.

Einsele, G., 1977, Range, velocity, and material flux of compaction flow in growing sedimentary sequences: Sedimentology, v. 24, p. 639–655.

Friedman, G. M., and Sanders, J. E., 1967, Origin and occurrence of dolostones, *in* Chilingar, G. V., Bissel, H. J., and Fairbridge, R. W., eds., Carbonate Rocks: Developments in Sedimentology 9A, Elsevier, Amsterdam, p. 267–348.

Garven, G., and Freeze, R. A., 1984, Theoretical analysis of the role of groundwater flow in the genesis of stratabound ore deposits. 1. Mathematical and numerical model: American Journal of Science, v. 284, p. 1085–1124.

Gordon, L., Salutsky, M. L., and Willard, H. H., 1959, Precipitation from homogeneous solution: Wiley and Sons, New York, 455 p.

Graf, D. L., and Goldsmith, J. R., 1956, Some hydrothermal synthesis of dolomite and protodolomite: Journal of Geology, v. 64, p. 173–186.

Gregg, J. M., and Hagni, R. D. (convenors), 1986, Symposium on the Bonneterre Formation (Cambrian), Southeastern Missouri. Department of Geology and Geophysics, University of Missouri, Rolla, Missouri, 24 p.

———, and Shelton, K. L., 1986, Minor and trace element distributions at the Bonneterre Dolomite/Lamotte sandstone contact (Cambrian), SE Missouri: Evidence for basinal fluid pathways: Geological Society of America, Abstracts with Programs, p. 621.

Hardie, L. A., 1987, Dolomitization: A critical view of some current views: Journal of Sedimentary Petrology, v. 57, p. 166–183.

Jacobson, R. L., and Usdowski, H. E., 1976, Partitioning of strontium between calcite, dolomite and liquids: Contributions to Mineralogy and Petrology, v. 59, p. 171–185.

Katz, A., and Matthews, A., 1977, The dolomitization of $CaCO_3$: An experimental study at 252–295°C: Geochimica et Cosmochimica Acta, v. 41, p. 297–308.

Kretz, R., 1982, A model for the distribution of trace elements between calcite and dolomite: Geochimica et Cosmochimica Acta, v. 46, p. 1979–1981.

Land, L. S., 1980, The isotopic and trace element geochemistry of dolomite: The state of the art, *in* Zenger, D. H., Dunham, J. B., and Ethington, R. L., eds., Concepts and Models of Dolomitization: Society of Economic Paleontologists and Mineralogists Special Publication 28, p. 87–110.

———, 1985, The origin of massive dolomite: Journal of Geological Education, v. 33, p. 112–125.

———, SALEM, M. R. I., AND MORROW, D. W., 1975, Paleohydrology of ancient dolomites: Geochemical evidence: American Association of Petroleum Geologists Bulletin, v. 59, p. 1602–1625.

MACHEL, H. G., 1984, Facies and dolomitization of the Upper Devonian Nisku Formation in the Brazeau, Pembina, and Bigoray areas, Alberta, Canada, *in* Eliuk, L. S., ed., Carbonates in Subsurface and Outcrop: Canadian Society of Petroleum Geologists Core Conference, p. 191–224.

———, 1985, Facies and diagenesis of the Upper Devonian Nisku Formation in the subsurface of central Alberta: Unpublished Ph.D. Dissertation, McGill University, Montreal, Canada, 392 p.

———, 1986a, Limestone diagenesis of Upper Devonian Nisku carbonates in the subsurface of central Alberta. Canadian Journal of Earth Sciences, v. 23, p. 1804–1822.

———, 1986b, Early lithification, dolomitization, and anhydritization of Upper Devonian Nisku buildups, subsurface of Alberta, Canada, *in* Schroeder, J. H., and Purser, B. H., eds., Reef Diagenesis: Springer-Verlag, Berlin, p. 336–356.

———, AND MOUNTJOY, E. W., 1986, Chemistry and environments of dolomitization–A reappraisal: Earth Science Reviews, v. 23, p. 175–222.

———, ———, 1987a, General constraints on extensive pervasive dolomitization–And their application to the Devonian carbonates of Western Canada: Bulletin of Canadian Petroleum Geology, v. 35, p. 143–158.

———, ———, 1987b, Chemistry and environments of dolomitization–A reappraisal (reply): Earth Science Reviews, v. 24, p. 213–215.

MORROW, D. W., 1982a, Diagenesis I. Dolomite–Part I: The chemistry of dolomitization and dolomite precipitation: Geoscience Canada, v. 9, p. 5–13.

———, 1982b, Diagenesis II. Dolomite–Part II: Dolomitization models and ancient dolostones: Geoscience Canada, v. 9, p. 95–107.

———, AND MAYERS, I. R., 1978, Simulation of limestone diagenesis–A model based on strontium depletion: Canada Journal of Earth Sciences, v. 15, p. 376–396.

MUELLER, H. W., 1975, Centrifugal progradation of carbonate banks: A model for deposition and early diagenesis, Ft. Terrett Formation, Edwards Group, Lower Cretaceous, central Texas: Unpublished Ph.D. Dissertation, University of Texas at Austin, Texas, 316 p.

PATTERSON, R. J., AND KINSMAN, D. J. J., 1982, Formation of diagenetic dolomite in coastal sabkha along the Arabian (Persian) Gulf: American Association of Petroleum Geologists Bulletin, v. 66, p. 28–43.

PINGITORE, N. E., 1982, The role of diffusion during carbonate diagenesis: Journal of Sedimentary Petrology, v. 52, p. 27–39.

RUDOLPH, K. W., 1978, Diagenesis of back-reef carbonates: An example from the Capitan complex: Unpublished M.A. Thesis, University of Texas at Austin, Texas, 285 p.

SASS, E., AND KATZ, A., 1982, The origin of platform dolomites: American Journal of Science, v. 282, p. 1184–1213.

SHINN, E. A., GINSBURG, R. N., AND LLOYD, R. M., 1965, Recent supratidal dolomite from Andros Island, Bahamas, *in* Pray, L. C., and Murray, R. C., eds., Dolomitization and Limestone Diagenesis: A Symposium: Society of Economic Paleontologists and Mineralogists Special Publication 13, p. 112–123.

SIMMS, M., 1984, Dolomitization by groundwater-flow systems in carbonate platforms: Gulf Coast Association of Geological Societies, Transactions, v. XXXIV, p. 411–420.

VEIZER, J., 1983, Chemical diagenesis of carbonates: Theory and application of trace element technique, *in* Arthur, M. A., Anderson, T. F., Kaplan, I. R., Veizer, J., and Land, L. S., eds., Stable Isotopes in Sedimentary Petrology: Society of Economic Paleontologists and Mineralogists Short Course No. 10, p. 3-1 to 3-100.

———, AND DEMOVIC, R., 1974, Strontium as a tool in facies analysis: Journal of Sedimentary Petrology, v. 44, p. 93–115.

WHITE, A. F., 1978, Sodium coprecipitation in calcite and dolomite: Chemical Geology, v. 26, p. 65–72.

SECTION V
GEOCHEMISTRY OF DOLOMITE TEXTURES AND FABRICS

SECTION V: INTRODUCTION
GEOCHEMISTRY OF DOLOMITE TEXTURES AND FABRICS

This section presents two papers which describe the geochemical variations in textures and fabrics of dolomites.

Cander and others describe three generations of dolomites with distinct geochemical and petrographic characteristics. This paper is one of two parts of a larger study; the other part is presented by Banner and others in Section IV.

Shukla presents data correlating Sr, Fe, Na, Mn, and B distributions in various dolostone fabrics. Some inductive explanations for these variations are also presented. This paper should be compared and contrasted with that by Sass and Bein (Section VII).

REGIONAL DOLOMITIZATION OF SHELF CARBONATES IN THE BURLINGTON-KEOKUK FORMATION (MISSISSIPPIAN), ILLINOIS AND MISSOURI: CONSTRAINTS FROM CATHODOLUMINESCENT ZONAL STRATIGRAPHY

HARRIS S. CANDER,[1] JONATHAN KAUFMAN, LAWRENCE D. DANIELS,[2] AND WILLIAM J. MEYERS
Department of Earth and Space Sciences
State University of New York
Stony Brook, New York 11794

ABSTRACT: Cathodoluminescence petrography defines three regionally extensive dolomite generations in the Mississippian Burlington-Keokuk Formation over an area greater than 100,000 km^2 in Illinois and Missouri. These dolomites represent widespread replacement of older dolomite generations by younger generations, each with distinct petrographic, geochemical, and distributional characteristics. Dolomite I is the oldest and most abundant generation and is characterized by fine-scale cathodoluminescent zoning that is correlative within a measured section but is not correlative between sections. Dolomite II is the second generation and is characterized by unzoned red to brown cathodoluminescence. Dolomite II replaced dolomite I and is primarily restricted to the lower part of Burlington-Keokuk strata. Dolomite III is the youngest generation and occurs as nonluminescent, syntaxial overgrowths on, and partial replacements of, the older two generations.

Dolomite I is pre-Pennsylvanian, on the basis of geometric relation with pre-Pennsylvanian calcite cements and with pre-Pennsylvanian cherts. The age of dolomite II is uncertain, but its occurrence within chert nodules suggests that it also is pre-Pennsylvanian. On the basis of geometric relations with pre-Pennsylvanian and post-Mississippian cements, dolomite III probably formed after the Mississippian but before late Pennsylvanian/early Permian time.

Dolomite I is proposed to have formed in a seawater-freshwater mixing environment associated with a regional meteoric groundwater system that developed beneath late Mississippian/early Pennsylvanian unconformities. This model is supported by the early timing, restriction of correlatable zonation to single measured sections, and by isotopic characteristics. Dolomite II is proposed to have formed as a replacement of dolomite I by warm, relatively Fe- and Mn-rich subsurface fluids. These fluids interacted with pre-Burlington rocks and invaded the Burlington-Keokuk sediments from below in "plumes," which had an irregular regional distribution. Dolomite III probably represents continued influx of progressively more Fe- and Mn-rich, warm subsurface fluids, which caused replacement and syntaxial overgrowth of the older two generations. These models imply that the Mg for dolomite I was derived from sea water and from intraformational skeletal Mg-calcites, whereas the Mg for dolomites II and III was derived mainly from precursor dolomites I and II.

INTRODUCTION

Regional dolomitization of shallow-water shelf and platform carbonates is an important yet poorly understood process. This lack of understanding is due in part to the absence of modern analogues for large-scale regional dolomitization, to the difficulty of synthesizing dolomite at low temperatures, and to the shortage of regional-scale studies that integrate detailed petrography with state-of-the-art geochemical and modeling approaches. This paper presents the detailed petrography of dolomites from the Mississippian Burlington-Keokuk Formation, which crops out in Illinois and northeastern and central Missouri (Fig. 1), and provides the framework for a major ongoing project on regional dolomitization and limestone diagenesis in the Burlington-Keokuk Formation. The project integrates a variety of geochemical approaches, including stable isotopes (C, O), radiogenic isotopes (Sr, Nd, Pb), and trace elements (Fe, Mn, Sr, REE, U, Th, Pb; Prosky and Meyers, 1985, Grams, 1987; Banner and others, in prep.; Harris and Meyers, in prep.; J. Hoff, pers. commun., 1987; K. Kohrt, pers. commun., 1987; Banner and others, this volume). The goals of the petrographic study are: (1) to identify the major generations of dolomite and calcite cements; (2) to determine their stratigraphic and regional distributions; (3) to constrain their timing and thus their diagenetic environments; (4) to identify and document replacement of early generations of dolomite by later generations; and (5) to provide a well-characterized sampling base for the geochemical work.

This paper documents several major discrete, yet regionally correlative, episodes of dolomitization of the Burlington-Keokuk Formation over about 100,000 km^2 of study area. Cathodoluminescence petrography established a zonal stratigraphy for the dolomites and associated calcite cements (Kaufman and others, 1988), which allowed integration of the dolomites into the overall diagenetic and burial history of the Burlington-Keokuk Formation. This paper provides the petrographic framework for the geochemical studies and should be viewed as a companion paper to that of Banner and others (this volume). Banner and others (this volume) show that each dolomite generation has a distinctive chemistry, and quantitatively model the dolomitizing fluids as well as replacement of early dolomite by later dolomites in terms of water-rock interactions.

METHODS

Our study is based on 60 measured sections from outcrops of the Burlington-Keokuk Formation in Missouri and Illinois (Fig. 1). About 650 thin sections were examined to assure adequate stratigraphic and geographic distribution throughout the study area. Cathodoluminescence petrography used a Nuclide Luminoscope and a Technosyn 3200 Mk II cathodoluminescence device, both operating in the 10- to 12-kv range. The procedures and precision for analyses of dolomites by mass spectrometry (C, O, Sr isotopes, Sr concentrations) are given in Banner (1986). Procedures

[1]Present addresses: Department of Geological Sciences, University of Texas, Austin, Texas 78712; and

[2]Dynamac Corporation, Fort Lee Executive Park, Fort Lee, New Jersey 07024

Sedimentology and Geochemistry of Dolostones, SEPM Special Publication No. 43

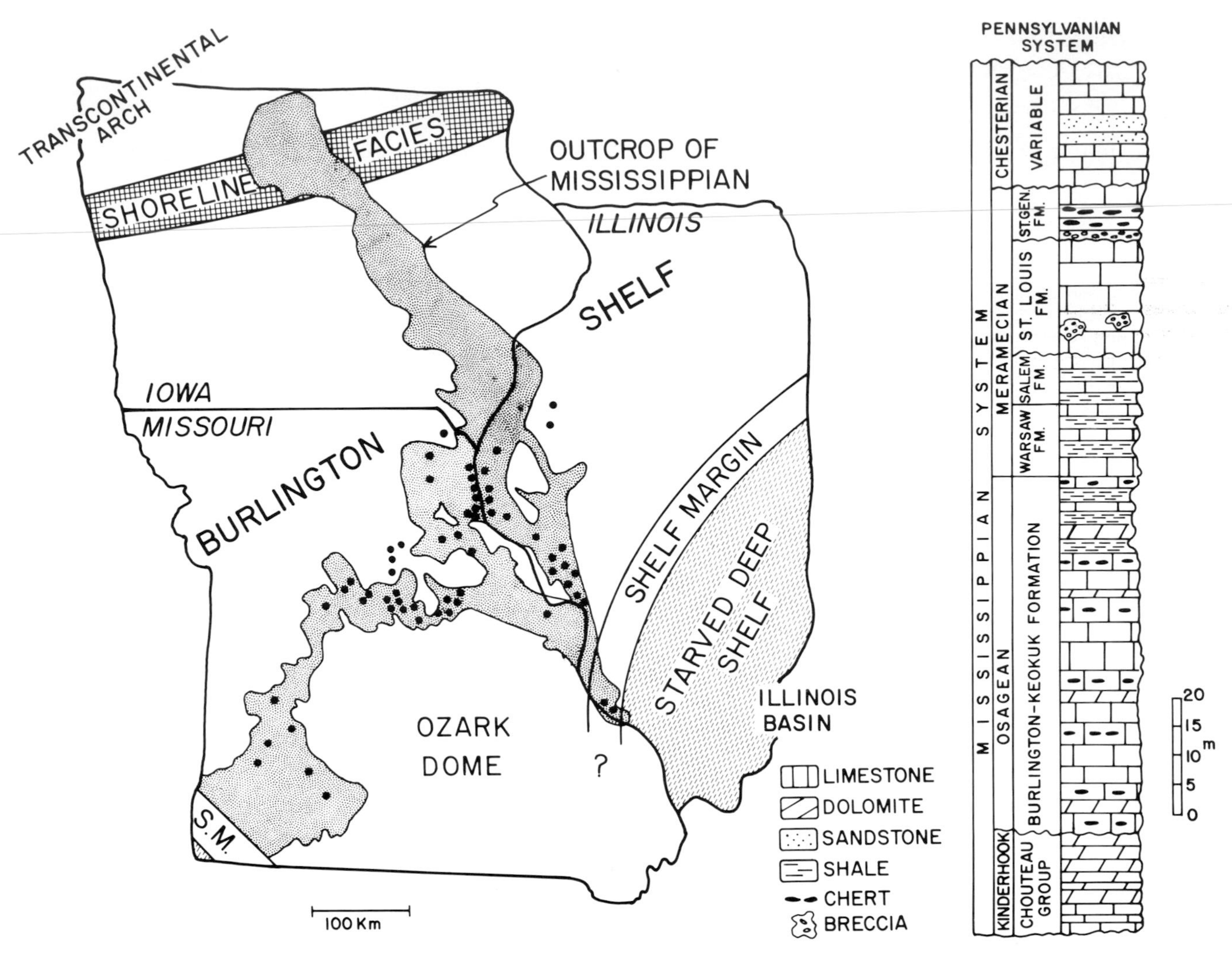

FIG. 1.—Locality map, generalized Osagean paleogeography (from Lane and DeKeyser, 1980; Sixt, 1983), and stratigraphy of study area. Stippled pattern indicates the Burlington-Keokuk Formation outcrop, Dots are measured sections used for this study.

and precision for analyses by electron microprobe (Fe, Mn, Ca, Mg) are given in Reeder and Prosky (1986).

STRATIGRAPHY AND DEPOSITIONAL FACIES

The Mississippian Burlington-Keokuk Formation (Osagean) consists of a 60- to 90-m sequence of crinoidal packstones and grainstones with interbedded dolomitized skeletal mudstones and wackestones (Fig. 1). Most mud-supported facies have been dolomitized in the study area. Although crinoids are the main skeletal component in both grain-supported and mud-supported rocks, fenestellid bryozoans are also common, and brachipods and corals are less common. The formation contains common nodular cherts and in its upper part thin (less than 0.3 m) shale beds. The formation also contains horizons rich in glauconite pellets and more rarely fish bones. Limestone and dolomite beds throughout the formation are generally less than a meter thick and often pinch out on the scale of an outcrop. The upper Burlington-Keokuk Formation is more thinly bedded, with shale beds becoming more numerous as the formation grades into the overlying Warsaw Shale.

The Burlington-Keokuk Formation unconformably overlies Kinderhookian strata throughout most of the study area. Underlying Kinderhookian strata include the Hannibal Shale in the northern part of the study area, the Meppen Limestone in the southern part, and the Chouteau Group in western Illinois and in eastern and central Missouri (Howe and Koenig, 1961; Atherton and others, 1975). Throughout much of the area, the Burlington-Keokuk Formation is overlain conformably by the Meramecian Warsaw Shale (Fig. 1), which in turn is overlain by shallow-water carbonates of the Salem Formation. The Salem Formation is overlain by the St. Louis Formation, and their contact has been interpreted by some workers to be a regional unconformity (Collinson, 1964; Harris and Parker, 1964; Avcin and Koch, 1979; Carlson, 1979; H. R. Lane, pers. commun. 1987). Throughout parts of northwestern Illinois and central Missouri, the Burlington-Keokuk Formation is overlain unconformably by Middle Pennsylvanian strata (Howe and Koenig, 1961; Atherton and others, 1975). This pre-Pennsylvanian unconformity surface contains well-developed paleokarst features in central Missouri (Bretz, 1950; Daniels, 1986).

The Burlington-Keokuk carbonates represent continuous, subtidal deposition of open-marine sediments on a broad, shallow, intracratonic shelf (Lane and DeKeyser, 1980; Gutschick and Sandberg, 1983). The Burlington shelf was bordered by the Transcontinental Arch to the northwest and by a starved deeper shelf to the south and southeast (Fig. 1; Lane and DeKeyser, 1980; Gutschick and Sandberg, 1983). Westward progradation of the Borden "delta" terrigenous silts and shales terminated carbonate production and resulted in deposition of the Warsaw Shale over the Burlington-Keokuk (Lineback, 1981). It is not clear if the Ozark Dome and the LaSalle Arch in northeastern Illinois were positive structural features during deposition of the Burlington-Keokuk Formation. These two features were high before the pre-Pennsylvanian erosion, however, as indicated by Pennsylvanian strata resting on pre-Mississippian rocks.

Burlington-Keokuk strata contain no peritidal facies nor physical evidence of intraformational subaerial unconformities that could have resulted in freshwater influx. Potentially, fresh water could have penetrated the formation during development of the pre-St. Louis unconformity, during the pre- and early Chesterian unconformities, and during the pre-Pennsylvanian unconformity.

CALCITE CEMENT STRATIGRAPHY OF THE BURLINGTON-KEOKUK LIMESTONES

The diagenetic history of the Burlington-Keokuk Formation includes multiple episodes of calcite cementation, dolomitization, chertification, compaction, and karstification (Harris, 1982; Cander, 1985; Kaufman, 1985; Daniels, 1986). The calcite cementation history is important because it serves as a temporal reference by which other diagenetic phases can be dated. In this section we present the main characteristics of the cements and evidence for their timing. Later, we present timing of the important generations of dolomites relative to calcite cementation episodes.

The calcite cement stratigraphy consists of seven regionally extensive cement zones. These zones are distinguished by their cathodoluminescence characteristics, Fe content, relative position in the sequence, interzonal dissolution surfaces and fracturing, and age relative to the pre-Pennsylvanian unconformity (Kaufman and others, 1988). Pre-Pennsylvanian cements are non-ferroan (i.e., they do not stain with K ferricyanide solution) and consist of two distinct, but age-equivalent stratigraphies. In the lower part of the formation, pre-Pennsylvanian calcite cements compose, from oldest to youngest, zones I through IV. Zones II and IV are nonluminescent, and zones I and III are luminescent. Early calcite cements in upper parts of the Burlington-Keokuk Formation consist of fully luminescent zone II′, which is interpreted as an age-equivalent "facies" of zones I through IV in underlying strata. Zone I is restricted mainly to Iowa, but zones II through IV and II′ extend throughout the study area in Iowa, Illinois, and Missouri. Post-Mississippian cements consist of ferroan zone V and younger non-ferroan zone VI. This cement stratigraphy is slightly modified from that established by Harris (1982) in Iowa.

Petrography of pre-Pennsylvanian paleokarst features in the Burlington-Keokuk Formation in central Missouri has shown that zones II–IV predate the karstification and zone V postdates karstification (Daniels, 1986; Kaufman and others, 1988). Specifically, paleokarst cavities are filled with sediment that contains reworked pre-zone V cements, and some of these sediments and overlying Middle Pennsylvanian sandstones are cemented with zone V. These timing constraints indicate that zones I through IV, and II′ precipitated under less than 500 m of burial on the basis of stratigraphic reconstructions (A. Cox, unpubl. data). Zones V and VI are crosscut and replaced by minor occurrences of Mississippi Valley-type mineralization. A late Pennsylvanian/early Permian age has been estimated for Mississippi Valley-type mineralization in the major ore deposits of the region (Wu and Beales, 1981; Wisniowiecki and others, 1983; Leach and Rowan, 1986). Although we cannot prove that the minor occurrences of Mississippi Valley-type (MVT) mineralization in our study area are the same age as those of the major ore deposits, we have assumed that they are the same age. If this assumption is correct, it implies a pre-late Pennsylvanian/early Permian age for zones V and VI, and less than about 1-km burial depth during their precipitation.

We point out that this interpretation of timing of cements differs from that of Harris (1982; Harris and Meyers, 1987), who suggested that all cement zones, including zone VI, predate the St. Louis Formation. This is based mainly on the absence of the Burlington-Keokuk cement stratigraphy in the St. Louis Formation grainstones.

All of the above zones are interpreted as non-marine cements on the basis of their petrography and geochemistry. They lack typical petrographic characteristics of shallow warm water marine cements, such as columnar or prismatic crystal morphologies, microdolomite inclusions, and post-cement marine internal sediment. Furthermore, Mg contents of all cements are low (mean Mg for zone I = 1,600 ppm; zone II = 725 ppm; zone III = 635 ppm; zone IV = 320 ppm; zone V = 890 ppm; zone VI = 600 ppm; Grams and Meyers, 1987). In addition, cement-rich rocks contain no intraclasts of grainstones, even within beds that experienced slow sedimentation, such as glauconite and fish bone horizons. The sequence of cathodoluminescent zones, which probably reflects fluctuating Eh in pore fluids during precipitation (Frank and others, 1982; Machel, 1985), is consistent over large areas and throughout relatively large stratigraphic intervals. These features are difficult to reconcile with a model of cementation by marine pore waters during progressive burial. These features, combined with the timing and burial-depth constraints, suggest zones I–IV are shallow burial, freshwater phreatic-calcite cements, and zones V and VI are deeper burial non-marine calcite cements.

DOLOMITE PETROGRAPHY

Dolostones compose 10 to 25 percent of the Burlington-Keokuk Formation and occur predominantly in former mud-

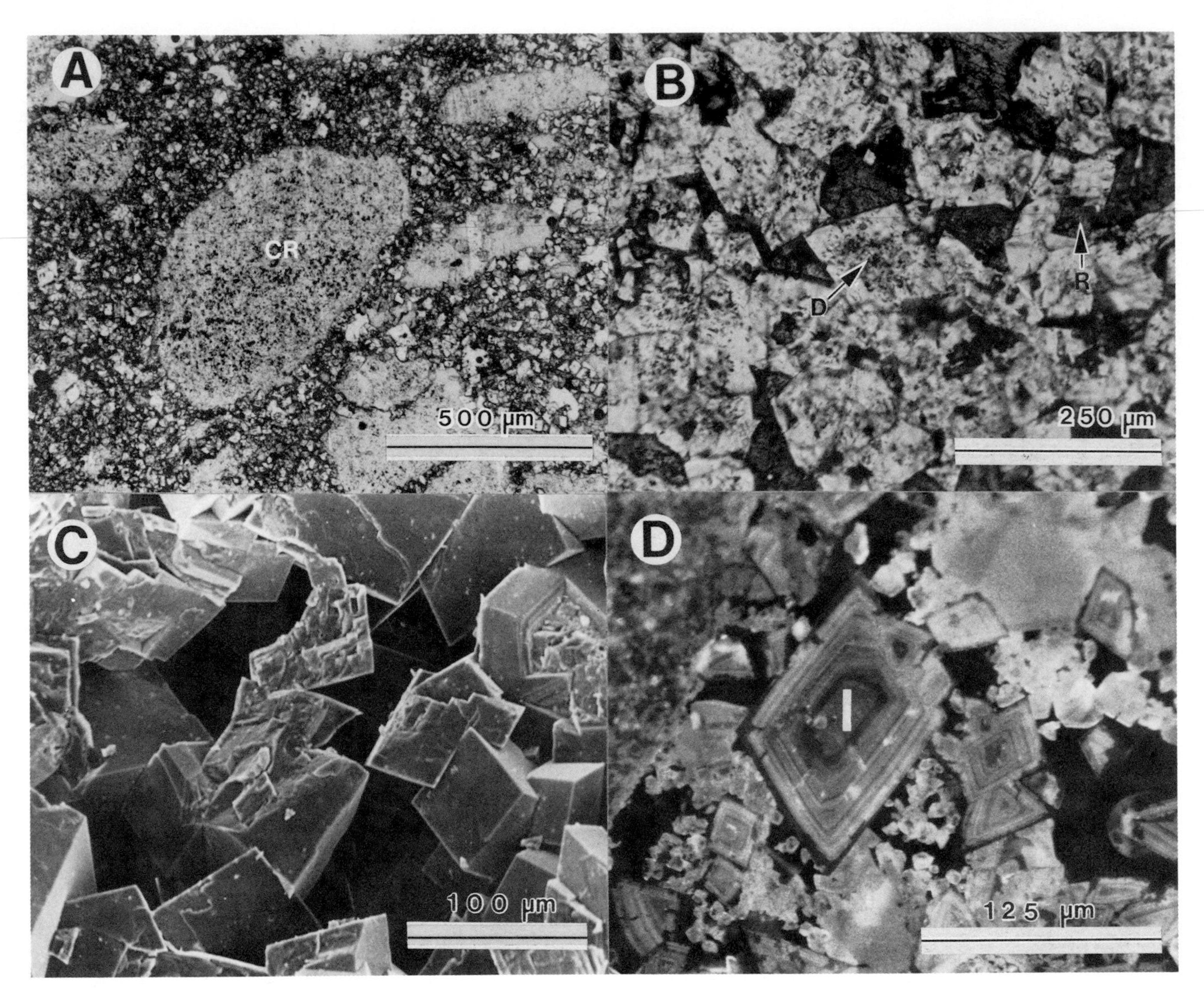

FIG. 2.—Petrography of Burlington-Keokuk Formation dolomites. (A) Dolomite has replaced all lime mud between crinoid grains in a skeletal wackestone. CR = crinoid. (B) Dolomite (D) occurs as loosely interlocking rhombs containing common inclusions that are concentrated in central part of rhombs. Dark gray is calcite cement between rhombs and replacing interior of some rhombs (R). (C) SEM photomicrograph of fractured surface of dolomite showing euhedral rhombs and intercrystalline porosity. (D) Cathodoluminescence of dolomite I (I) showing fine-scale zoning.

supported beds that are intercalated with crinoidal packstone and grainstone beds. Only in southwestern Missouri is dolomite rare (Daniels, 1986). Dolomite has replaced most of the lime mud in the Burlington-Keokuk Formation, and although it is concentrated in mud-supported rocks, it also occurs commonly in packstones and grainstones.

Plane-Light Petrography

Dolomite has selectively replaced former lime mud in the Burlington-Keokuk Formation (Cander, 1985; Kaufman, 1985; Harris and Meyers, 1987). Skeletal grains in dolomitized, former lime mudstone/wackestone are preserved as calcite grains (Fig. 2A), molds, or silicified grains. The abundance of dolomite, therefore, is primarily a function of original lime mud content.

Burlington-Keokuk dolomite rhombs are euhedral to subhedral and typically are 50 to 75 μm in size but range from 10 to 150 μm (Fig. 2B). Inclusion density is variable in individual rhombs; identified inclusions consist of calcite, clays, sulfides (Cander, 1985; Banner, 1986) and two-phase fluid inclusions (Smith, 1984). Solid inclusions are concentrated near dolomite cores, which results in a cloudy-core, clear-rim appearance (Fig. 2B). Dolomite occurs as loosely intergrown rhombs, locally with significant intercrystalline porosity (Fig. 2C), some of which has been filled with calcite cement (Fig. 2B). Dolomite in lime packstones and grainstones occurs as individual rhombs or clusters of rhombs in geopetal sediments or in non-geopetal distributions surrounded by calcite cements. In the latter case, dolomite may have replaced peloids or some other microcrystalline grain or could be cement. In argillaceous layers, dolomite is fine-grained (approximately 20 μm) and euhedral to subhedral.

Coarsely crystalline, ferroan, saddle dolomite occurs in rare Mississippi Valley-type mineralized vugs and postdates the previously described cements and the regional dolomites, which are the main topic of this paper. These saddle dolomites will not be discussed further.

Cathodoluminescent Zonal Stratigraphy of Dolomite

Burlington-Keokuk Formation dolomite can be divided, using cathodoluminescence, into three regionally correlative generations. These generations, from oldest to youngest, are: luminescently zoned dolomite I; luminescently unzoned dolomite II; nonluminescent dolomite III.

Dolomite I consists of cloudy-core, clear-rim rhombs with fine-scale concentric luminescent zoning (Fig. 2D). The luminescence varies from yellow to brown within a single rhomb, and zone width also varies. Cores of rhombs often show irregular luminescence. Cathodoluminescence reveals that crystal faces are commonly etched, and Reeder and Prosky (1986) have documented sector zoning in dolomite I rhombs. Geochemically, dolomite I is Ca rich (54.5 to 56.5 mole percent $CaCO_3$) and contains 800–1,500 ppm Mn, 400–8,000 ppm Fe, the brighter luminescent zones containing the lower Fe and Mn (J. L. Prosky, pers. commun., 1987). Dolomite I has initial $^{87}Sr/^{86}Sr$ values which range from Mississippian sea water (0.7076) to more radiogenic values (0.7081). Stable isotopes are relatively heavy, with mean $\delta^{18}O$ of −0.4‰ PDB (n = 33, range = −2.2 to +2.5‰), and mean $\delta^{13}C$ of +2.5‰ PDB (n = 33, range = −0.9 to +4.0‰; Table 1; Banner and others, this volume).

Dolomite I is the most abundant of the three dolomite generations, accounting for about two-thirds of all Burlington-Keokuk dolomites. It can be correlated throughout the study area. In spite of this widespread correlation, internal zonal patterns in dolomite I vary between localities. Dolomite I zonal patterns *are* correlatable throughout single measured sections, however, (Fig. 3A, B). This consistency of zonation within a single measured section is important for constraining the models for dolomite I formation, as will be discussed later.

Dolomite II is the second most abundant dolomite generation in the study area and has unzoned red to brown cathodoluminescence (Fig. 3C). Dolomite II most commonly occurs as complete rhombs, but it also commonly occurs in the centers of rhombs that have zoned dolomite I rims (Fig. 3D). Dolomite II also occurs as irregular patches within predominantly dolomite I rhombs (Fig. 4A). Where dolomite II occurs as irregular patches or as centers of rhombs, the dolomite I-dolomite II contact typically cross-cuts the fine zoning of dolomite I (Fig. 3D). In these cases, zoned dolomite I rims could be interpreted as syntaxial overgrowths on corroded dolomite II rhombs, or conversely, zoned dolomite I rims could be interpreted as the chemically more resistant relicts after replacement of cores by dolomite II (as proposed by Banner, 1986; J. L. Prosky, pers. commun., 1987). We favor the second interpretation, that dolomite II replaced dolomite I, for the following reasons: dolomite II rhombs are about the same size as dolomite I rhombs, even where they occur in the same sample (Fig. 3D), and dolomite II dolostones have porosities that are statistically the same as those for dolomite I dolostones (A. Cox, unpubl. data). The presence of rhombs composed entirely of dolomite I in a dolomite II dolostone implies that dolomite I fluids moved through these rocks for the same length of time as through dolomite I dolostones. If, in the same sample, dolomite I rims were syntaxial overgrowth on corroded dolomite II cores, these rims should be equal in thickness to about one-half the width of the coexisting totally dolomite I rhombs. In such cases, the dolomite II plus dolomite I overgrowths should be significantly larger than the coexisting dolomite I rhombs, and the dolostones should have lower intercrystalline porosities than dolomite I dolostones. Therefore, the fact that dolostones dominated by dolomite II have about the same rhomb sizes and porosities as those dominated by dolomite I is best explained by dolomite II being a replacement. In the above cases, most dolomite I rhombs cannot be slices through the outer part of rhombs, because such slices (either parallel or oblique to a face) would result in a two-dimensional image that would be unzoned in the center or would have equivalent zones of unequal apparent thicknesses on opposing sides of the crystal. Most dolomite I crystals are zoned throughout and have equivalent zones of equal thicknesses on opposing sides of the rhombs (Fig. 3D).

A second argument for the dolomite II having replaced dolomite I is that dolomite I is more Ca rich than dolomite II (Prosky and Meyers, 1985), and in some cases dolomite I rhombs are more Ca rich in their centers than in their margins. This implies that dolomite I is less stable than dolomite II, and the irregular geometries are thus more reasonably interpreted as dolomite II being the younger. Furthermore, intra-rhomb pores (hollow cores and selectively dissolved zones) occur predominantly in dolomite I; they are rare in dolomite II. This is additional evidence that dolomite I is the less stable and is evidence that fluids have penetrated the interior of rhombs without leaving petrographically visible traces of their pathways.

TABLE 1.—SUMMARY OF GEOCHEMISTRY OF DOLOMITES

	Dolomite I	Dolomite II	Dolomite III
$\delta^{18}O$ PDB	−2.2 to +2.5‰	−0.2 to −6.6‰	−3.5 to −5.9‰
$\delta^{13}C$ PDB	−0.9 to +4.0‰	−1.0 to +4.1‰	+2.6 to +3.1‰
$^{87}Sr/^{86}Sr$	0.7076–0.7081	0.7089–0.7094	—
Fe	400–8,000 ppm	1,600–43,000 ppm	52,800–89,400 ppm
Mn	800–1,500 ppm	1,000–2,600 ppm	2,500–8,200 ppm
$CaCO_3$	54.5–56.5 mole percent	51–52 mole percent	53 mole percent

Values shown are ranges. For more detailed summary, see Banner and others (this volume).
Isotope data are from Banner (1986); trace- and major-element data are from J. L. Prosky (pers. commun., 1987).

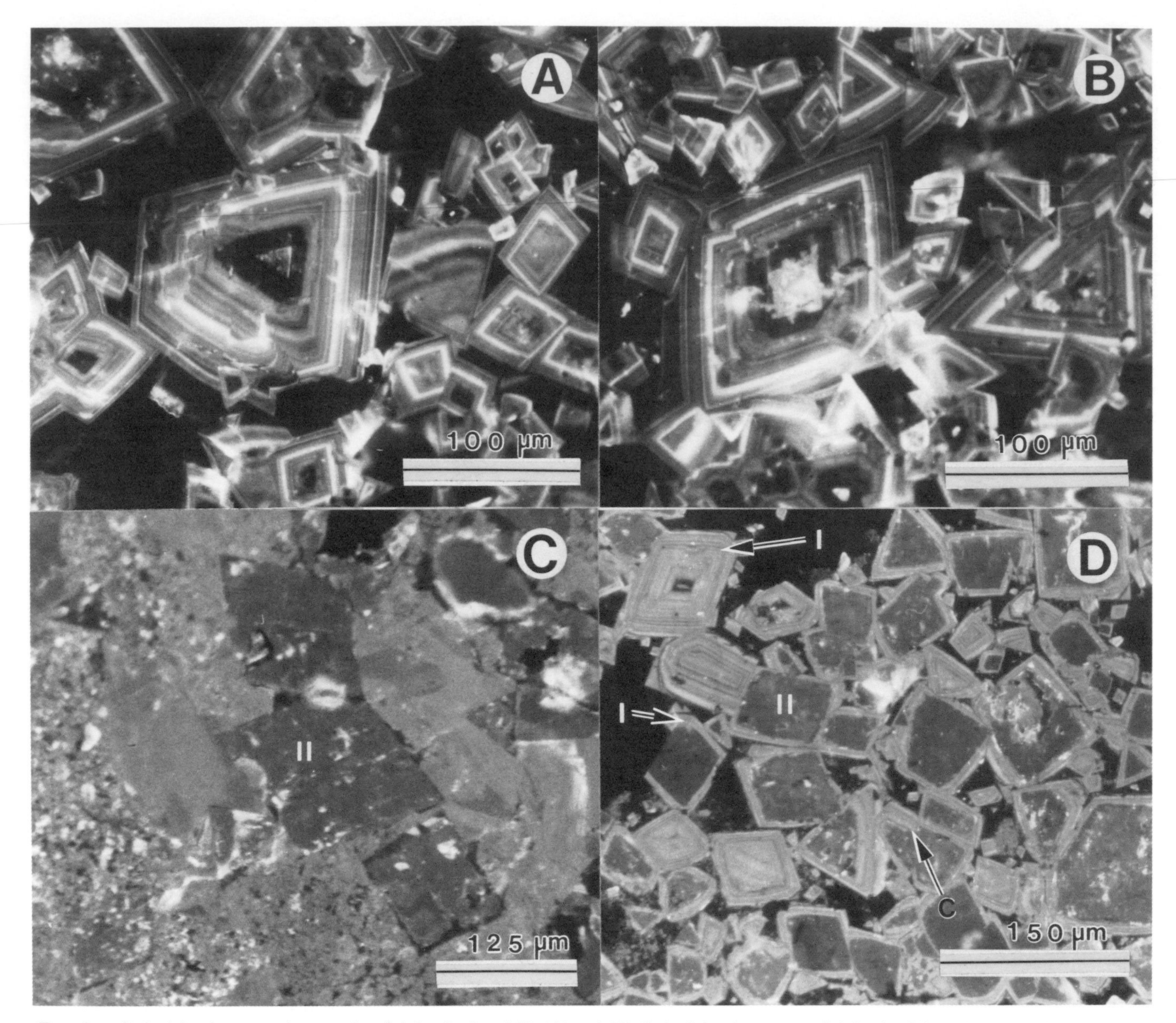

FIG. 3.—Cathodoluminescent photographs of dolomite I and II. (A) and (B) Cathodoluminescence of dolomite I from 2 m and 10 m above base of Burlington Formation within the same measured section showing correlation of fine-scale zoning within a measured section. Note the difference in zoning between these rhombs and those in other sections, such as in Figures 3D and 4B. (C) Dolomite II (II) showing unzoned dull red cathodoluminescence. Small bright patches within some rhombs are interpreted as remnants of dolomite I. (D) Dolomite II (II) forming interiors of rhombs rimmed with dolomite I (I). Dolomite I also forms entire rhombs (I). Note irregular contact (C) between some dolomite I rims and dolomite II interiors. Dolomite II is interpreted as a replacement of dolomite I. Black = intercrystalline pores.

In the cases where rhombs consist entirely of dolomite II, it is unclear whether dolomite II was the initial dolomite phase or whether it completely replaced precursor dolomite I. These "pure" dolomite II rhombs generally coexist with dolomite II rhombs containing bright patches interpreted as relicts of dolomite I (Fig. 3C). Thus, the pure red luminescent dolomite II rhombs are interpreted as replacements of dolomite I.

In addition to the typical red luminescing dolomite II, there is a dark brown variety (dolomite II′), which occurs as entire rhombs or as patches coexisting with the red variety within rhombs and with dolomite I. These patches are interpreted as a replacement of dolomite I and red dolomite II (Banner, 1986; J. L. Prosky, pers. commun., 1987). Dolomite II′ is of limited regional extent.

Geochemically, dolomite II differs markedly from dolomite I. Dolomite II is nearly stoichiometric (51 to 52 mole percent $CaCO_3$) and contains 1,000 to 2,600 ppm Mn and 1,600 to 43,000 ppm Fe, the darker luminescing varieties having the greater Fe and Mn contents (J. L. Prosky, pers. commun., 1987). Dolomite II contains radiogenic Sr (initial $^{87}Sr/^{86}Sr$ = 0.7089 to 0.7094), mean $\delta^{18}O$ of −3.8‰

FIG. 4.—Cathodoluminescent photographs of dolomites I, II, and III. (A) Irregular patch of dolomite II (II) in center of dolomite I rhomb (I). Dolomite II is interpreted as a replacement of dolomite I. (B) Dolomite III (DIII) forms a syntaxial rim on dolomite I (DI). Dolomite III occurs only where rhomb is not encased in pre-Pennsylvanian calcite cement zones II, III, IV (II, III, IV), and dolomite III is encased in calcite cement zone V (V). (C) Dolomite III (III) occurs as an extensive replacement of dolomite I (I). Contacts (C) between the two are generally irregular and cut across zoning of dolomite I. (D) Dolomite III rims (III) on dolomite II (II) cores. Note irregular contacts between the two, interpreted as dolomite III having replaced dolomite II.

PDB (n = 31, range = −0.2 to −6.6‰), and mean $\delta^{13}C$ of +2.7‰ PDB (n = 30, range = −1.0 to +4.1‰; Table 1; Banner and others this volume).

The third major generation of dolomite, dolomite III, is nonluminescent and highly ferroan. Dolomite III typically occurs as syntaxial rims on Dolomites I and II (Fig. 4B). The contacts between dolomite III syntaxial overgrowths and dolomite I substrates may be conformable (Fig. 4B) or irregular, in which case they cut across zoning in dolomite I (Fig. 4C). The conformable geometries are interpreted as dolomite III passively overgrowing precursor rhombs and the unconformable geometries as dolomite III having replaced the precursor rhombs. Similar irregular geometries between dolomite III rims and dolomite II substrates are interpreted as dolomite III being the younger replacement (Fig. 4D). In the cases of irregular contacts, we cannot rule out that there was a significant time gap between dissolution of the older generations and "healing" by dolomite III.

Dolomite III is the most Fe and Mn rich of the three dolomite types (52,800 to 89,400 ppm Fe; 2,500 to 8,200 ppm Mn), and its $CaCO_3$ content (53 mole percent) is intermediate between dolomites I and II (J. L. Prosky, pers.

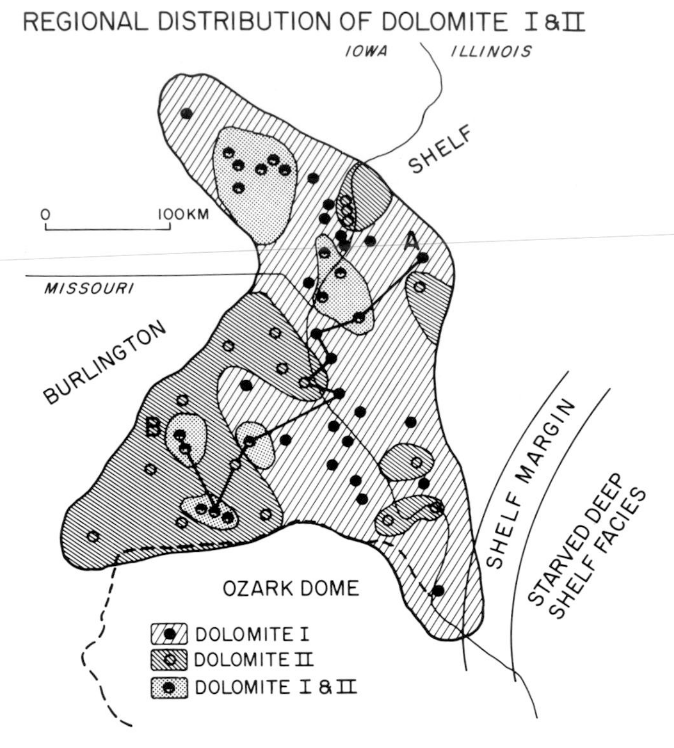

FIG. 5.—Map of distribution of dolomite types based on measured sections shown in Figure 1, plus those of Harris (1982). Most sections do not span entire Burlington-Keokuk Formation, so map represents our best interpretation. Note patchy regional distribution of dolomite II, and its apparent increase in abundance in central Missouri. Dolomite III occurs throughout the region. Line of section, A-B, is that shown in Figure 6.

commun., 1987). Stable isotopes of dolomite III ($\delta^{18}O$ = -3.5 to -5.9‰ PDB; $\delta^{13}C$ = $+2.6$ to $+3.1$‰ PDB, n = 3) fall within the range of those of dolomite II (Table 1; Daniels, 1986).

Regional Distribution of Dolomite Generations

Dolomites I, II, and III have been correlated throughout the study area, covering a region over 100,000 km^2. Regionally, dolomite II occurs in patchy distribution throughout the study area (Figs. 5, 6). Dolomite II may increase in abundance in west central Missouri (Figs. 5, 6); however, incomplete exposure and preservation of the Burlington-Keokuk Formation make this uncertain. Dolomite III occurs in virtually all measured sections in which dolomite occurs. In localities where dolomites I and II coexist, dolomite II is usually limited to the lower Burlington-Keokuk Formation, and dolomite I is unreplaced in the upper strata (Fig. 6). This concentration of dolomite II in the lower part of the formation is a pattern that has also been recognized in Iowa (J. L. Prosky, pers. commun., 1987). As will be discussed later, the concentration of dolomite II in the lower part of the formation is an important constraint on models for dolomitization.

In addition to the complex regional distribution of dolomite generations, cathodoluminescent zoning in dolomite I exhibits two regional patterns. First, there is little variation in luminescent zoning in rhombs at individual localities, regardless of stratigraphic position (Fig. 3A, B). Second, great variability in cathodoluminescent zoning exists between dolomite I rhombs at different localities. In other words, cathodoluminescent zoning in dolomite I rhombs varies regionally between measured sections but not within individual sections of the Burlington-Keokuk Formation.

TIMING OF DOLOMITE GENERATIONS

Dolomite-Calcite Cement Relations

The relative timing of Burlington-Keokuk dolomite generations was constrained by geometric relations with the calcite cement zones. One of the most important of these relations is the pinching out of pre-Pennsylvanian cement zones II through IV in proximity to dolomite I rhombs (Fig. 7A). We interpret this geometry to indicate that the presence of the dolomite rhomb inhibited the growth of the cement zones and therefore predated the cements. Alternatively, the dolomite may have replaced some $CaCO_3$ grains, such as peloids or bryozoans, after the precipitation of calcite cement zone IV. According to this interpretation, these peloids or bryozoans adjacent to crinoids could have inhibited growth of the calcite cement zones. The absence of relicts of the precursor $CaCO_3$ grain, however, and the perfect fit of the dolomite rhomb to the re-entrant in the cement rim (Fig. 7A) argue against this latter interpretation.

The geometric relations between dolomite I and calcite cement zone I are more equivocal because petrographic data support dolomite I both predating and postdating calcite cement I. Specifically, in rare cases, calcite cement zone I pinches out adjacent to dolomite I (Fig. 7B), although the pinchouts are not as clearly related to inhibition by dolomite I as are the pinchouts in calcite zones II through IV. The geometric relations shown in Figure 7B are best interpreted as dolomite I predating calcite cement zone I. Conversely, Figure 7C shows subzones within dolomite I pinching out against the adjacent calcite cement zone I, a geometric relation best interpreted as calcite cement I predating dolomite I. These petrographic relations may indicate that dolomite I and calcite cement zone I precipitated during the same time interval (although they need not be strictly co-precipitates).

Dolomite II shows the same geometric relations with the pre-Pennsylvanian calcite cements as dolomite I. In addition, dissolution voids in dolomite II are filled with cement zone II (Fig. 7D). The simplest interpretation of these features is that dolomite II predated calcite cement zone II. Since dolomite II is a replacement of dolomite I, however, the geometric relations with calcite cements could be inherited from dolomite I. Where dolomite II is not encased in pre-Pennsylvanian cements (zones II–IV), it is encased in calcite zone V. We therefore cannot rule out dolomite II replacing dolomite I after cement zone IV and before zone V.

Dolomite III occurs as syntaxial overgrowths on dolomite I and dolomite II rhombs only where rhombs are not in contact with pre-Pennsylvanian cements (Figs. 4B, 7A). Faces of dolomite I or II rhombs in contact with pre-Pennsylvanian cements do not have dolomite III rims. Dolomite III growth, however, was not inhibited by calcite cement

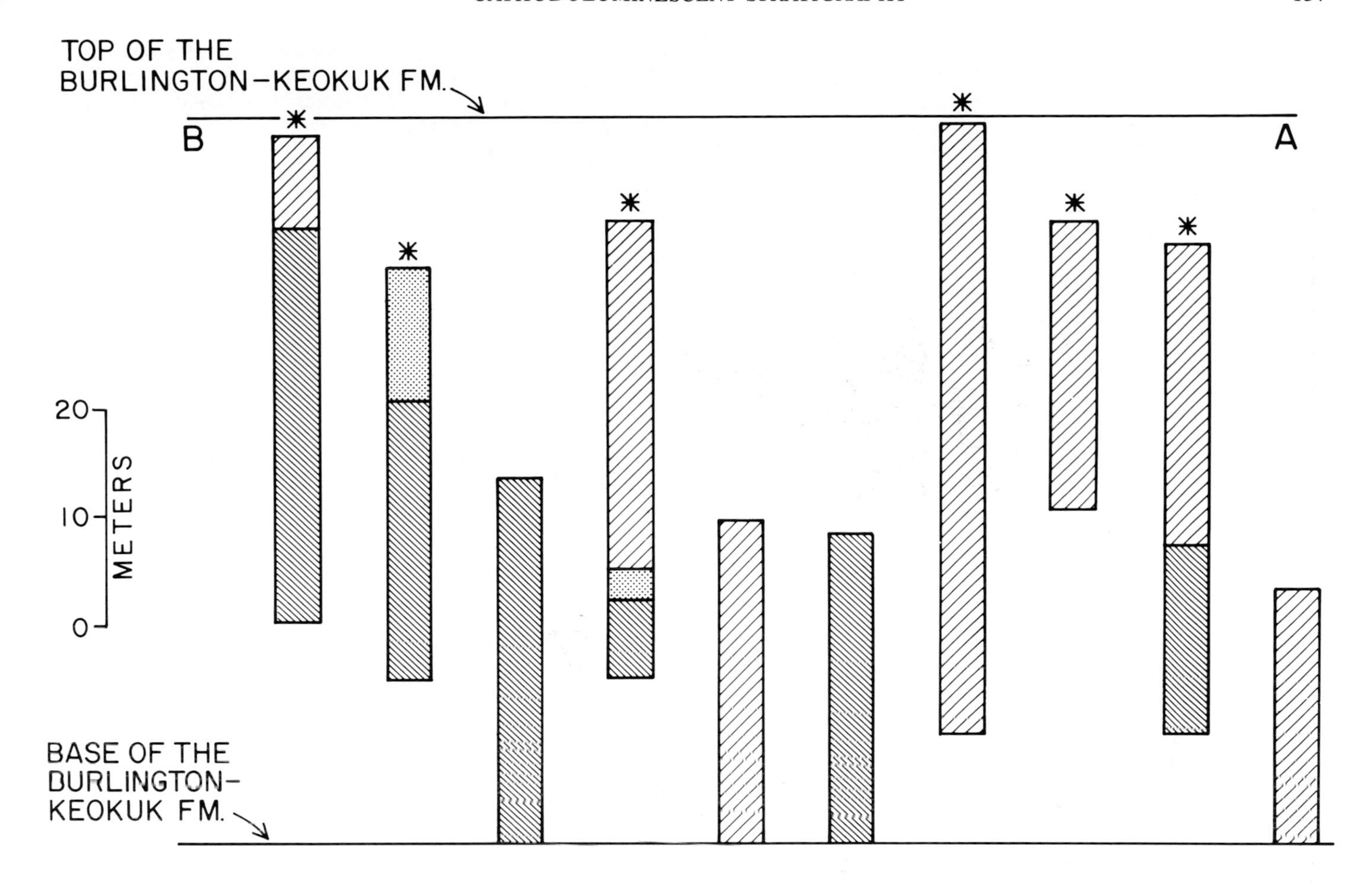

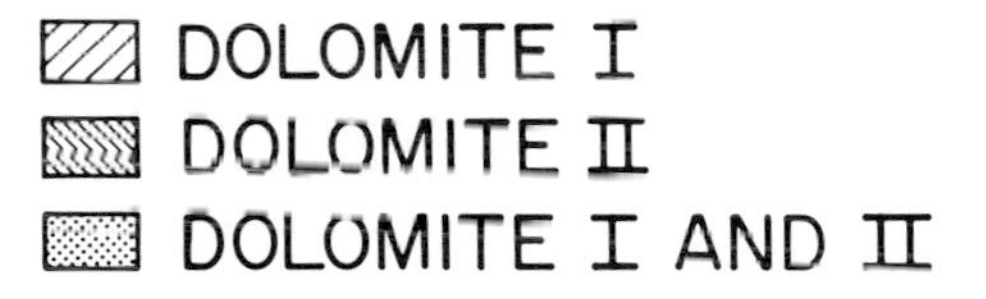

FIG. 6.—Cross section from Figure 5 showing stratigraphic distribution of dolomites I and II. Note that, where both are present within a measured section, dolomite II occurs in the lower part of the section.

zone V, which grew around dolomite III (Fig. 4B). These features indicate that dolomite III postdates calcite cement zone IV and predates cement zone V.

Calcite cement zone V also often fills intercrystalline and moldic porosity in dolomites and partially replaces dolomite rhombs (Fig. 2B). This is true for dolostones dominated by dolomite I, II, or III and is consistent with the previously presented timing relations interpreted from grain-supported rocks.

Dolomite-Chert Relations

A second line of evidence for timing of dolomitization is the age relation between chertification and dolomitization. This is useful for constraining dolomitization, because chert nodules occur as residua at the pre-Pennsylvanian unconformity on the Burlington-Keokuk Formation in rare outcrops in northeastern Missouri (Kaufman, 1985). Furthermore, chert nodules derived from weathering of the Burlington-Keokuk Formation are major components of the fills of pre-Pennsylvanian karst cavities in central Missouri (Daniels, 1986). These features indicate that nearly all nodular cherts formed before the pre-Pennsylvanian unconformity.

In northeastern Missouri, dolomite I and II rhombs occur within chert nodules where they are deeply embayed by microquartz that composes the bulk of the nodules (Fig. 8). This is interpreted as partial replacement of dolomites by microquartz. Accepting a pre-Pennsylvanian age for the cherts, this implies a pre-Pennsylvanian age for dolomites I and II. Dolomite III has not been observed partly replaced by microquartz within chert nodules.

Summary of Paragenetic Sequence

Figure 9 summarizes our interpretation of the sequence of the three major dolomite types (I, II, III), and Figure 10 summarizes the present understanding of the overall para-

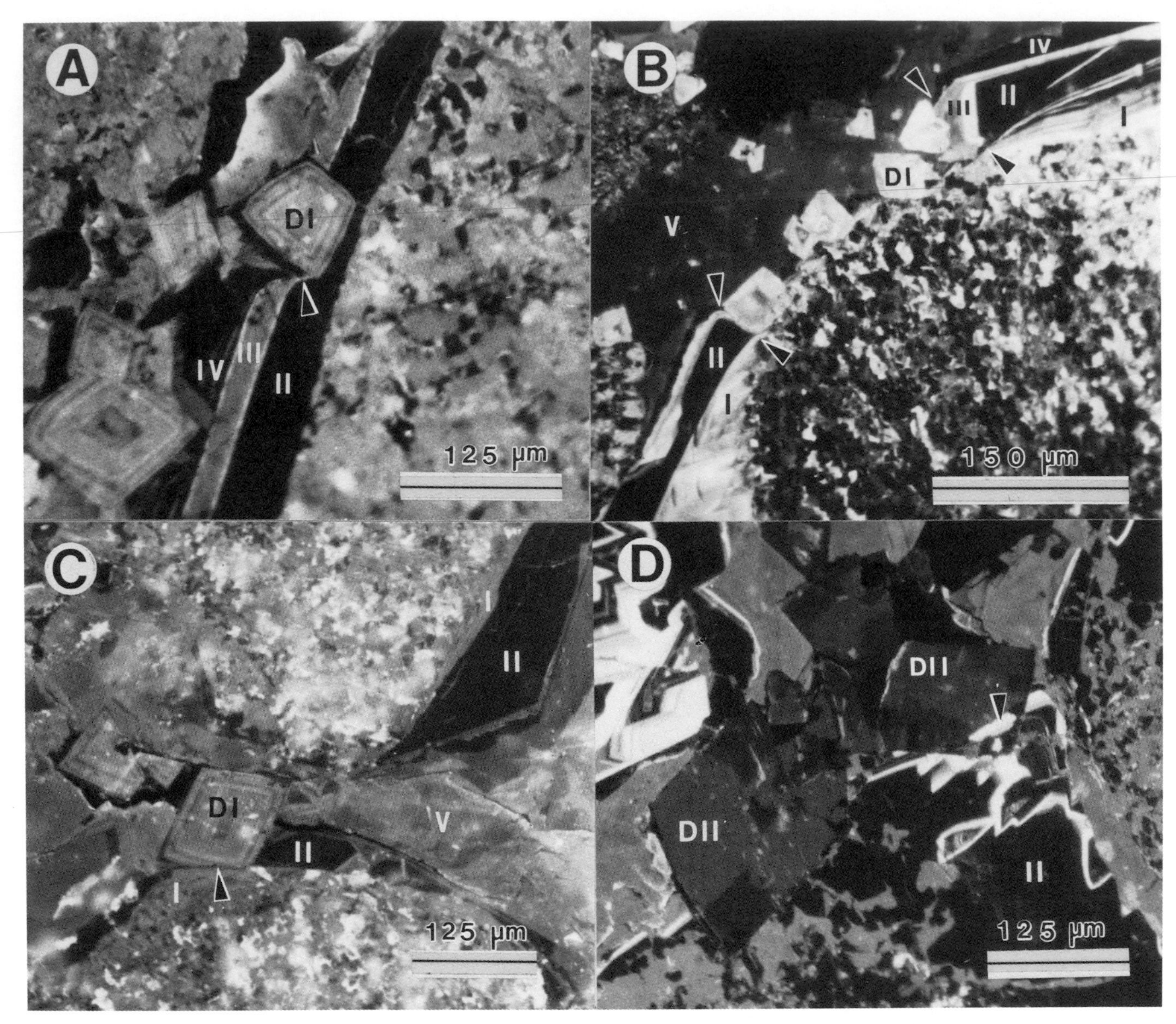

FIG. 7.—Cathodoluminescent photographs of geometric relations between dolomites and calcite cements. (A) Pinchout of cement zones II, III, and IV adjacent to dolomite I rhomb (arrow), indicating that dolomite I predates the cements. II, III, IV = calcite cement zones; DI = dolomite I. (B) Pinchout of cement zones I, II, III, and IV adjacent to dolomite I rhomb (arrows), indicating that dolomite I predates cement zone I (from southeast Iowa). I, II, III, IV, V = cement zones; DI = dolomite I. (C) Pinchout of zone within dolomite I against calcite cement zone I (arrow), indicating that dolomite I postdates cement zone I (from southeast Iowa). I, II, V = cement zones; DI = dolomite I. (D) Calcite cement zone II (II) fills dissolution hole in dolomite II rhomb (arrow). DII = dolomite II.

genetic sequence. Dolomite I predates calcite cement zones II through IV and some nodular chertification and therefore formed before the sub-Pennsylvanian unconformity. Thus, initial dolomitization of most lime muds by dolomite I, and dolomite I precipitation in coarse crinoidal grainstones and packstones, were among the earliest diagenetic events, being predated by some nodular chertification and by some meteoric zone I calcite cementation. These timing relations imply burial depths of less than 500 m during dolomite I precipitation.

Dolomite II formation, mainly as a replacement of dolomite I, also probably predated the sub-Pennsylvanian unconformity, as indicated by its predating some nodular chertification and by its having the same geometric relations with zones II–IV as dolomite I. This suggests burial depths of less than 500 m for dolomite II formation. Dolomite III, on the other hand, probably postdated the sub-Pennsylvanian unconformity, because it has never been observed partly replaced by microquartz within chert nodules. It did, of course, predate calcite cement zone V, the minimum age of which is constrained by Mississippi Valley-type mineralization of probable late Pennsylvanian/early Permian age.

Our interpretation of the pre-Pennsylvanian ages for dolomite I and II is consistent with that of Harris (Harris and Meyers, 1987), who interpreted these dolomites as pre-St.

FIG. 8.—Cathodoluminescent photograph of dolomite-chert relations. Dolomite II (II) rhombs occur encased in microquartz of chert nodule (CH). Note that dolomite II is deeply embayed by microquartz (arrows), indicating replacement by microquartz.

Louis Formation. Our interpretation of dolomite III as post-dating the pre-Pennsylvanian unconformity differs from that of Harris, who interpreted dolomite III as pre-St. Louis Formation.

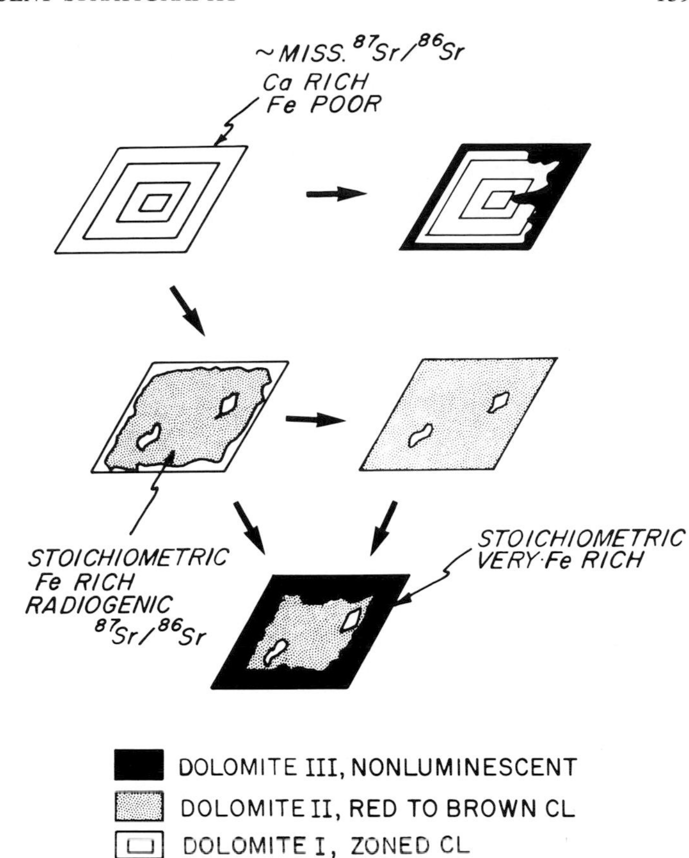

FIG. 9. Summary of geometric relations and geochemistry of three major dolomite types in the Burlington-Keokuk Formation. Dolomite I was commonly replaced by dolomite II, with replacement often selective to interiors of rhombs leaving rims and small remnants of dolomite I. Dolomite I was also completely replaced with dolomite II in many rocks. Dolomite III syntaxially overgrew and replaced dolomite I and dolomite II. In rare cases, all three dolomite types are present in a rhomb.

DOLOMITE MODELS

In this section we discuss the geologic constraints on models for dolomitization as a companion to the discussion by Banner and others (this volume), which summarizes the dolomite geochemistry and presents quantitative geochemical modeling of the data. We discuss models invoking sea

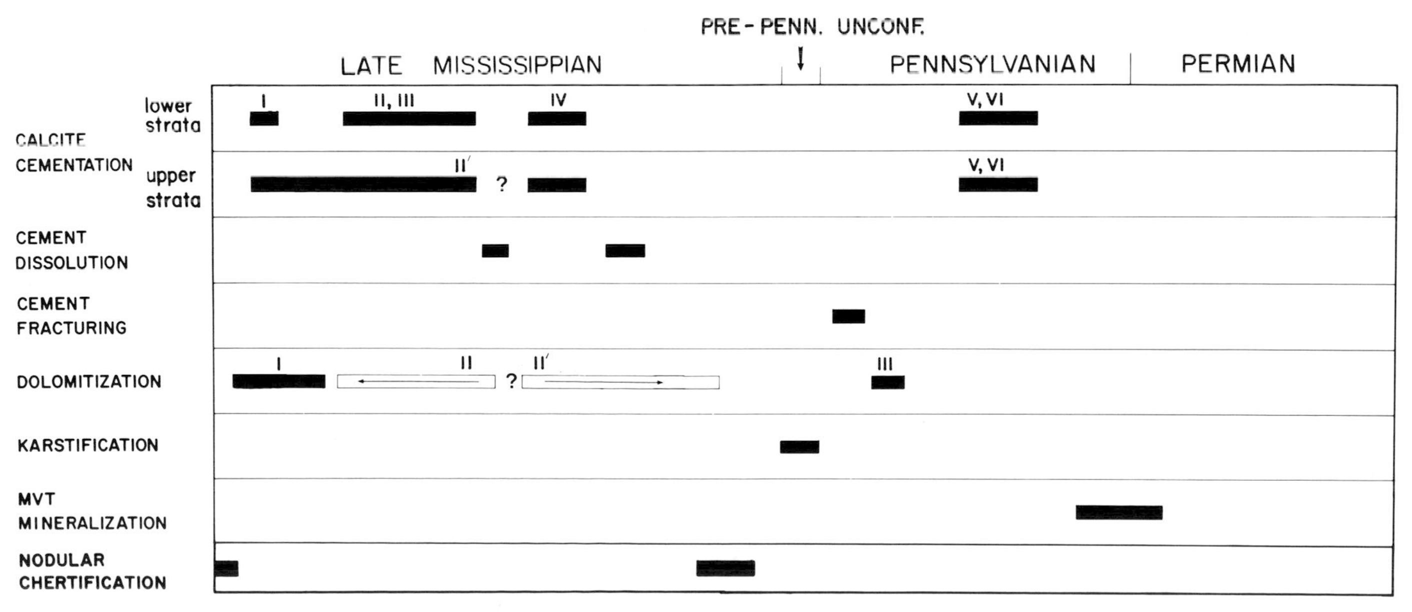

FIG. 10.—Summary of paragenetic sequence for diagenetic components relevant to constraining timing of major Burlington-Keokuk dolomite generations.

water, mixed fresh and sea water, hypersaline waters, and subsurface fluids as possible dolomitizing solutions.

An important consideration in any model for dolomitization is the source for the Mg. A rough estimate of the potential intraformational source can be made by assuming that the Burlington-Keokuk sediments originally composed 100% high-Mg calcites containing 15 mole percent $MgCO_3$, which was redistributed to make dolomite. If 100% of the Mg was incorporated into dolomite, the resultant dolomite would compose about 27% of the total rock volume. Based on our measured sections, the Burlington-Keokuk Formation contains about 14% dolomite by volume, and therefore theoretically intraformational sources would be adequate. It is highly unlikely that 100% of the Mg was used in formation of dolomite, however. Studies in crinoidal shelf limestones of the Lake Valley Formation (Osagean), New Mexico, show that about 75% (by volume) of the original skeletal Mg was retained within the echinoderm grains as microdolomite inclusions (Leutloff and Meyers, 1984). If comparable values are applicable to the Burlington-Keokuk rocks, in which the main nondolomite rocks are crinoidal calcarenites, it would suggest that *available* intraformational Mg could account for about 6 to 7% (by volume) dolomite, or about half of that present. Therefore, the formation of dolomite I probably required import of significant amounts of extraformational Mg. The fact that many of the dolostones have low porosities implies that even greater amounts of Mg had to be imported than in the above mass balance model.

Our interpretation of dolomite II as a replacement of dolomite I implies that the Mg for dolomite II was largely derived from within the formation and therefore required virtually no import of Mg. Since dolomite II is more nearly stoichiometric than dolomite I, a mole-for-mole replacement would require small amounts of nondolomite I Mg to be added. Similarly, dolomite III being partly a syntaxial cement on older dolomites would also require small amounts of Mg to be added from outside the Burlington-Keokuk.

Dolomite I

Seawater dolomitization.—

Precipitation of dolomite I at or near the depositional interface from sea water of normal salinity and Mg content is supported by its early timing, widespread occurrence, and is consistent with the Mississippian seawater Sr isotopes. If dolomite I did predate calcite cement zone I (an interpreted meteoric cement), it would be consistent with a seawater model. Furthermore, the correlation of dolomite I throughout the 100,000 km^2 of study area indicates that the chemistry of the dolomitizing fluids was similar over this region, a feature compatible with seawater precipitation. A seawater model is also attractive because sea water is a prime source for "extraformational" Mg.

Arguing against a simple seawater model, the similarity of cathodoluminescent zoning in dolomite I rhombs throughout the vertical section at any single locality implies that dolomite I precipitated at the same time throughout the thickness of the Burlington-Keokuk strata. This argues against a synsedimentary seafloor origin for dolomite I. Synsedimentary dolomitization by sea water may have occurred, but there is no petrographic record of such a phase. In other words, a seawater dolomite phase was either completely replaced by dolomite I after deposition of the Burlington-Keokuk Formation, or synsedimentary dolomitization by sea water never occurred. In regard to timing, if dolomite I precipitated during or after cement zone I, it is difficult to invoke seafloor seawater dolomitization. Furthermore, the range of oxygen isotope values, many depleted in ^{18}O and ^{13}C relative to expected Mississippian seawater dolomites, also is difficult to reconcile with a seafloor seawater model.

An additional problem with a seawater model is the difficulty of driving sea water through a sediment pile 90+ m thick. It is difficult to envision a process that would provide the flux of sea water needed to provide the necessary allochthonous Mg. A thermally driven Kohout-type convection model (Kohout, 1967) seems unlikely considering the relatively thin stratigraphic section and thus small temperature changes downward (probably less than 15°C temperature range *could,* however, account for the range in $\delta^{18}O$ values for dolomite I, which amounts to about 3‰ for most samples. A Kohout model also requires an intake region for cold sea water, such as the steep margins of high-relief carbonate platforms. The relatively low-relief, gentle and distant (from most of the shallow-shelf area) margin of the Burlington-Keokuk shelf renders this model unlikely.

Another possible model is one invoking refluxing of evaporated sea water from post-Keokuk restricted marine depositional settings, a model patterned after that of Simms (1984). Under this model, sea water with elevated salinities, but below saturation with respect to evaporite minerals, could have been generated during deposition of the Salem, St. Louis, or Ste. Genevieve formations and, due to their excess densities, migrated down through Burlington-Keokuk strata. Dolomites in the Salem Formation have similar cathodoluminescence and stable isotope values to dolomite I, supporting this model (Banner, 1986). This model, as with the synsedimentary sea water model, is difficult to reconcile with the wide range of oxygen isotope values, most of which are depleted in ^{18}O relative to expected marine dolomite values. Assuming dolomitization during St. Louis time, the basal part of the Burlington would have been buried to about 200 m maximum, which could result in a range of burial temperatures of as much as about 6°C (at 10°C/300 m). This would produce a 1.5‰ range of $\delta^{18}O$ in the dolomites, about half the 3‰ range seen in dolomite I values.

Dolomitization by hypersaline evaporated sea water.—

We rule out synsedimentary evaporative reflux models based on the entirely open-marine facies of the Burlington-Keokuk Formation, the absence of intercalated shoreline/peritidal or restricted marine facies, and the absence of evidence of former evaporites within the Burlington-Keokuk strata. The closest possible shoreline facies is in northern Iowa, where Sixt (1983) has recently interpreted the Humboldt beds of the Gilmore City Formation as age equivalent to the Burlington-Keokuk Formation. If this interpretation is correct, there were peritidal and restricted marine conditions in northern Iowa. If synsedimentary hypersaline reflux from shoreline settings was important, we would expect extensive dolomitization of the Humboldt beds, yet they are nearly all limestones, with rare dedolomites.

Dolomitization by mixed meteoric and sea water.—

The early, but post-depositional timing of dolomite I is consistent with mixed-water dolomitization, particularly if it was precipitated over the same time interval as the first meteoric cements (zone I). This is the timing that would be expected for mixing zone dolomitization, that is, during the earliest phases of establishment of a regional fresh-groundwater system, while the lime muds had high porosities and metastable mineralogies. This model can also account for leached crinoids and other skeletal molds, found only in the dolomites, features that are difficult to account for by seawater or hypersaline dolomitization. In addition, the range of stable isotopes, including the ^{18}O-depleted values, is more easily explained by this model than by the above marine and hypersaline models. In regard to the Sr isotopes, the Mississippian marine and the slightly radiogenic values are consistent with this model, as they are with the above seawater models (Banner and others, this volume). The proposed allochthonous Mg in the mixed-water model could have derived from sea water, the "pump" being the gravity-driven groundwater system.

In spite of the many attractive features of the mixed-water model, it is not clear what caused the fine-scale zoning in dolomite I, although it could be due to slight fluctuations in Eh or in crystallization rates as a result of fluctuations of freshwater input. It also is not clear what processes account for the variations in zoning between measured sections.

Subsurface-fluid model.—

A fourth major dolomitization model invokes warm subsurface basinal fluids that migrated into the Burlington-Keokuk strata at shallow burial depths. There are no petrographic features that rule this out for dolomite I. The subsurface-fluid model, however, is not consistent with the Mississippian marine and nearly marine Sr isotopes, since most subsurface fluids are markedly radiogenic (Chaudhuri, 1978; Sunwall and Pushkar, 1979; Steuber and others, 1984). In addition, in the context of the range of oxygen isotope values for dolomite I and dolomite II, the oxygen isotopes of dolomite I are not particularly depleted in ^{18}O and are markedly enriched relative to dolomite II (Banner and others, this volume).

Summary of dolomite I.—

In summary, the petrography and geochemistry of dolomite I are most compatible with a mixed meteoric-seawater model in which the mixing zone was established beneath late Mississippian or early Pennsylvanian subaerial exposure surfaces (Fig. 11A). It is unclear if the main episode of dolomitization predated the St. Louis Formation, as suggested by Harris (Harris and Meyers, 1987), because we have not studied the post-Warsaw formations in any detail. Similarly, although we favor the mixed-water model, the model invoking reflux from post-Warsaw depositional settings requires further testing through more detailed diagenetic studies of the younger Mississippian strata.

Dolomite II and III

In applying the above models to the origin of dolomite II, the highly radiogenic Sr (Banner and others, this volume) and the strongly depleted $\delta^{18}O$ values argue against marine seafloor, hypersaline reflux, or mixing zone models, all of which would predict marine (or nearly marine) oxygen and Sr isotope values. At the same time, there is nothing inherent in the petrographic characteristics that rule out these models. Of particular importance is the concentration of dolomite II in the lower part of the Burlington-Keokuk strata and its patchy regional distribution. These distributional features, coupled with their geochemistry, suggest dolomitization by an extraformationally derived, warm, subsurface fluid that migrated through pre-Burlington strata and penetrated the Burlington-Keokuk strata from below. Under this model, the radiogenic Sr derived from extraformational sources by interaction of the fluids with pre-Burlington sedimentary rocks or with basement rocks. Our petrographic data do not clearly define flow directions, which could have been from the north (Transcontinental Arch) or from the south. Dolomite II and III seem to be more abundant in central Missouri, however, and this, combined with the northward increase in $\delta^{18}O$ in dolomite II (suggesting northward decrease in fluid temperatures or increase fluid-rock interaction; Banner and others, this volume), suggests a component of fluid flow from the south (Fig. 11B).

Subsurface fluids are generally Mg poor, have low Mg/Ca ratios, and therefore have been questioned as widespread dolomitizing fluids (Land, 1983, 1985). Dolomite II and some of III, however, are replacements of dolomite I, a feature which makes a subsurface-fluid model more attractive, because the fluid would not have had to import large quantities of Mg. We therefore envision dolomite II as having formed mainly from an externally derived warm, saline(?), subsurface fluid which became more Fe and Mn rich with time, and possibly culminated in the very Fe-rich dolomite III. We have no Sr isotope data on dolomite III, but the few stable isotope values are similar to those of dolomite II.

Another consideration in evaluating the subsurface-fluid model is the recognition of two-phase fluid inclusions in dolomite III that have final melting temperatures and pressure-corrected homogenization temperatures that indicate high salinities (about 19 to 20 weight percent total dissolved salt), and warm entrapment temperatures (mean = 109°C; Smith, 1984). Similar two-phase fluid inclusions are found in dolomite I, however, and inclusions with similar salinities but slightly lower temperatures (mean of 85°C) are found in both the pre-Pennsylvanian calcite cements (zones I through IV) and post-Mississippian cements (zones V and VI). Considering the differences in timing, cathodoluminescence, and geochemistry between many of these phases, it raises the possibility that the fluids in the inclusions, and their homogenization temperatures, are second-

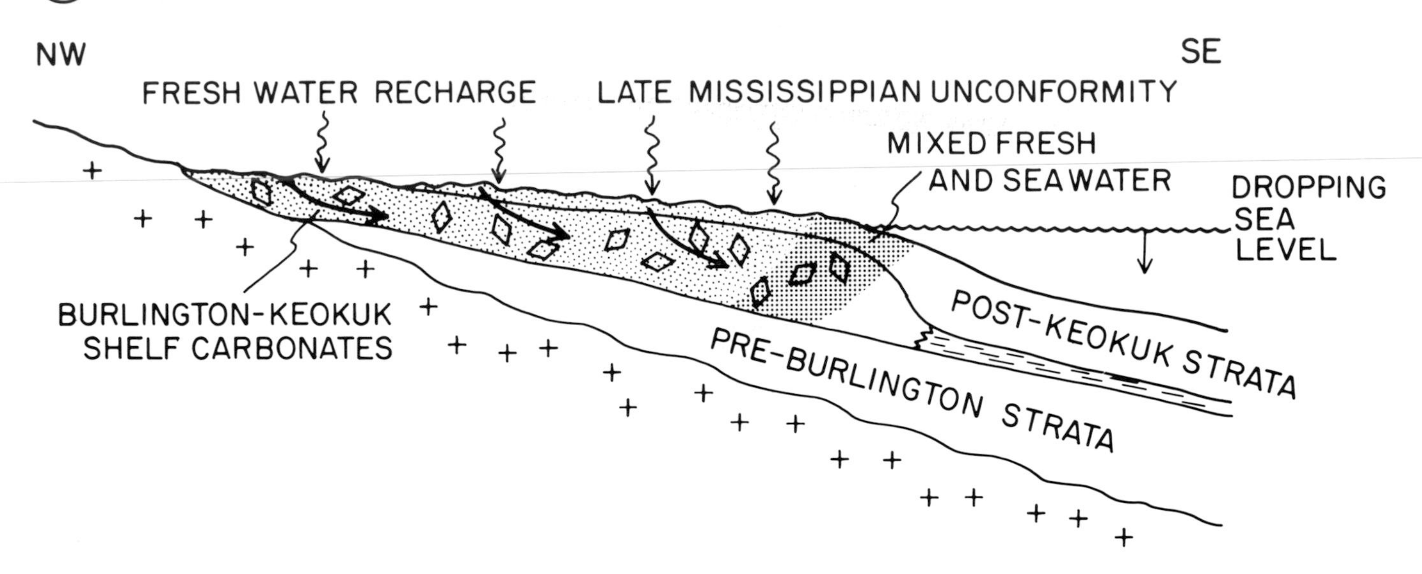

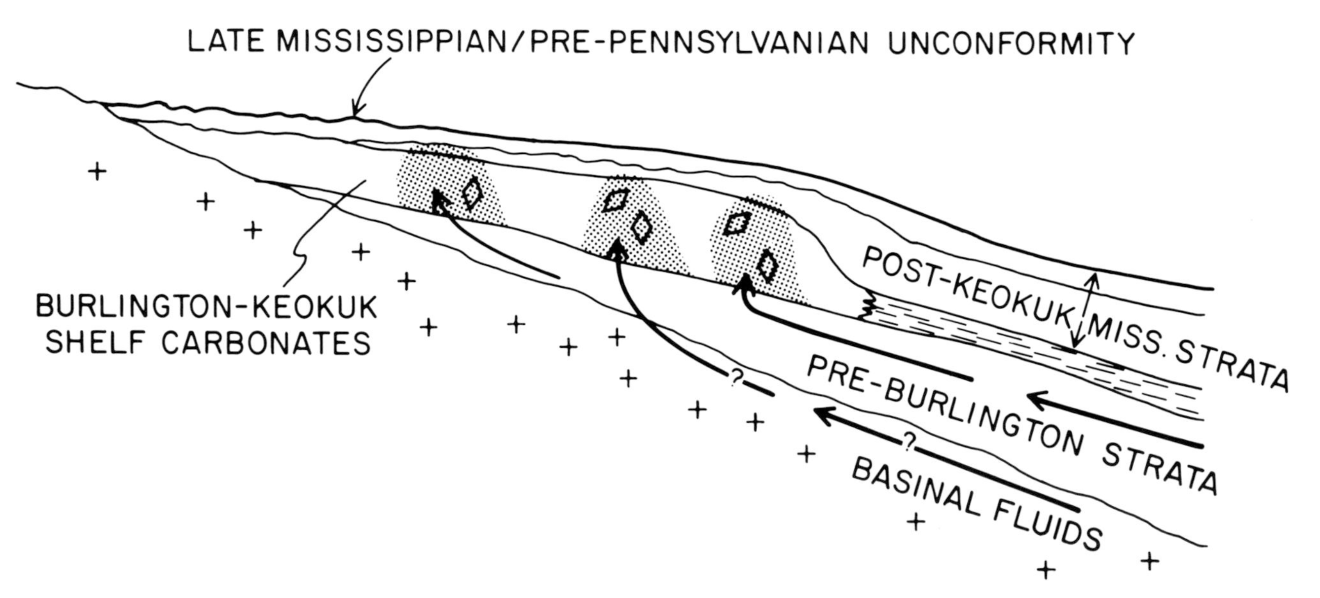

FIG. 11.—Regional models for formation of Burlington-Keokuk dolomites I and II. (A) Dolomite I is interpreted to have formed in a freshwater-seawater mixing zone that moved through the formation during one or more late Mississippian subaerial exposures. Fresh waters were likely recharged over large areas of the subaerial surface but probably had a general southward flow direction. (B) Dolomite II is interpreted to have formed, probably during the late Mississippian, from warm subsurface fluids that migrated through sub-Burlington strata and into the Burlington-Keokuk strata from below. These subsurface fluids probably interacted with pre-Burlington strata and possibly with basement rocks. Although subsurface fluids are shown flowing from the south, major flow components also could have been derived from the north (Transcontinental Arch).

ary. If this is the case, the inclusion brines may represent the fluids that precipitated the youngest cement (zone VI) or may represent younger brines that precipitated the minor occurrences of Mississippi Valley-type ores in the study area. In either case, the fluids in the dolomite inclusions, and in cements I–V, would comprise contaminations by younger brines. In summary, fluid inclusions provide evidence that high-temperature saline brines moved through the Burlington-Keokuk sediments, but it is still an open question whether these are samples of the same fluids that formed the calcite cements and dolomites.

CONCLUSIONS

(1) Cathodoluminescent petrography defines three regionally extensive dolomite generations (dolomites I, II, III) in the Burlington-Keokuk Formation that are correlative throughout an approximately 100,000 km^2 study area in western Illinois and northeastern and central Missouri. Their widespread occurrence attests to the truly regional scale of the dolomitizing paleohydrologic systems.

(2) The cathodoluminescent petrography documents the widespread and extensive replacement of earlier dolomite generations (dolomites I, II) by later generations (dolomites II, III). Geochemical studies (Prosky and Meyers, 1985; Banner, 1986; Banner and others, this volume) demonstrate that this involved the replacement of an early nonstoichiometric dolomite (dolomite I) by younger more stoichiometric dolomites (dolomites II, III), all three generations having distinct trace-element and isotopic chemistries.

(3) Dolomite I, the oldest and volumetrically most abundant generation, is finely cathodoluminescently zoned and replaced most of the lime mud in the study area. Its zonation is consistent within a measured section but not correlatable between measured sections.

(4) Dolomite II, the second most abundant generation, is unzoned and has red to brown cathodoluminescence. Dolomite II is interpreted as a replacement of dolomite I.

(5) Dolomite III, the youngest generation, is nonluminescent and occurs as syntaxial overgrowths on, and replacements of, dolomites I and II. It is a widespread but volumetrically minor type of dolomite.

(6) Dolomite II is largely restricted to the lower part of the Burlington-Keokuk strata and has a patchy regional distribution.

(7) On the basis of geometric relations with pre-Pennsylvanian meteoric calcite cements and with pre-Pennsylvanian cherts, dolomite I formed before calcite cement zones II through IV and therefore is pre-Pennsylvanian. The age of dolomite II is uncertain since it is a replacement of dolomite I, but it too is probably pre-Pennsylvanian. Dolomite III postdates the pre-Pennsylvanian cements but predates late-stage, post-Mississippian/pre-Permian cements.

(8) Integration of dolomite distribution, timing, petrographic, and geochemical characteristics suggests that dolomite I formed in a seawater-freshwater mixing environment associated with a regional meteoric-groundwater system established beneath late Mississippian/early Pennsylvanian unconformities. The requisite Mg probably was derived from intraformational Mg calcites and from sea water.

(9) Dolomite II represents widespread replacement of dolomite I in warm, relatively Fe- and Mn-rich subsurface fluids. These fluids migrated to shallow burial depths through pre-Burlington strata and possibly basement rocks, penetrating the Burlington-Keokuk strata from below in "plumes" having a patchy regional distribution. Most of the Mg for dolomite II was probably derived from dolomite I precursors.

(10) Dolomite III probably represents continuation of the influx of progressively more Fe- and Mn-rich subsurface fluids replacing and overgrowing older dolomites.

ACKNOWLEDGMENTS

This paper is taken from the master's theses of H. Cander, J. Kaufman, and L. Daniels, completed at the State University of New York at Stony Brook under the direction of W. J. Meyers and G. N. Hanson. We are grateful to J. Grams for providing unpublished microprobe data on Burlington-Keokuk cements. Financial support for this research was provided by U.S. Department of Energy contract DE-AC02-83ER-13112A001 held by G. N. Hanson and W. J. Meyers. Additional support was provided by Texaco, Exxon, Chevron, Amoco, Marathon, and Gulf. We thank Steve Dorobek and Vijai Shukla for their thoughtful reviews, which significantly improved the manuscript, and J. Banner and E. Oswald for making numerous helpful suggestions.

REFERENCES

ATHERTON, E., COLLINSON, C., AND LINEBACK, J. A., 1975, Mississippian System, *in* Williams, H. B., Elwood Atherton, T. C. Buschbach, Charles Collinson, J. C. Frye, M. E. Hopkins, J. A. Lineback, and J. A. Simon, eds., Handbook of Illinois Stratigraphy: Illinois State Geological Survey, Bulletin 95, Urbana, Illinois, 261 p.

AVCIN, M. J., AND KOCH, D. L., 1979, The Mississippian and Pennsylvanian (Carboniferous) Systems in the United States Iowa: U.S. Geological Survey Professional Paper No. 1110-M, 13 p.

BANNER, J. L., 1986, Petrologic and geochemical constraints on the origin of regionally extensive dolomites of the Burlington-Keokuk Formations of Iowa, Illinois, and Missouri: Unpublished Ph.D. Dissertation, State University of New York, Stony Brook, New York, 368 p.

BRETZ, D. H., 1950, Origin of the filled sink structures and circle deposits of Missouri: Geological Society of America Bulletin, v. 61, p. 789–934.

CANDER, H. S., 1985, Petrology and diagenesis of the Burlington-Keokuk Limestone, Illinois: Unpublished M.S. Thesis, State University of New York, Stony Brook, New York, 406 p.

CARLSON, M. P., 1979, The Nebraska-Iowa Region, *in* Craig, L. C., and Varnes, K. L., eds., Paleotectonic Investigations of the Mississippian System in the United States, Part I: Introduction and Regional Analyses of the Mississippian System: U. S. Geological Survey Professional Paper No. 1010-F, p.107–114.

CHAUDHURI, W., 1978, Strontium isotopic composition of several oil-field brines from Kansas and Colorado: Geochimica et Cosmochimica Acta, v. 42, 329–331.

COLLINSON, C., 1964, Western Illinois: 28th Annual Tri-State Field Conference, Quincy, Illinois, Illinois Geological Survey Guidebook Series 6, 30 p.

DANIELS, L. D., 1986, Diagenesis and paleokarst of the Burlington-Keokuk Formation (Mississippian), central and southwetern Missouri: Unpublished M.S. Thesis, State University of New York, Stony Brook, New York, 403 p.

FRANK, M. H., AND LOHMANN, K. C., 1982, Cathodoluminescent and isotopic analysis of diagenetically altered dolomite. Bonneterre Formation, southeast Missouri: Geological Society of America, Annual Meeting Abstracts, p. 490–491.

GRAMS, J. C., 1987, Trace element geochemistry of calcite cement in the Burlington-Keokuk Formation, southeastern Iowa: Unpublished M.S. Thesis, State University of New York, Stony Brook, New York, 193 p.

———, AND MEYERS, W. J., 1987, Trace-element variation in calcite cements of the Burlington-Keokuk Limestone: Society of Economic Paleontologists and Mineralogists Annual Midyear Meeting, Austin, Texas, Abstracts, v. IV, p. 31.

GUTSCHICK, R. C., AND SANDBERG, C. A., 1983, Mississippian continental margins of the conterminous U.S., *in* Stanley, D. J., and Moore, G. T., eds., The Shelfbreak: Critical Interface on Continental Margins: Society of Economic Paleontologists and Mineralogists Special Publication 33, p. 79–96.

HARRIS, D. C., 1982, Carbonate cement stratigraphy and diagenesis of the Burlington Limestone (Miss.), southeast Iowa, western Illinois: Unpublished M.S. Thesis, State University of New York, Stony Brook, New York, 296 p.

HARRIS, D. C., AND MEYERS, W. J., 1987, Regional dolomitization of subtidal shelf carbonates: Burlington and Keokuk Formations (Mississippian), Iowa and Illinois, in Marshall, J. D., ed., Diagenesis of Sedimentary Sequences: Geological Society Special Publication No. 36, p. 237–258.

HARRIS, S. E., AND PARKER, M. C., 1964, Stratigraphy of the Osage Series in southeastern Iowa: Iowa Geological Survey Report of Investigations, 52 p.

HOWE, W. B., AND KOENIG, J. W., 1961, The Stratigraphic Succession in Missouri: Missouri Geological Survey and Water Resources, second series, v. 40, 185 p.

KAUFMAN, J., 1985, Diagenesis of the Burlington-Keokuk Limestones (Miss.), eastern Missouri: Unpubl. M.S. Thesis), State University of New York, Stony Brook, New York, 326 p.

KAUFMAN, J., CANDER, H. S., DANIELS, L., AND MEYERS, W. J., 1988, Calcite cement stratigraphy and cementation history of the Burlington-Keokuk Formation (Mississippian), Illinois and Missouri: Journal of Sedimentary Petrology, v. 58, p. 312–326.

KOHOUT, F. A., 1967, Groundwater flow and the geothermal regime of the Floridan Plateau: Gulf Coast Association of Geological Societies, Transactions, v. 17, p. 339–354.

LAND, L. S., 1983, Dolomitization: American Association of Petroleum Geologists Education Course Note Series No. 24, 20 p.

LAND, L. S., 1985, The origin of massive dolomite: Journal of Geological Education, v. 33, p. 112–125.

LANE, H. R., AND DEKEYSER, T. L., 1980, Paleogeography of late early Mississippian (Tournasian 3) in the central and southwestern United States, *in* Fouch, T. D., and Magathan, E. G., eds., Paleozoic Paleogeography of the West-Central United States: West-Central U.S. Paleogeography Symposium: Rocky Mountain Section, Society Economic Paleontologists and Mineralogists, p. 149–162.

LEACH, D. L., AND ROWAN, L., 1986, Genetic link between Ouachita foldbelt tectonism and the Mississippi Valley-type lead-zinc deposits of the Ozarks: Geology, v. 14, p. 931–935.

LEUTLOFF, A. H., AND MEYERS, W. J., 1984, Regional distribution of microdolomite inclusions in Mississippian echinoderms from southwestern New Mexico: Journal of Sedimentary Petrology, v. 54, p. 432–446.

LINEBACK, J. A., 1981, The eastern margin of the Burlington-Keokuk carbonate bank in Illinois: Institute of Natural Resources, State Geological Survey Division, Circular 520, 24 p.

MACHEL, H. G., 1985, Facies and diagenesis of the Upper Devonian Nisku Formation in the subsurface of central Alberta: Unpublished Ph.D. Dissertion, McGill University, Montreal, Canada, 300 p.

PROSKY, J. L., AND MEYERS, W. J., 1985, Nonstoichiometry and trace element geochemistry of the Burlington-Keokuk dolomites: Society of Economic Paleontologists Mineralogists Annual Midyear Meeting, Golden, Colorado, Abstracts, v. II, p. 73.

REEDER, R. J., AND PROSKY, J. L., 1986, Compositional sector zoning in dolomite: Journal of Sedimentary Petrology, v. 56, p. 237–247.

SIMMS, M., 1984, Dolomitization by groundwater-flow-systems in carbonate platforms: Gulf Coast Association of Geological Societies, Transactions, v. 34, p. 411–420.

SIXT, S. C. S., 1983, Depositional environments, diagnesis and stratigraphy of the Gilmore City Formation (Mississippian) near Humboldt, north-central Iowa: Unpublished M.S. Thesis, University of Iowa, Iowa City, Iowa, 164 p.

SMITH, F. S., 1984, A fluid inclusion study of the Burlington Limestone (Mississippian), southeastern Iowa and western Illinois: Unpublished M.S. Thesis, State University of New York, Stony Brook, New York, 219 p.

STEUBER, A. M., PUSHKAR, P., HETHERINGTON, E. A., 1984, A strontium isotopic study of Smackover brines and associated solids, southern Arkansas: Geochimica et Cosmochimica Acta, v. 48, p. 1063–1077.

SUNWALL, M., AND PUSHKAR, R., 1979, The isotopic composition of strontium in brines from petroleum fields of southeastern Ohio: Chemical Geology, v. 24, p. 189–197.

WISNIOWIECKI, M. J., VAN DER VOO, R., MCCABE, C., AND KELLY, W. C., 1983, A Pennsylvanian paleomagnetic pole from the mineralized Late Cambrian Bonneterre Formation, southeast Missouri: Journal of Geophysical Research, v. 88, p. 6540–6548.

WU, Y., AND BEALES, F. W., 1981, A reconnaissance study by paleomagnetic methods of the age of mineralization along the Viburnum Trend, southeastern Missouri: Economic Geology, v. 76, p. 1879–1894.

SEDIMENTOLOGY AND GEOCHEMISTRY OF A REGIONAL DOLOSTONE: CORRELATION OF TRACE ELEMENTS WITH DOLOMITE FABRICS

VIJAI SHUKLA[1]

Texaco U.S.A., Exploration and Production Technology Division, 3901 Briarpark, Houston, Texas 77042

ABSTRACT: The Interlake Formation (Silurian) in North Dakota is a 366-m-thick dolostone unit representing dolomitization of subtidally deposited lime grainstones and minor lime mudstones. Dolomitization was early and resulted from: (1) hypersaline brines in supratidal settings; and (2) mixing of marine fluids with hypersaline brines and/or meteoric water. Burial dolomites are quantitatively minor. Three types of dolostone fabrics are described: grain-supported fabrics (GSF), pervasive-dolostone fabrics (PDF), and particle-relict fabrics (PRF). Grain-supported fabric (GSF) dolostones result from replacement of lime grainstones or packstones, whereas PDF dolostones lack evidence of precursor depositional textures and fabrics; PRF dolostones have intermediate fabrics consisting of ghosts and relicts of particles in a mosaic of dolomite crystals.

Grain-supported fabric (GSF), PDF, and PRF dolostones were analyzed for Sr, Mn, Fe, Na, and B. In general, GSF have elevated Mn and Na, whereas Fe, Sr and B are depressed. In contrast, PDF have elevated Fe, Sr and B, whereas Mn and Na are depressed. A more complicated picture emerges if PRF dolostones are considered. These dolostones have elevated B and Sr, whereas Mn, Fe, and Na are depressed. In contrast to fabric, dolomite texture does not show systematic trends in element distribution. These variations cannot be easily explained, because factors controlling the incorporation of these trace elements into various dolomite fabrics are poorly understood. Nevertheless, significant differences between PDF and GSF dolostones indicate that concentrations of some trace elements (Mn and Na) in dolostones may be influenced by the concentration of those elements in precursor minerals.

INTRODUCTION

Inorganic geochemical concepts are being routinely used in studying dolomites and dolomitization models (Morrow, 1982a, b; Arthur and others, 1983; Land, 1985; Machel and Mountjoy, 1986). Studies linking geochemistry with major fabrics of dolostones, however, are uncommon (e.g., Veizer and others, 1978). Geochemical modeling is a very common approach to determining dolomitization chemohydrology (Hardie, 1987), but rarely are the following questions addressed: (1) can different fabrics in dolostones be correlated with variations in trace-element concentrations? (2) how can such correlations be explained? and (3) why do dolostones containing precursor relicts have significant differences in trace-element concentrations compared to pervasive dolostones?

This paper addresses these questions by describing trace-element characteristics of the Interlake dolostone (Silurian), Williston basin, North Dakota. Research in this aspect of dolostone geochemistry is in the inductive stage (Johnson, 1933; Matthews, 1986). Accordingly, this study offers a preliminary hypothesis (Johnson, 1933), rather than a unique solution. More research is required to advance the inductive stage to a deductive mode. The terms "fabric" and "texture" are used as defined by Friedman (1965, p. 646): "fabric" refers to size and mutual relations of crystals, whereas "texture" refers to shape of crystals.

GEOLOGIC HISTORY AND PREVIOUS WORK

The Williston basin is located in southern Canada and the northern United States and is one of the largest cratonic basins in North America (Fig. 1A; Sloss, 1985; Johnson and Lescinsky, 1986). The Middle Cambrian to Middle Ordovician section is composed of mostly siliciclastic rocks, whereas the Upper Ordovician to Upper Mississippian section is mostly carbonates and evaporites. The remainder of the Paleozoic sequence is siliciclastic rock (Peterson and Maccary, 1985).

The depositional center of the basin has migrated through geologic time (Fig. 1A). During deposition of Interlake carbonates (29 Ma, from Early to Middle Silurian), the depocenter was located in McKenzie County, where approximately 4,880 m of sedimentary rock (including the Interlake) overlie the Precambrian basement (Carlson and Anderson, 1965). Subsequently, the entire area underwent prolonged erosion for approximately 38 Ma (Early Silurian to Early Middle Devonian).

The Interlake Formation (Silurian) was named by Baillie (1951) for rocks present between the Ashern (Middle Devonian) and the Stony Mountain (Upper Ordovician) Formation in Manitoba (Fig. 1B). The Interlake Formation varies in thickness from approximately 366 m in the basin center to less than 3 m at the erosional edge near the eastern limit of the basin. The formation has a simple structure with deepest portions present in McKenzie County, where the Nesson anticline is prominent.

Previous workers have interpreted the Interlake Formation as low-energy, open-marine carbonate deposits, reflecting water salinities ranging from normal saline to penesaline. Carlson and Eastwood (1962) interpreted "middle" Interlake strata as deposits of penesaline seas, which had not reached the stage of evaporite precipitation. This interpretation was based on the presence of medium crystalline dolomite, which was implied to have been a primary precipitate. In contrast, they interpreted "upper" Interlake strata as representing deposition near wave base in shallow seas with high-energy conditions (Carlson and Eastwood, 1962). Roehl (1967) arrived at similar conclusions based on the fine crystalline nature of the dolomite. Lobue (1983) interpreted the dolomite as secondary, presenting evidence that it replaced precurser carbonate sediments deposited in twelve lithotopes. Lobue (1982, 1983) described three dolomitizing processes: (1) hypersaline brines; (2) mixing of

[1]Present address: SCI (Shukla Consultants, Inc.), P.O. Box 2158, Portland, Oregon 97208-2158.

Note: The following valences are used throughout this paper: Sr^{2+}, Mn^{2+}, Fe^{2+}, Na^{+}, and B^{3+}.

Sedimentology and Geochemistry of Dolostones, SEPM Special Publication No. 43

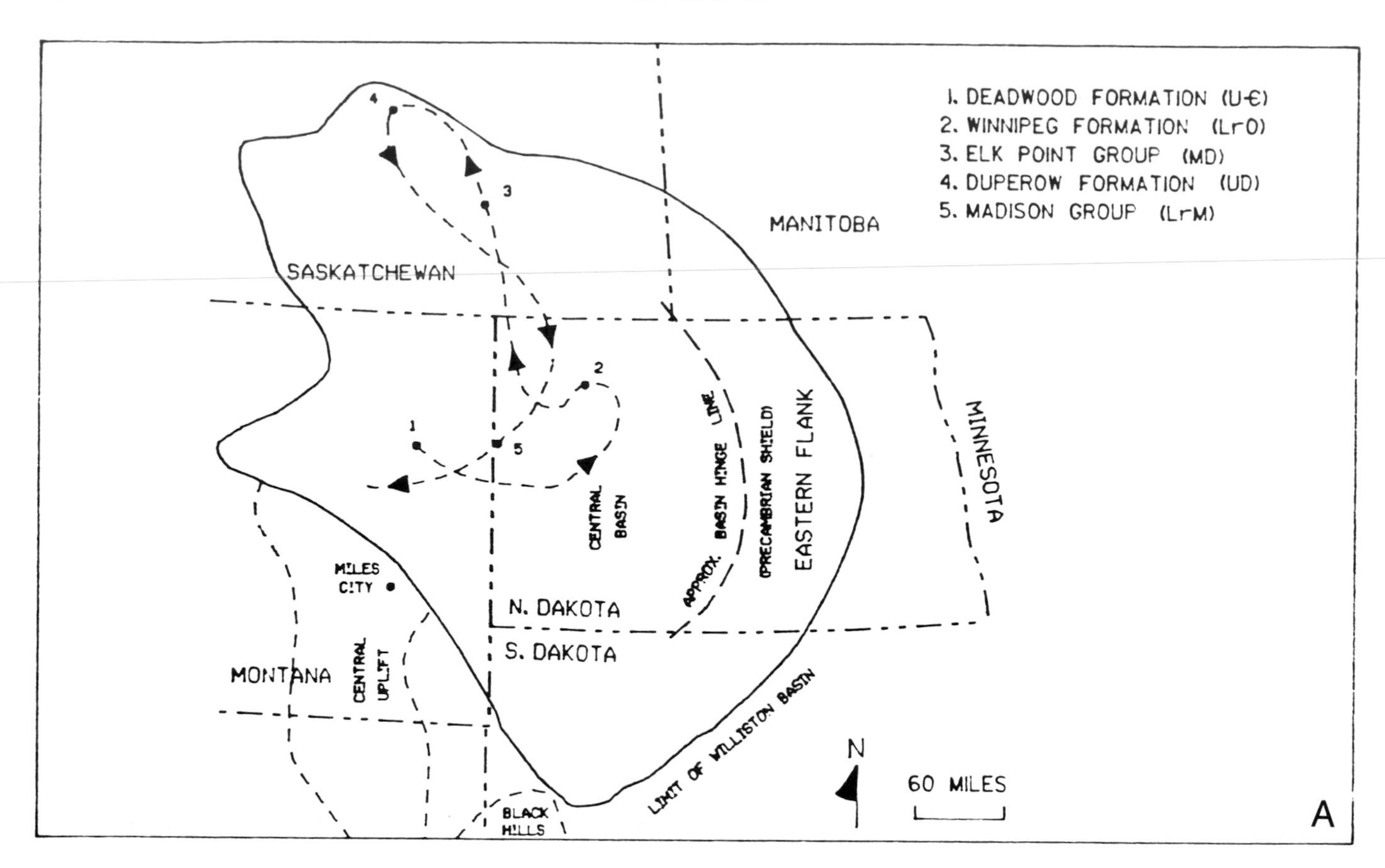

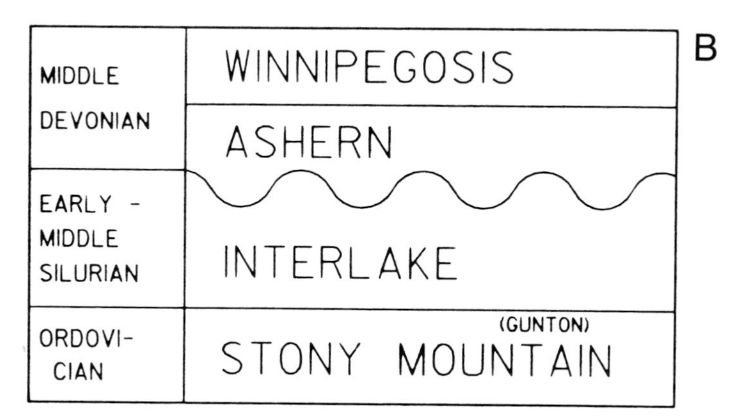

FIG. 1.—(A) Location of the Williston basin in southern Canada and northern United States and migration of the depocenter during the Paleozoic (compiled from Ballard, 1963, and Lobue, 1983). (B) Stratigraphy of the Interlake Formation.

seawater and freshwater in the shallow subsurface; and (3) pressure solution during deep burial.

Shukla (1985) described regional cathodoluminescence and geochemistry and interpreted the Interlake dolomite as a replacement of lime sediment and/or limestone. Dolomitization occurred early in the geologic history of the Interlake from two main processes: hypersaline brines in supratidal settings, and mixing of marine fluids with hypersaline brines and/or meteoric water. Burial dolomites are quantitatively minor. Shukla (1986) made a preliminary correlation of variation in trace elements with dolomite fabric and texture.

METHODS

Core samples and geophysical logs were obtained for nine wells in North Dakota (Fig. 2). Conventional sedimentologic analyses of the samples were complemented with cathodoluminescence, blue-light fluorescence, and trace-element and stable isotope (O, C) analyses.

Cathodoluminescence was studied in a Nuclide ELM-2B Luminoscope®. Operating conditions included: 100-millitorr vacuum, 15-KV voltage, 0.5-MA current, and 20 units of focus.

Fluorescence was studied using a Hg-sourced, blue-light,

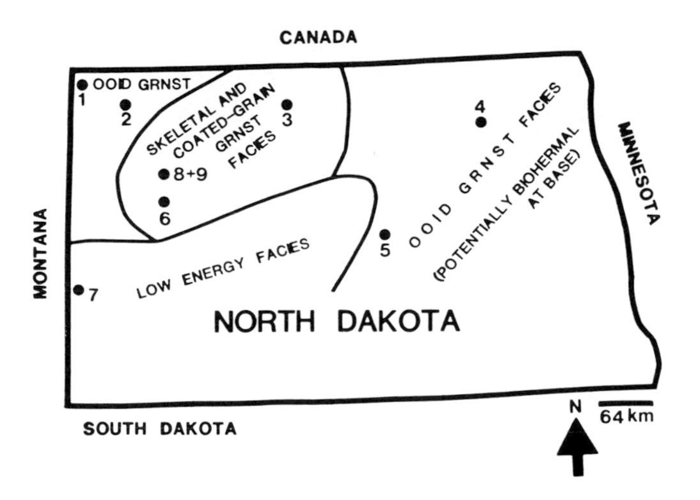

FIG. 2.—Map of North Dakota showing locations of wells used for this study and distribution of depositional facies (GRST = grain-support fabric): (1) North Dakota Geological Survey (NDGS) No. 2010, Carter D. Moore #1, Divide County, 7-T163N-R102W, 9,612–9,640 ft. (2) NDGS No. 548, Oil Development Corporation, Texas O. Gunderson #1, Divide County, 11-T160N-R98W, 10,930–11,126 ft. (3) NDGS No. 38, California Company B. Thompson #1, Bottineau County, 31-T160N-R81W, 6,559–6,750 ft. (4) NDGS No. 20, Union Oil Company Aanstaad Stratigraphic Test #1, Ramsey County, 29-T158N-R62W, 2,148–2,213 ft. (5) NDGS No. 207, Continental Oil Company J. Lueth #1, Wells County, 27-T146N-R73W, 4,663–4,832 ft. (6) NDGS No. 33, S & J Operating Company B. Risser #1, McKenzie County, 12-T149N-R96W, 12,626–12,689 ft. (7) NDGS No. 470, Blackwood and Nichols M. E. Gilman #1, Golden Valley County, 15-T140N-R105W, 11,020–11,065 ft. (8) Tiger Oil Siguardson Trust 11–25, McKenzie County, 25-T153N-R95W, 12,096–12,170 ft. (9) Hunt Oil Haugen #1, McKenzie County, 22-T153N-R95W, 12,045–12,140 ft.

epi-illumination system (Nikon) mounted on a polarizing microscope.

Trace elements (Sr, Mn, Fe, Na, and B) were analyzed using an inductively coupled plasma spectrometer by Instrumental Laboratory Plasma (100 ICP) at an operating temperature of approximately 15,000°C. Standards were prepared from Fisher 75 AA materials. Machine error (relative standard deviation) is based on the atomic structure of elements and is reported in decreasing order: Na = 1.61%, Fe = 1.35%, Mn = 1.30%, B = 0.6%, and Sr = 0.5%.

Stable isotopes ($\delta^{18}O$, $\delta^{13}C$) were analyzed in a Micromass 602-E by VG Isogas. Powdered samples were reacted with 100% phosphoric acid in a vacuum maintained at 29°C (McCrea, 1950). As recommended by Land (1980) δO^{18} results were not corrected for fractionation factors for phosphoric acid-dolomite and calcite-dolomite. Standard deviation in reported results is 0.3 per mil.

DOLOMITE FABRICS AND TEXTURES

The general sedimentology and geochemistry of the Interlake dolostone are summarized in Table 1. The Interlake Formation represents mostly subtidally deposited lime grainstones (even though mudstones are also present), which were later dolomitized. For this study, Interlake dolostone fabrics were divided into three major categories: grain-supported fabrics (GSF), pervasive-dolomite fabrics (PDF), and particle-relict fabrics (PRF).

TABLE 1.—SUMMARY OF INTERLAKE (SILURIAN) DOLOMITES, WILLISTON BASIN, NORTH DAKOTA

Characteristic	Evidence
Early Dolomitization:	
1. Pre-compaction	Fractured dolomite-crusts.
	Isopachous cements dolomitized before compaction.
	Polygonal interparticle boundaries.
2. Pre-pressure solution	Dolomite rhombs truncated by pressure-solution seams.
Supratidal (penecontemporaneous dolomite)	Presence of dolomite crusts.
	Dolomite crystal size 100 μm
	$\delta^{18}O$ = +0.45 to +3.25 per mil PDB
	$\delta^{13}C$ = +2.5 to +3.0 per mil PDB
Eogenetic dolomite (mixing between marine water, and meteoric water or brines	Dolomite crystals in mosaics
	Dolomite crystal size 100–500 μm
	$\delta^{18}O$ = 0.0 to −3.4 per mil PDB
	$\delta^{13}C$ = 0.0 to +2.7 per mil PDB

Grain-supported fabrics (GSF) of dolostones result from replacement of lime grainstones or lime packstones. Absence or presence of predolomite lime mud cannot be established with certainty in many dolostones that have dolomitized particles in a grain-supported framework. Thus, these dolomites cannot be termed "dolomite grainstones" or "dolomite packstones" (Dunham, 1962). Therefore, the general term "grain-supported" is used for these fabrics.

Examples of GSF dolostones are: coated-grain GSF, including distinct ooid GSF, and skeletal GSF (Fig. 3). Ooid GSF dolostones developed in northwestern and eastern North Dakota, whereas skeletal and other coated-grain GSF dolostones developed in the depocenter.

Pervasive-dolomite fabric (PDF) dolostones are similar to "crystalline carbonates" of Dunham (1962). This is an umbrella term, which was proposed for carbonate rocks lacking evidence of depositional textures and fabrics. Figure 4 shows examples of PDF Interlake dolostones.

Intermediate between GSF and PDF are those dolostone fabrics that contain relicts of dolomitized particles which are not in a grain-supported framework. These are termed particle-relict fabrics (PRF) and are shown in Figure 4.

Dolomitization of Interlake calcium carbonate occurred prior to significant burial, even though burial-related dolomites are present (Table 1). Evidence for early dolomitization includes: (1) pre-compaction dolomitization of isopachous submarine cements; (2) pre-compaction dolomitization of originally nondolomite allochems; (3) fractured dolomite-crusts; and (4) truncation of dolomite crystals by pressure-solution seams. Thus, Interlake dolomitization was an early process and was essentially completed prior to the Siluro-Devonian unconformity.

RESULTS

Correlation of Trace Elements with Dolomite Fabric and Texture

Figure 5 shows downcore trends in concentration of Sr, Mn, Fe, Na, and B with respect to dolomite fabrics (GSF,

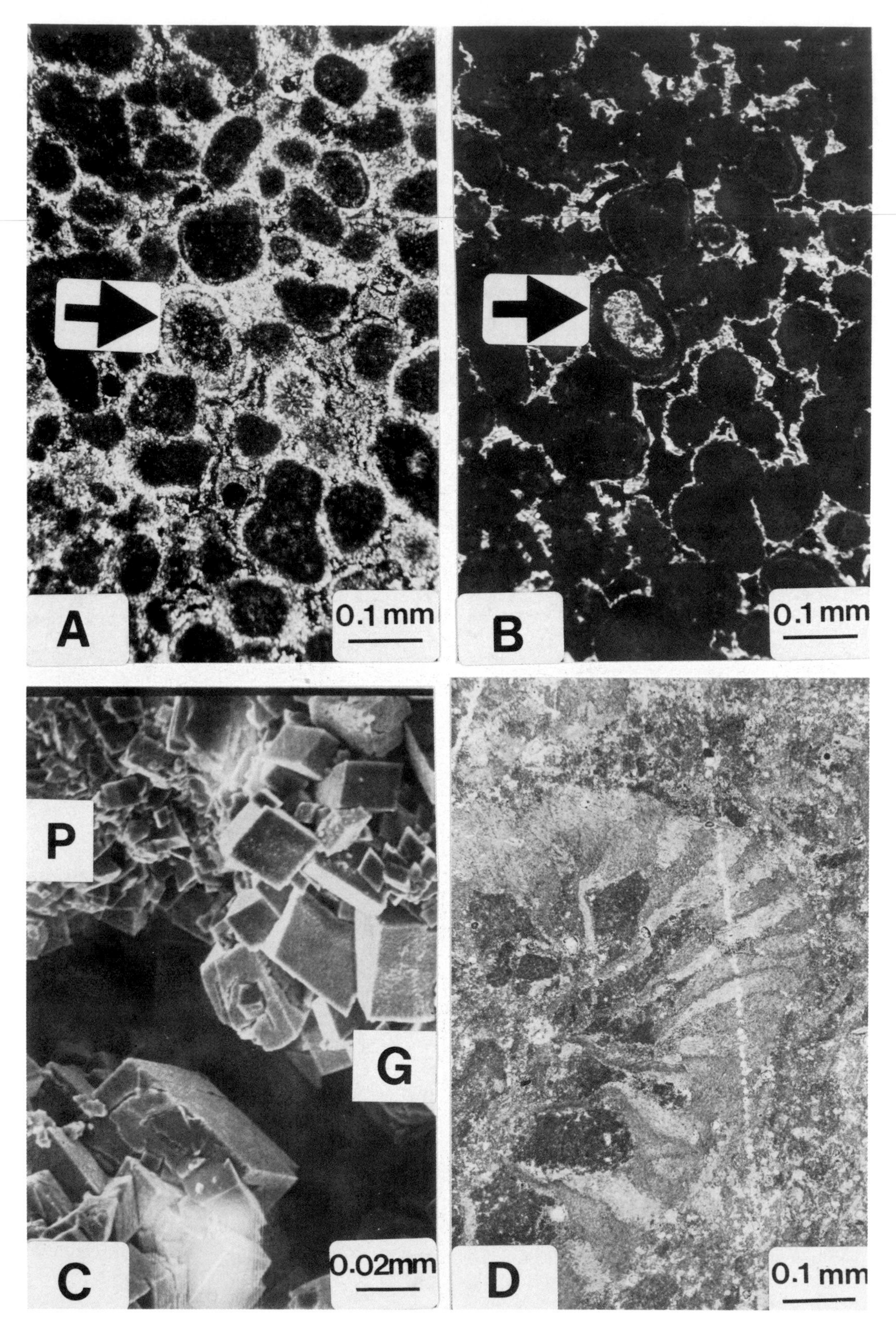

FIG. 3.—Grain-supported fabrics in Interlake dolostones. (A) Dolomitized oolite showing grain-supported framework of ooids (arrow). Isopachous submarine cements were dolomitized prior to significant compaction. This is indicated by presence of dolomite at point contacts and plane contacts between ooids. Uncrossed polars. Conoco Lueth, 1,441 m. (B) Same field in cathodoluminescence. Note the presence of preburial luminescing dolomite rims on nonluminescent ooids. (C) SEM view showing dolomite crystal size difference between particle (P) and groundmass (G). Hunt Haugen, 3,701 m. (D) Dolomitized coral fragment in a grain-support fabric. Uncrossed polars. Conoco Lueth, 1,461 m.

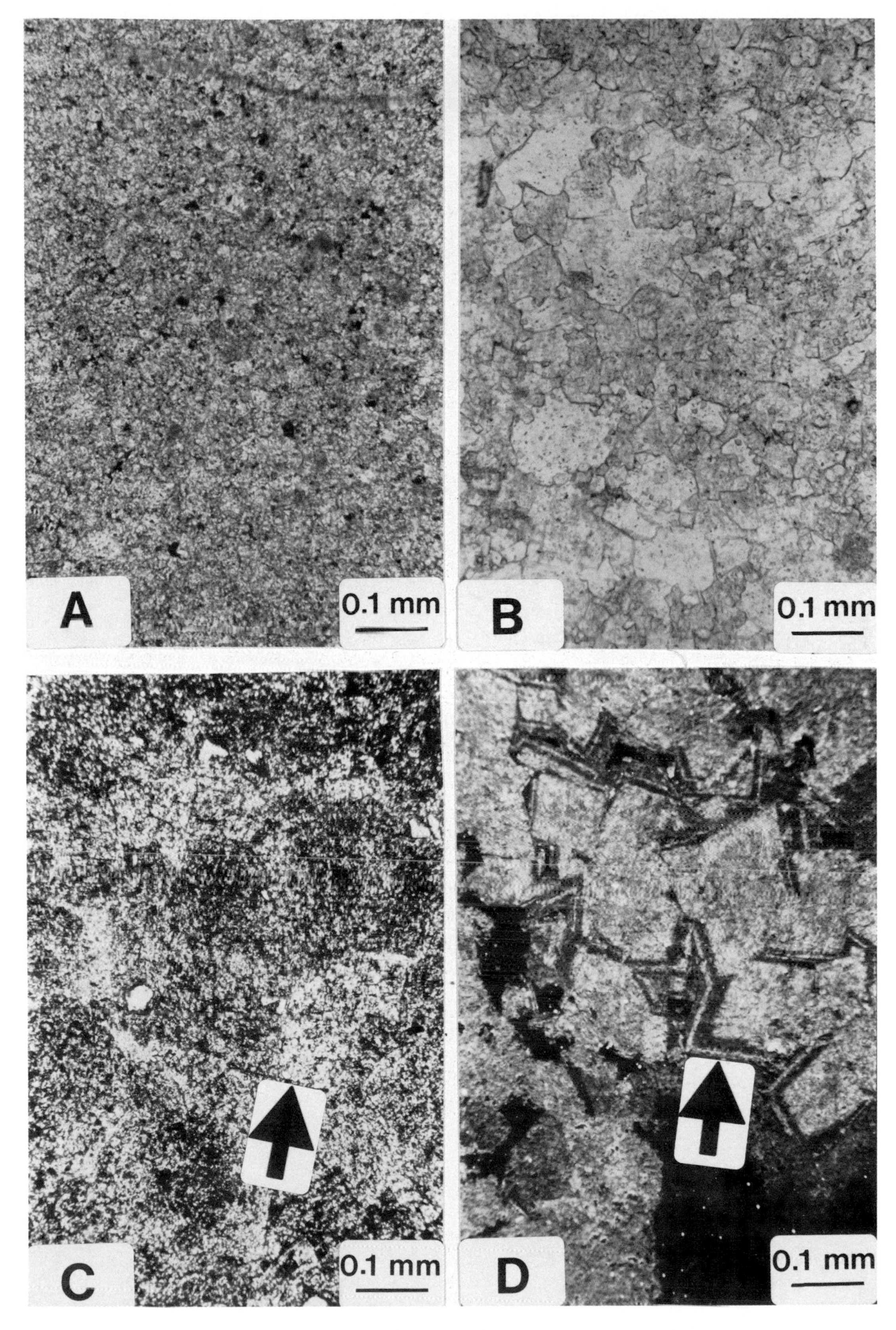

FIG. 4.—Pervasive-dolostone fabrics (PDF) and particle-relict fabrics (PRF) in Interlake dolostones. (A) Fine crystalline PDF dolostone. Uncrossed polars. Conoco Lueth, 1,441 m. (B) Medium crystalline PDF dolostone. Uncrossed polars. California Company Thompson, 2,000 m. (C) Coarse crystalline PRF dolostone. Uncrossed polars. California Company Thompson, 2,010 m. (D) Same field as (C) in cathodoluminescence. Arrows in (C) and (D) point to the same rhomb.

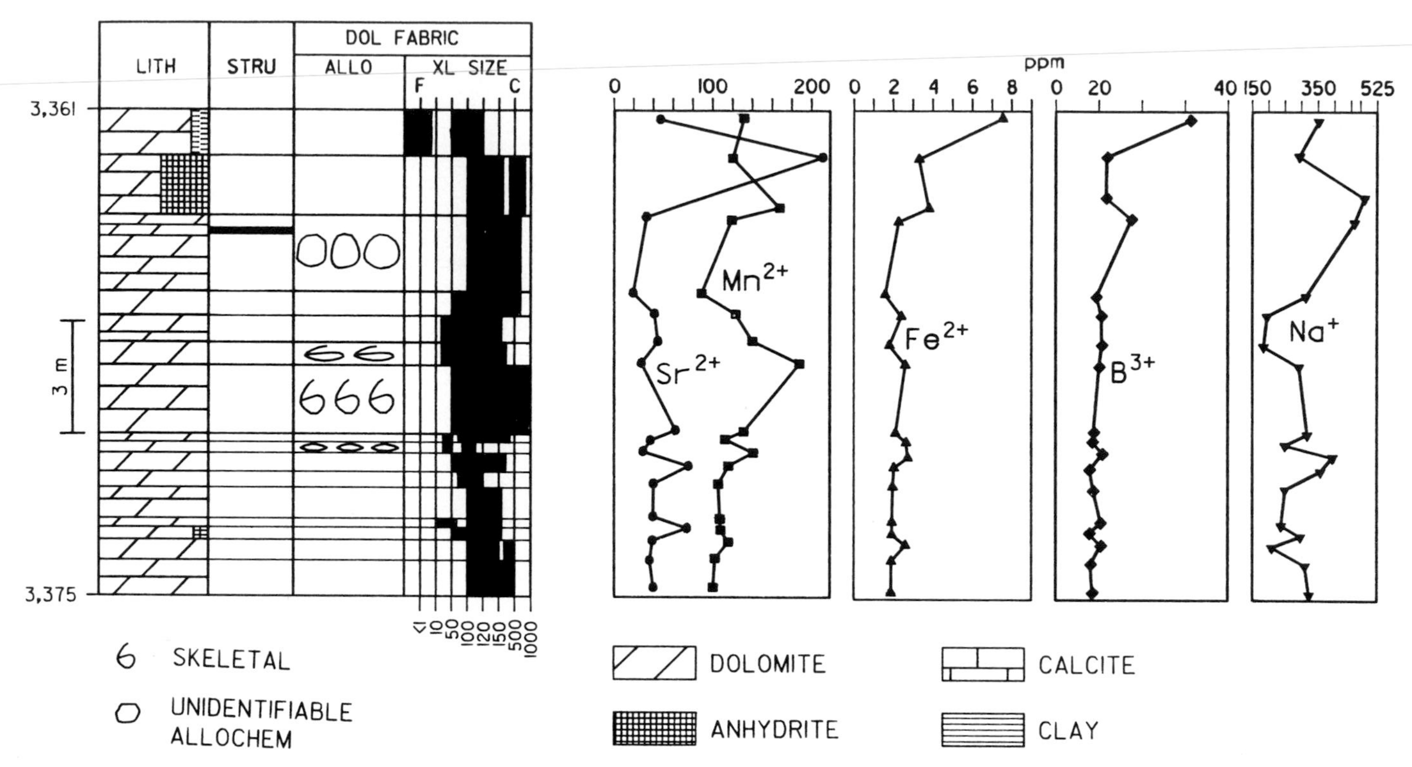

FIG. 5.—Example of a typical downcore trend in concentration of trace elements with GSF, PDF, and PRF dolostones in the Interlake Formation (Silurian), North Dakota; GSF is indicated by the shaded horizontal bar, PDF by absence of depositional fabrics, and PRF by presence of allochems, which are not in grain-supported framework. (LITH = lithology; STRUC = structures; ALLO = allochems; XL = crystal; F = fine; C = coarse.)

PDF, and PRF), and Table 2 shows elemental data for GSF, PDF and PRF dolomites. Figures 6–10 show variations in trace elements with various dolostone fabrics. In general, grain-supported fabrics (GSF) have elevated Mn and Na, whereas Fe, Sr, and B are depressed (Table 3). In contrast, pervasive-dolomite fabrics (PDF) have elevated Fe, Sr, and B, whereas Mn and Na are depressed. A more complicated picture emerges if particle-relict fabrics (PRF) are considered. Particle-relict fabric dolostones have elevated B and Sr, whereas Mn, Fe, and Na are depressed (Table 3). In contrast to fabric, dolomite texture does not show systematic trends in element distribution.

Interpretation

Strontium.—

Of all the trace elements in dolomite, Sr has received the most attention in recent years (Land, 1980, 1985; M'Rabet, 1981; Veizer, 1983). Much of this attention has focused on three areas:

(1) kinetic explanations of Sr incorporation into the dolomite lattice by replacement of Ca in dolomite cation planes; as shown in Figure 11, early (penecontemporaneous) dolomites are enriched in Sr compared to late (diagenetic) dolomites (Veizer and others, 1978; Land, 1985). These differences are also reflected in dolomite fabric because penecontemporaneous dolomites are finer crystalline than diagenetic dolomites. Our knowledge of kinetic aspects of Sr incorporation in the dolomite lattice has not made substantial progress since reviews by Land (1980) and Veizer (1983).

(2) behavior of Sr in mixed-water (dorag) systems; according to Veizer and others (1978), large variations in meteoric-water input will not significantly change the Sr content of dolomites. This is especially true if seawater constitutes $\geq 15\%$ of the mixture. Dolomitization of Interlake sediments involved a high water:rock ratio (open system), as indicated by Mg requirements for hundreds of meters of dolostone. This, in addition to mostly eogenetic dolomitization, indicates that the principal source of Mg was seawater modified by evaporation or mixing with meteoric water and/or brines. The resulting dolomites have significant differences in Sr content, despite the preponderance of seawater. This interpretation is in sharp contrast to that of Veizer and others (1978) as noted above.

(3) influence of the Sr content of the replaced (precursor) mineral upon the Sr content of dolomite; Veizer and others (1978) regarded the "precursor influence" to be significant.

TABLE 2.—TRACE-ELEMENT CONCENTRATIONS (IN PPM) OF VARIOUS DOLOSTONE FABRICS, INTERLAKE FORMATION (SILURIAN), NORTH DAKOTA (GSF–GRAIN-SUPPORTED FABRIC; PDF–PERVASIVE-DOLOSTONE FABRIC; PRF–PARTICLE-RELICT FABRIC)

Mn GSF	Mn PRF	Mn PDF	Sr GSF	Sr PRF	Sr PDF	Fe GSF	Fe PRF	Fe PDF	Na GSF	Na PRF	Na PDF	B GSF	B PRF	B PDF	Well Name
68	77	–	25	26	–	3158	2623	–	241	227	–	10	12	–	Carter
57	56	–	60	27	–	1951	2406	–	280	209	–	9	12	–	Moore
49	–	–	–	–	–	–	–	–	–	–	–	–	–	–	NDGS 2010
150	108	182	51	32	228	1911	1014	9776	432	288	613	16	15	26	Union
–	197	137	–	46	73	–	1549	1872	–	337	464	–	32	15	Aans-
–	167	189	–	57	43	–	1674	27780*	–	461	1170*	–	58	93	taad
–	127	140	–	56	90	–	660	1701	–	407	396	–	35	14	
–	69	–	–	46	–	–	1297	–	–	276	–	–	32	–	NDGS 20
–	73	–	–	57	–	–	778	–	–	431	–	–	60	–	*clay
–	74	–	–	48	–	–	2226	–	–	181	–	–	22	–	rich
267	164	95	56	60	55	1933	1053	1092	898	148	467	38	36	77	
281	95	–	69	55	–	4311	1092	–	1390	467	–	30	77	–	
134	161	–	46	58	–	9088*	1636	–	4014*	808	–	73	25	–	
82	171	–	68	44	–	1677	1615	–	312	500	–	49	33	–	Conoco
30	198	–	67	202	–	601	7168	–	272	239	–	91	30	–	Lueth
–	108	–	–	55	–	–	1581	–	–	500	–	–	53	–	
–	131	–	–	44	–	–	2131	–	–	491	–	–	58	–	NDGS 207
–	140	–	–	55	–	–	1955	–	–	332	–	–	45	–	
–	109	–	–	48	–	–	18890*	–	–	496	–	–	43	–	
–	124	–	–	60	–	–	986	–	–	3189*	–	–	105	–	
–	111	–	–	55	–	–	1023	–	–	267	–	–	73	–	*clay
–	58	–	–	50	–	–	630	–	–	390	–	–	71	–	rich
–	64	–	–	110	–	–	1021	–	–	277	–	–	52	–	
–	–	–	–	–	–	–	–	–	–	791	–	–	–	–	
159	–	153	50	–	49	2587	–	10540	124	–	542	13	–	36	
91	–	162	80	–	41	16970	–	12230	204	–	439	8	–	40	S & J
140	140	–	71	71	–	2724	2724	–	656	656	–	13	13	–	Oper. Co.
142	142	–	76	76	–	2934	2934	–	329	329	–	13	13		Risser
122	–	–	49	–	–	2769	–	–	2235	–	–	14	–	–	
146	146	141	45	45	54	4603	4603	9062	235	235	432	19	19	36	NDGS 33
133	133	–	76	76	–	3869	3869	–	243	243	–	13	13	–	
169	138	130	82	42	41	4008	1789	8201	515	170	368	11	10	32	
118	189	118	32	37	458	2380	1829	3503	–	279	390	18	9	12	Blackwood
–	110	87	27	30	18	–	2813	–	473	396	316	–	10	9	& Nichols
–	142	127	–	79	40	–	2846	1800	227	349	184	–	8	10	Gilman
	112	–	–	40	62	–	2880	2416	–	229	312	–	6	8	
–	99	–	–	38	–	–	2041	2371	–	211	–	–	8	–	NDGS 470
–	104	–	–	76	–	–	2103	–	–	281	–	–	9	–	
–	104		–	38	–	–	2133	–	–	191	–	–	7	–	
–	117	–	–	35	–	–	2081	–	–	296	–	–	8	–	
–	97	–	–	–	–	–	2576	–	–	–	–	–	6	–	
63	46	76	76	50	43		–		1256	719	990	9	9	15	
97	41	75	41	52	65	–	–	–	1340	1906	1007	11	8	9	Hunt
58#	107	49	64#	34	86	–		–	1160#	1154*	1182	10#	9	7	Haugen
64	75	76	51	65	49	–	–	–	1250	1007	1030	11	9	8	
82	49	45#	76	86	50#		–	–	1070	1182	1170#	17	7	7#	*contains
–	76	46#	–	49	57#	–	–	–	–	1030	1130#	–	8	7#	anhydrite
–	70	–	–	63		–	–	–	–	2641	–	–	14	–	
–	51	–	–	64		–	–	–	–	920			8	–	#calcian
–	57#	–	–	76#	–	–	–	–	–	2640#	–	–	11	–	dolostone

Thus, penecontemporaneous dolomite (high Sr) was interpreted to have replaced aragonite. In contrast, diagenetic dolomite (low Sr) was interpreted to have replaced sediment that lacked high-Sr aragonite (Veizer and others, 1978). Katz and Matthews (1977) concluded that Sr partitioning was not significantly affected by changes in temperature or type of reactant, i.e., aragonite or calcite. The Interlake Formation commonly has alternating sequences of GSF, PDF and PRF dolostone relicts of particles. What, then, produces these variations? Some possibilities are discussed ahead.

Strontium concentration in the Interlake dolostone varies systematically with dolomite fabric (Fig. 6): GSF and PRF dolostones are Sr poor whereas PDF dolostones are Sr rich. The most reasonable interpretation of these differences is that Sr content in Interlake dolostones was not influenced by the Sr content of the precursor carbonate. If the Sr in the precursor carbonate had influenced the incorporation of Sr into dolomite, then GSF and PRF dolostones should have

TABLE 3.—VARIATIONS IN TRACE-ELEMENT CONCENTRATION IN DOLOSTONES WITH GSF, PDF, AND PRF FABRICS, INTERLAKE FORMATION (SILURIAN), NORTH DAKOTA (SEE TABLE 2 FOR EXPLANATION OF GSF, PDF, AND PRF)

Element Fabric	Elevated	Depressed
Grain-supported fabric (GSF)	Mn Na	Fe Sr B
Pervasive-dolostone fabric (PDF)	Fe Sr B	Mn Na
Particle-relict fabric (PRF)	Sr B	Mn Fe Na

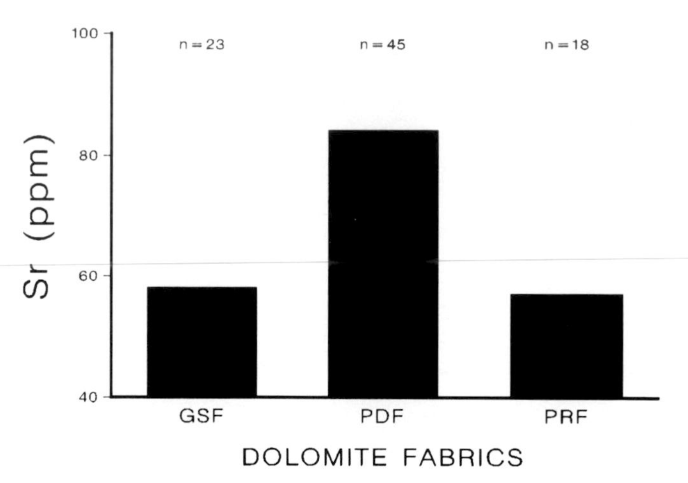

FIG. 6.—Histogram showing Sr distribution in Interlake dolostones with grain-supported fabrics (GSF), pervasive-dolostone fabrics (PDF), and particle-relict fabrics (PRF); "n" indicates number of different data points (analyses) from which the bar was calculated.

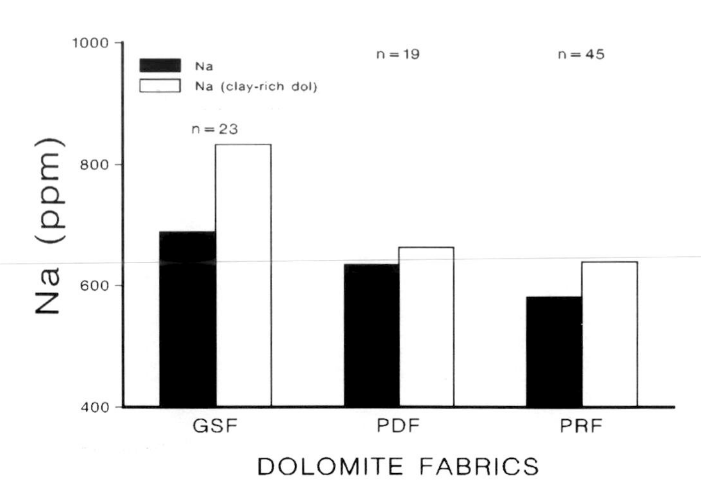

FIG. 7.—Histogram showing Na distribution in clay-poor and clay-rich Interlake dolostones with grain-supported fabrics (GSF), pervasive-dolostone fabrics (PDF), and particle-relict fabrics (PRF); "n" indiciates number of different data points (analyses) from which the bars were calculated.

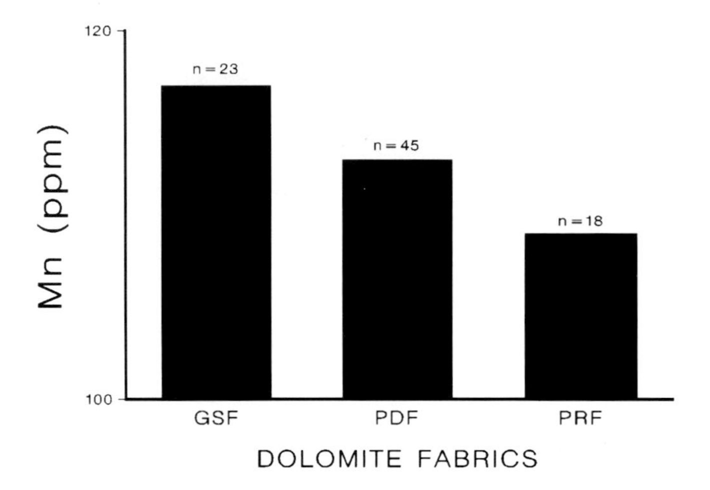

FIG. 8.—Histogram showing Mn distribution in Interlake dolostones with grain-supported fabrics (GSF), pervasive-dolostone fabrics (PDF), and particle-relict fabrics (PRF); "n" indicates number of different data points (analyses) from which the bar was calculated.

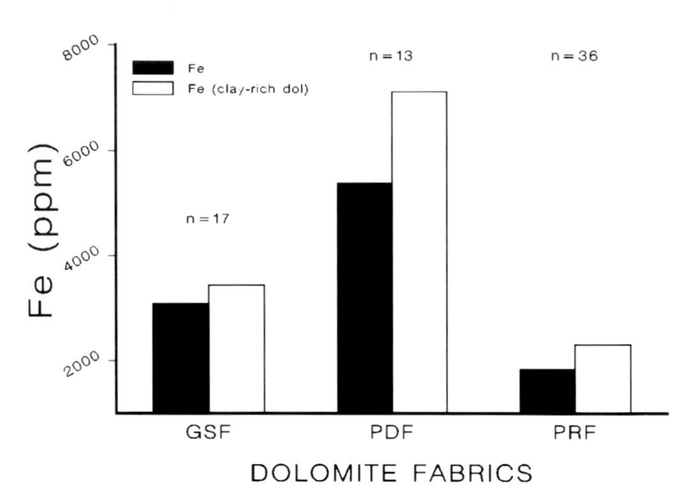

FIG. 9.—Histogram showing Fe distribution in clay-poor and clay-rich Interlake dolostones with grain-supported fabrics (GSF), pervasive-dolostone fabrics (PDF), and particle-relict fabrics (PRF); "n" indicates number of different data points (analyses) from which the bars were calculated.

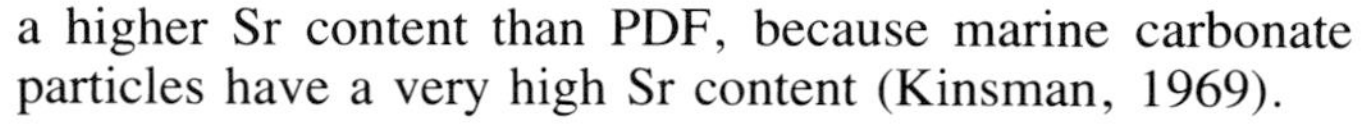

a higher Sr content than PDF, because marine carbonate particles have a very high Sr content (Kinsman, 1969).

Sodium.—

Sodium is the most abundant cation in seawater (10,565 ppm; compare with Mg, which has the next highest value at 1,269 ppm; values from Pytkowicz, 1983). Accordingly, its utility as a paleosalinity indicator has been widely examined since Weber (1964) first mentioned the possibility (Fritz and Katz, 1972; Land and Hoops, 1973; Veizer and others, 1977). These studies can be applied to dolostones only if the location the Na ion can be determined with certainty. Sodium can be present as solid inclusions (NaCl) or, upon combining with H, may replace Ca (Fritz and Katz, 1972), or both Ca and Mg (Land and Hoops, 1973).

Sodium in the Interlake dolostone shows the following trends (Fig. 7): (1) Na values are generally elevated in GSF dolostones compared to PDF and PRF dolostones; (2) PDF dolostones are Na rich compared to PRF dolostones; and (3) clay-rich dolostones are Na rich compared to clay-poor dolostones.

These patterns indicate that Na in the Interlake dolostone was certainly influenced by precursor mineralogy. In addition, Interlake Na is interpreted to be lattice-bound, because Na distribution shows no correlation with evaporites. Some correlation should be present if the NaCl inclusions, derived from evaporating brines, supplied Na during dolomitization. In contrast, high Na dolomites in places contain relicts of normal marine fauna.

Manganese.—

Concentration of Mn shows trends similar to those of Na (Fig. 8): (1) GSF dolostones are Mn rich compared to PDF and PRF dolostones; and (2) PDF dolostones are Mn rich

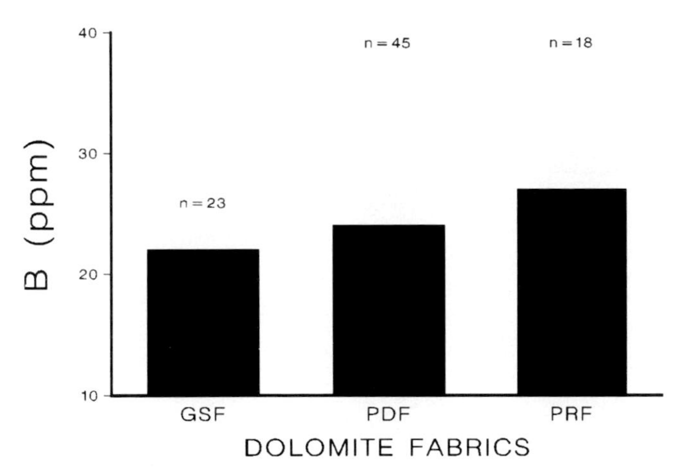

FIG. 10.—Histogram showing B distribution in Interlake dolostones with grain-supported fabrics (GSF), pervasive-dolostone fabrics (PDF), and particle-relict fabrics (PRF); "n" indicates number of different data points (analyses) from which the bar was calculated.

compared to PRF dolostones. As in the case of Na, the manganese distribution is interpreted to reflect the influence of precursor-derived Mn.

Iron.—

Concentration of Fe shows trends similar to those of Sr (Fig. 9): (1) GSF dolostones are Fe poor compared to PDF dolostones; (2) PDF dolostones are Fe rich; and (3) clay-rich dolostones are Fe rich compared to clay-poor dolo-

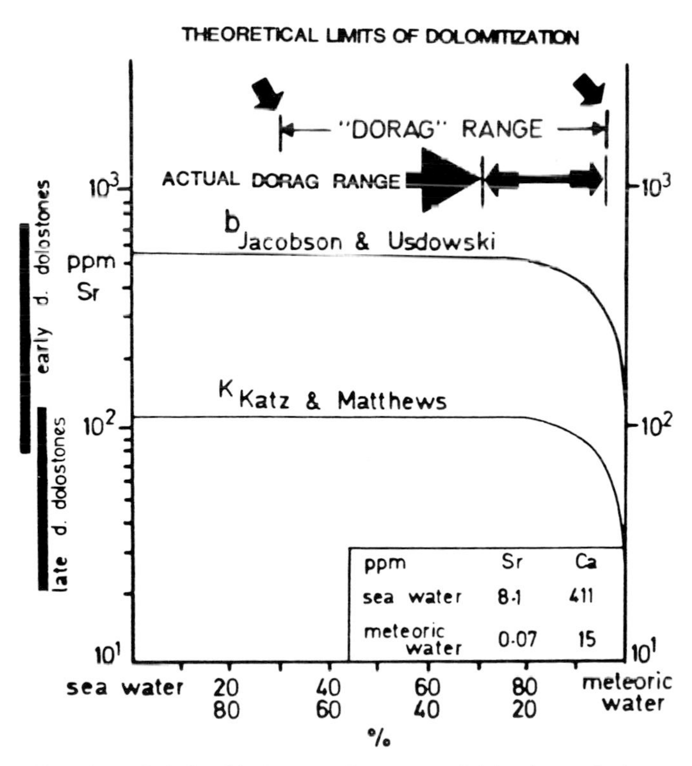

FIG. 11.—Relationship between Sr content of dolomites and mixtures ("dorag") of seawater and meteoric water. Note that the "dorag" range shown by Veizer and others (1978) denotes the theoretical limits of dolomitization as defined by Badiozamani (1973) for a mixed-water system. The actual "dorag zone" is much narrower, between 5 and 30% sea water (modified from Veizer and others, 1978).

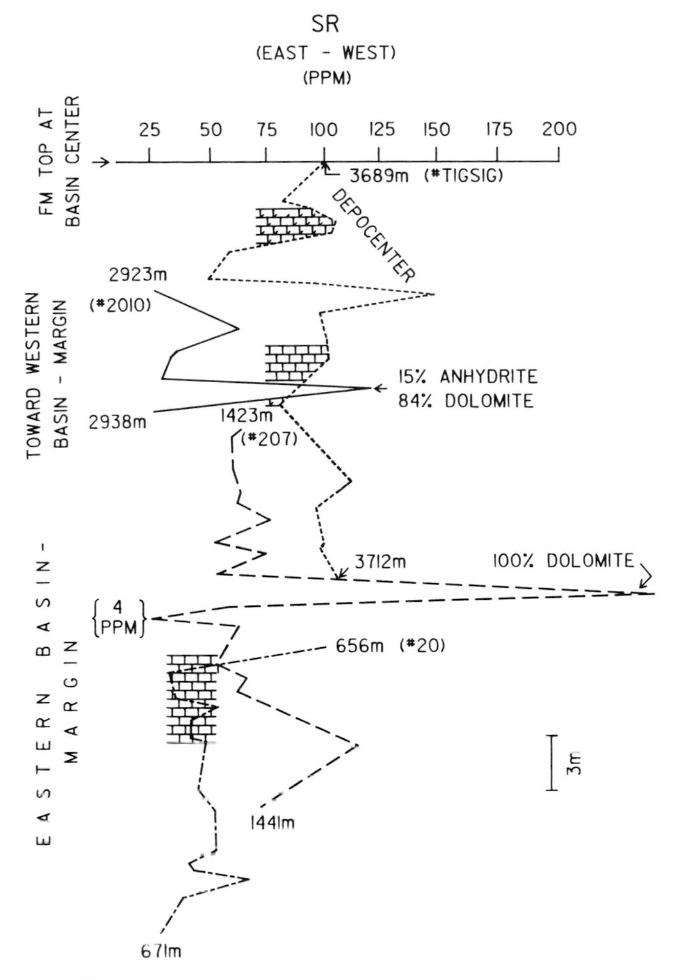

FIG. 12.—Composite vertical section of the Interlake Formation showing the vertical distribution of Sr in the Interlake dolostone. Each curve is a different well, and together they represent a vertical composite of the Interlake Formation (see text for details). Note that Sr increases toward the top of the Interlake Formation by a factor of two (values in the depocenter have a mean = 100 ppm, compared to the eastern and western basin margins where the mean = 50 ppm).

stones. As in the case of Sr, the iron distribution in the Interlake dolostone was not influenced by precursor-derived Fe.

Boron.—

Distribution of B in the Interlake dolostone shows the most unusual trends (Fig. 10): GSF and PDF dolostones are similar, and both are only slightly B-poor compared to PRF dolostones. This is unusual because in the case of Sr, Na, Mn, and Fe, either GSF or PDF are relatively enriched. The unusual trends of B in the Interlake dolostones cannot, at present, be interpreted with certainty.

CHEMOSTRATIGRAPHY AND THE REGIONAL DISTRIBUTION OF TRACE ELEMENTS

Distributions of Sr, Mn, and Na in the Interlake dolostones also reveal vertical and regional trends (Figs. 12–14). The top of the Interlake Formation is an erosional boundary, which has two effects on the formation: the basin center has the thickest Interlake sequence, whereas the ba-

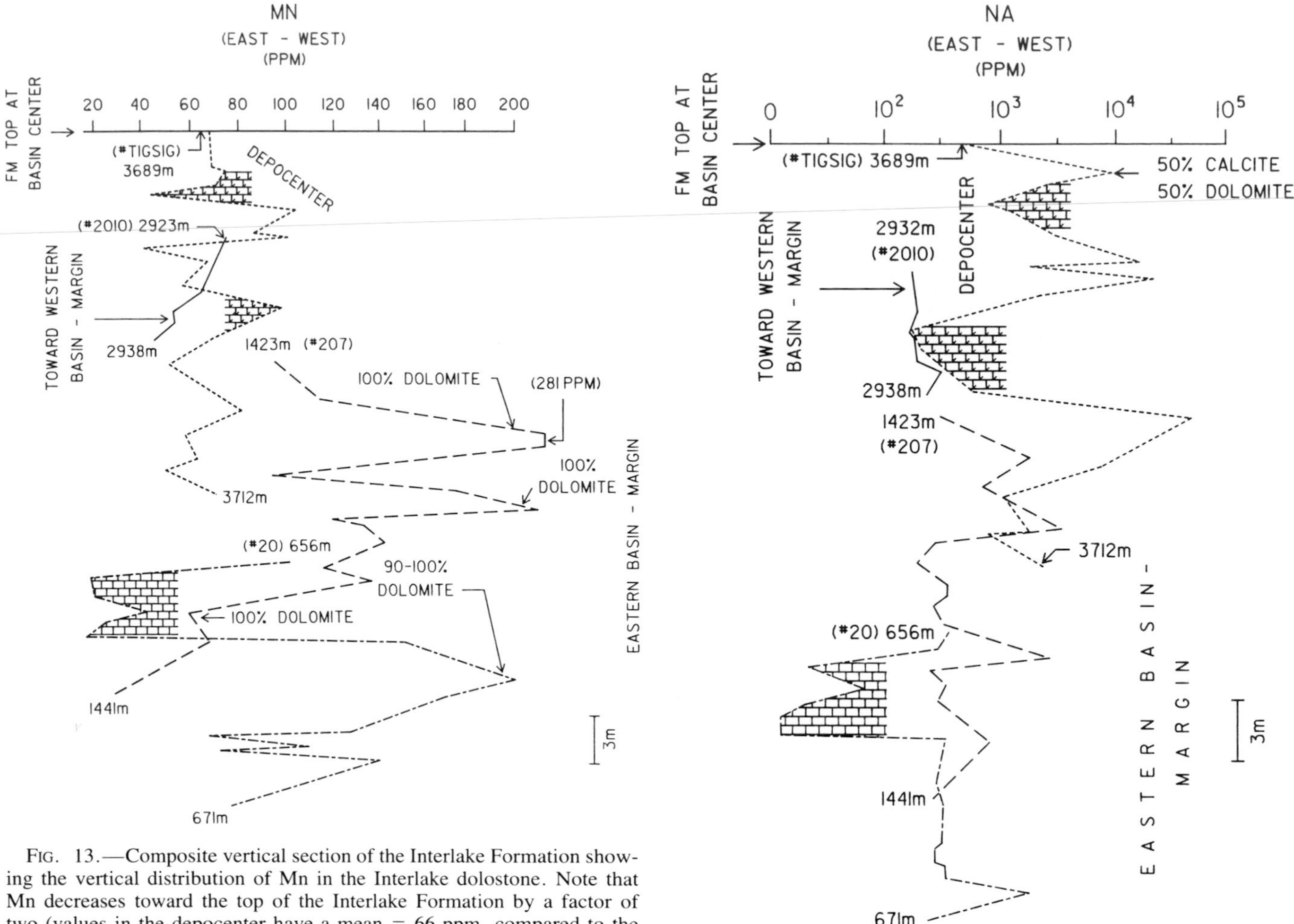

FIG. 13.—Composite vertical section of the Interlake Formation showing the vertical distribution of Mn in the Interlake dolostone. Note that Mn decreases toward the top of the Interlake Formation by a factor of two (values in the depocenter have a mean = 66 ppm, compared to the southeastern basin margin where the mean = 130 ppm).

FIG. 14.—Composite vertical section of the Interlake Formation showing the vertical distribution of Na in the Interlake dolostone. Note that Na increases toward the top of the Interlake Formation by more than one order of magnitude (values in the depocenter have a mean = 5,000 ppm, compared to the basin margins where the mean = 300 ppm).

sin margins have progressively thinner sequences; and wells drilled toward the basin margins encounter progressively older Interlake rocks, i.e., the top of the formation is a diachronous surface, becoming older toward the basin margins. These effects have to be corrected before attempting any stratigraphic or chemostratigraphic analysis. The curves in Figures 12–14 show various wells (#2010, 207, 20, and TIGSIG) placed in their corrected stratigraphic position, thus making a vertical composite of the Interlake Formation. This was achieved by the following steps: (1) the top of the Interlake in a basin-center well (TIGSIG) was chosen as the datum, because the rock record is most complete in the basin center; (2) the TIGSIG geophysical log was correlated with that from the Interlake discovery well (1951) Amerada Inverson No. 1, which penetrated approximately 305 m (1,000 ft) of Interlake dolostone (Carlson and Eastwood, 1962); and (3) the geophysical logs from the other wells were then correlated with the logs from the TIGSIG well. The results of this stratigraphic correlation are shown in Figures 12–14, which represent vertical composites of the Interlake Formation and show a number of features: the eastern basin margin has the lower (oldest) Interlake rocks, whereas the western basin margin has the middle and upper Interlake rocks; and a distinct chemostratigraphy exists between the oldest (eastern basin margin) and the youngest (depocenter) Interlake rocks. This chemostratigraphy has the following general trends: (1) basin-center dolostones are enriched in Sr compared to basin-margin dolostones (Fig. 12). In addition, dolostones are enriched in Sr compared to limestones; and, anomalously high values of Sr are present in anhydrite-rich dolomites; (2) basin-center dolostones are depleted in Mn compared to basin-margin dolostones (Fig. 13); and (3) regional trend of Na distribution is similar to that of Sr (Fig. 14), i.e., basin-center dolostones are enriched in Na compared to basin-margin dolostones. (This is in accordance with the distribution coefficients of Sr and Na, both less than one; Veizer, 1983).

DISCUSSION

Dolomite Fabric and Trace Elements

Previous studies have generally dealt with three principal factors that may influence the trace-element composition of dolomites: stoichiometry and kinetics of dolomitization (see review by Hardie, 1987); dolomitizing fluids and rock porosity and permeability (Ward and Halley, 1985; Lloyd and others, 1986); geochemistry of the replaced (precursor) mineral (Veizer and others, 1978; Veizer, 1983).

Stoichiometry and kinetics.—

This aspect of both Ca-rich and almost stoichiometric dolomites has received much attention (Hsu, 1967; Land, 1980; Machel and Mountjoy, 1986; Hardie, 1987). As yet, no consensus exists regarding the solubility constants of dolomite, which may differ by approximately two orders of magnitude (Hardie, 1987). Sulfate reduction (along with a concomitant increase in alkalinity) remains the most recently identified kinetic control over dolomite precipitation in many hydraulic settings (Baker and Kastner, 1981). They cite (p. 215) dolomite in organic-rich sediments and supratidal sabkhas as the principal examples of dolomite formation induced by sulfate reduction; however, a thick sequence of dolostone does not form in either instance. In a later study, Baker and Burns (1985) argued that the fundamental control over dolomite formation in organic-rich sediments is the availability of Ca (through dissolution of precursor calcium carbonate). They also note (p. 1919) that dolomite ". . . is by no means pervasive or complete." In contrast, examples abound of dolomite forming in marine waters that have not undergone sulfate reduction (e.g., Saller, 1984; Mazzullo and others, 1987) or have undergone only slight sulfate reduction (Carballo and others, 1987). Therefore, while sulfate reduction may abet dolomitization, it cannot by itself produce thick dolostones.

As noted previously, the Interlake dolomites formed early during diagenesis by processes utilizing modified seawater as the Mg source. Variations in trace elements with dolomite fabrics (GSF, PDF and PRF) indicates that Sr in Interlake dolomites is not influenced by the Sr in precursor minerals. This conclusion is in accord with that of Veizer and others (1978), who noted that in seawater-dominated dolomitization, Sr should not vary greatly (Fig. 11).

Dolomitizing fluids and rock properties.—

The Mg required for generating ≥350 m of Interlake dolostone makes modified seawater the only reasonable dolomitizing fluid. This is further confirmed by characteristics indicative of mostly preburial dolomitization. Moreover, dolomitization was essentially completed prior to the Siluro-Devonian unconformity, implying rapid dolomitization. Precise rates cannot be determined because the quantity of eroded Interlake is unknown. Present thickness yields abnormally low rates of dolomitization.

Movement of large quantities of seawater requires great porosity and permeability in the carbonate sediment. Because Interlake sediments were deposited in an epeiric sea, tidal pumping and other coastal phreatic-lens dynamics cannot be invoked (e.g., Saller, 1984; Ward and Halley, 1985; Carballo and others, 1987). Instead, modification of seawater through mixing (with either meteoric water or brines) may have occurred around isolated highs. These highs are indicated by presence of localized shoals of ooids and skeletal debris in northern North Dakota (Fig. 2).

Geochemistry of the precursor mineral.—

Trace-element trends in various fabrics of the Interlake dolostone indicate that the Na and Mn content of the precursor carbonate mineral influenced the Na and Mn content in the dolomite. This could have occurred by rapid incorporation of Na and Mn in the dolomite during the precipitation phase of the dissolution-precipitation reaction (Katz and Matthews, 1977). Even though it cannot be definitively proven at present, it is possible that the Na and Mn released by the precursor mineral are locally available for some time before being exported in an open-system setting. This is similar to the suggestion by Lippmann (1973, p. 150) that ". . . the only conceivable mechanism leading from ionized dissolved species to an ionic crystal can be:

$$Ca + Mg + 2CO_3^{2-} \rightarrow CaMg(CO_3)_2 \text{ (solid)."}$$

Thus, in addition to Ca and Mg, sodium and manganese also become briefly available as ionic species. In summary, distinct differences in Na and Mn concentration of GSF and PDF dolostones most likely reflect precursor influences.

Boron has received extensive attention as a paleosalinity indicator of clays and cherts (Kolodny and others, 1980). Graf (1960, p. 18) reviewed early work on B in carbonates, including ". . . a marine dolomite of Carboniferous age . . ." that contained 25 ppm boron. Graf (1960) also cited published values of B from the Lockport dolomite (Silurian), New York (20 ppm; Table 1) and dolomite from Illinois (10 ppm; Table 27A). Weber (1964) noted that dolostones associated with evaporites are B rich (98.7 ppm) compared to other dolostones (68.1 ppm B). The Interlake dolostone ranges from 0 to 91 ppm (Table 2). This value is high, especially because B is a minor solute in seawater (4.5 ppm; Veizer, 1983, Table 3-2). In summary, B values of dolostones cannot be precisely interpreted at present beyond describing overall distributions and variations. In addition, clays did not affect the B content of Interlake dolostone, because clay-rich dolostones do not have a different B content compared to clay-poor dolostones.

In summary, the Interlake dolostones display pronounced variations in Sr, Mn, Fe, Na, and B concentrations with dolostone fabric. The most likely explanation for this is the incorporation of some trace elements into a local cation "pool" rather than being immediately exported.

CONCLUSIONS

(1) The Interlake dolostone (Silurian) in North Dakota represents replacement of subtidally deposited lime grainstones (mudstones were minor).

(2) Dolomitization was early and resulted from: (1) hypersaline brines in supratidal settings, and (2) mixing of marine fluids with hypersaline and/or meteoric water.

(3) Burial dolomites are quantitatively minor.

(4) Dolostone fabrics are of three major types, grain-supported fabrics (GSF), pervasive-dolomite fabrics (PDF), and particle-relict fabrics (PRF).

(5) In general, GSF dolostones have elevated Mn and Na, whereas Fe, Sr, and B are depressed; PDF dolostones have elevated Fe, Sr, and B, whereas Mn and Na are de-

pressed. Trace-element concentration does not vary systematically with dolomite texture.

(6) Strontium and iron content in Interlake dolostones was not influenced by precursor-derived elements. In contrast, Mn and Na content in Interlake dolostones was influenced by precursor-derived elements. Boron content of Interlake dolostones cannot, at present, be interpreted with certainty.

ACKNOWLEDGMENTS

I thank Wm. C. Dawson, Texaco USA (EPTD), B. H. Wilkinson, and J. Veizer for their critique of the manuscript; however, the views expressed in this paper remain the author's responsibility. I also acknowledge Texaco for permission to publish.

REFERENCES

ARTHUR, M. A., ANDERSON, T. F., KAPLAN, I. R., VEIZER, J., AND LAND, L. S., 1983, Stable Isotopes in Sedimentary Geology: Society of Economic Paleontologists and Mineralogists Short Course No. 10, p. A-1 to 5–54.

BADIOZAMANI, KHOSROW, 1973, The Dorag dolomitization model–Application to the Middle Ordovician of Wisconsin: Journal of Sedimentary Petrology, v. 43, p. 965–984.

BAILLIE, A. D., 1951, Silurian geology of the Interlake area, Manitoba: Manitoba Department of Mines and Natural Resources, Mines Branch Publication 50-1, 82 p.

BAKER, P. A., AND BURNS, S. J., 1985, Occurrence and formation of dolomite in organic-rich continental margin sediments: American Association of Petroleum Geologists Bulletin, v. 69, p. 1917–1930.

———, AND KASTNER, M., 1981, Constraints on the formation of sedimentary dolomite: Science, v. 213, p. 214–216.

BALLARD, F. V., 1963, Structure and Stratigraphic relationships in the Paleozoic rocks of eastern North Dakota: North Dakota Geological Survey Bulletin, v. 40, 42 p.

CARBALLO, J. D., LAND, L. S., AND MISER, D. E., 1987, Holocene dolomitization of supratidal sediments by active tidal pumping, Sugarloaf Key, Florida: Journal of Sedimentary Petrology, v. 57, p. 153–165.

CARLSON, C. G., AND ANDERSON, S. B., 1965, Sedimentary and tectonic history of North Dakota part of Williston basin: American Association of Petroleum Geologists Bulletin, v. 49, p. 1833–1846.

———, AND EASTWOOD, W. P., 1962, Upper Ordovician and Silurian rocks of North Dakota: North Dakota Geological Survey Bulletin, v. 38, 52 p.

DUNHAM, R. J., 1962, Classification of carbonate rocks according to depositional texture, *in* Ham, W. E., ed., Classification of Carbonate Rocks: American Association of Petroleum Geologists Memoir 1, p. 108–121.

FRIEDMAN, G. M., 1965, Terminology of crystallization textures and fabrics in sedimentary rocks: Journal of Sedimentary Petrology, v. 35, p. 643–655.

FRITZ, PETER, AND KATZ, AMITAI, 1972, The sodium distribution of dolomite crystals: Chemical Geology, v. 10, p. 237–244.

GRAF, D. L., 1960, Geochemistry of carbonate sediments and sedimentary carbonate rocks, Part III: Minor element distribution: Illinois State Geological Survey, Circular 301, 71 p.

HARDIE, L. A., 1987, Dolomitization: A critical view of some current views: Journal of Sedimentary Petrology, v. 57, p. 166–183.

HSU, K. J., 1967, Chemistry of dolomite formation, *in* Chilingar, G. V., Bissell, H. J., and Fairbridge, R. W., eds., Carbonate Rocks: Developments in Sedimentology 9B: Elsevier, New York, p. 169–191.

JOHNSON, DOUGLAS, 1933, Role of analysis in scientific investigation: Geological Society of America Bulletin, v. 44, p. 461–494.

JOHNSON, M. E., AND LESCINSKY, H. L., 1986, Depositional dynamics of cyclic carbonates from the Interlake Group (Lower Silurian) of the Williston basin: PALAIOS, v. 1, p. 111–121.

KATZ, AMITAI, AND MATTHEWS, ALAN, 1977, The dolomitization of $CaCO_3$: An experimental study at 252–295°C: Geochimica et Cosmochimica Acta, v. 41, p. 279–308.

KINSMAN, D. J. J., 1969, Interpretation of $Sr^0U^{+20}D$ concentrations in carbonate minerals and rocks: Journal of Sedimentary Petrology, v. 39, p. 486–508.

KOLODNY, YEHOSHUA, TARABOULOS, ALBERT, AND FRIESLANDER, URI, 1980, Participation of fresh water in chert diagenesis: Evidence from oxygen isotopes and boron α-track mapping: Sedimentology, v. 27, p. 305–316.

LAND, L. S., 1980, The isotopic and trace element geochemistry of dolomite: The state of the art, *in* Zenger, D. H., Dunham, J. B., and Ethington, R. C., eds., Concepts and Models of Dolomitization: Society of Economic Paleontologists and Mineralogists Special Publication 28, p. 87–110.

———, 1985, The origin of massive dolomite: Journal of Geological Education, v. 33, p. 112–125.

———, AND HOOPS, G. K., 1973, Sodium in carbonate sediments and rocks: A possible index to the salinity of diagenetic solutions: Journal of Sedimentary Petrology, v. 43, p. 614–617.

LIPPMANN, FRIEDRICH, 1973, Sedimentary Carbonate Minerals: Springer-Verlag, New York, 228 p.

LLOYD, J. M., RAGLAND, P. C., AND PARKER, W. M., 1986, Diagenesis of Jurassic Smackover Formation, Jay Field, Florida (abst.): American Association of Petroleum Geologists Bulletin, v. 70, p. 1185.

LOBUE, CHARLES, 1982, Depositional environments and diagenesis of the Silurian Interlake Formation, Williston basin, western North Dakota, *in* Christopher, J. E., and Kaldi, J., eds., Fourth International Williston Basin Symposium, Regina, Saskatchewan, p. 29–42.

———, 1983, Depositional environments and diagenesis, Interlake Formation (Silurian), Williston basin, North Dakota: Unpublished M.S. Thesis, University of North Dakota, Grand Forks, North Dakota, 233 p.

MACHEL, H.-G., AND MOUNTJOY, E. W., 1986, Chemistry and environments of dolomitization–A reappraisal: Earth Science Reviews, v. 23, p. 175–222.

MATTHEWS, R. K., 1986, The role of paradigms in sedimentary geology: PALAIOS, v. 1, p. 433.

MAZZULLO, S. J., REID, A. M., AND GREGG, J. M., 1987, Dolomitization of Holocene Mg-calcite supratidal deposits, Ambergris Cay, Belize: Geological Society of America Bulletin, v. 98, p. 224–231.

MCCREA, J. M., 1950, On the isotopic chemistry of carbonates and a paleotemperature scale: Journal of Chemical Physics, v. 18, p. 849–857.

MORROW, D. W., 1982a, Diagenesis 1. Dolomite, part 1: The chemistry of dolomitization and dolomite precipitation: Geoscience Canada, v. 9, p. 5–13.

———, 1982b, Diagenesis 2. Dolomite, part 2: Dolomitization models and ancient dolostones: Geoscience Canada, v. 9, p. 95–107.

M'RABET, ALI, 1981, Differentiation of environments of dolomite formation, Lower Cretaceous of central Tunisia: Sedimentology, v. 28, p. 331–352.

PETERSON, J. A., AND MACCARY, L. M., 1985, Regional stratigraphy and general petroleum geology, Williston basin (abst.): American Association of Petroleum Geologists Bulletin, v. 69, p. 859.

PYTKOWICZ, R. M., 1983, Equilibria, Nonequilibria, and Natural Waters, Volume 1: John Wiley, New York, 351 p.

ROEHL, P. O., 1967, Stony Mountain (Ordovician) and Interlake (Silurian) facies analogs of Recent low-energy marine and subaerial carbonates, Bahamas: American Association of Petroleum Geologists Bulletin, v. 51, p. 1979–2032.

SALLER, A. H., 1984, Petrologic and geochemical constraints on the origin of subsurface dolomite, Enewetak Atoll: An example of dolomitization by normal seawater: Geology, v. 12, p. 217–220.

SHUKLA, VIJAI, 1985, Cathodoluminescence and geochemistry of a regional dolostone, Williston basin: Society of Economic Paleontologists and Mineralogists Midyear Meeting, Golden, Colorado, Abstracts, v. II, p. 82.

———, 1986, Sedimentology and geochemistry of a regional dolostone: Correlation of trace elements with dolomite fabric and texture: Society of Economic Paleontologists and Mineralogists Midyear Meeting, Raleigh, North Carolina, Abstracts v. III, p. 102.

SLOSS, L. L., 1985, Williston basin in the family of cratonic basins (abst.): American Association of Petroleum Geologists Bulletin, v. 69, p. 867.

VEIZER, JAN, 1983, Chemical diagenesis of carbonates: Theory and application of trace element technique, *in* Arthur, M. A., Anderson,

T. F., Kaplan, I. R., Veizer, Jan, and Land, L. S., eds., Stable Isotopes in Sedimentary Geology: Society of Economic Paleontologists and Mineralogists Short Course No. 10, p. 3-1 to 3–100.

———, Lemieux, Jean, Jones, Brian, Gibling, M. R., and Savelle, Jim, 1977, Sodium: Paleosalinity indicator in ancient carbonate rocks: Geology, v. 5, p. 177–179.

———, ———, ———, ———, and ———, 1978, Paleosalinity and dolomitization of a Lower Paleozoic carbonate sequence, Somerset and Prince of Wales Islands, Arctic Canada: Canadian Journal of Earth Sciences, v. 15, p. 1448–1461.

Ward, W. C., and Halley, R. B., 1985, Dolomitization in a mixing zone of near-seawater composition, Late Pleistocene, northern Yucatan Peninsula: Journal of Sedimentary Petrology, v. 55, p. 407–420.

Weber, J. N., 1964, Trace element composition of dolostones and dolomites and its bearing on the dolomite problem: Geochimica et Cosmochimica Acta, v. 28, p. 1817–1868.

SECTION VI
DOLOMITE DIAGENESIS

SECTION VI: INTRODUCTION
DOLOMITE DIAGENESIS

This section contains four papers dealing with various aspects of alteration of dolomite.

Zenger and Dunham describe replacement dolomite with a chemistry suggesting influence of hot fluids in the subsurface. It is unclear, however, whether these dolomites resulted from mesogenetic replacement or by burial neomorphism of early-formed dolomite.

Moore and others present mutually contrasting sedimentological and geochemical data. Smackover (Jurassic) sedimentology suggests dolomitization by refluxing evaporative brines, whereas geochemical data suggest possibility of recrystallization of reflux dolomite in a meteoric water system.

Holail and others describe dolomitization and dedolomitization. Dolomitization was complex and may have involved dolomite recrystallization. In contrast, calcite precipitation and dedolomitization occurred in various stages of diagenesis.

Fischer describes various dolomite fabrics some of which form by neomorphism of earlier-formed dolomite. The dolomite fabrics are related to major diagenetic environments.

DOLOMITIZATION OF SILURO-DEVONIAN LIMESTONES IN A DEEP CORE (5,350 M), SOUTHEASTERN NEW MEXICO

DONALD H. ZENGER AND JOHN B. DUNHAM

Geology Department, Pomona College, Claremont, California 91711-6339; and Unocal Research Center, Brea, California 92621

ABSTRACT: One hundred thirty-two feet (40 m) of continuous, conventional core are divisible into an upper part (71 ft; 21.5 m) of limestone, dolomitic limestone, and minor intervals of dolomite sharply separated from a lower (61 ft; 18.5 m) completely dolomitized sequence. Limestone lithotypes, such as stromatoporoid rudstone and peloidal skeletal packstone to grainstone, indicate a generally shallow-water depositional setting of near-normal salinity, whereas laminated to massive, fenestral, peloidal mudstone and packstone, suggest a tidal flat environment. Intraclast breccia consisting of rip-up clasts of this latter lithology, commonly occurring above that *in situ* sequence, and of skeletal grainstone also associated with laminated fenestral units are interpreted as supratidal deposits analogous to Holocene deposits in Shark Bay, Australia.

Obliterative replacement dolomite in the upper part occurs as concentrated seams to thicker bands focused along stylolites, disseminated rhombs, and completely replaced intervals of various unfossiliferous dolomite types, such as brecciated (intraclastic?), laminated, fenestral, burrowed, and stylolitic, which also compose the completely dolomitized lower 61-ft (18.5 m) sequence. Near-stoichiometric, non- to very slightly luminescent dolomite crystal size ranges from mud to millimeter-size saddle void-filling cement, but the main mode is coarse decimicron- to fine centimicron-size. Average $\delta^{13}C$ for calcite = −1.13‰, for dolomite = +0.51‰; average $\delta^{18}O$ for calcite = −7.43‰, for dolomite = −6.69‰ (PDB). Average trace-element content (in ppm) for dolomites is Fe = 313, Na = 985, Sr = 31, and Mn = 73. Average homogenization temperatures (pressure uncorrected) of fluid inclusions for selected groups of dolomite crystals suggest a general direct relation to crystal size and range from 130°C for fine centimicron-size crystals to 193°C for saddle dolomite.

A sequence containing dolomitized tidal flat features and intimate associations of calcite and dolomite intraclasts suggests early dolomitization. Some deep burial dolomitization is indicated by dolomite growth along stylolites; more pervasive late dolomitization is suggested by broader bands of dolomite, whose geometry suggests stylolite control. Coarse crystallinity, xenotopic fabric, relatively depleted $\delta^{18}O$ values for all dolomite types, trace-element content, and limited fluid inclusion data strongly suggest the influence of hot and deep subsurface solutions, but it is unclear whether the dolomite resulted from mesogenetic replacement, early dolomitization followed by neomorphism in the burial environment, or some combination of those two "end-member" models.

INTRODUCTION

Unocal's Red Hills Unit #3 well was drilled in Section 5, Township 26S, Range 33E, in Lea County, southeastern New Mexico (Fig. 1). The principal objective of this paper is to describe and interpret the dolomitization in the 132 ft (40 m) of this deep, conventional core (17,465 ft; 5,323.3 m–17,597 ft; 5,363.6 m) of mixed carbonates of Siluro-Devonian age. Interestingly, dolomitization is incomplete; depositional textures and sedimentary structures are generally well preserved in the limestones.

The Silurian and Devonian rock units of West Texas were deposited in the early to medial Paleozoic Tobosa Basin (McGlasson, 1967). The core under consideration includes a carbonate sequence consisting of limestone, dolomitic limestone, calcareous dolomite, and dolomite immediately underlying the Devonian Woodford Shale. This carbonate section beneath the Woodford has been termed the "Siluro-Denovian" unit due both to uncertainty concerning the exact age of the interval and to the fact that the sequence does not crop out, thus precluding formal stratigraphic designation (McGlasson, 1967). This "Siluro-Devonian" unit is lithologically distinct from, though possibly correlative with, the shaly facies in the "Upper Silurian" unit and the entire "Devonian" unit (chert and siliceous limestone) southeast of Lea County in a different part of the Tobosa Basin.

METHODS

The study included description of the slabbed core and examination of about 90 large (2″ × 3″; 5 cm × 7.6 cm) stained (alizarin red-S and potassium ferricyanide) thin sections of carboniates impregnated with blue-dyed epoxy. Selected samples, including all those of mixed calcite/dolomite mineralogy chosen for stable isotope analysis, given sufficient material, were analyzed by X-ray diffraction (XRD) primarily for dolomite stoichiometry and calcite/dolomite ratios. A Norelco vertical goniometer was used in a 2θ-step scan from 24°–33° at a scan rate of 0.02°/s/step (quartz was used as an internal standard). Calcite:dolomite ratios were determined by relating peak-area ratios to weight percent dolomite. Dolomite stoichiometry was determined using the relation between calcium content and d_{104} spacing (Goldsmith and Graf, 1958). Forty-five samples, predominantly of dolomite and dolomitic limestone (n = 34), but also including limestone and dolomitic limestone (n = 11), were analyed for $\delta^{13}C$ and $\delta^{18}O$; samples were selected on the basis of petrography. For example, calcite types include stromatoporoids, groundmass (peloidal material and some early spar), and late cement. In addition to dolomite disseminated along stylolites, other dolomites were analyzed, with textures ranging from medium decimicron-size through millimeter-size saddle crystals, to test whether there might be a trend with respect to $\delta^{18}O$. Sampling was done with a hand drill to permit selectivity on a reasonably small scale. On the basis of their ratios, calcite-dolomite mixtures were leached with the appropriate amount of 1N acetic acid so as to minimize the possibility of inclusion of any CO_2 gas evolved from calcite in the dolomite analyses. The remainder of the procedure follows that of McCrea (1950). Trace elemental content for Fe, Na, and Mn was determined by inductively coupled plasma (ICP) atomic emission spectrophotometry, and Sr by flame atomic emission spectrophotometry. Selected samples were observed by cathodoluminescence (CL) and scanning electron microscopy (SEM). In a limited study, three dolomite samples were chosen for determination of temperatures of homogenization of gas-

Sedimentology and Geochemistry of Dolostones, SEPM Special Publication No. 43

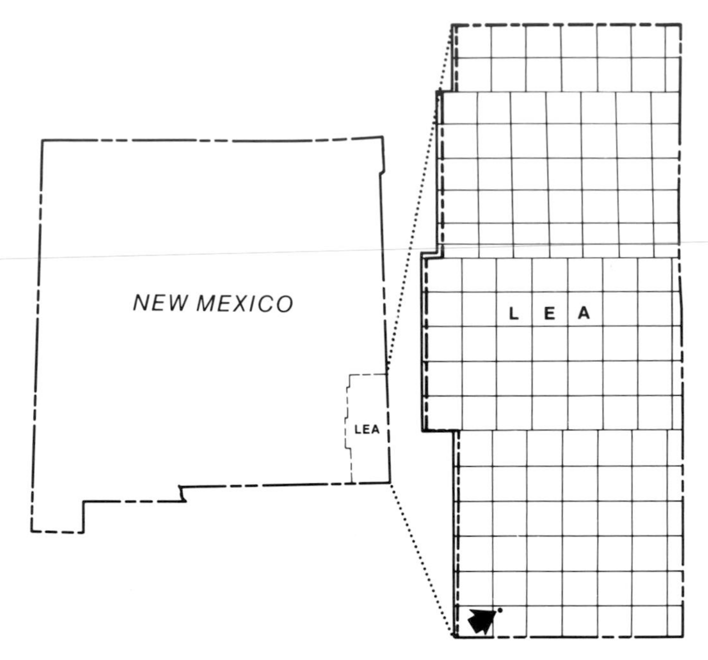

FIG. 1.—Location map of Unocal Red Hill Unit #3 well.

fluid inclusions; in order of increasing crystal size there are: coarse decimicron to fine centimicron size; fine to medium centimicron size; and millimeter size (saddle).

CORE DESCRIPTION

The general core lithologies are shown in Figure 2. The upper part of the cored interval (71 ft; 21.5 m), from 17,465 ft (5,323.3 m) to 17,536 ft (5,345 m), contains marine fossils, is of mixed calcite-dolomite mineralogy, and comprises several carbonate lithotypes. Contrastingly, the lower part (61 ft; 18.5 m), from 17,536 ft (5,345 m) to 17,597 ft (5,363.6 m), consists completely of dolomite, lacks discernible fossils, and is much less diverse in lithotypes. In the following descriptions, we follow generally the carbonate classification of Dunham (1962) as modified by Embry and Klovan (1971).

Stratigraphy and Petrography of the Upper Sequence (above 17,536 ft; 5,345 m)

With the exception of a few dolomite beds and some very dolomitic limestone, low insoluble limestone is by far the dominant lithology in the uppermost 33 ft (10 m) of this upper sequence. Most limestone contains significant dolomite throughout, however. There are roughly equal amounts of limestone and dolomite from 17,498 ft (5,333.4 m) to the lower contact with the lower sequence (Fig. 2). Intraclasts, commonly in breccia, are the dominant grain type in both limestone and dolomite. Although fossils and peloids are very abundant in the limestones and dolomitic limestones, they are not observed in the dolomites. Stromatoporoids, including bulbous, nodular, tabular, and dendroid types (e.g., amphiporids and stachyodids) are the most common fossils followed by pelmatozoans (probably crinoids), corals (tabulates and rugosans), brachiopods, ostracodes, and gastropods. Most fossils occur as skeletal fragments in the matrix or in intraclasts, which consist of limestone, dolomite, or mixtures of calcite and dolomite; some exceed 10 cm in greatest dimension.

Limestone lithotypes.—

There are four major limestone lithotypes present in the upper sequence:

(1) Stromatoporoid, peloidal rudstone to floatstone (Fig. 3A); neomorphosed stromatoporoids range from millimeter to centimeter size. The matrix itself consists of a grain-supported fabric of peloids and other skeletal fragments. Calcite spar, interpreted as cement, ranges from fine to medium decimicron size. Stylolites are common and the facies is generally dolomitic (to be discussed later).

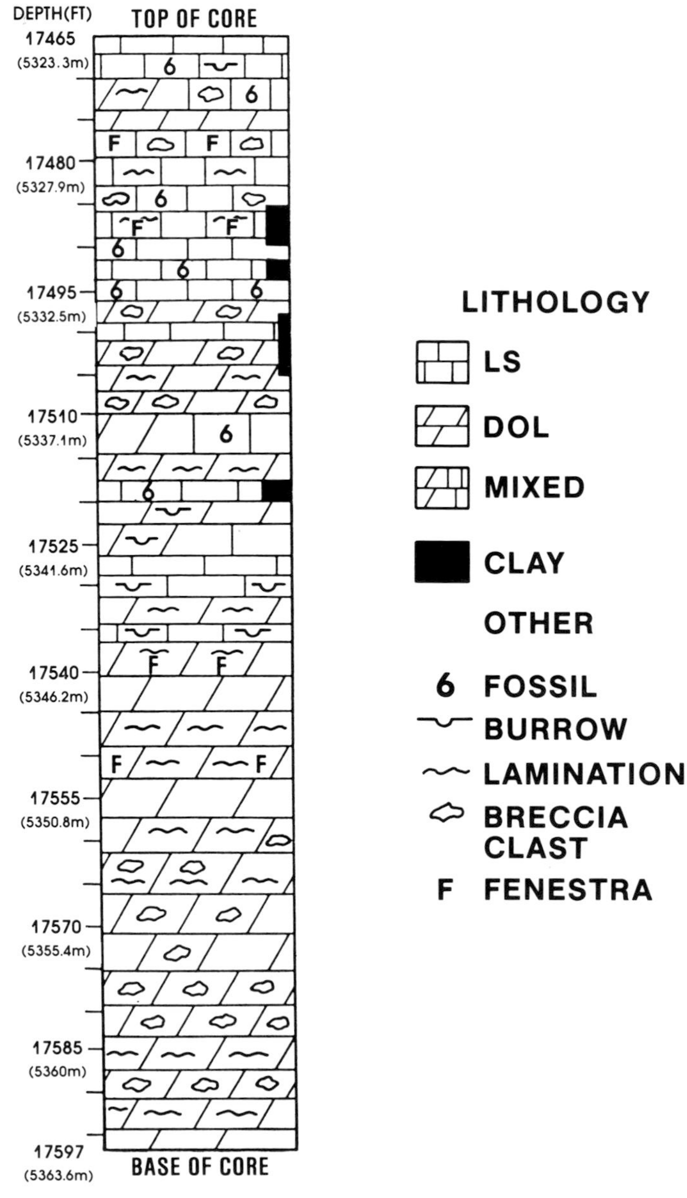

FIG. 2.—Stratigraphic section of Siluro-Devonian unit, Unocal Red Hills Unit #3 well.

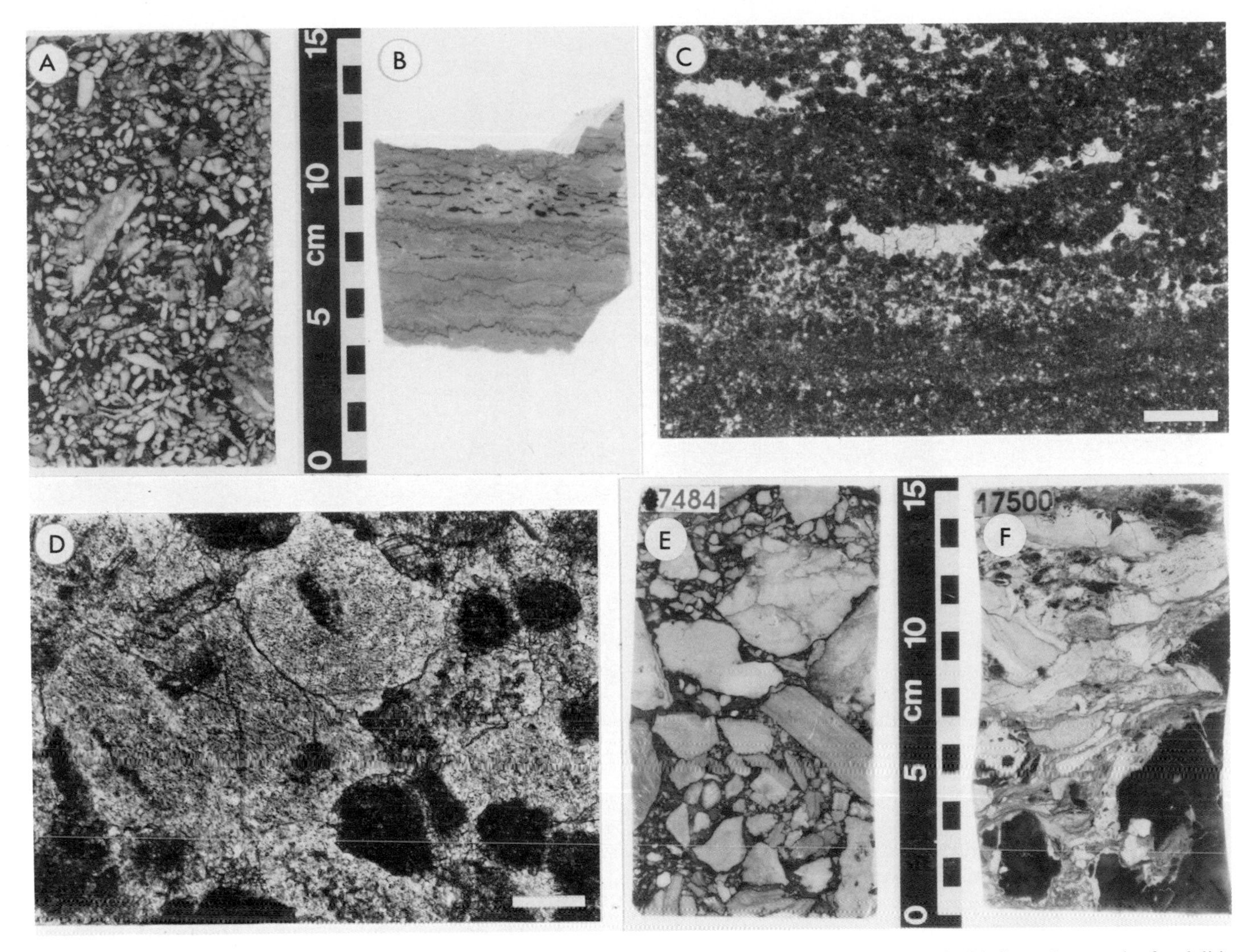

FIG. 3.—Limestone lithotypes. (A) Core photograph of stromatoporoid (amphiporid) rudstone, upper interval. (B) Core photograph of stylolitic, fenestral, laminated to massive mudstone, upper interval. (C) Photomicrograph of laminated, peloidal mudstone to packstone with fenestrae; plane polarized light; scale bar 500 μm. (D) Photomicrograph of peloidal skeletal (pelmatozoans) grainstone, plane polarized light; scale bar 500 μm. (E) Core photograph of intraclast breccia. (F) Core photograph of intraclast breccia with mixture of stylolite-bounded dolomite (lighter and more abundant) and limestone clasts (stained and darker).

(2) Laminated to massive, peloidal mudstone to packstone (Fig. 3B, C); peloids and sand-size mudstone intraclasts are the dominant grains. It is possible that some laminae represent smooth algal mats. Skeletal fragments are relatively uncommon. Certain sequences are fenestral but generally small, spar-filled mesointerparticle pores (Choquette and Pray, 1970) do not resemble fenestrae. Stylolites contribute to the laminated appearance. This lithotype contains little dolomite, although clasts of similar lithology are commonly dolomitic (see lithotype 4).

(3) Peloidal, skeletal grainstone (Fig. 3D); this relatively minor facies, which grades to the packstones of lithotype 2, contains little dolomite. Dominant grains are peloids and skeletal fragments, including pelmatozoans, stromatoporoids, corals, and gastropods, many of which are micritized to varying degrees. Pressure solution is not as common in this facies.

(4) Intraclast breccia (Fig. 3E); this most common facies, particularly above 17,510 ft (5,337 m), has a distinctly brecciated appearance. The contacts between generally irregular and angular clasts are typically stylolitic, resulting in an overly close packing of clasts to the exclusion of matrix. Limestone intraclasts consist primarily of lithotype 1, lithotype 3, and less commonly, lithotype 2. Partial dolomitization is characteristic of this lithotype, which grades into the dolomite intraclast breccia lithotype, described later, by way of mixtures of limestone and dolomite clasts (Fig. 3F).

Dolomite occurrences.—

A very significant quantity of dolomite occurs in partially dolomitized limestone. The sequences designated dolomite and calcareous dolomite, more common between 17,498 ft (5,333.4 m) and 17,536 ft (5,345 m), are roughly com-

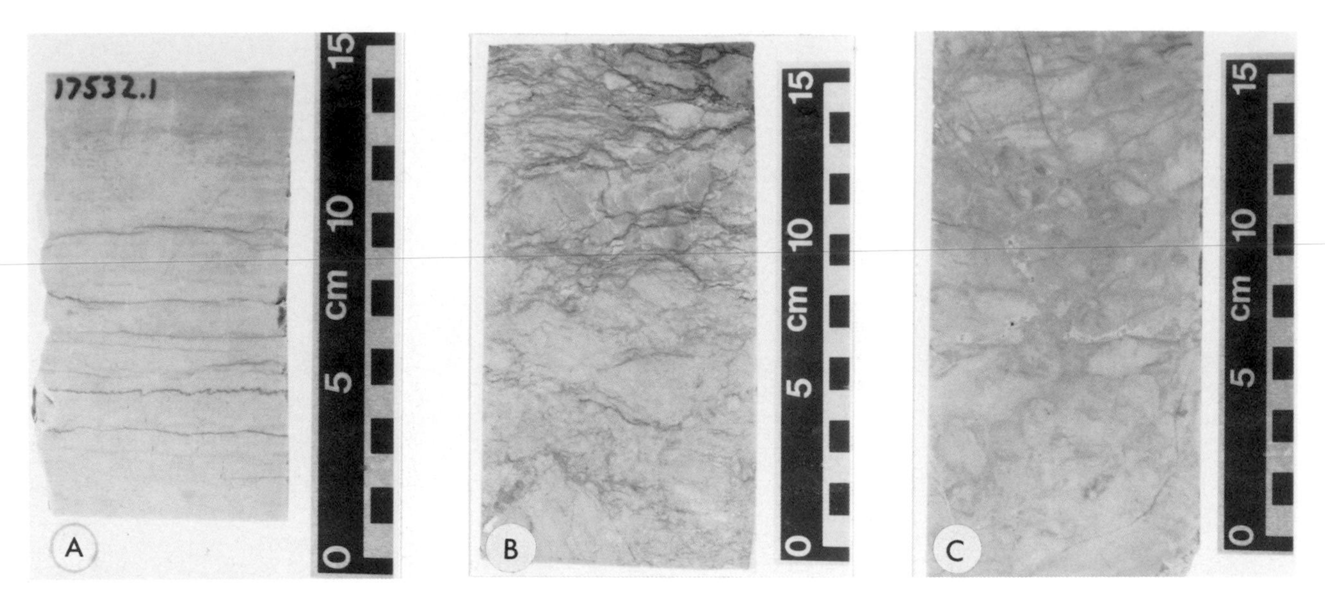

FIG. 4.—Dolomite lithotypes. (A) Core photograph of stylolitic, laminated, mosaic dolomite, upper interval. (B) Core photograph of dolomite intraclast breccia with relatively distinct clasts, upper interval. (C) Core photograph of dolomite intraclast breccia, with vague clasts, lower interval.

parable to two of the limestone lithotypes. Laminated mosaic dolomite (Fig. 4A) appears to resemble the laminated, peloidal mudstone to packstone, whereas dolomite intraclast breccia (Fig. 4B, C) is equivalent to its limestone counterpart. There are, however, some essential differences other than the main mineralogical distinction—most important, there is no recognizable skeletal or peloidal material in the dolomite.

Laminated mosaic dolomite consists of vague, relatively even to irregular, continuous to broken, thin to thick, subtle laminae of medium to coarse decimicron-size crystals alternating with coarse decimicron- to fine centimicron-size dolomite (Fig. 5A). More rarely, laminae are also identified by alternations of seams of cloudy crystals with clearer dolomite and also by remnants of incompletely replaced peloidal packstone and grainstone. The fabric of the groundmass dolomite is xenotopic (Fig. 5B). Also common in this facies are thin, wispy seams of finer decimicron-size dolomite. Relict stylolites and microstylolites are possibly represented by these and/or similar undulatory seams with brownish interstitial material padding dolomite crystals. The dolomite is "cloudy" with unstained inclusions, and sporadic subhedral to euhedral crystals possess clear rims. Large mesopores (Choquette and Pray, 1970), varying in shape from equidimensional to elongate in a direction parallel to stratification, are commonly filled primarily by saddle dolomite (Fig. 5C). In some thin intervals, such filled pores resemble fenestrae (Fig. 5D), but generally they are more regular in shape and less uniformly distributed; possibly, these represent burrows. Healed fractures <50 μm in width are relatively common (Fig. 5E). In the rare, incompletely dolomitized laminae referred to earlier, euhedral, fine centimicron-size dolomite crystals truncate peloidal material. Normally, the inclusions in these cloudy dolomite crystals are uniformly distributed, and the truncated peloids have lost their identity. All samples examined in CL showed no luminescence except one specimen of saddle dolomite with a moderately bright zone, the outer margin of which coincides with the contact between a cloudy core and a clear rim. Distinct stylolites and microstylolites transecting the dolomite (Fig. 5F) are not as common as in the limestones.

Distinct to vague clasts in the dolomite intraclast breccias range from millimeter size to 15 cm in greatest dimension, the greatest bulk being between 2 and 5 cm. Clast outlines are generally very angular and irregular, as in their limestone counterparts, and the lithology is similar to that of the laminated mosaic dolomite. Of possible significance is the presence of rare, small patches of unreplaced limestone in some intraclasts; one such pocket consists of a portion of a stromatoporoid.

Dolomite in partially dolomitized limestone occurs in three settings which appear to be intergradational:

(1) focused along definite stylolites or zones of closely spaced microstylolites (sutured seam and non-sutured seam, respectively, of Wanless, 1979); at one end of the spectrum is a scattering of dolomite crystals along and truncating a single stylolite; the dolomite crystals commonly coalesce into thicker seams laterally (Fig. 6A, B). This fabric, in one direction, grades into disseminated to locally concentrated crystals associated with microstylolite swarms (Fig. 6C). Another fabric that could well be related to either, or both, of the above is a concentration of dolomite in seams as thick as one mm or more, which have the styluslike geometry noticeably similar to stylolites (Fig. 6D) and, in addition, commonly show some argillaceous residue of the pressure-solution process. These dolomites are typically medium to coarse decimicron size but range upward through fine centimicron-size to millimeter-size saddle crystals, particularly along the margins of the more concentrated seams.

(2) in bands of pure or slightly calcareous, xenotopic,

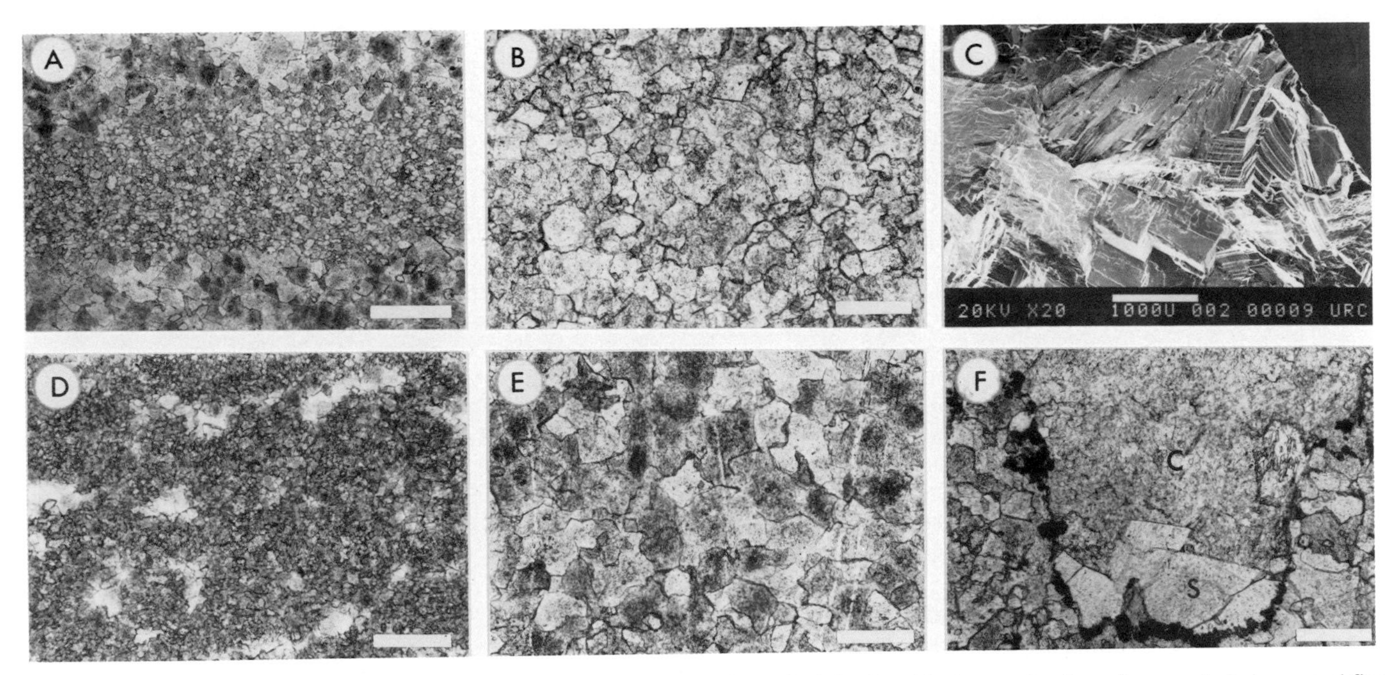

FIG. 5.—Dolomite fabrics and textures. (A) Photomicrograph of laminated mosaic dolomite with alternate laminae of coarse decimicron- and fine centimicron-size crystals; plane polarized light; scale bar 500 μm. (B) Photomicrograph of xenotopic dolomite; plane polarized light; scale bar 500 μm. (C) Scanning electron micrograph of void-filling saddle dolomite. (D) Photomicrograph of dolospar-filled fenestrae in coarse decimicron- to fine centimicron-size dolomite; plane polarized light; scale bar 500 μm. (E) Photomicrograph of predominantly fine centimicron-size, xenotopic dolomite with healed fractures; plane polarized light; scale bar 200 μm. (F) Photomicrograph of post-dolomite stylolite; note truncation of saddle dolomite (s) that is replacing a calcitic stromatoporoid (c); plane polarized light; scale bar 200 μm.

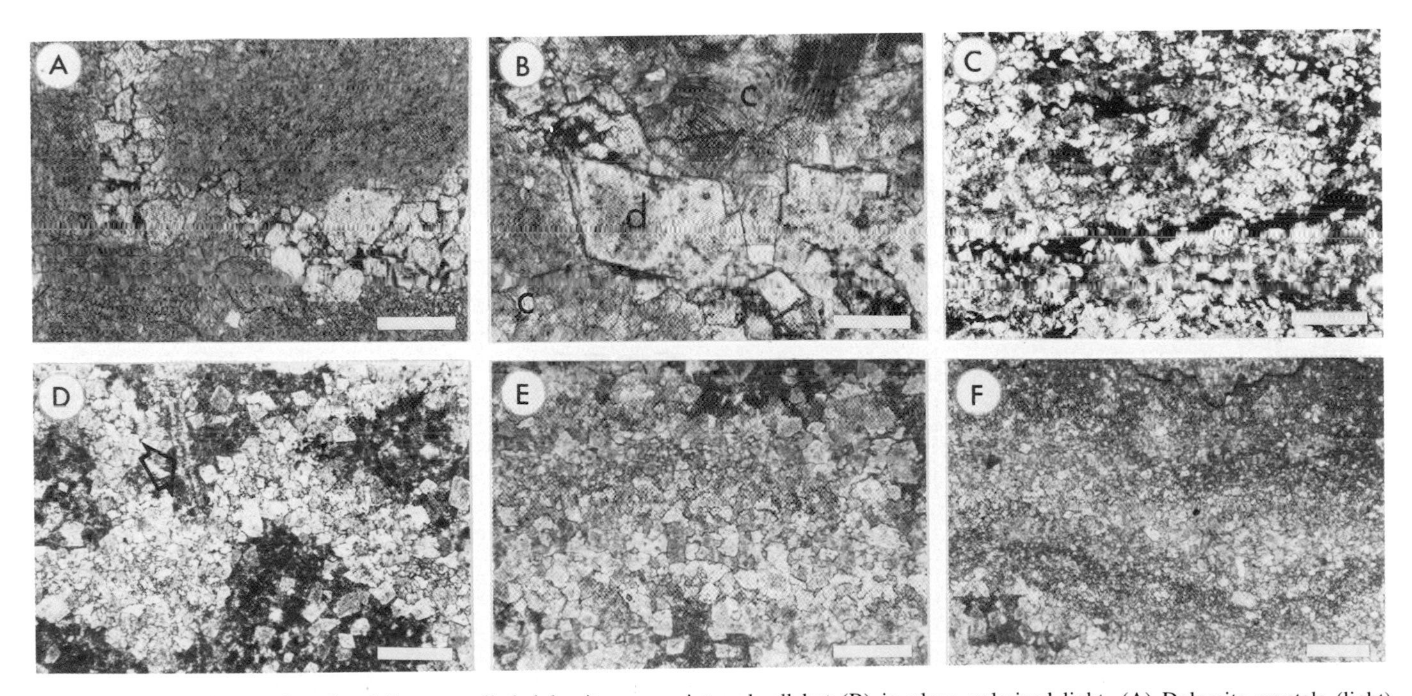

FIG. 6.—Photomicrographs of stylolite-controlled dolomite, upper interval; all but (B) in plane polarized light. (A) Dolomite crystals (light) truncating microstylolite that separates limestone clasts (dark); note thickening of dolomite band on left and right; scale bar 150 μm. (B) Dolomite rhombs (d) truncating stylolite that separates limestone clasts (c), one in upper right showing calcite twinning; ordinary light; scale bar 150 μm. (C) Disseminated, pressure-solution–controlled (dark stringers) dolomite (light) replacing calcite (darker) groundmass; scale bar 500 μm. (D) Probable stylolite-controlled seams of dolomite (light) resembling net fabric (Zenger, 1983) and disseminated dolomite crystals replacing peloidal packstone; note truncation of calcite-filled fracture, upper left center (arrow); scale bar 500 μm. (E) Relatively wide band of probably stylolite-controlled dolomite (light) replacing peloidal mudstone (dark) note seam of finer crystalline dolomite in upper part of band; scale bar 200 μm. (F) "Wispy," decimicron-size dolomite (darker) in coarser laminated mosaic dolomite (lighter); small angular pocket of incompletely replaced lime mudstone (dark), lower left; note post-dolomite stylolite at top; scale bar 500 μm.

fine to coarse decimicron-size dolomite as thick as several millimeters (Fig. 6E); at the narrower end of the spectrum, there appears to be a gradation to the stylolite-controlled seams of the first type (Fig. 6D). The thicker bands, although generally lacking the revealing styluslike geometry of distinct stylolites, do occupy a similar position between intraclasts, commonly pass laterally into stylolites or distinct stylolite-controlled seams, and commonly contain irregular stringers of very fine crystalline dolomite (Fig. 6E) and/or argillaceous matter that could represent relict stylolites.

(3) disseminated in limestone (Fig. 6C, D); these crystals are characteristically euhedral, although locally "ragged" in their detailed outlines, and fine centimicron size. They occur in the matrix, but also in intraclasts, replacing primarily peloidal material. In most instances, this type is spatially related to types 1 and 2; it is usually most abundant in the proximity of dolomitized stylolites and bands of concentrated dolomite described earlier. The occurrences described in 1–3 commonly resemble various stages of the "net fabric" described in the Devonian Lost Burro Formation of California (Fig. 6D; see also Zenger, 1983).

Stratigraphy and Petrography of the Lower Sequence (below 17,536 ft; 5,345 m)

At precisely 17,536 ft (5,345 m), a sharp contact separates burrowed(?), slightly dolomitized peloidal grainstone from an underlying, entirely dolomite sequence to the base of the core. The diversity of recognizable textures and fabrics is much more limited in this lower sequence. These dolomites bear a strong resemblance to the pure dolomites in the upper interval in which precursor calcite fabrics have been obliterated. Fossils appear to be completely lacking, and only uncommon late-stage calcite is present. Abundant voids occur as intercrystalline pores and incompletely filled meso- to megavugs (some representing burrows; some probably fenestrae) and fractures. With the exception of void-filling dolospar, dolomite is xenotopic (Fig. 5B); essentially all dolomite, whether groundmass or void-filling, is cloudy. Stylolites and microstylolites postdating dolomite are relatively common (Fig. 6F), as are healed fractures.

The two most prevalent lithotypes, very similar and corresponding to those completely dolomitized types in the upper sequence, are laminated mosaic dolomite and dolomite intraclast breccia (Fig. 4). The former is most common in the upper 20 ft (6.1 m), whereas essentially the whole interval below 17,560 ft (5,352.3 m) consists of the latter.

As in the upper sequence, laminae in the laminated mosaic dolomite are defined by alternations of medium to coarse decimicron-size and coarse decimicron-to fine centimicron-size dolomite (Fig. 5A), both being relatively inequigranular. Characteristic are thin, wispy seams of very fine crystalline dolomite ranging in appearance from partings to stylolites (Fig. 6F). Mudstone intraclasts, much smaller than those in the intraclast breccia, are sparsely distributed. Large mesovugs to small megavugs are partially to completely filled with as large as millimeter-size, idiotopic, predominantly saddle dolomite (Fig. 5C). Pores, or locally late-stage calcite, occupy the void center. These generally rounded pockets are interpreted to represent burrows, although there are some intervals which appear to be fenestrate.

Clasts in the dolomite intraclast breccias are ill-defined in the slabbed core (Fig. 4C) and even less distinguishable in thin section; their character is nearly identical to that of the breccias with poorly defined intraclast boundaries in the upper sequence. Although the intraclasts are commonly angular, many intervals have more rounded clasts exhibiting a compacted appearance. Except for void-filling dolomite, the fabric in both intraclasts and matrix is relatively inequigranular and xenotopic. Generally, the large intraclasts are coarser crystalline (average fine centimicron size) than the matrix (average coarse decimicron size), although there are exceptions where crystal sizes are reversed. Smaller neomorphosed(?) mudstone intraclasts are locally present. Rare mesovugs have a druse of cloudy-centered, clear-rimmed dolospar with void space toward the vug center; fracture porosity and permeability are more typical of this lithotype.

ANALYTICAL RESULTS

Table 1 includes the analytical data for various groups of limestone and dolomite samples. Generally, the dolomites are strikingly near-stoichiometric; the mean of 26 samples is 50.9 mole percent of $CaCO_3$. Only three samples are enriched in $CaCO_3$ more than 2 mole percent; 85% are within 1.5 mole percent of perfect stoichiometry. The average values of mole percent $CaCO_3$ for dolomites (>96.0% dolomite, $n = 10$) = 50.1, for calcareous dolomites (50.0–96.0% dolomite, $n = 7$) = 51.7, and for dolomitic limestone (4.0–50.0% dolomite, $n = 9$) = 51.2. There is a lack of correlation between mole percent $CaCO_3$ and dolomite content ($r = -0.1893$), which is in accordance with the findings of Lumsden and Chimahusky (1980) on the basis of a much larger sample. The mole percent $MgCO_3$ in calcite is extremely low, all but one of the 21 samples analyzed (not shown in Table 1) being $\leq 0.9\%$.

There is a noticeable homogeneity in the stable isotopic values (relative to PDB), except for a few samples with primarily unexplained depleted $\delta^{13}C$ (Table 1, Fig. 7). The mean $\delta^{13}C$ for the dolomites = +0.51‰ (s = 1.42‰), whereas $\delta^{18}O$ is depleted, averaging −6.69‰ (s = 1.81‰); the mean $\delta^{13}C$ for calcite = −1.13‰ (s = 1.96‰), whereas $\delta^{18}O$ = −7.43‰ (s = 1.11‰). There is no significant correlation between $\delta^{13}C$ and $\delta^{18}O$ for either dolomite ($r = -0.09$) or calcite ($r = -0.05$). For the two instances of coexisting calcite and dolomite, $\Delta^{18}O$ is about at the lower limit of the 1‰–7‰ range predicted by Land (1980) for dolomite and its calcite precursor. Figure 7 shows the distribution of stable isotopes in textural and compositional groups. Conspicuously variant groups, e.g., disseminated dolomite, coarse decimicron-size dolomite, and late calcite, result from one or two samples in each group that are discrepant for either or both $\delta^{13}C$ or $\delta^{18}O$ with regard to the general range (Fig. 7, Table 1); a re-examination of the petrography sheds no light on these "mavericks." There appears to be no consistent trend of more depleted $\delta^{18}O$ with increase in crystal size. For example, although $\delta^{18}O$ for medium to coarse centimicron-size dolomite ($\delta^{18}O$ = −9.1‰; $n = 4$) averages slightly more than 2 per mil more depleted

TABLE 1.—ANALYTICAL RESULTS (AS AVERAGES OF CARBONATE TYPES)

Carbonate Type	Wt. Percent Dolomite (of Carbonate Fraction) (max. error = 1.6%)	Dolomite Stoichiometry ($CaCO_3/MgCO_3$) (max. error = 0.7%)	Stable Isotopes (‰, PDB) $\delta^{13}C$ (±0.1‰)	$\delta^{18}O$ (±0.2‰)	Trace Elements (ppm) Fe (±10%)	Na (±10%)	Sr (±20%)	Mn (±10%)
Dolomite (Saddle)	D = 100 (n = 5)	50.2/49.8 (n = 1)	−0.02 (n = 5)	−7.80 (n = 5)	250 (n = 1)	759 (n = 1)	28 (n = 1)	117 (n = 1)
Dolomite (m.c.s.–c.c.s.)	D = 100 (n = 4)	49.6/50.4 (n = 1)	−0.01 (n = 4)	−9.06 (n = 4)	134 (n = 1)	884 (n = 1)	24 (n = 1)	75 (n = 1)
Dolomite (f.c.s.)	D = 100 (n = 4)	50/50 (n = 2)	+0.53 (n = 4)	−7.13 (n = 4)	154 (n = 2)	832 (n = 2)	38 (n = 2)	74 (n = 2)
Dolomite (c.d.s.–f.c.s.)	D = 99 (n = 4)	50.6/49.4 (n = 3)	+3.15 (n = 3)	−6.30 (n = 3)	593 (n = 2)	1165 (n = 2)	34 (n = 2)	52 (n = 2)
Dolomite (m.d.s.–c.d.s.)	0.98.6 (n = 5)	50.3/49.7 (n = 5)	+1.14 (n = 5)	−6.96 (n = 5)	314 (n = 4)	1053 (n = 4)	29 (n = 4)	73 (n = 4)
Dolomite (Along stylolites)	D = 62.3 (n = 10)	51.2/48.8 (n = 10)	+0.77 (n = 10)	−6.20 (n = 10)	—	—	—	—
Dolomite (Disseminated)	D = 22.7 (n = 2)	53.2/46.8 (n = 2)	−0.85 (n = 2)	−3.35 (n = 2)	—	—	—	—
Calcite (Late cement)	D = 5.6 (n = 3)	49.7/50.3 (n = 1)	−3.0 (n = 3)	−7.8 (n = 3)	—	—	—	—
Calcite (Stromatoporoid)	D = 0.4 (n = 5)	—	−0.6 (n = 4)	−7.13 (n = 4)	91 (n = 1)	995 (n = 1)	170 (n = 1)	11 (n = 1)
Calcite (Groundmass)	D = 1 (n = 5)	52.7/47.3 (n = 1)	−0.34 (n = 4)	−7.32 (n = 4)	196 (n = 2)	751 (n = 2)	130 (n = 2)	12 (n = 2)

Note: m.d.s., c.d.s. = medium, coarse decimicron size; f.c.s., m.c.s., c.c.s. = fine, medium, coarse centimicron size. (Readers may request data for individual samples.)

than medium to coarse decimicron-size dolomite ($\delta^{18}O$ = −7.0‰; n = 5), saddle dolomite, the coarsest dolomite ($\delta^{18}O$ = −7.8‰; n = 5), averages 1.3‰ more enriched. Interestingly, a test for correlation between mole percent $CaCO_3$ in dolomite vs. $\delta^{13}C$ and $\delta^{18}O$ gave r = −0.38713 and +0.84185, respectively (n = 24); thus, there is a significant positive correlation (99% confidence level) between $CaCO_3$ in dolomite and $\delta^{18}O$.

For 10 selected dolomites (Table 1), mean trace-element contents (ppm) are Fe = 313 (s = 213), Na = 985 (s = 288), Sr = 31 (s = 7), and Mn = 73 (s = 19); for three calcite samples, the corresponding values are Fe = 161 (s = 134), Na = 832 (s = 144), Sr = 143 (s = 38), and Mn = 11 (s = 1.5). In contrast to other studies in this volume (Machel; Shukla), no obvious trend or pattern was detected in the admittedly small number of dolomite samples, either with regard to depth or to crystal size. The one sample of saddle dolomite is relatively low in Fe (250 ppm) and high in Mn (117 ppm). The correlation coefficient for Fe vs. Mn in dolomite is −0.615, just insignificant at the 95% confidence level; this relation contrasts with the positive relation reported by Pierson (1981).

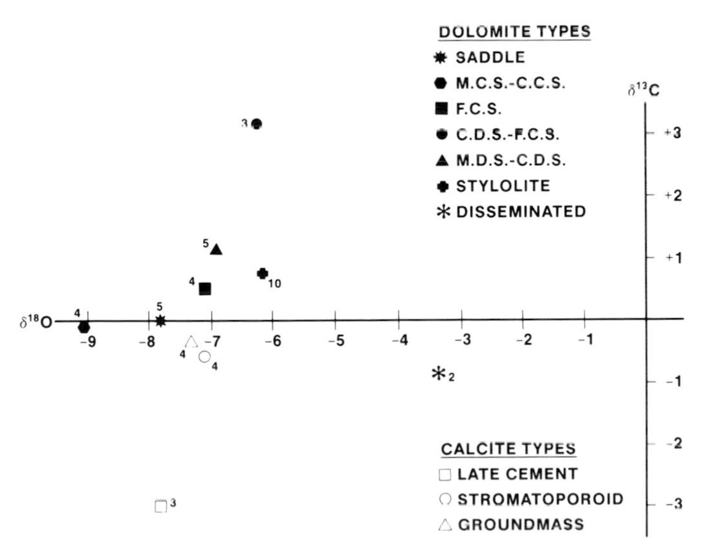

FIG. 7.—Scattergram of carbon and oxygen isotope values for mineralogical (i.e., calcite, dolomite) and textural groups. Abbreviations for crystal size as in footnote for Table 1.

Given the low Mn content, it is not surprising that nonluminescent dolomite predominates. One sample of saddle dolomite mentioned earlier exhibited a moderately bright luminescent zone; this sample, referred to immediately above, possesses the highest Mn content determined, 117 ppm. No quenching was noted in the limited samples (n = 5), in accord with the Fe content, which is well below that reported necessary for inhibition (Pierson, 1981).

Fluid inclusions analyzed appeared to be primary, varied in shape, and ranged in greatest dimension from 3 to 8 μm, most being small and in the lower end of that range, 3 to 4 μm. No fluorescence was observed in the inclusions. Given the vagaries and pitfalls of fluid inclusion study (e.g., Hanor, 1980; Bodnar and Bethke, 1984; Prezbindowski and Larese, 1987), we wish to be cautious about overextending the significance of the determinations, but there may be a direct relation between crystal size and homogenization temperatures (Fig. 8), mean values (pressure uncorrected) for the three groups being as follows: coarse decimicron- to fine centimicron-size dolomite = 130.4°C (n = 17, s = 14.7°); fine to medium centimicron-size dolomite = 144.9°C (n =

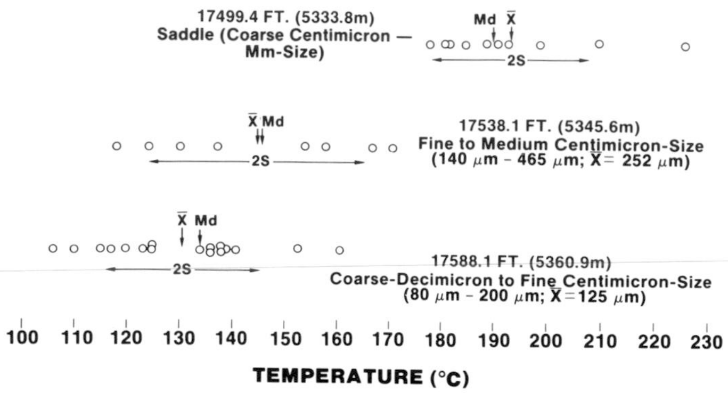

17499.4 FT. (5333.8m)
Saddle (Coarse Centimicron — Mm-Size)
Md X̄
2S

X̄ Md
2S
17538.1 FT. (5345.6m)
Fine to Medium Centimicron-Size
(140 μm - 465 μm; X̄= 252 μm)

X̄ Md
2S
17588.1 FT. (5360.9m)
Coarse-Decimicron to Fine Centimicron-Size
(80 μm - 200 μm; X̄=125 μm)

100 110 120 130 140 150 160 170 180 190 200 210 220 230

TEMPERATURE (°C)

FIG. 8.—Homogenization temperatures for three groups of varying dolomite crystal size.

8, s = 20.2°); millimeter-size saddle dolomite = 193.4°C (n = 10, s = 14.9°). The saddle dolomite is significantly separated from the fine to medium centimicron-size dolomite ("t" test; 99% probability level); the latter, however, barely fails statistical distinction from the finer dolomite at the 95% probability level.

DISCUSSION

Environment of Deposition

Insight into the environment of deposition as well as the nature of dolomitization is limited to a considerable extent by the lack of control adjacent to this one core, which precludes a more reliable interpretation of lithotypes depending in part on their shape and extent. In spite of these limitations, however, many conclusions can be drawn based on the mineralogy, textures, fabrics, and small-scale structures observed. The common presence of invertebrates, particularly corals and stromatoporoids, in the limestones in the upper sequence strongly suggests the occurrence of near-normal subtidal conditions for a significant part of the depositional interval. Of the four major limestone lithotypes, that of the laminated to massive, peloidal mudstone to packstone contains the least abundant and varied fauna, consisting primarily of pelmatozoan and ostracode fragments. Practically all the fossil material has undergone at least some transport. Possibly, some rudstone bioclasts represent bioherm detritus, although there is no definite evidence of any such structures. Possible low-angle burrows suggest subtidal depositional conditions for the peloidal, skeletal grainstone lithotype.

In fact, the most abundant and striking lithotype is the intraclast breccia. Certainly, the clasts are similar to closely associated carbonates, as described earlier. For example, there are intervals in which most of the clasts consist of laminated to massive, peloidal lime mudstone to packstone and which grade downward to that *in situ* lithotype (Fig. 9). This lithology, including the presence of some fenestral horizons, may well represent the tidal flat environment; the few characteristic fossils present could have been washed into this site of more restricted circulation during storms or during brief pulses of relatively rare sea-level rise, although the calcispheres and ostracodes present could be euryhaline. Probably, the clasts represent true intraclasts resulting from "rip up" of lithified crusts, as is occurring today along the southern margin of Shark Bay, western Australia (observed by JBD). Some of the more fossiliferous clasts in certain breccias also represent a similar ripping up of originally subtidal sediments in the tidal flat environment, or, alternatively, in the subtidal zone itself. The lack of any deep-water interbeds (e.g., dark, thinly laminated, and so on) and fauna precludes a slope depositional environment for the clasts. It is very unusual to observe clasts in "normal" contact with the groundmass. Stylolite-bounded clasts are the rule and commonly have a fitted fabric (Logan and Semeniuk, 1976) with little to no matrix. The variance in clast orientation and their sharp distinction across stylolitic boundaries is interpreted to result from storm or wind tide "rip up" and subsequent deposition, followed much later by pressure-solution welding. Locally, however, there are some breccias that may have a tectonic or solution-collapse origin.

FIG. 9.—Core photograph of sequence of *in situ* laminated peloidal lime mudstone (below) with overlying intraclast limestone breccia consisting of clasts of same lithology in fine calcareous groundmass (dark), upper interval.

No recognizable fossils were found in any of the dolomites. Such a situation could result were dolomitization penecontemporaneous, in which case conditions in the depositional environment (e.g., hypersalinity) enhancing dolomitization could have stifled faunal development. Small laths of anhydrite or carbonate-filled anhydrite molds occur rarely in limestone and dolomite. Much of the anhydrite appears to be later void filing (Fig. 10). Possibly some void-filling dolospar represents space provided by the solution of earlier anhydrite. Thus, the lower, completely dolomi-

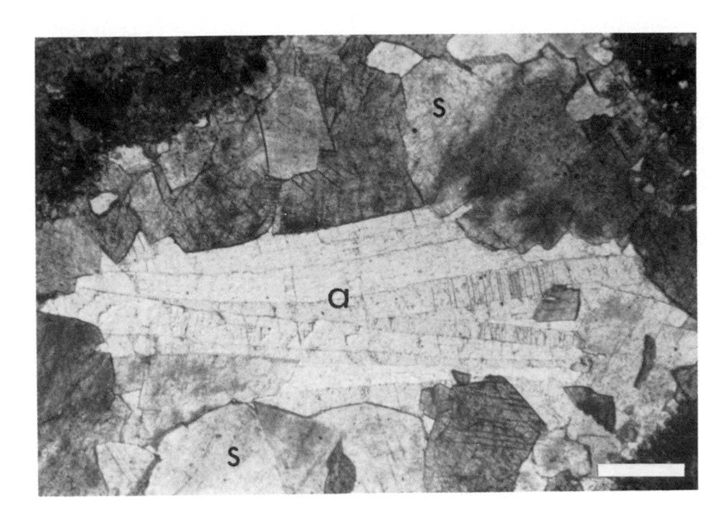

FIG. 10.—Photomicrograph of peloidal limestone groundmass (dark, all corners but lower left) showing lath-shaped anhydrite (a) filling center of void and surrounded by coarse saddle dolomite (s); plane polarized light; scale bar 500 μm.

tized interval could represent more extended periods of restricted conditions than the upper sequence. On the other hand, dolomitization could have been obliterative (Figs. 6, 11); the fact that the latter is the more likely situation will be discussed in the subsequent section on dolomitization. Given such a paucity of original fabrics and textures in the dolomites, it is difficult to make any reliable environmental interpretation for them. As a general statement, we suggest that the depositional environments of the laminated mosaic dolomite and the dolomite intraclast breccia are equivalent to their limestone "counterparts"—tidal flat for the laminated to massive, peloidal mudstone to packstone and subtidal to tidal flat for the intraclast breccia, respectively. One anomaly here is that, of the partially dolomitized limestone lithologies, one of the least dolomitized is the laminated to massive, peloidal lime mudstone to packstone.

Dolomitization

Except for later void fill, dolomite is replacive, as conclusively shown by disseminated crystals that truncate peloids, early spar, fossils, and stylolites (Figs. 6A, B, 11). Dolomite is most common in the peloidal matrix between bioclasts and intraclasts in the stromatoporoid rudstones and intraclast breccias (Fig. 11A), which reflects its association with pressure-solution seams. Dolomite is rarer in intraclasts in the limestones, particularly those composed of peloidal mudstone and packstone, but again is related to stylolites that truncate the clasts. Dolomite is not a significant component in two limestone facies—laminated, peloidal packstone to grainstone and peloidal, skeletal grainstone. Generally, both bioclasts and intraclasts resisted dolomitization; as observed where dolomitization is partial, the replacement is selective, that is, mudstone and peloidal material are preferentially replaced (Figs. 6D, 11A) with respect to early spar, and particularly skeletal debris, which is the most resistant. Where dolomite does rarely replace portions of fossils (generally in the proximity of stylolites), the replacement is nonmimic (Bullen and Sibley, 1984; Fig. 11B), and, in fact, appears to be obliterative (Figs. 6, 11). Dolomite crystals, especially disseminated in peloidal fields, are strikingly cloudy with included matter, although rarely can the outlines of the truncated portion of peloids be recognized; cloudy crystals with clear rims are relatively common. As a rule, disseminated dolomite crystals are coarser than those along stylolites or in argillaceous seams (Fig. 6D). This relation argues against a residue origin (i.e., stylocumulate; Logan and Semeniuk, 1976) for dolomite focused along stylolites. There is more compelling evidence for a reactate origin such as truncation of stylolites (Fig. 6A, B) and innumerable instances of much greater concentrations of dolomite crystals immediately along the stylolite, with a sharp drop in number into the adjacent groundmass, which, in fact, may be dolomite-free. It seems very unlikely that pressure dissolution of such groundmass could have provided the dolomite as a residue. Saddle dolomite is ubiquitous in terms of distribution at various depths, but it is not generally abundant. It is replacive (Fig. 11C) along the margins of dolomitized stylolite zones and occurs as late-stage void-filling cement in fractures and pores (burrows?).

Petrographically, the paragenetic sequence seems to be: original carbonate; replacive dolomite; void-filling dolomite; later stage pore- and fracture-filling quartz, calcite, anhydrite, and barite (it is difficult to determine the temporal relations of these late cements).

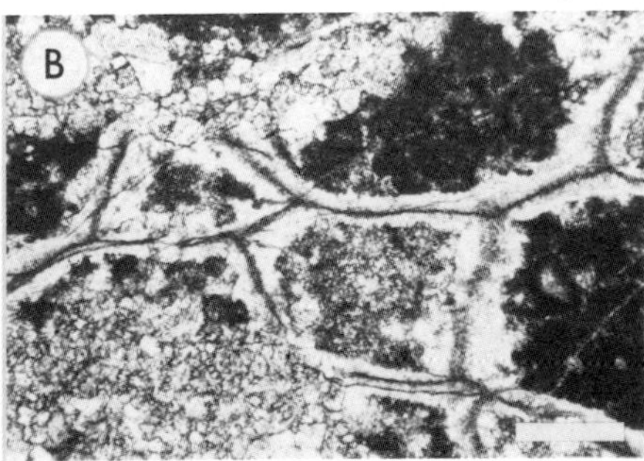

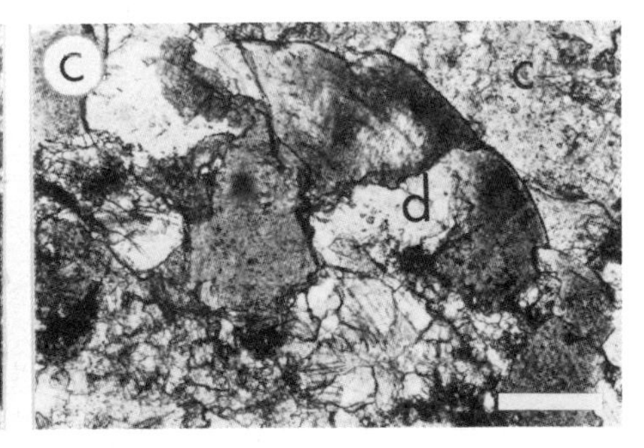

FIG. 11.—Photomicrographs of dolomite replacement fabrics; plane polarized light. (A) Disseminated dolomite crystals (light) selectively replacing peloidal groundmass (dark) vs. coral (on left); scale bar 250 μm. (B) Thin seams of stylolite-controlled dolomite (light, top and bottom) nonmimetically replacing corallite walls and peloidal grainstone filling (dark) of coral *(Thamnopora?);* scale bar 500 μm. (C) Saddle dolomite crystals (d; note undulating extinction) oriented along stylolite and replacing calcite groundmass (c); cross-polarized light; scale bar 200 μm.

There seems to be no easily recognized cyclicity or pattern of lithotypes in the limestones and dolomites in the upper sequence; as was mentioned above, the only generalization possible for the lower, completely dolomitized sequence is that the laminated mosaic dolomite is more common above and the dolomite intraclast breccia below.

Alternative models of dolomitization.—

Controversy continues over the relative responsibility of dolomitization models in accounting for extensive ancient dolomites. In determining the origin of a particular dolomitic unit, especially given our still incomplete state of knowledge, it is generally necessary to examine the *collective* evidence, from the field to the laboratory, as rarely do one or a few criteria prove to be reliable. In fact, as will be shown by our efforts here, an unequivocal conclusion may not be forthcoming, even when appealing to a wide range of available information.

Basically, two "end-member" possibilities are presented for replacement dolomitization of these Siluro-Devonian carbonates. These models are briefly presented, followed by a discussion of the evidence for each.

(1) The major dolomitization was penecontemporaneous to very early diagenetic and was originally associated with surface-derived solutions. Such dolomite is forming today in numerous Holocene peritidal environments. Small amounts of late diagenetic dolomite occur along stylolites (which the dolomite transects) in the upper interval of the core. The geochemistry indicates, however, that the early diagenetic dolomite has experienced neomorphism in the deeper burial environment. According to this model, limestone-dolomite distribution bears a relation to the paleogeography and depositional setting. For instance, the gross difference between the mixed carbonates in the upper interval and dolomite in the lower sequence relates to environmental parameters. The completely dolomitized lower interval contains no fossils and represents restricted hypersaline conditions that promoted dolomitization; fossiliferous limestones above record periods of more normal salinity. The extended residence time of the lower carbonates in and near the active diagenetic setting of the supratidal zone is the factor responsible for the complete dolomitization of that interval. By contrast, sediments in the upper portion were incompletely dolomitized because of a greater frequency of relative sea-level rise, which interrupted establishment of long-term supratidal conditions during sedimentation. Recently, Machel and Mountjoy (1986) have argued that there is a continuous spectrum from normal marine subtidal to hypersaline-subaerial dolomitization.

(2) The major dolomitization is later diagenetic and deeper burial. Not only did the dolomite that focused along discrete "sutured-seam stylolites" (Wanless, 1979) form subsequent to these structures, it also evolved into the complete gamut of dolomites, through the narrower seams and bands (Fig. 6) to the more pervasive sequences of dolomite or slightly calcareous dolomite in the upper interval and ultimately to the completely dolomitized lower interval. Thus, dolomitization was not effected by solutions related to the contemporary surface and accordingly, limestone-dolomite distribution is not directly a reflection of the depositional environment.

Of course, there is the possibility of an intermediary model involving either some early diagenetic dolomite that acted as seeds for subsequent later diagenetic replacive overgrowths, or a scenario in which some of the dolomite at certain stratigraphic intervals (e.g., those in the upper interval) follows one model and some the other.

Evidence for effect of burial environment.—

Common to both these proposed models is the significant role played by diagenesis in the burial environment, particularly as suggested by the geochemical evidence, considering the apparent lack of significant recrystallization in the unreplaced limestone. *Much of the following evidence could be accounted for by either burial neomorphism of an earlier replacement dolomite or by late diagenetic, burial replacement:*

(1) Crystal size. Other than relatively minor fine decimicron-size dolomite in the argillaceous and/or organic (stylocumulate?) seams in dolomitic limestone and in wispy seams in dolomite (Fig. 6F), crystal size ranges from medium decimicron-size to millimeter-size saddle dolomite, the great bulk being in the coarse decimicron to fine centimicron size, which greatly exceeds that of modern tidal flat dolomite and many ancient analogs (Zenger, 1972). Coarse crystal size is certainly not definitive of late diagenesis but is suggestive of higher temperature (Zenger, 1983).

(2) Xenotopic texture. Gregg and Sibley 1984) proposed that dolomite forming above its "critical roughening temperature" of >50°C tends to be xenotopic, but Shukla (1986) has suggested a revision downward to about 35°C. In these dolomites the texture is xenotopic (Fig. 5B), suggesting an effect of temperature possibly well above that of the earth's surface.

(3) Stable isotopes. Although $\delta^{13}C$ values (Table 1, Fig. 7) for both limestone and dolomite are generally within those of normal marine carbon, $\delta^{18}O$ values for dolomite are depleted, averaging −6.69‰ PDB. Similar depleted values have been recorded for purported burial dolomites in the geologic record (e.g., Mattes and Mountjoy, 1980; Zenger, 1983; Gregg and Sibley, 1984). Modern hypersaline dolomite formed by evaporative brines is isotopically heavy (Botz and von der Borch, 1984; Pierre and others, 1984). As described earlier, there is no consistent trend of more depleted oxygen values with increasing crystal size (Fig. 7); apparently there has been an isotopic "homogenization" among dolomites of various textures; presumably this reflects an exchange with subsurface solutions. The histogram of $\delta^{18}O$ for Devonian dolomites (Veizer and Hoefs, 1976, p. 1391) is bimodal, the more depleted mode of about −8 to −9‰ per mil probably representing late diagenesis (Fritz, 1971); however, recent stable isotopic studies of well-preserved Devonian calcitic skeletal elements (Veizer and others, 1986; Popp and others, 1986) strongly suggest lower values of $\delta^{18}O$ for Devonian oceans relative to present sea water. Consequently, pending further evaluation, their work questions the tacit assumption that it was only deep burial that contributed to depleted Devonian $\delta^{18}O$.

(4) Homogenization temperatures of fluid inclusions. Aware of the various problems that beset fluid inclusion studies, we are cautious in suggesting a direct relation between homogenization temperatures and crystal size (Fig. 8). Recently, Goldstein (1986) has proposed that during deep burial heating, most low-temperature fluid inclusions reequilibrate, some repeatedly, by leaking, exchanging with pore fluids and rehealing. At the least, the range of average values (pressure uncorrected) from 130°C (coarse decimicron- to fine centimicron-size crytals) to 193°C for saddle dolomite certainly suggests the deep burial environment.

(5) Stoichiometry. The near-stoichiometry of the dolomites (Table 1) most likely reflects burial conditions and either neomorphism and stabilization of metastable "protodolomite," the typical phase forming in modern tidal flats, or replacive crystalline dolomite forming initially in the mesogenetic zone (see also Lumsden and Chimahusky, 1980).

(6) Trace elements. Because of a number of factors, including uncertainty about appropriate distribution coefficients, it is difficult to interpret the trace-element content of ancient dolomite quantitatively (Land, 1980; Mattes and Mountjoy, 1980; Veizer, 1983; Hardie, 1987), although general statements may be made. Overall, the trace-element content (Sr, Fe, Mn, and Na) of these Siluro-Devonian rocks (Table 1), as in most ancient carbonates, indicates the effects of late diagenesis; accordingly, the trace-element content of *both* calcite and dolomite differs from that of modern marine, particularly supratidal, dolomite. The average Sr contents of calcite (143 ppm) and dolomite (31 ppm) are vastly depleted compared to their Holocene counterparts (Land and Hoops, 1973; Land, 1980; Veizer, 1983). Mattes and Mountjoy (1980) recorded an average of 50 ppm for their Miette (Devonian) dolomites, many of which were judged to be burial.

Iron tends to be enriched in later diagenetic dolomites, especially those formed from saline formation waters, compared to its extremely low concentration in modern marine sediments (Land, 1980; Veizer, 1983). Our average Fe values for dolomite (313 ppm) fall within the range of those of other deep burial and/or late diagenetic dolomites (Mattes and Mountjoy, 1980; Budai and others, 1984; Morrow and others, 1986).

Because its distribution coefficient is greater than one, Mn is generally concentrated in crystal phases during diagenesis (Land, 1980), but it is in low concentration in these Siluro-Devonian dolomites as it is in other diagenetic dolomites (e.g., Mattes and Mountjoy, 1980; Budai and others, 1984), although it is relatively enriched compared to calcite (Table 1). The low Mn content accounts for the nonluminescent character of most of the dolomites—iron content is too low to serve as a quencher. Morrow and others (1986) felt their similarly low content of Sr, Fe, and Mn in the Manetoe dolomite facies (Denovian) is consistent with its high degree of major-element stoichiometry and cation ordering, which are characteristics of the rocks described here.

The bulk of Na in Recent carbonate sediments is rarely less than 1,000 ppm and is commonly more than 3,000 ppm (Land and Hoops, 1973), whereas ancient dolomites generally show a range of 114–982 (Baum and others, 1985). Fritz and Katz (1972) pointed out that possible modes of occurrence of Na in dolomite crystals are in high-salinity fluid inclusions, related clay minerals, and the carbonate lattice. The fluid inclusion study here was limited by the rarity and small size of the inclusions, which would have made determination of their freezing temperatures difficult; consequently, there are no data on what part of the total Na content might be so bound. These Siluro-Devonian rocks include little insoluble material except the stylocumulate along stylolites, so it is questionable whether much Na would be contributed from that source (illite was the only clay mineral detected in an XRD analysis of an argillaceous seam). Sodium content in both calcite and dolomite (832 ppm and 985 ppm, respectively) is higher than that reported for many ancient dolomites, some of which have been interpreted as having been formed by hypersaline solutions (Veizer and others, 1977; Mattes and Mountjoy, 1980). We conclude that our higher than normal average reflects the effect of either initial or diagenetic brines.

Except for obvious recrystallization of stromatoporoids, limestone fabrics suggest little severe neomorphism; the geochemical data, however, attest to rather impressive, later, burial diagenetic effects, but it must be emphasized that in *this* core the geochemistry generally is biased toward *the later stages* of diagenesis. It is necessary to invoke petrology to attempt to help distinguish between early and late dolomitization.

Arguments for alternative models.—

Certain lines of evidence and reasoning tend to support a very early replacement origin for the dolomite, whereas others suggest a replacement at depth. These arguments are summarized in Table 2. From this, and the geochemical evidence, it is tenuous to distinguish between the two alternatives (early replacement followed by neomorphism in the deep subsurface vs. burial replacement) or some combination. For example, very early diagenetic dolomite could have provided "seeds" for later diagenetic overgrowths, and/or dolomite in certain stratigraphic intervals could represent earlier replacement, whereas in other sequences, the dolomitization was late. A general lack of correlation between crystal size and both $\delta^{18}O$ and trace-element content, coupled with a lack of "petrologic" zonation (as represented by the general nonluminescent character), suggests that in this study, the nature of single crystals is not helpful in our interpretation.

SUMMARY AND CONCLUSIONS

(1) Textures, fabrics, and structures in the upper interval of mixed carbonates suggest an association of shallow subtidal and tidal flat environments. Comparison of the much less variable, lower, completely dolomitized sequence with dolomites and their relation to limestones in the upper interval suggests the possibility that the lower part of the section was more intensely influenced by a hypersaline tidal flat environment.

(2) The dolomite is replacive. Evidence in the dolomite,

TABLE 2.—EVIDENCE AND REASONING FOR EARLY AND LATE DIAGENETIC (BURIAL) REPLACEMENT

Early Diagenetic Replacement	Late Diagenetic (Burial) Replacement
1. Features, such as intraclasts, fenestrae, algal laminae(?), and minor evaporites, characteristic of supratidal zone.	1. Apparent gradation, in upper interval, from minor dolomite truncating discrete stylolite seams or disseminated in zones related to them to thicker seams and zones (Fig. 6) to thicker dolomite sequences and possibly whole lower interval; thin, irregular, wispy seams of finer dolomite (Fig. 6F) might represent relict stylolites.
2. Absence of fossils in dolomitized zones in upper sequence and entire lower interval, suggestive of restricted, hypersaline depositional conditions that would have promoted dolomitization and inhibited development of diverse benthonic fauna.	2. Common presence of saddle dolomite (Radke and Mathis, 1980).
3. Selective dolomitization of fine mud matrix and peloids vs. skeletal fragments (Fig. 11A) more likely in early diagenesis than in tighter burial environment.	3. Coarseness of dolomite crystals; much of their volume may represent later diagenetic replacive overgrowths.
4. Association of limestone and dolomite intraclasts in some breccias (Fig. 3F) suggests dolomitization prior to "rip up" and redeposition; preservation of limestone clasts difficult to explain with late diagenetic dolomitization.	4. Dolomite transecting calcite-filled fractures (Figs. 6D, 12) and presence of healed fractures in dolomite (Fig. 5E).
5. Presence in upper interval of considerable undolomitized or slightly dolomitized limestone; late diagenetic dolomitization should be more pervasive.	5. Obliterative dolomitization; disseminated crystals transecting grains contain no relict structure; suggests lack of environmental control on unfossiliferous dolomite.
6. Numerous stylolites in upper interval have no associated dolomite; post-dolomitization stylolites (Fig. 5F) common in both upper and lower intervals.	6. Most evidence for early evaporites (e.g., anhydrite laths, calcite pseudomorphs) in *limestone*; anhydrite in dolomite mainly as later stage void filling (Fig. 10).
7. Ready supply of Mg^{2+} ions in environment bathed in essentially seawater; provision of Mg^{2+} in burial environment may be a limiting factor (Morrow, 1982; Land, 1985).	7. Fenestrae more common in limestone; laminated, peloidal, fenestrate limestone, a likely supratidal lithotype, contains relatively little dolomite.
	8. Generalizations that association of dolomite and limestone intraclasts and dolomitization controlled by primary textures and structures indicate early dolomitization not without exception; later dolomitization could be selective.
	9. Possibility of significant fluid flow in burial environment (Zenger, 1983; Land, 1985; Gregg, 1985), such as along stylolites, which may serve as efficient permeability conduits (Koepnick, 1985; Dawson, 1986; von Bergen and Carozzi, 1986); possible provision of Mg^{2+} ions by post-dolomitization stylolitization acting in nearby subsurface.

such as relatively coarse crystallinity, xenotopic texture, depleted oxygen isotopes, high temperatures of homogenization, trace-element content, and near-stoichiometry, collectively attest to the effect of deep burial diagenesis; however, such a conclusion allows for two possible modes of dolomitization: (1) very early diagenetic replacement effected by hypersaline brines related to a tidal flat environment followed by neomorphism, during which the dolomite assumed many characteristics, particularly geochemical, of the mesogenetic environment, and (2) late diagenetic, deep burial replacement during which the dolomite developed its mesogenetic characteristics as it formed.

(3) Supratidal textures and structures, absence of fossils, selective dolomitization of primary features, association of limestone and dolomite clasts, nonpervasive dolomitization, post-dolomite stylolites, and the problem of magnesium supply in a burial environment are arguments in support of the first model, whereas collective evidence and reasoning in favor of burial replacement include: a suggested evolution from dolomite localized along distinct stylolites to thicker but geometrically similar seams to possibly thick intervals; saddle dolomite; truncation of calcite-filled fractures by dolomite; obliterative dolomitization; minor evaporites in calcite and late-stage anhydrite in pores in dolomite; fenestrae most commonly in limestones; little dolomite in some supratidal lithotypes; underestimated possibility of selective late diagenetic replacement of primary features; and likelihood of availability of Mg^{2+} ions from post-dolomitization stylolitization. We suggest that sea water provided Mg^{2+} ions for any early dolomitization; post-dolomite stylolitization in the Tobosa Basin may have provided Mg^{2+} ions for burial dolomitization.

FIG. 12.—Photomicrograph of disseminated dolomite crystals (lighter, mostly rhomb-shaped) replacing peloidal groundmass and truncating calcite-filled fracture (arrow); plane polarized light; scale bar 500 μm.

(4) Finally, in this study (and very likely in others), it is difficult to distinguish conclusively between these two alternatives or some combination thereof. Our geochemical data are not conclusive in such a distinction. Disconcerting as such an indefinite conclusion may be, we are heartened by the recent admonitions of Hardie (1987, p. 180), who argued that one should not be hesitant to conclude ". . . that no conclusion is possible with the present state of knowledge" regarding dolomitization. A particularly fruitful area for future investigation would be the study of such petrologic and geochemical criteria that might be diagnostic in distinguishing between early, but later neomorphosed, dolomite and that formed during late diagenesis in the deep subsurface.

ACKNOWLEDGMENTS

We are grateful to Unocal for their support and permission to publish this paper. We thank Jean MacKay for typing the manuscript and Gerhard Ott for help with some of the photography. We are also grateful to Ann Zenger for her help with figure layout. John Cooper and David Lumsden read early drafts of the manuscript and we appreciate their thoughtful comments, as well as the valuable suggestions of reviewers Gerald Friedman, Vijai Shukla, and Maurice Tucker.

REFERENCES

BAUM, G. R., HARRIS, W. B., AND DREZ, P. E., 1985, Origin of dolomite in the Castle Hayne Limestone, North Carolina: Journal of Sedimentary Petrology, v. 55, p. 506–517.

BODNAR, R. J., AND BETHKE, P. M., 1984, Systematics of stretching of fluid inclusions 1: Fluorite and sphalerite at 1 atmosphere confining pressure: Economic Geology, v. 79, p. 141–161.

BOTZ, R. W., AND VON DER BORCH, C. C., 1984, Stable isotope study of carbonate sediments from the Coorong area, South Australia: Sedimentology, v. 31, p. 837–849.

BUDAI, J. M., LOHMANN, K. C., AND OWEN, R. M., 1984, Burial dedolomite in the Mississippian Madison Limestone: Journal of Sedimentary Petrology, v. 54, p. 276–288.

BULLEN, S. B., AND SIBLEY, D. F., 1984, Dolomite selectivity and mimic replacement: Geology, v. 12, p. 655–658.

CHOQUETTE, P. W., AND PRAY, L. C., 1970, Geologic nomenclature and classification of porosity in sedimentary carbonates: American Association of Petroleum Geologists Bulletin, v. 54, p. 207–250.

DAWSON, W. C., 1986, Stylolite-associated porosity: Society of Economic Paleontologists and Mineralogists, Third Annual Meeting, Raleigh, North Carolina, Abstracts, p. 27.

DUNHAM, R. J., 1962, Classification of carbonate rocks according to depositional texture, *in* Ham, W. E., ed., Classification of Carbonate Rocks: American Association of Petroleum Geologists Memoir 1, p. 108–121.

EMBRY, A. F., AND KLOVAN, J. E., 1971, A late Devonian reef tract on northeastern Banks Island, Northwest Territories: Canadian Bulletin of Petroleum Geology, v. 19, p. 730–781.

FRITZ, PETER, 1971, Geochemical characteristics of dolomites and the $\delta^{18}O$ content of Middle Devonian oceans: Earth and Planetary Science Letters, v. 11, p. 277–282.

———, AND KATZ, AMITAI, 1972, The sodium distribution of dolomite crystals: Chemical Geology, v. 10, p. 237–244.

GOLDSMITH, J. R., AND GRAF, D. L., 1958, Relation between lattice constants and composition of the Ca-Mg carbonates: American Mineralogist, v. 43, p. 84–101.

GOLDSTEIN, R. H., 1986 Reequilibration of fluid inclusions in low-temperature cement: Geology, v. 14, p. 792–795.

GREGG, J. M., 1985, Regional epigenetic dolomitization in the Bonneterre Dolomite (Cambrian), southeastern Missouri: Geology, v. 13, p. 503–506.

———, AND SIBLEY, D. F., 1984, Epigenetic dolomitization and the origin of xenotopic dolomite texture: Journal of Sedimentary Petrology, v. 54, p. 907–931.

HANOR, J. S., 1980, Dissolved methane in sedimentary brines: Potential effect on the PVT properties of fluid inclusions: Economic Geology, v. 75, p. 603–609.

HARDIE, L. A., 1987, Dolomitization: A critical view of some current views: Journal of Sedimentary Petrology, v. 57, p. 166–183.

KOEPNICK, R. B., 1985, Impact of stylolites on carbonate reservoir continuity: Example from Middle East: American Association of Petroleum Geologists Bulletin, v. 69, p. 274.

LAND, L. S., 1980, The isotopic and trace element geochemistry of dolomite: The state of the art, *in* Zenger, D. H., Dunham, J. B., and Ethington, R. L., eds., Concepts and Models of Dolomitization: Society of Economic Paleontologists and Mineralogists Special Publication 28, p. 87–110.

———, 1985, The origin of massive dolomite: Journal of Geological Education, v. 33, p. 112–125.

———, AND HOOPS, G. K., 1973, Sodium in carbonate sediments and rocks: A possible index to the salinity of diagenetic solutions: Journal of Sedimentary Petrology, v. 43, p. 614–617.

LOGAN, B. W., AND SEMENIUK, VIC, 1976, Dynamic metamorphism: Processes and products in Devonian carbonate rocks, Canning Basin, western Australia: Geological Society of Australia Special Publication 6, 138 p.

LUMSDEN, D. N., AND CHIMAHUSKY, J. S., 1980, Relationship between dolomite nonstoichiometry and carbonate facies parameters, *in* Zenger, D. H., Dunham, J. B., and Ethington, R. L., eds., Concepts and Models of Dolomitization: Society of Economic Paleontologists and Mineralogists Special Publication 28, p. 123–137.

MACHEL, H.-G., AND MOUNTJOY, E. W., 1986, Chemistry and environments of dolomitization–A reappraisal: Earth Science Reviews, v. 23, p. 175–222.

MATTES, B. W., AND MOUNTJOY, E. W., 1980, Burial dolomitization of the Upper Devonian Miette buildup, Jasper National Park, Alberta, *in* Zenger, D. H., Dunham, J. B., and Ethington, R. L., eds., Concepts and Models of Dolomitizaton: Society of Economic Paleontologists and Mineralogists Special Publication 28, p. 259–297.

MCCREA, J. M., 1950, On the isotopic chemistry of carbonates and a paleotemperature curve: The Journal of Chemical Physics, v. 18, p. 849–857.

MCGLASSON, E. H., 1967, The Siluro-Devonian of west Texas and southeast New Mexico: Tulsa Geological Society Digest, v. 35, p. 148–164.

MORROW, D. W., 1982, Diagenesis 2. Dolomite, part 2: Dolomitization models and ancient dolostones: Geoscience Canada, v. 9, 95–107.

———, CUMMING, G. L., AND KOEPNICK, R. B., 1986, Manetoe facies–A gas-bearing, megacrystalline, Devonian dolomite, Yukon and Northwest Territories, Canada: American Association of Petroleum Geologists Bulletin, v. 70, p.702–720.

PIERRE, CATHERINE, ORTLIEB, LUC, AND PERSON, ALAIN, 1984, Supratidal evaporitic dolomite at Ojo de Liebre Lagoon: Mineralogy and isotopic arguments for primary crystallization: Journal of Sedimentary Petrology, v. 54, p. 1049–1061.

PIERSON, B. J., 1981, The control of cathodoluminescence in dolomite by iron and manganese: Sedimentology, v. 28, p. 601–610.

POPP, B. N., ANDERSON, T. F., AND SANDBERG, P. A., 1986, Textural, elemental, and isotopic variations among constituents in Middle Devonian limestones, North America: Journal of Sedimentary Petrology, v. 56, p. 715–727.

PREZBINDOWSKI, D. R., AND LARESE, R. E., 1987, Experimental stretching of fluid inclusions in calcite–Implications for diagenetic studies: Geology, v. 15, p. 333–336.

RADKE, B. M., AND MATHIS, R. L., 1980. On the formation and occurrence of saddle dolomite: Journal of Sedimentary Petrology, v. 50, p. 1149–1168.

SHUKLA, VIJAI, 1986, Epigenetic dolomitization and the origin of xenotopic dolomite texture–Discussion: Journal of Sedimentary Petrology, v. 56, p. 733–734.

VEIZER, JÁN, 1983, Chemical diagenesis of carbonates: Theory and application of trace element technique, *in* Arthur, M. A., ed., Stable Isotopes in Sedimentary Geology: Society of Economic Paleontologists and Mineralogists Short Course No. 10, Tulsa, Oklahoma, p. 1–100.

———, FRITZ, PETER, AND JONES, BRIAN, 1986, Geochemistry of brachiopods: Oxygen and carbon isotopic records of Paleozoic oceans: Geochimica et Cosmochimica Acta, v. 50, p. 1679–1696.

———, AND HOEFS, JOCHEN, 1976, The nature of O^{18}/O^{16} and C^{13}/C^{12} secular trends in sedimentary carbonate rocks: Geochimica et Cosmochimica Acta, v. 40, p. 1387–1395.

———, LEMIEUX, JEAN, JONES, BRIAN, GIBLING, M. R., AND SAVELLE, JIM, 1977, Sodium: Paleosalinity indicator in ancient carbonate rocks: Geology, v. 5, p. 177–179.

VON BERGEN, DONALD, AND CAROZZI, A. V., 1986, Stylolitic porosity as a critical factor for deep gas production, Atokan limestone, northern Delaware Basin, Reeves County, Texas, Society of Economic Paleontologists and Mineralogists Third Annual Meeting, Raleigh, North Carolina, Abstracts, p. 113.

WANLESS, H. R., 1979, Limestone responses to stress: Pressure solution and dolomitization: Journal of Sedimentary Petrology, v. 49, p. 437–462.

ZENGER, D. H., 1972, Significance of supratidal dolomitization in the geologic record: Geological Society of America Bulletin, v. 83, p. 1–11.

———, 1983, Burial dolomitization in the Lost Burro Formation (Devonian) east-central California and the significance of late diagenetic dolomitization: Geology, v. 11, p. 519–522.

UPPER JURASSIC SMACKOVER PLATFORM DOLOMITIZATION, NORTHWESTERN GULF OF MEXICO: A TALE OF TWO WATERS

CLYDE H. MOORE AND AHAD CHOWDHURY

Applied Carbonate Research Program, Basin Research Institute, Louisiana State University, Baton Rouge, Louisiana 70803

AND

LUI CHAN

Department of Geology, Louisiana State University, Baton Rouge, Louisiana 70803

ABSTRACT: The Upper Jurassic Smackover Formation of eastern Texas and western Arkansas exhibits a classic example of platform dolomitization. The close stratigraphic proximity of dolomitized Smackover strata, immediately below thick Buckner evaporites, combined with common vertical dolomitization gradients increasing toward the formational contact, strongly suggests an evaporative-reflux origin for the dolomite. The geologic setting of the Buckner, interpreted as a long-lived evaporative lagoon, supports this model. Chemical composition of the dolomite, however, such as depleted ^{18}O (average − 2.7‰), low Sr and Na content (average 40 and 157 ppm, respectively) suggest meteoric-water influences. Elevated ^{87}Sr, and high Fe and Mn compositions of some platform interior dolomites indicate that extremely large volumes of meteoric water, passing first through intercalated siliciclastics, were involved.

Two models generally satisfy the necessity for involvement of both refluxed evaporative waters and meteoric waters: (1) an evaporative reflux-meteoric water-mixing model, where refluxing brines from the overlying Buckner lagoon mix with an active meteoric-water system in the upper Smackover, providing the Mg for dolomitization; (2) a recrystallization model, where the upper Smackover was initially dolomitized by reflux of evaporative brines from the overlying Buckner lagoon and subsequently recrystallized in a major meteoric-water system. The authors do not unanimously support either model. Chowdhury and Chan prefer the recrystallization model, because trace-element and Sr isotope data seem to preclude involvement of marine waters during final dolomitization. Moore favors the evaporative reflux-meteoric mixing model, because it is the simplest model that satisfies most constraints, the Sr isotope data supporting recrystallization is limited, and because there is no clear petrographic evidence for recrystallization.

INTRODUCTION

Upper Jurassic Smackover ooid-dominated platform sequences rim the northern Gulf of Mexico in the subsurface from Texas to Florida. In the central area of the trend (Arkansas, Louisiana, and Mississippi), the ooid grainstones are dominantly limestone (Moore, 1984; Fig. 1). On opposing ends of this arc (east Texas and Alabama-Florida; Fig. 1), upper smackover ooid grainstones exhibit pervasive dolomitization. In Alabama and Florida, dolomitized ooid grainstones are localized above isolated salt-related anticlines, surrounded by mud-dominated limestones. These dolomites have been interpreted as having been formed by evaporative reflux from the overlying Buckner evaporites (Koepnick and others, 1985). Bradford (1984) called upon Buckner-related fluids at elevated temperatures. Alternatively, mixed meteoric-marine dolomites formed as a result of local shoal exposure (Vinet, 1984). In contrast, the east Texas-west Arkansas dolomite province is a classic example of regional platform dolomitization. In this area, the upper Smackover is a blanketlike ooid grainstone covering a 100-km-wide platform. The upper 30 m of Smackover grainstone is pervasively dolomitized across an area totaling over 3,000 km^2 (Fig. 1B) and forms one of the major reservoir facies of the Smackover along the western extent of its subcrop.

This paper deals solely with the northwestern part of the trend, encompassing the east Texas and west Arkansas platform dolomite province. Pervasive platform dolomitization along this trend has long been thought to be the result of early brine reflux from extensive Buckner evaporites immediately above the upper Smackover (Dickinson, 1969; Moore, 1984). We will, for the first time, present a review of the geologic, petrographic, and geochemical framework for dolomitization of the upper Smackover in this part of the trend. We will then present and discuss new geochemical and geologic data, which suggest that, whereas these regional platform dolomites undoubtly involved refluxing evaporative brines, meteoric waters played a major role in their formation,

METHODS

Data from 20 cores were utilized in the present study Twelve cores (wells 1–12, Table 1, Fig. 1A) sampled most areas where the upper Smackover is pervasively dolomitized across east Texas. Seven additional cores from Texas and Arkansas (wells 13–20, Table 1, Fig. 1A) represent the geochemical and petrographic data base from an earlier study of upper Smackover dolomitization (Stamatedes, 1982). New geochemical data generated by the present study (Table 3) are clearly differentiated from Stamatedes' earlier data (Table 2).

Several hundred new thin sections were prepared from cores 1–12 for comparison with Stamatedes' original petrographic observations concerning the timing of dolomitizing events in western Arkansas. Samples were carefully chosen for geochemical analysis using light microscopy combined with X-ray diffraction to ensure that samples consisted of a single generation of dolomite and that other phases did not affect elemental compositions. Elemental (Ca, Mg, Sr, Na, Fe, and Mn) analyses were performed by inductively coupled plasma (ICP) atomic emission spectrometry. Precision and accuracy of the ICP analyses were 1.5%. Oxygen and carbon stable isotope analyses were performed at Coastal Science Laboratories in Austin, Texas, using standard techniques. Precision is 0.2‰. All values are reported relative to PDB.

Strontium isotopes were determined at Louisiana State University on a 12-in. Nuclide mass spectrometer with a single Faraday cup collector. Carbonate samples were processed by standard techniques with Sr separation by ion exchange chromatography using AG 50 × 8 cation exchange resin (Biorad). Reproducibility determined by rep-

Sedimentology and Geochemistry of Dolostones, SEPM Special Publication No. 43

TABLE 1.—WELL CORES USED IN STUDY

Well # Operator	Well Name	County	Core Footage
1 Schneider, et al.	#1 Hargraves	Texas, Hopkins	9924-9461
2 Humble	#1 Beltex	Texas, Bowie	9905-10805
3 Humble	#2 Moody	Texas, Kaufman	9830-10190
4 Humble	#3C Calfe	Texas, Titus	8895-955; 9514-10998
5 Hamon	#1 Chitsey	Texas, Franklin	13061-260
6 Getty	#7-1-D Elledge	Texas, Franklin	11968-12406
7 Sun	#1 Travis	Texas, Van Zandt	13010-531
8 Lone Star	#B-1 Allyn	Texas Henderson	13584-996
9 Pan Am	#1 Brown	Texas, Wood	12870-985
10 Texaco	#1 WV Moore	Texas, Rains	13127-177
11 Texaco	#1 Lynch	Texas, Rains	13079-135
12 Getty	#6-2-D Evers and Rhodes	Texas, Franklin	11809-12302
13 Murphy	#1 Giffco	Texas, Bowie	7734-783
14 Guardian	#1 Cox Meek	Ark., Miller	7785-808
15 Pennzoil	#1 Allen	Ark., Lafayette	10809-867
16 Penzoil	#1 Taylor	Ark., Lafayette	10790-865
17 Pennzoil	#1 Alford	Ark., Lafayette	?
18 Feazel	#1 Fouke	Ark., Miller	9628-680
19 Tlapek & Cardell	#2 Charles Nix	Ark., Columbia	7965-8005
20 Lear	#1 Harold Clements	Ark., Miller	10728-810

Notes.—Wells 1 through 12 were used by authors to develop east Texas data base (Table 3). Wells 13–20 were used by Stamatedes (1982; Table 2). Thin sections from all wells were used in the present study. Well locations, by number, are shown in Figure 1.

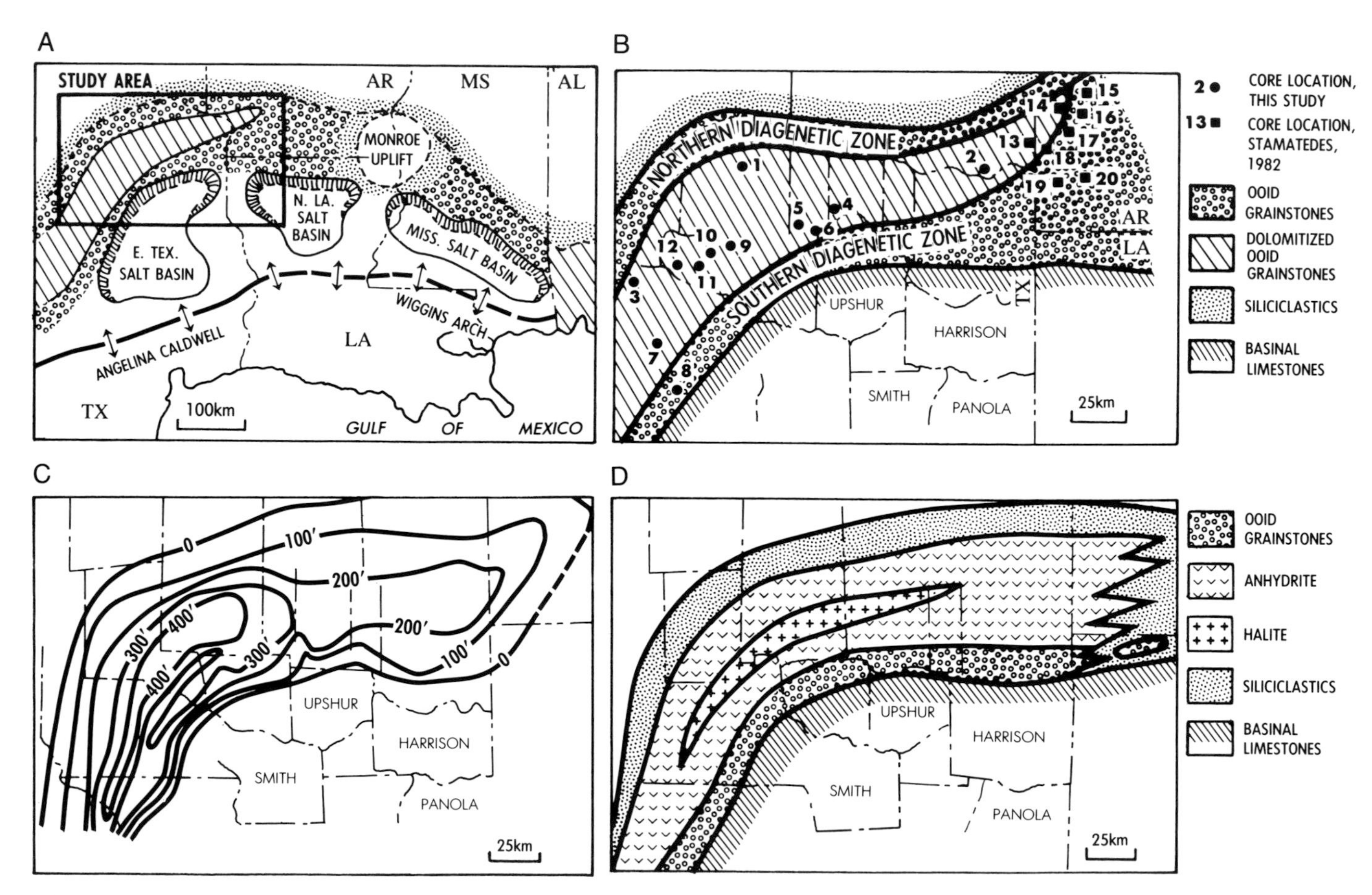

FIG. 1.—Geologic setting, Upper Jurassic Smackover Formation and Buckner Member, Haynesville Formation, northern Gulf of Mexico. (A) Regional setting, showing structural features affecting Upper Jurassic sedimentation: upper Smackover high-energy platforms (ooid pattern) and area of upper Smackover regional dolomitization (cross-hatched pattern). Study area is outlined in black. (B) Study area showing well locations, limits of upper Smackover lime grainstones, and distribution of pervasive dolomitization in the upper Smackover. Diagenetic zones of Moore and Druckman (1981) are extended through east Texas by McGillis (1984), Stewart (1984), Wilkinson (1984). Well numbers are keyed to Table 1. (C) Isopach map of the Buckner evaporite member of the Haynesville Formation in the study area based on 308 control points. Data from Dickinson (1969), McGillis (1984), Stewart (1984), and Wilkinson (1984). (D) General lithofacies distribution of the Buckner Evaporite across the study area. Compiled from Dickinson (1969), Moore and Druckman (1981), Stamatedes (1982), McGillis (1984), Stewart (1984), and Wilkinson (1984).

TABLE 2.—GEOCHEMICAL DATA FOR UPPER SMACKOVER AND BUCKNER DOLOMITES, WESTERN ARKANSAS

Well	County	Sample	Dolomite type	Sr(ppm)	Na(ppm)	$\delta^{18}O$ PDB	$\delta^{13}C$ PDB	Mole % excess Ca
13	Bowie	7770	Early, oomoldic	50	201	−1.45	5.77	2.2
		7779	Early, oomoldic	52	155	−2.7	6.2	1.4
		7788	Early, oomoldic	55	140	−2.18	5.65	2.52
		7793	Early, oomoldic			−2.3	5.86	
		7842	Early, oomoldic	59	340	−2.62	5.1	3.5
		7851	Early, oomoldic	56	134	−2.5	5.2	5.7
14	Miller	7734	Early, oomoldic	1124*	142	−2.29	5.45	1
		7737	Early, oomoldic	1007*	195	−3.3	5.6	1
		7744	Early, oomoldic	127*	170	−2.97	5.61	1.4
		7752	Early, oomoldic	69	177	−2.53	5.45	0.8
		7755	Early, oomoldic	72	263	−1.71	5.52	0.2
		7764	Early, oomoldic	371*	138	−2.6	5.4	1.9
		7770	Early, oomoldic	168*	169	−2.6	5.6	1.7
18	Miller	9628	Early, oomoldic			−3.3	5.3	
		9631	Early, oomoldic	61	163	−3.4	5.3	2.33
		9661	Post-compaction			−4.6	4.6	
		9676	Post-compaction	59	135	−4	5	1.48
15	Lafayette	10850	Post-compaction	61	160	−4.3	5.3	0
		10856	Post-compaction			−3.9	5.3	
20	Miller	10747	Post-compaction	58	121	−4.9	5	0.02
		10751	Post-compaction	60	161	−4.2	4.5	0
16	Lafayette	10794	Post-compaction	66	163	−3.4	3.4	1.47
20	Miller	10732	Smackover, saddle			−4.88	1.56	
		10748				−6.6	3.1	
16	Lafayette	10792	Smackover, saddle	169	1267	1.4	2.9	5
17	Lafayette	8388		489	512	2.4	4.8	4.4
19	Columbia	7969	Buckner dolomite	262	512	1.3	4.6	4.8
		7979	Buckner dolomite			1.2	4.3	
			Buckner dolomite					
			Buckner dolomite					

Data from Stamatedes, 1982

*Contaminated by Celestite

Notes: Well locations are shown in Figure 1, well names in Table 1. Sample relates to core footage from which sample was taken. Excess Ca determined by X-ray diffraction.

licate analyses was within 0.00010‰. The $^{87}Sr/^{86}Sr$ ratio of the Eimer and Amend Sr CO_3 standard was determined to be 0.70802 ± 0.00005.

GEOLOGIC SETTING

The Upper Jurassic Smackover Formation was deposited on the flanks of a series of interior salt basins rimming the present northern Gulf of Mexico (Fig. 1A). These salt basins were partially isolated from the main Gulf basin by marginal highs such as the Wiggins Arch (Moore, 1984; Fig. 1A). The area of interest for the present paper includes east Texas, southwestern Arkansas, and northwestern Louisiana (Fig. 1B). In this area, the upper Smackover, deposited during a late Oxfordian sea-level highstand (Moore, 1984; Fig. 2) consists of a wedge of grainstones ranging in thickness from 50 m updip to over 300 m downdip. Lithofacies within this wedge grade from ooid and pellet grainstones in the platform interior (along the updip limits of the Smackover, Fig. 1B) to ooid grainstones containing algae-coated grains and bioclasts along the platform margin (just shoreward of the basinal limestones indicated on Fig. 1B; McGillis, 1984; Stewart, 1984; Moore, 1984). The dominance of ooids to the exclusion of all biological components except crustacean pellets in the platform interior, combined with the occurrence of normal marine biota (coral, algae, echinoids, and so on) toward the platform margin, suggests a platformwide salinity gradient in the upper Smackover, with increasing salinity toward the platform interior. Presumed high salinities in the study area were no doubt related to the partial isolation of the east Texas salt basin from the Gulf of Mexico basin by a marginal high (Angelina-Caldwell Flexure) to the south (Moore, 1984; Fig. 1A). This presumed salinity gradient may have been a causative factor in the apparent change in original ooid mineralogy across the platform in the study area from aragonite in the interior to magnesian calcite at the platform margin (Moore and others, 1986).

A rimmed Haynesville shelf developed in the study area during the subsequent Kimmeridgian sea-level rise (Figs. 2, 3; Moore, 1984). This shelf was marked by a high-energy carbonate-shelf margin that acted as a partial barrier to marine circulation across the platform (Dickinson, 1969; Harwood and McGillis, 1984). Consequently, an extensive, shallow, evaporative lagoon, represented by the Buckner Member of the Haynesville Formation, was developed immediately above the porous upper Smackover ooid grainstones in east Texas and southwest Arkansas (Dickinson, 1969; Moore, 1984). The depositional setting for the Buckner lagoon is directly analogous to that of the Ferry Lake Anhydrite, described by Loucks and Longman (1982) as a widespread, shallow, subtidal lagoon developed shoreward of middle Glen Rose shelf margin reefs. Buckner evaporites attain thicknesses of over 140 m, are partly of subsqueous origin, and contain significant halite (over

TABLE 3.—GEOCHEMICAL DATA FOR UPPER SMACKOVER AND BUCKNER DOLOMITES (PRESENT STUDY)

Well	Country	Sample #	Dolomite type	Mean XTL size (mm)	Sr (ppm)	Na (ppm)	Fe (ppm)	Mn (ppm)	$\delta^{18}O$ PDB	$\delta^{13}C$ PDB	$^{87}Sr/^{86}Sr$
3	Kaufman	9892	Buckner, evap. dolomite	—	261	—	—	—	0.3	3.8	
3	Kaufman	9901	Buckner, evap. dolomite	—	333	512	416	90	−0.4	3.6	—
2	Bowie	10230	Buckner, evap. dolomite	—	490	517	517	92	−2.2	4.6	—
7	Van Zandt	13109	Buckner, evap. dolomite	—	340	—	—	—	0.5	4.6	—
7	Van Zandt	13112	Buckner, evap. dolomite	—	486	990	889	89	1	4.8	—
7	Van Zandt	13134	Buckner, evap. dolomite	—	440	989	576	92	1.2	4.6	0.70726
7	Van Zandt	13154	Buckner, evap. dolomite	—	273	621	1200	172	−0.3	4.5	—
7	Van Zandt	13186	Buckner, evap. dolomite	—	383	—	—	—	0.3	4.4	—
12	Franklin	11819	Early, oom., coarse xtl., S.	0.14	66	169	96	42	−2.58	5.3	—
12	Franklin	11826	Early, oom., coarse xtl., S.	0.14	49	143	81	72	−2.89	5.1	—
12	Franklin	11844	Early, oom., coarse xtl., S.	—	43	133	106	110	−3.2	4.9	0.70744
12	Franklin	11854	Early, oom., coarse xtl., S.	—	—	—	—	—	−3.32	5	—
12	Franklin	11863	Early, oom., coarse xtl., S.	—	50	129	57	69	−3.34	5.2	—
12	Franklin	11882	Early, oom., coarse xtl., S.	—	—	—	—	—	−4.01	4.7	—
12	Franklin	11891	Early, oom., coarse xtl., S.	—	—	—	—	—	−3.08	4.9	—
12	Franklin	11900	Early, oom., coarse xtl., S	0.14	62	148	113	81	−3.15	5.1	—
12	Franklin	11903	Early, oom., coarse xtl., S.	0.14	—	—	—	—	−2.9	4.9	—
12	Franklin	11991	Early, oom., coarse xtl., S.	—	42	121	113	126	−3	5.3	0.70734
6	Franklin	12018	Early, oom., coarse xtl., S.	0.09	47	117	53	63	−3.09	5.1	—
6	Franklin	12024	Early, oom., coarse xtl., S.	—	52	137	71	80	−3.71	5.31	—
6	Franklin	12030	Early, oom., coarse xtl., S.	—	—	—	—	—	−2.89	5.41	—
6	Franklin	12054	Early, oom., coarse xtl., S.	—	42	122	51	72	−3.25	5.08	—
6	Franklin	12070	Early, oom., coarse xtl., S.	—	48	185	67	92	−3.43	4.79	—
9	Wood	12073	Early, oom., coarse xtl., S.	0.09	57	184	73	96	−2.5	4.9	0.70714
9	Wood	12886	Early, oom., coarse xtl., S.	0.11	63	178	83	125	−2.33	4.8	—
9	Wood	12898	Early, oom., coarse xtl., S.	—	52	165	90	82	−2.5	4.8	—
9	Wood	13040	Early, oom., coarse xtl., S.	—	—	—	—	—	−2.99	4.69	—
10	Rains	13075	Early, oom., coarse xtl., S	—	48	169	114	137	−2.4	5.3	—
10	Rains	13134	Early, oom., coarse xtl., S	—	69	182	82	86	−3	4.9	—
11	Rains	13120	Early, oom., coarse xtl., S	—	67	158	90	92	−2.18	5.3	—
11	Rains	13130	Early, oom., coarse xtl., S	—	56	143	86	111	−2.38	4.6	—
7	Van Zandt	13288	Early, oom., coarse xtl., S	—	—	—	—	—	−2.99	5.09	—
7	Van Zandt	13296	Early, oom., coarse xtl., S	—	52	143	79	93	−2.7	5.31	—
7	Van Zandt	13306	Early, oom., coarse xtl., S	0.11	—	—	—	—	−2.57	5.46	—
7	Van Zandt	13420	Early, oom., coarse xtl., S	0.11	55	154	85	84	−3.2	5.1	0.70712
4	Titus	8896	Early, oom., fine xtl., N.	0.04	48	187	1675	610	−2.2	5.9	—
1	Hopkins	9411	Early, oom., fine xtl., N.	0.03	54	166	2184	560	−2.17	5.54	0.70794
1	Hopkins	9427	Early, oom., fine xtl., N.	0.03	77	180	2663	421	−2.48	5.38	0.70812
1	Hopkins	9434	Early, oom., fine xtl., N.	0.03	—	—	—	—	−1.96	5.65	—
1	Hopkins	9440	Early, oom., fine xtl., N.	—	65	197	1984	484	−2.02	5.54	—
3	Kaufman	9914	Early, oom., fine xtl., N.	0.05	56	143	1362	221	−2.1	5.3	0.70716
3	Kaufman	9925	Early, oom., fine xtl., N.	—	89	212	1280	166	−2.3	5.8	—
2	Bowie	10246	Early, oom., fine xtl., N.	0.03	40	—	2460	640	−2.27	5.3	0.70837
2	Bowie	10251	Early, oom., fine xtl., N.	—	47	—	2050	710	−2.43	5.4	—
5	Franklin	13062	Early, oom., fine xtl., N.	—	54	173	58	59	−1.8	4.9	—
5	Franklin	13066	Early, oom., fine xtl., N.	—	50	129	67	74	−1.9	5.3	0.70727
5	Franklin	13068	Early, oom., fine xtl., N.	—	—	—	—	—	−2.1	4.8	—
5	Franklin	13075	Early, oom., fine xtl., N.	—	58	154	96	65	−2.2	4.9	0.7073
5	Franklin	13090	Early, oom., fine xtl., N.	—	—	—	—	—	−2.03	5.2	—
5	Franklin	13098	Early, oom., fine xtl., N.	—	44	143	124	72	−2.3	4.9	—
6	Franklin	12088	Post compact., zoned repl.	—	56	112	156	131	−4.46	4.3	0.7074
6	Franklin	12090	Post compact., zoned repl.	—	—	—	—	—	−4.9	5.1	—
6	Franklin	13034	Post compact., zoned repl.	—	83	154	216	137	−5	4.4	—
7	Van Zandt	13321	Post compact., zoned repl.	—	—	—	—	—	−4.2	4.9	—
7	Van Zandt	13459	Post compact., zoned repl.	—	—	—	—	—	−4.91	3.79	—
7	Van Zandt	13524	Post compact., zoned repl.	—	65	183	173	110	−5.02	3.47	—
8	Henderson	13903	Post compact., zoned repl.	—	53	192	4120	689	−4.7	4.2	0.7076
12	Franklin	11884	Post compact., saddle dolo	—	132	399	36294	1571	−7.8	2.6	0.71022
12	Franklin	11889	Post compact., saddle dolo	—	139	256	3631	614	−6.8	3.1	0.70946
8	Henderson	13809	Post compact., saddle dolo	—	93	256	3683	418	−4.7	4.5	—

Notes—Well locations are shown in Figure 1, well names in Table 1. Sample relates to core footage from which sample was taken. Dolomite crystal size is mean value of 40 measurements. Abbreviations for dolomite types as follows; oom. (oomoldic); xtyl. (crystalline); N. (north, platform interior); S (south, platform margin); evap. (evaporite-related); compact. (compaction); repl. (replacement).

30 m) associated with a distinct evaporite depocenter located in east Texas (Dickinson, 1969; Moore, 1984; Stewart, 1984; Wilkinson, 1984; McGillis, 1984; Fig. 1C, D).

Figure 3 illustrates the basic sedimentologic setting of the lagoon and postulated paths taken by both marine and evaporated marine waters moving through the lagoon and interacting with the porous Smackover Formation below. Of critical importance is the implication that the locus of gypsum precipitation and gypsum sedimentation was located initially along the interior (northern) margin of the lagoon as marine surface waters moving across the lagoon finally reached gypsum saturation in response to evaporation (Fig. 3). A return of concentrated brines, moving south toward the shelf margin under less dense marine surface waters, would result in a density stratification of the lagoon, effectively precluding further evaporation of the dense bot-

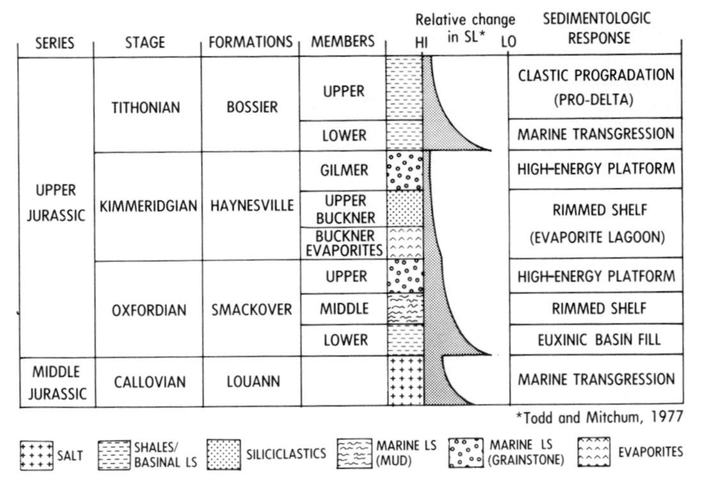

FIG. 2.—Stratigraphic setting, Middle Upper Jurassic, Gulf of Mexico. Modified from Moore (1984).

tom brines and further precipitation of evaporites such as gypsum or halite (Sloss, 1969). Thus, the porous limestones of the Smackover below would be exposed to dense evaporative brines of elevated Mg/Ca ratios across a wide (40–60 km) swath of the shelf (Fig. 3). Under the conditions of rising sea level, as found in the Haynesville, this geometry could have been maintained for tens of thousands of years, while enormous volumes of brine, driven by density differences, could have seeped through and dolomitized the upper Smackover. The depth of penetration of these brines can be estimated from the distribution of dolomite within the upper Smackover Formation. Figure 4 summarizes data from a core through an 18-m sequence of upper Smackover ooid grainstones beneath the Buckner evaporite that exhibits a distinct dolomitization gradient, from 100% dolomite at the contact with the Buckner, to less than 10% dolomite 15 m downcore. A depth of 15 m, therefore, is probably close to the penetration depth of the refluxing brines. All cores studied containing oomolodic-related early dolomite exhibited similiar dolomite distribution patterns.

This refluxing was probably shut down by evaporative drawdown triggered by the estabishment of a more effective shelf edge barrier during the slowing of sea-level rise in middle and upper Haynesville time (Fig. 2), leading to the precipitation of halite in the lagoon center and the final sealing of the upper Smackover from the evaporative fluids above. Evaporites in the Haynesville lagoon thin and disappear to the east in central Arkansas, giving way to siliciclastics (Fig. 1D). These shales, silts, and sands were shed from the Monroe uplift (Fig. 1A; Moore, 1984). The facies change from evaporites to siliciclastics in the Buckner coincides with the end of platform dolomitization in the upper Smackover (compare Figs. 1B–D).

DIAGENETIC FRAMEWORK FOR SMACKOVER PLATFORM DOLOMITIZATION

Ooid grainstones of the upper Smackover exhibit a striking diagenetic gradient across southern Arkansas and northern Louisiana, documented by Moore and Druckman in 1981 (Fig. 1B). Extensive oomoldic porosity and widespread intergranular calcite cement are developed in the platform interior facies across central Arkansas (Northern Diagenetic Zone). An intermediate area, in central south Arkansas, is characterized by moderate volumes of bladed circumgranular calcite cement and partially dissolved ooids (Transitional Diagenetic Zone). At the platform margin, in north Louisiana, well-preserved ooids occur associated with early fibrous circumgranular cement, spotty replacement dolomite, moderate volumes of post-compaction poikilitic calcite, and saddle dolomite cements (Southern Diagenetic Zone). In west Arkansas, and east Texas, however, the transitional diagenetic zone of Moore and Druckman (1981) is the locus of the massive platform dolomites that are the subject of this paper (Fig. 1).

In the area straddling east Texas and west Arkansas, Stamatedes (1982) recognized three petrographically distinct

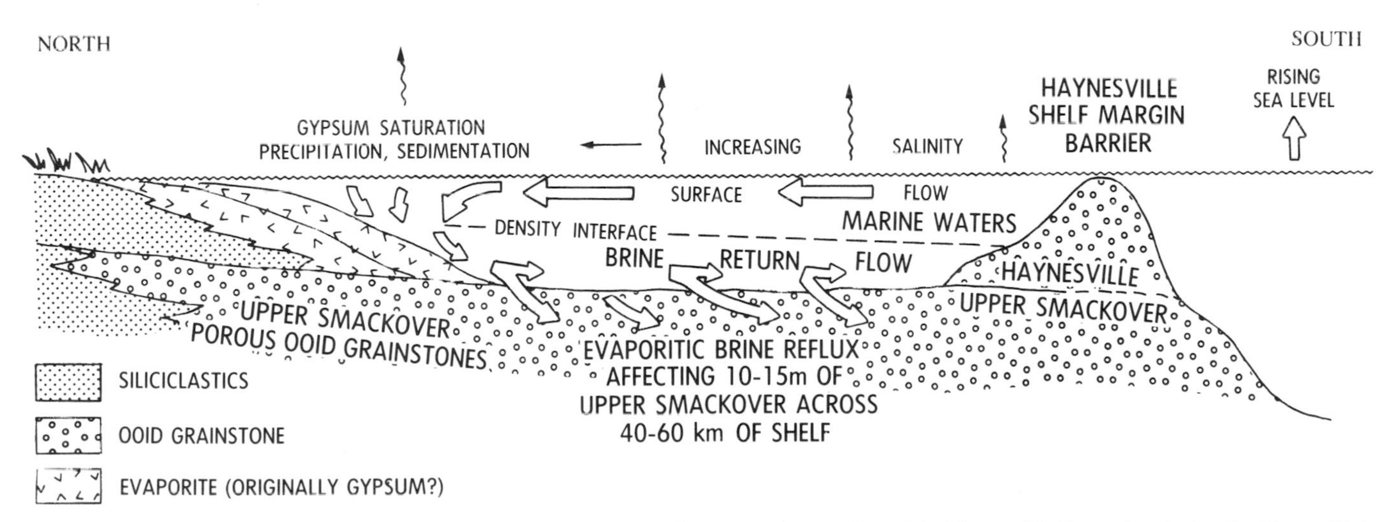

FIG. 3.—Schematic sedimentologic and hydrologic setting of the Buckner evaporite member of the Haynesville Formation during the Kimmeridgian sea-level rise (see Fig. 2). Concept of a Haynesville lagoon and a Haynesville shelf margin barrier first developed by Dickinson (1969). No vertical or horizontal scale intended.

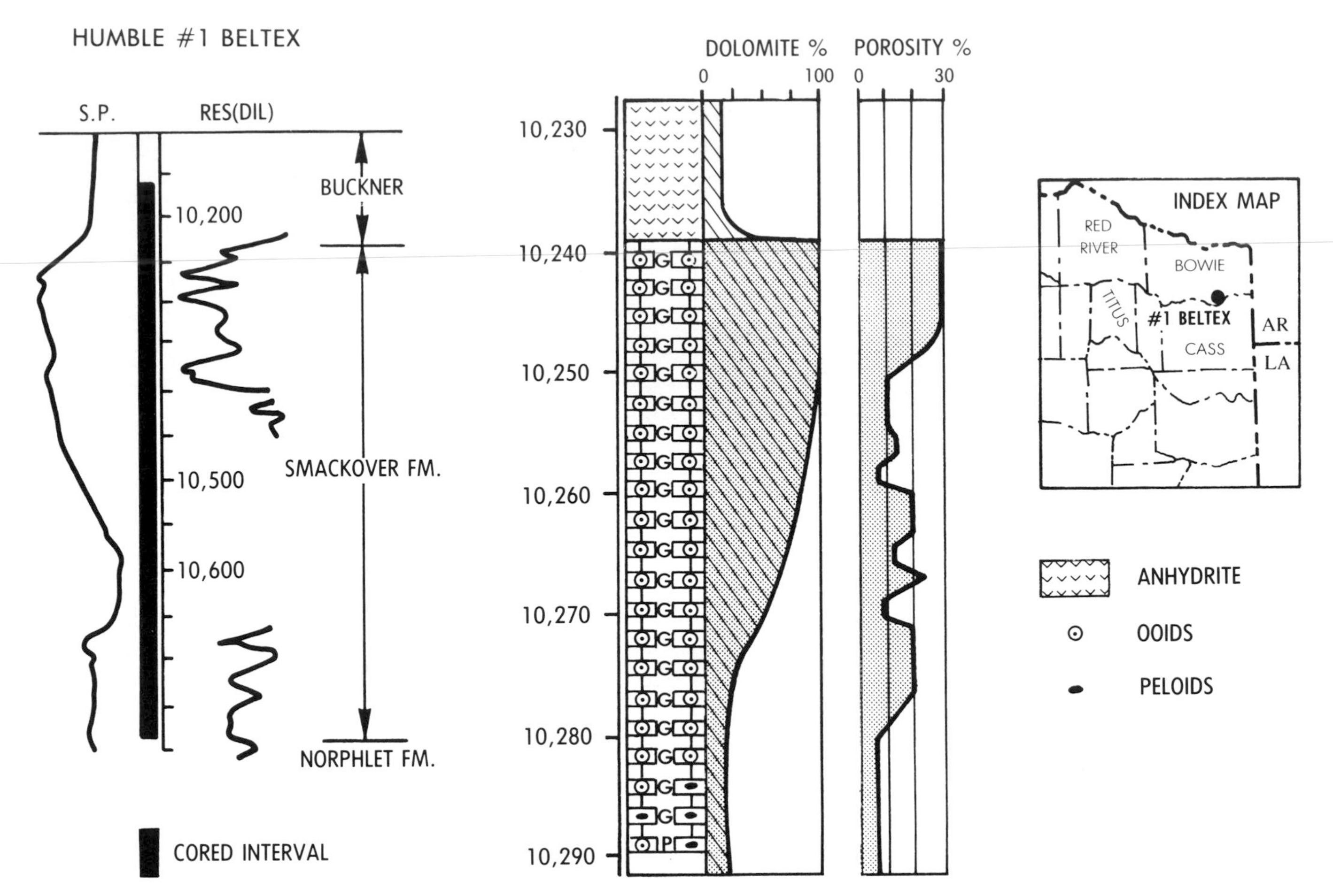

FIG. 4.—Dolomite distribution in the Humble #1 Beltex, Bowie County, Texas (Well 2 on Fig. 1 and Table 1). Note that dolomite percentage decreases vertically away from Buckner anhydrite, even though lithology throughout is well-sorted ooid grainstone.

populations of dolomite in the upper Smackover: (1) pre-compaction, coarse to medium crystalline, fabric-selective dolomite associated with an oomoldic fabric; (2) post-compaction, medium to fine crystalline, nonfabric-selective, zoned (Fe, Mn, and cathodoluminescence) dolomite; and (3) coarse crystalline, post-compaction saddle dolomite. Stamatedes (1982) also found that these three dolomite populations could be separated on the basis of their stable isotope compositions (Table 2, Fig. 5). The pre-compaction dolomites are heaviest with respect to $\delta^{18}O$ and $\delta^{13}C$, and the saddle dolomites have significantly lighter stable isotope ratios. The zoned dolomites are intermediate between the oomoldic and saddle dolomites. The stable isotope trends of Stamatedes' data set appear to be consistent with episodic dolomitization during progressive burial associated with increasing temperature and/or changing water composition. Stamatedes (1982) interpreted the early oomoldic dolomites of east Texas and west Arkansas as being of mixed meteoric-marine origin because of their light isotopic composition relative to superjacent marine dolomites associated with massive Buckner evaporites (Fig. 5). Furthermore, Loucks and Budd (1981) interpreted platform dolomitization of the Smackover in south Texas as mixed-meteoric marine in origin, but had no supporting geochemical data. O'Hearn (1984) confirmed the relative timing of each of these dolomite populations by a study of two-phase fluid inclusions that utilized Stamatedes' (1982) material. Only single-phase inclusions were found in Stamatedes' oomoldic dolomites, suggesting low temperatures of formation. On the other hand, the zoned dolomites exhibited two-

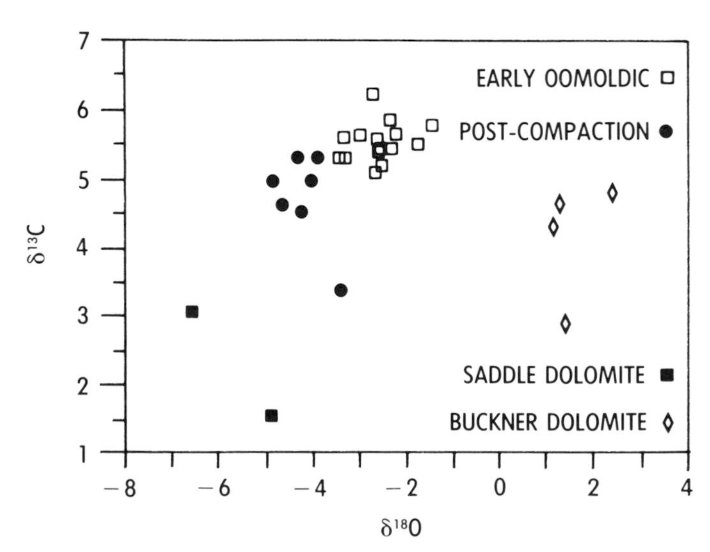

FIG. 5.—Cross-plot of $\delta^{18}O$ vs. $\delta^{13}C$ of upper Smackover early oomoldic dolomites, western Arkansas. Data from Stamatedes (1982). See Table 2, this paper.

phase fluid inclusions with homogenization temperatures averaging 120°C; those of the saddle dolomites averaged 150°C. These results support Stamatedes' paragenetic sequence and interpretations.

The general diagenetic trends developed in Arkansas and Louisiana (Moore and Druckman, 1981) were extended into and through east Texas by McGillis (1984), Stewart (1984), Wilkinson (1984), Harwood and Fontana (1984), and Harwood and Moore (1984; Figs. 1B, 6). Most important, these workers were able to document detailed geographic patterns of dolomite distribution that related early, oomoldic Smackover dolomites directly to the distribution of overlying massive Buckner evaporites (Fig. 1C, D). In addition, they were able to confirm that the vertical dolomitization gradients observed by Stamatedes (1982) in west Arkansas, that pointed to the Buckner as a source for dolomitizing fluids, were a general pattern across east Texas (see Fig. 4 for an example).

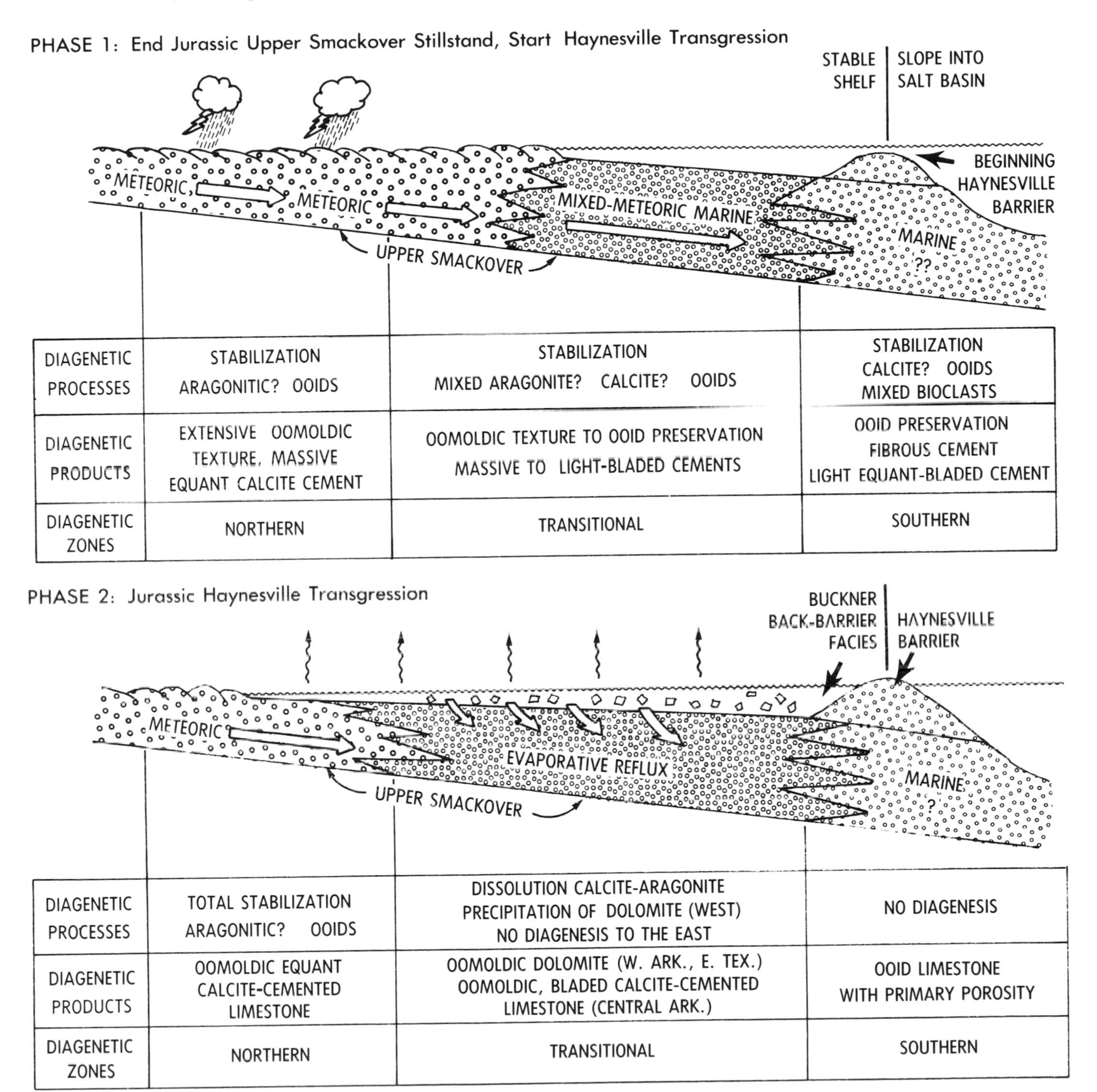

FIG. 6.—Conceptual model of early diagenetic environments, processes, and products in upper Smackover ooid grainstones, northwestern Gulf of Mexico. Diagrams modified from Moore (1984). Diagenetic zones from Moore and Druckman (1981), extended through east Texas by McGillis (1984), Stewart (1984), and Wilkinson (1984), as shown in Figure 1B.

These studies led to the development of a general two-phase diagenetic model for the upper Jurassic in the study area (Moore, 1984), as shown schematically in Figure 6. The first two diagenetic phases occurred early (pre-burial), were related to sea-level fluctuations between Oxfordian and Kimmeridgian sequences, and form the basic diagenetic framework for the present study. The extensive moldic porosity, cementation, and the development of the general regional diagenetic patterns seen across the study area (Fig. 6) probably resulted from a major meteoric-water system, with input along the platform inner margins at the end of the Oxfordian. These patterns were originally thought to be the result of a north-to-south gradient from fresh to marine water in the Smackover during early diagenesis (Moore and Druckman, 1981). The diagenetic fluids may have been dominantly meteoric across the entire trend, however, and the north-to-south change in primary mineralogy of ooids from aragonite to calcite may have been a major cause for the observed diagenetic gradient (Moore and others, 1986). The second phase involved dolomitization of the upper Smackover by reflux of evaporative waters from the Buckner lagoon through porous underlying upper Smackover grainstones and mixing with meteoric waters during the early Kimmeridgian sea-level rise. The final phase took place during burial. The relative timing of major diagenetic events in the upper Smackover in the area of interest is shown in Figure 7, with dolomites highlighted.

The geologic and general diagenetic framework of the oomoldic-related early platform dolomites of the upper Smackover, as reviewed earlier, clearly supports a reflux origin for this important early diagenetic event. The stable isotopic compositions of these dolomites in west Arkansas, however, are at variance with such as origin and are more compatible with a dolomitizing model involving either meteoric or mixed meteoric-marine water (Stamatedes, 1982). We are left with an apparent conflict between geochemical and geologic evidence for the origin of upper Smackover platform dolomites. The present investigation extends and expands the geochemical data base of the upper Smackover platform dolomites to include the rest of east Texas and provides not only new stable isotope data but also more extensive trace-element data, as well as radiogenic isotope analyses, in order to reconcile geologic and geochemical evidence for the origin of these dolomites.

RESULTS

Additional petrographic observations.—

The early upper Smackover platform dolomites formed prior to compaction and were emplaced during or shortly

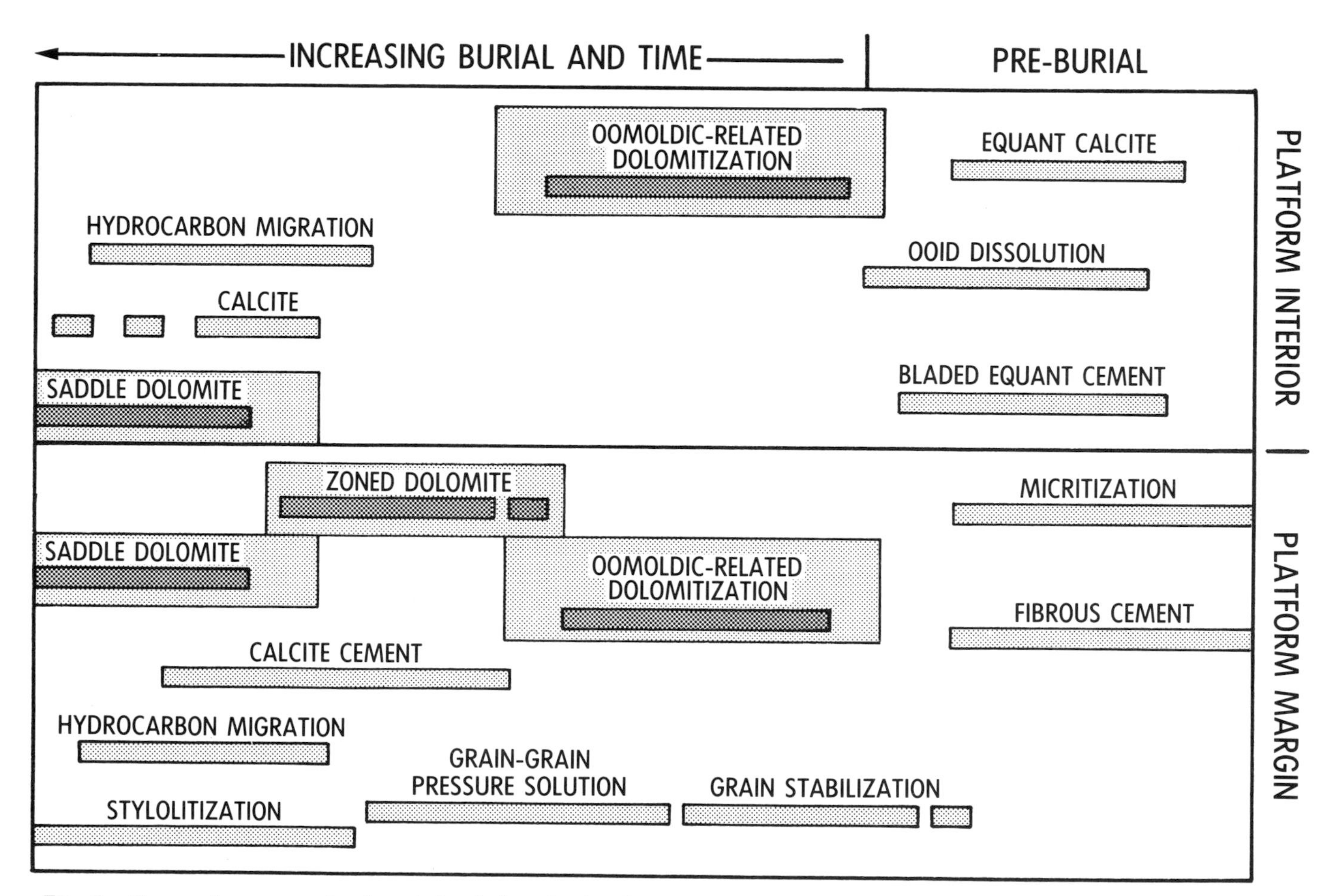

FIG. 7.—Paragenetic sequence showing relative timing of major diagenetic events in the upper Smackover, east Texas and west Arkansas. Data from Moore and Druckman (1981), Stamatedes (1982), O'Hearn (1984), McGillis (1984), Stewart (1984), and Wilkinson (1984).

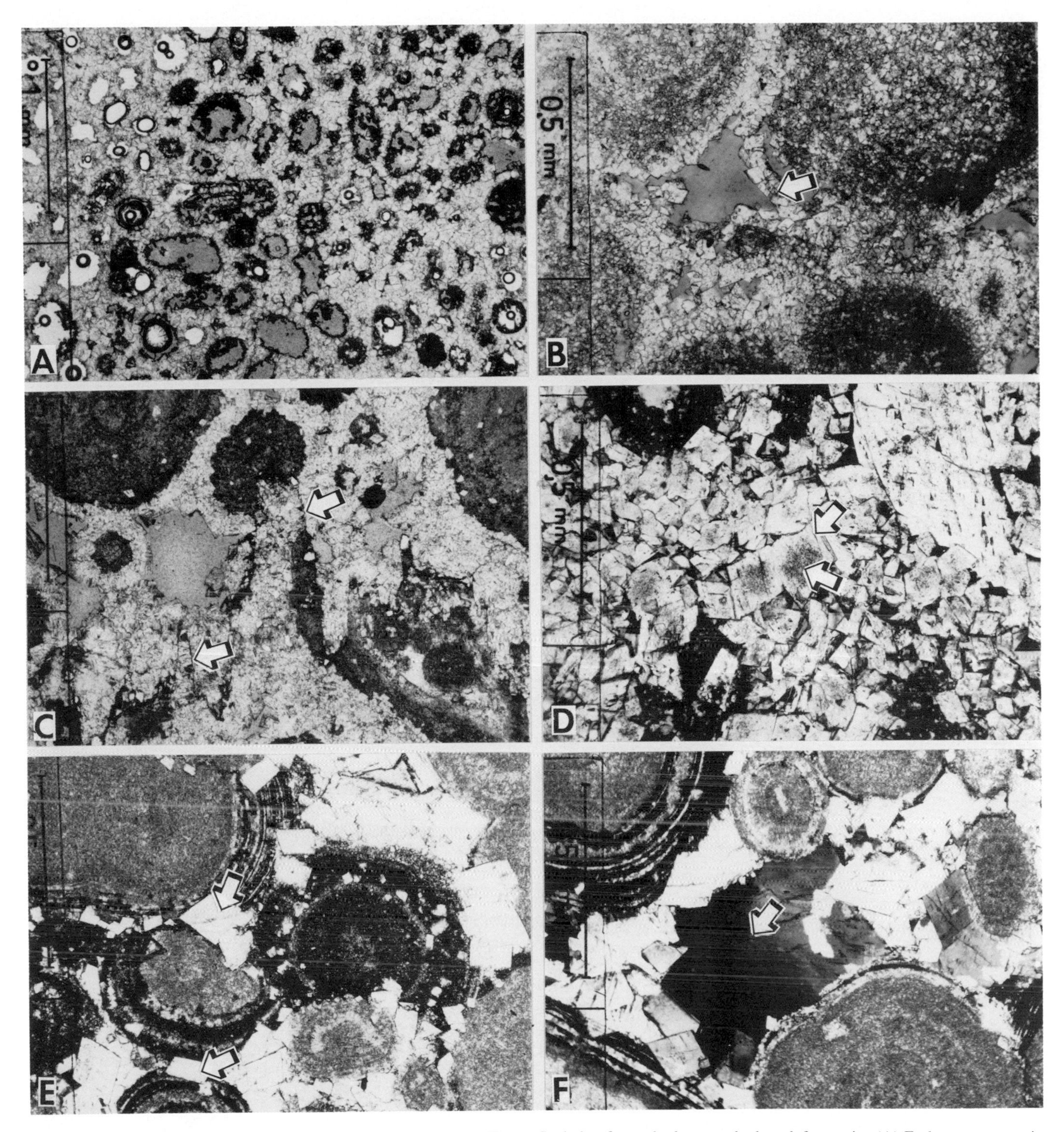

FIG. 8.—Photomicrographs of upper Smackover dolomites, east Texas. Scale bar for each photograph along left margin. (A) Early pre-compaction platform interior oomoldic dolomite. Some oomolds show evidence of compaction. Black material in oomolds in pyrobitumen. Unpolarized light; scale bar 1 mm. Schneider and others #1 Hargraves, Hopkins County, Texas, 9424′ (Well 1, Table 1). (B) Early pre-compaction platform interior oomoldic dolomite showing compaction-induced spalling of dolomitized ooid cortex or dolomitized calcite circumgranular cement (arrow). Unpolarized light; scale bar 0.5 mm. Schneider and others #1 Hargraves, Hopkins County, Texas, 9427′ (Well 1, Table 1). (C) Early pre-compaction platform margin dolomite showing stylolite cutting dolomite fabric (arrow). Unpolarized light; scale bar 0.5 mm. Getty #6-2-D Evers and Rhodes, Franklin County, Texas, 11,819′ (Well 12, Table 1). (D) Early pre-compaction platform margin dolomite showing clear overgrowths over cloudy core (arrows). Unpolarized light; scale bar 0.5 mm. Sun #1 Travis, Van Zandt County, Texas, 13,305′ (Well 7, Table 1). (E) Post-compaction dolomite crosscutting grain-to-grain pressure-solution contacts (arrows). Unpolarized light; scale bar 0.5 mm. Sun #1 Travis, Van Zandt County, Texas, 13,280′ (Well 7, Table 1).(F) Post-compaction saddle dolomite void fill, showing sweeping extinction (arrow). Polarized light; scale bar 0.5 mm. Sun #1 Travis, Van Zandt County, Texas, 13,321′ (Well 7, Table 1).

after mineral stabilization (Stamatedes, 1982; Harwood and Fontana, 1984; Harwood and Moore, 1984; Fig. 8A-D). Dolomite crystal size varies widely from 0.005 mm to 0.2 mm. There is a crystal size gradient within the oomoldic dolomites, with the finer sizes more prevalent toward the interior of the platform and the coarser sizes concentrated toward the platform margin (Table 3, Figs. 8A, D, 11). The dolomites of the platform interior show no zoning of any kind. The coarse dolomites of the platform margin, however, do show clear overgrowths around a cloudy core (Fig. 8D). Luminescence characteristics vary across the oomoldic dolomite facies, with the finer crystalline dolomites of the platform interior luminescing a uniform bright brick red (Fig. 9A, B), and the coarser crystals toward the platform margin generally showing a dull red color to nonluminescent with a single, dark, final zone present in about 10% of the rhombs (Fig. 9C, D).

Post-compaction replacement dolomite consists of scattered euhedral rhombs in compacted grainstones and packstones. The rhombs are commonly associated with stylolites and randomly crosscut pressure-solution contacts and grain-cement boundaries (Fig. 8E). Crystals are generally compositionally zoned and average 0.1 mm in size. Luminescent zoning is common with a dull red core (zone 1) followed by a dark nonluminescent zone (zone 2), a dull red zone (zone 3), and a brightly luminescing final zone (zone 4; Fig. 9E, F). Zones 1 and 2 of these post-compaction dolomites are similiar to, and may represent in part, the two zones noted in the coarse, pre-compaction platform margin dolomites (Fig. 9C, D). These dolomites occur across the platform but are most common along the platform margin.

Post-compaction saddle dolomite occurs as pore-filling cement across the platform but is most common along the platform margin. Crystals are usually large (as much as 0.5 mm) with crystal faces projecting into and partially to fully infilling pores (Fig. 8F). These dolomites are commonly nonluminescent and unzoned.

Trace-element composition.—

Table 3 summarizes the trace-element composition of the three general populations of dolomites that occur in the upper Smackover, as well as fine crystalline dolomite encased in the evaporites of the Buckner Formation above. The main dolomite populations of the upper Smackover and Buckner can be differentiated on the basis of their trace-element composition, particularly Sr, Na, and Fe, suggesting that the dolomitizing waters responsible for these dolomite populations were chemically distinct.

The Sr composition of all Smackover dolomites is relatively low, ranging from 40 to 139 ppm. The early oomoldic and zoned dolomites show the lowest levels (55 and 64 ppm average, respectively) whereas the saddle dolomite contains significantly more Sr with an average of 121 ppm for three samples. Buckner dolomite is considerably enriched in Sr relative to Smackover dolomites, below, with an average of 376 ppm Sr and a range from 261 to 490 ppm Sr.

Sodium composition tracks that of Sr, with the early oomoldic and zoned dolomites averaging 157 and 160 ppm, respectively; saddle dolomites at higher levels, with 304 ppm average; and Buckner dolomites significantly enriched, at 726 ppm average. The sympathetic relation between Na and Sr in Smackover and Buckner dolomites is shown in Figure 10. These results generally parallel those of Stamatedes (1982; Table 2).

Iron composition is more complex. Oomoldic dolomite of the platform interior contains significant Fe, averaging 1,334 ppm. Coarser oomoldic dolomite toward the platform margin is depleted relative to the platform interior, averaging just over 84 ppm, whereas Buckner dolomite averages 720 ppm Fe (Fig. 8). Post-compaction zoned and saddle dolomites average 1,166 and 14,500 ppm Fe, respectively. These averages, however, are difficult to assess due to compositional zoning and the large standard deviation of the data. Manganese varies sympathetically with Fe, with highest Mn compositions in saddle dolomites and lowest Mn compositions in platform margin oomoldic dolomites.

Dolomite stoichiometry.—

Most upper Smackover dolomites analyzed by Stamatedes (1982) are nearly stoichiometric, with excess $CaCO_3$ seldom over 2 mole percent and most commonly less than 0.5 mole percent (Table 2). Buckner dolomites, however, commonly exhibit 4 to 5 mole percent excess $CaCO_3$ and are clearly nonstoichiometric relative to Smackover dolomites. A preliminary survey of east Texas dolomites yielded similiar values, and as a result, stoichiometry was not routinely determined during the present study.

Stable isotopes of oxygen and carbon.—

Table 3 lists the stable isotopic ratios of upper Smackover and Buckner dolomites determined in the present study. The cross-plot of all data (Fig. 12) indicates that the above three dolomite populations are indeed viable for the entire study area (compare with Stamatedes' data, Fig. 5). The Buckner dolomite is significantly enriched in ^{18}O relative to all upper Smackover dolomites. Both carbon and oxygen isotopes of the upper Smackover show a distinct trend toward more depleted values, from the early pre-compaction oomoldic dolomites to the post-compaction saddle dolomites. In detail, the early oomoldic-related dolomites can be separated into two distinct subpopulations (Fig. 13), with the fine crystalline platform interior dolomites distinctly enriched in ^{18}O relative to their coarser platform margin counterparts.

Strontium isotopes.—

Ratios of $^{87}Sr/^{86}Sr$ from upper Jurassic dolomites are listed in Table 3. Whereas most dolomites analyzed cluster around Upper Jurassic seawater values of 0.7070 (Burke and others, 1982), two populations are more radiogenic: late post-compaction saddle dolomites, and iron-rich oomoldic platform interior dolomites (Fig. 14).

DISCUSSION OF EARLY UPPER SMACKOVER PLATFORM DOLOMITIZATION

This discussion centers on the origin of the early, pre-compaction, oomoldic platform dolomites of the upper

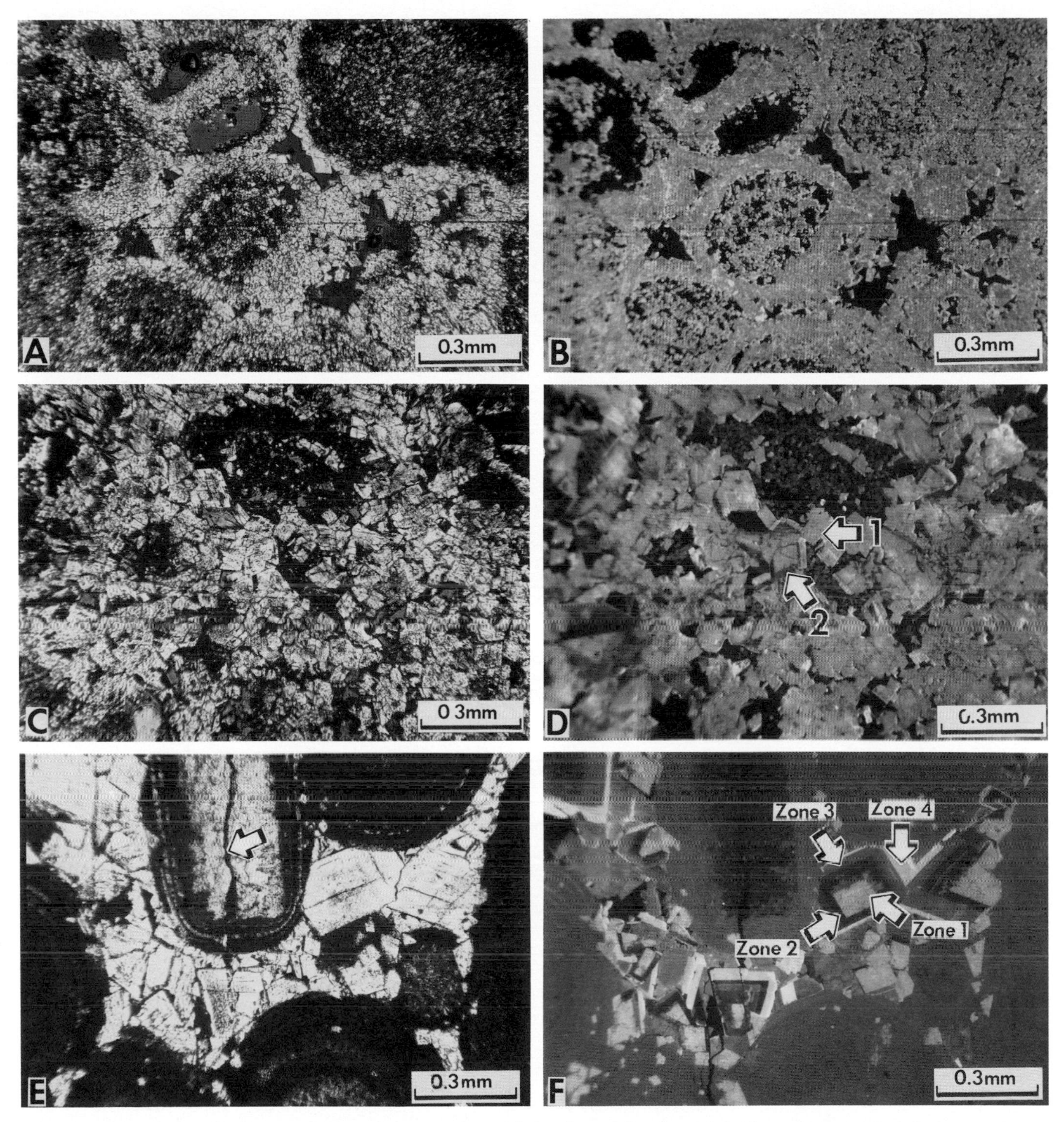

FIG. 9.—Photomicrographs showing cathodoluminescence characteristics of upper Smackover dolomites, east Texas. Scale is the same for each photograph and is shown in the lower right corner. (A) Early pre-compaction, platform interior, oomoldic dolomite. Unpolarized light. Schneider and others #1 Hargraves, Hopkins County, Texas, 9427′ (Well 1, Table 1). (B) Luminescent microscopy of sample shown in (A). Luminescence is a bright uniform brick red with no apparent zones. (C) Early pre-compaction platform margin dolomite. Unpolarized light. Sun #1 Travis, Van Zandt County, Texas, 13305′ (Well 7, Table 1). (D) Luminescent microscopy of sample shown in (C). Luminescence is dull red (arrow 1) with an occasional outer dark zone (arrow 2). (E) Post-compaction replacement dolomite. Ooid-coated echinoid fragment (arrow) in photo center. Unpolarized light. Sun #1 Travis, Van Zandt County, Texas, 13,321′ (Well 7, Table 1). (F) Luminescent microscopy of sample shown in (E). Zone 1 (arrow) similiar to general dolomite luminescence in pre-compaction dolomite shown in Figure 9D (arrow 1). Zone 2 (arrow) similar to final zone present in pre-compaction dolomites shown in Figure 9D (arrow 2). Zones 3 and 4 (arrows) are unique to post-compaction dolomites.

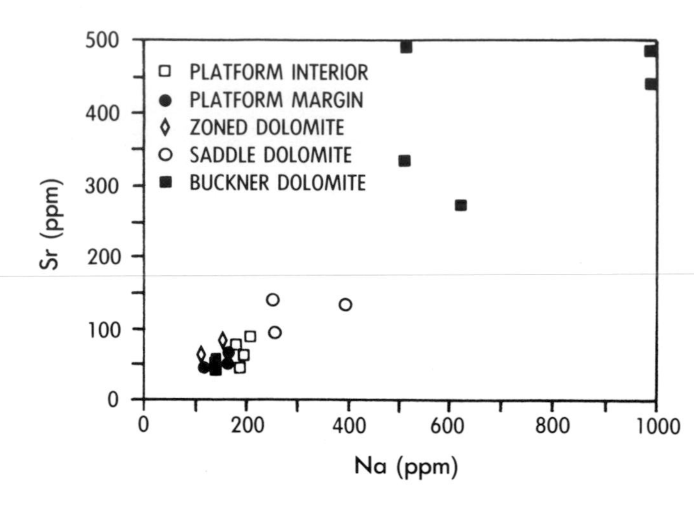

FIG. 10.—Cross-plot of Sr vs. Na for all dolomites, study area, showing that Buckner and saddle dolomites are distinctly relative to both Sr and Na.

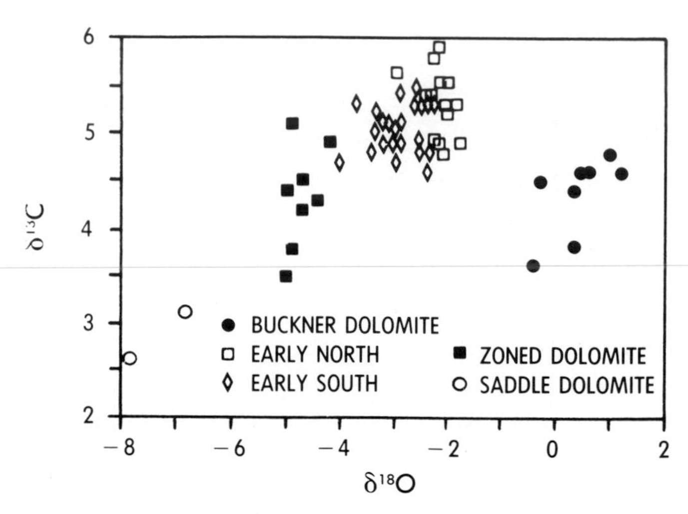

FIG. 12.—Cross-plot $\delta^{18}O$ vs. $\delta^{13}C$ for all dolomites, study area. Uncorrected data, expressed relative to PDB. Four major dolomite types are isotopically distinct.

Smackover in east Texas. The key question is: what model will satisfy the geochemical and geologic constraints outlined for these dolomites?

Even though dolomitization patterns in the upper Smackover strongly point to involvement of evaporative brines from the Buckner above (Figs. 1, 3, 4), no consistent vertical geochemical gradients have been observed in these dolomites in east Texas (Table 3). Indeed, there is a clear separation in composition of stable isotopes (Fig. 12), Sr, and Na (Fig. 10) between Buckner dolomites and the Smackover platform dolomites. There is, however, a distinct regional geographical Fe gradient in Smackover platform dolomites, with high Fe dolomites along the platform interior and low Fe dolomites toward the platform margin (Fig. 11). The upper Smackover interfingers with siliciclastics along its updip limits all across east Texas (McGillis, 1984; Stewart, 1984; Wilkinson, 1984). These siliclastics are a logical source for the Fe and ^{87}Sr in the early dolomites of the updip Smackover. Stable and radiogenic-isotope data as well as trace-element data, then, would seem to indicate that fluids responsible for upper Smackover dolomitization came from the interior margins of the platform rather than from the Buckner above.

The $\delta^{18}O$ composition of the platform dolomites (average -2.66‰, with a range from -1.8 to -4.9‰) could reflect a number of diagenetic environments, depending on the temperature and isotopic composition of the diagenetic fluid at the time of precipitation, or recrystallization, of the dolomite (Land, 1980; Moore, 1985). These dolomites petrographically predate compaction, however, and contain no two-phase fluid inclusions to indicate elevated temperatures. One could reasonably assume near-surface diagenetic temperatures. Using a 3‰ fractionation between calcite and dolomite (Land, 1980), a 25°C diagenetic temperature, and the temperature equation of O'Neil and others (1969), one could estimate that the $\delta^{18}O$ composition of the dolomitiz-

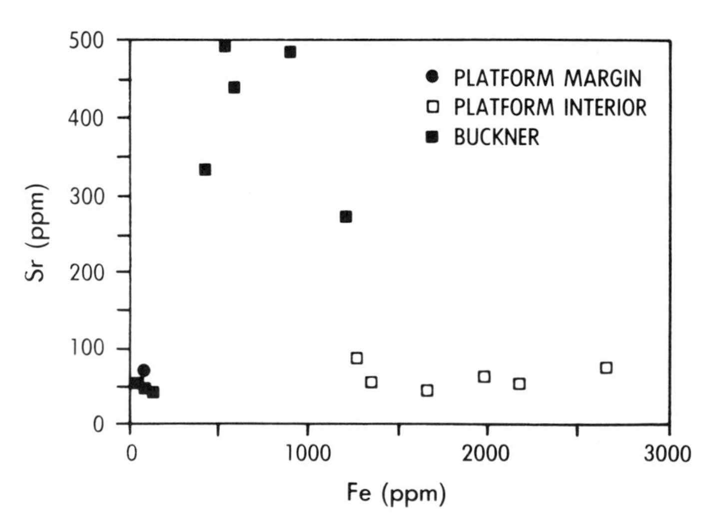

FIG. 11.—Cross-plot of Sr vs. Fe for platform and Buckner dolomites, study area. Platform interior dolomites are enriched in iron relative to platform margin and Buckner dolomites.

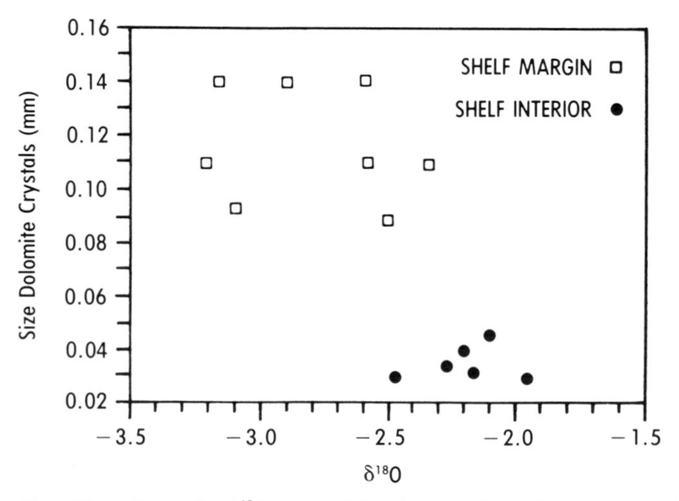

FIG. 13.—Cross-plot $\delta^{18}O$ versus dolomite crystal size for early, oomoldic platform dolomites, upper Smackover, study area. Uncorrected data, expressed relative to PDB. Data show shift in crystal size and $\delta^{18}O$ composition from platform interior to platform margin.

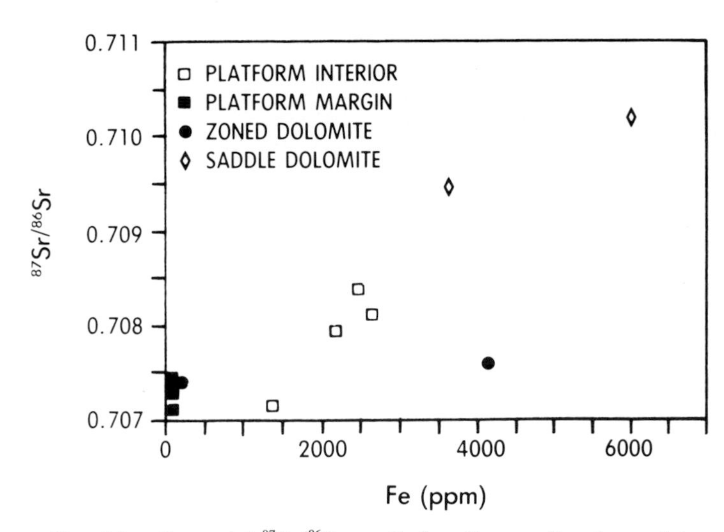

FIG. 14.—Cross-plot $^{87}Sr/^{86}Sr$ vs. Fe for all upper Smackover dolomites, study area. Upper Jurassic sea water $^{87}Sr/^{86}Sr$ is 0.7070 (Burke and others, 1982). Platform interior and saddle dolomites show increase in ^{87}Sr relative to Jurassic sea water.

ing fluid had an average composition of −3.0‰, with a range from −1.5 to −4.5‰. The average calculated $\delta^{18}O$ value is near the average isotopic composition of recent Yucatan coastal meteoric waters (−3.7‰) reported by Ward and Halley (1985). Finally, the $\delta^{13}C$ composition of these early east Texas dolomites averages +5.25‰ and probably reflects the carbon-isotopic composition of the precursor limestones and/or dolomites (Moore and Druckman, 1981).

The very low Sr (average 55 ppm) and Na (average 157 ppm) compositions of these east Texas platform dolomites are compatible with involvement of meteoric waters during the dolomitizing or recrystallizing event (Kinsman, 1969; Land and Hoops, 1973; Land and others 1975).

Whereas the geochemical data certainly implicate a meteoric-water system in the diagenetic history of upper Smackover dolomites, they do not necessarily support a conventional meteoric-marine water-mixing dolomitization model. Large areas of the upper Smackover subcrop in nearby Arkansas were probably under the influence of mixed meteoric-marine fluids for a significant period of time (Moore and Druckman, 1981) but were not pervasively dolomitized. Indeed, the elevated $^{87}Sr/^{86}Sr$ of some updip dolomites indicates that very large volumes of meteoric water must have been involved during dolomitization or recrystallization in order to overcome the buffering effects of marine-derived Jurassic sea water and limestones (Moore, 1985). In a conventional meteoric-marine mixing situation, the small levels of radiogenic Sr derived from siliciclastics by interaction with meteoric fluid would have been swamped by the greater volume of Sr derived from Jurassic limestones and sea water (Banner and others, 1986). In a conventional mixed meteoric-marine-water system then, the resulting dolomite should have a marine Jurassic $^{87}Sr/^{86}Sr$ signature (0.7070). The elevated $^{87}Sr/^{86}Sr$ compositions of some of the northern platform dolomites, therefore, when combined with all other geochemical data, strongly suggest that an active meteoric-water system was responsible for dolomitization of pre-existing limestones or recrystallization of pre-existing dolomites. The blanket nature of upper Smackover ooid grainstones and their intercalation updip with coarse siliciclastics are compatible with the development of such a major meteoric-water system.

Our study clearly indicates that dolomitization of the upper Smackover involved both refluxed evaporative brines as well as meteoric waters. Two models generally satisify the necessity for involvement of both water types: (1) an evaporite reflux-meteoric water-mixing model, where refluxing brines from the overlying Buckner lagoon mix with an active meteoric-water system in the upper Smackover. The refluxing evaporative brines provide the Mg necessary for dolomitization. As indicated earlier, this model was originally proposed by Moore (1984) for the study area. The Sr, Na, and Sr isotope data suggest, however, that the system must have been dominated by meteoric rather than evaporative waters; (2) a recrystallization model, where the upper Smackover was initially dolomitized by reflux of evaporative brines from the overlying Buckner lagoon and subsequently recrystallized in major meteoric-water system. This model requires the chemical reequilibration of the dolomites, destruction of the vertical geochemical gradients originally produced by refluxing brines, and the establishment of the present platform-wide horizontal geochemical gradients by intensive meteoric-water flushing.

As is common in all discussions of dolomite and dolomite-related phenomena, the geologists involved in this study do not unanimously support any one of the above models. Chowdhury and Chan prefer the recrystallization model, because they feel that the Sr, Na, and Sr isotope data preclude any involvement of marine waters during the final dolomitization event. Moore supports the reflux-meteoric water-mixing model, because it is the simplest model that satisifies most of the geologic and geochemical constraints. In addition, he feels that the Sr isotope data are too limited to be definitive, and most important, that there is no clear petrographic evidence for recrystallization. The reader is invited to assess each model on the merits of the data presented.

Finally, we must deal with the observed subtle shift in the stable isotopic compositions of these early Smackover platform dolomites toward more depleted values from platform interior toward platform margin (Fig. 13). The isotopic shift is paralled by a shift from fine crystal size in the platform interior to coarser crystal size toward platform margin, as well as a change from unzoned bright red cathodoluminescence in the interior to occasionally zoned dull or nonluminescence toward the platform margin (Figs. 9A–D, 13). These relations could well reflect greater burial effects toward the basin, suggesting the addition of isotopically depleted material as overgrowths during early burial phases. Indeed, the coarse platform margin dolomites commonly exhibit clear overgrowths around a cloudy core (Fig. 8D), and as noted earlier, an occasional dull luminescent outer zone, suggesting addition of a later dolomite phase (Fig. 9D). The change in luminescence characteristics may be related to the Fe-Mn gradient observed in the upper Smackover early dolomites. The elevated Mn levels of the platform interior dolomites (Table 3), derived from interbedded siliciclastics, could trigger luminescence. Whereas the Fe levels of these platform interior dolomites are high (Table 3), they are not elevated enough to quench the luminescence (Fairchild, 1983; Machel, 1985). The low Mn compositions (Table 3) found in the dolomites farthest away from the siliciclastics, at the platform margin, would lead to dull luminescence or nonluminescence in these dolomites (Fairchild, 1983; Machel, 1985). The zoned post-compaction replacement dolomites overlap slightly in isotopic composition with the early coarse dolomites (Fig. 13) and seem to share their first two zones with the earlier pre-compaction dolomites (Fig. 9E, F). These relations suggest that the overgrowths formed on early dolomites related to the platform margin during the initial development of the post-compaction zoned dolomites in the subsurface but ceased growth prior to the development of the bright luminescent outer zones characteristic of these post-compaction dolomites (Fig. 9D, E).

CONCLUSIONS

(1) There are three distinct dolomite populations in the upper Smackover in east Texas and Arkansas that can be distinguished on the basis of petrography and stable isotopic signature: (a) pervasive, early, pre-compaction, oomoldic-related dolomite; this dolomite forms the major petroleum reservoir facies in east Texas; (b) subordinate, post-compaction replacement dolomite associated with stylolites and pressure solution; and (c) subordinate, late, post-compaction, pore-filling saddle dolomite.

(2) There is a direct link between early pervasive upper Smackover dolomites and the distribution of evaporite facies in the overlying Buckner Formation.

(3) A widespread, shallow, density-stratified Buckner lagoon, developed behind a shelf margin barrier, provides an ideal setting for relatively long-term evaporite brine reflux into underlying porous upper Smackover sequences. This model is consistent with observed sedimentologic and stratigraphic relations within the Smackover and overlying Buckner formations.

(4) Strontium, Na, stable isotopes, and radiogenic isotopes suggest dolomite formation in an environment dominated by meteoric waters.

(5) Two models generally satisify the necessity for involvement of both evaporative reflux as well as meteoric water in the early dolomitization of the upper Smackover: (a) an evaporative reflux-meteoric water-mixing model, where refluxing brines from the overlying Buckner lagoon mix with a meteoric-water system in the underlying upper Smackover; and (b) a recrystallization model, where the upper Smackover was initially dolomitized by reflux from the overlying Buckner lagoon and subsequently recrystallized in a major meteoric-water system.

(6) The authors do not unanimously support either model. Chowdhury and Chan prefer the recrystallization model, because trace-element and Sr isotope data seem to preclude involvement of marine waters during final dolomitization. Moore favors the evaporative reflux-meteoric mixing model, because it is the simplest model that generally satisfies the constraints of the data, the Sr isotope data that are crucial to the recrystallization thesis are limited and are not definitive, and finally because there is no compelling petrographic evidence for recrystallization.

(7) The shift in the stable isotope compositions of the early pervasive upper Smackover dolomites across the platform is though to reflect greater burial effects toward the basin. An increase in crystal size toward the platform margin, the occurrence of clear overgrowths, and the presence of cathodoluminescent zones in platform margin early dolomites that seem to be shared with later, post-compaction zoned dolomites, support this conclusion.

ACKNOWLEDGMENTS

This project was funded by the Louisiana State University Applied Carbonate Research Program (ACRP). The associates are thanked for their continued enthusiastic support for carbonate studies at Louisiana State University. In addition, we thank Ron Snelling for his help with the ICP analyses, Sam Reed for his continuous help in the rock laboratory through all phases of the project, Mary Lee Eggart for drafting, Ezat Heydari for cathodoluminescence assistance, Paul Aharon for insights on problems concerning stable isotopes, and Ellen Tye and Hans Machel for their help with the manuscript. Thoughtful reviews by Phil Choquette, John Humphries, and V. Shukla made this a much better paper. We especially thank those ACRP alumni who furnished the foundation for this study and without whose hard work this paper could not have been written: Mike Stamatedes, Terry O'Hearn, Chris Fontana, and the members of the "East Texas Gang," Gil Harwood, Kathy McGillis, Sara Stewart, and Scott Wilkinson.

REFERENCES

Banner, J. L., Hanson, G. N., and Meyers W. J., 1986 Determination of initial Sr-isotope compositions of dolostones from the Burlington-Keokuk Formations (Mississippian) of Iowa, Illinois, and Missouri: American Association of Petroleum Geologists Bulletin, v. 70, p. 562.

Bradford, C. A., 1984, Transgressive-regressive carbonate facies of the Smackover Formation, Escambia County, Alabama, *in* Ventress, W. P. S., Bebout, D. G. Perkins, B. F., Moore, C. H., eds., The Jurassic of the Gulf Rim: Gulf Coast Section, Society of Economic Paleontologists and Mineralogists Foundation, Third Annual Research Conference, Proceedings, p. 27–39.

Burke, W. H., Denison, R. E., Hetherington, E. A., Koepnick, R. B., Nelson, H. F., and Otto, J. B., 1982, Variation of seawater $^{87}Sr/^{86}Sr$ throughout Phanerozoic time: Geology, v. 10, p. 516–519.

Dickinson, K. A., 1969, Upper Jurassic rocks in northeastern Texas and adjoining parts of Arkansas and Louisiana: Gulf Coast Association of Geological Societies, Transactions, v. 19, p. 175–187.

Fairchild, K. J., 1983, Chemical controls of cathodoluminescence of natural dolomites and calcites: New data and review: Sedimentology, v. 30, p. 579–583.

Harwood, G., and Fontana, C., 1984, Smackover deposition and diagenesis and structural history of the Bryan's Mill area, Cass and Bowie Countries, Texas, *in* Ventress, W. P. S., Bebout, D. G., Perkins, B. F., and Moore, C. H., eds., The Jurassic of the Gulf Rim: Gulf Coast Section, Society of Economic Paleontologists and Mineralogists Foundation, Third Annual Research Conference, Proceedings, p. 135–147.

Harwood, G. M., and McGillis, K. A., 1984, Location of Gilmer shelf margin, Upper Jurassic, east Texas basin: American Association of Petroleum Geologists Bulletin, v. 68, p. 484.

———, and Moore, C. H., 1984, Comparative sedimentology and diagenesis of Upper Jurassic ooid grainstone sequences, east Texas basin, *in* Harris, P. M., ed., Carbonate Sands–A Core Workshop: Society of Economic Paleontologists and Mineralogists Core Workshop No. 5, p. 176–232.

KINGSMAN, D. J. J., 1969, Interpretation of Sr^{+2} concentrations in carbonate minerals and rocks: Journal of Sedimentary Petrology, v. 39, p. 486–508.

KOEPNICK, R. B., EBY, D. E., AND KING, K. C., 1985, Controls on porosity and dolomite distribution in upper Smackover Formation (Upper Jurassic), southwestern Alabama and western Florida: American Association of Petroleum Geologists Bulletin, v. 69, p. 274.

LAND, L. S., 1980, The isotopic and trace element geochemistry of dolomite: The state of the art, *in* Zenger, D. H., Dunham, J. B., and Ethington, R. L. eds., Concepts and Models of Dolomitization: Society of Economic Paleontologists and Mineralogists Special Publication 28, p. 87–110.

———, AND HOOPS, G. K., 1973, Sodium in carbonate sediments and rocks: A possible index to the salinity of diagenetic solutions: Journal of Sedimentary Petrology, v. 43, p. 614–617.

———, SALEM, M. R. I., AND MORROW, D. W., 1975, Paleohydrology of ancient dolomites: Geochemical evidence: American Association of Petroleum Geologists Bulletin, v. 59, p. 1602–1625.

LOUCKS, R. G., AND BUDD, D. A., 1981, Diagenesis and reservoir potential of the Upper Jurassic Smackover Formation of south Texas: Gulf Coast Association of Geological Societies, Transactions, v. 31, p. 339–346.

———, AND LONGMAN, M. W., 1982, Lower Cretaceous Ferry Lake Anhydrite, Fairway Field, east Texas: Product of shallow-subtidal deposition: *in* Society of Economic Paleontologists and Mineralogists Core Workshop No. 3, Calgary, Canada, p. 130–173.

MACHEL, H. G., 1985, Cathodoluminesence in calcite and dolomite and its chemical interpretation: Geoscience Canada, v. 12, p. 139–147.

MCGILLIS, K. A., 1984, Upper Jurassic stratigraphy and carbonate facies, northeastern east Texas basin, *in* Presley M. W., ed., The Jurassic of East Texas: East Texas Geological Society, p. 63–66.

MOORE, C. H., 1984, The upper Smackover of the Gulf Rim: Depositional systems, diagenesis, porosity evolution and hydrocarbon production, *in* Ventress, W. P. S., Bebout, D. G., Perkins, B. F., and Moore, C. H., eds., The Jurassic of the Gulf Rim: Gulf Coast Section, Society of Economic Paleontologists and Mineralogists Foundation, Third Annual Research Conference, Proceedings, p. 283–308.

———, 1985, Upper Jurassic subsurface cements: A case history, *in* Schneidermann, N., and Harris, P. M., eds., Carbonate Cements: Society of Economic Paleontologists and Mineralogists Special Publication 36, p. 291–308.

———, AND DRUCKMAN, Y., 1981, Burial diagenesis and porosity evolution, Upper Jurassic Smackover, Arkansas and Louisiana: American Association of Petroleum Geologists Bulletin, v. 65, p. 597–628.

———, CHOWDHURY, A., AND HEYDARI, E., 1986, Variation of ooid mineralogy in Jurassic Smackover limestones as control of ultimate diagenetic potential: American Association of Petroleum Geologists Bulletin, v. 70, p. 622–623.

O'HEARN, T. C., 1984, A fluid inclusion study of diagenetic mineral phases, Upper Jurassic Smackover Formation, southwest Arkansas and northeast Texas: unpublished M.A. Thesis, Louisiana State University, Baton Rouge, Louisiana, 190 p.

O'NEIL, J. R., CLAYTON, R. N. AND MAYEDA, T. K., 1969, Oxygen isotope fractionation factors in divalent metal carbonates: Journal of Chemical Physics, v. 51, p. 5547–5558.

SLOSS, L. L., 1969, Evaporite deposition from layered solutions: American Association of Petroleum Geologists Bulletin, v. 53, p. 776–789.

STAMATEDES, M. R., 1982, Dolomitization of the upper Smackover in Miller County, Arkansas, and adjacent areas: unpublished M.S. thesis, Louisiana State University, Baton Rouge, Louisiana, 164 p.

STEWART S. K., 1984, Smackover and Haynesville facies relationships in north-central east Texas, *in* Presley, M. W., ed., The Jurassic of East Texas: East Texas Geological Society, p. 56–62.

TODD, R. G., AND MITCHUM, R. M., 1977, Seismic stratigraphy and global changes of sea level, part 8: Identification of Upper Triassic, Jurassic, and Lower Cretaceous seismic sequences in Gulf of Mexico and offshore West Africa *in* Seismic Stratigraphy–Applications to Hydrocarbon Exlporation: American Association of Petroleum Geologists Memoir 26, p. 145–163.

VINET, M. J., 1984, Geochemistry and origin of Smackover and Buckner dolomites (Upper Jurassic), Jay Field area, Alabama, *in* Ventress, W. P. S., Bebout, D. G., Perkins, B. F., and Moore C. H., eds., The Jurassic of the Gulf Rim: Gulf Coast Section, Society of Economic Paleontologists and Mineralogists Foundation, Third Annual Research Conference, Proceedings, p. 365–374.

WARD, W. C., AND HALLEY, R. B., 1985, Dolomitization in a mixing zone of near seawater composition, Late Pleistocene, northeast Yucatan Peninsula: Journal of Sedimentary Petrology, v. 55, p. 407–420.

WILKINSON, S., 1984, Upper Jurassic facies relationships and their interdependence on salt tectonism in Rains, Van Zandt, and adjacent counties, east Texas, *in* Presley M. W., ed., The Jurassic of East Texas: East Texas Geological Society, p. 153–156.

DOLOMITIZATION AND DEDOLOMITIZATION OF UPPER CRETACEOUS CARBONATES: BAHARIYA OASIS, EGYPT

HANAFY HOLAIL AND KYGER C. LOHMANN

Department of Geological Sciences, The University of Michigan, Ann Arbor, Michigan 48109-1063

AND

IVAN SANDERSON

Department of Geological and Atmospheric Sciences, Purdue University, West Layfayette, Indiana 47907

ABSTRACT: The Upper Cretaceous shallow-marine sequence at Bahariya Oasis, northern Egypt, is composed of medium-grained dolomite of the El Heiz Formation. (Upper Cenomanian) and the El Hefhuf Formation (Campanian). Early formed crystal cores possess chemical signatures and cathodoluminescent (CL) microfabrics characteristic for each formation. In the El Heiz Formation, dolomite crystal cores are nonstoichiometric, ferroan, and possess relatively high Na and Mn contents and depleted $\delta^{18}O$ values. Under CL, these dolomites exhibit a mottled microfabric, indicating partial dissolution and replacement of a precursor dolomite phase. In contrast, dolomite crystal cores of the El Hefhuf Formation are nearly stoichiometric, possess low Fe, Mn, and Na contents, are relatively enriched in $\delta^{18}O$, and exhibit a uniformly nonluminescent fabric.

Despite differences in the character of the dolomite crystal cores among these two formations, the overall paragenetic sequence is quite similar. Dolomite core crystals are overgrown by a brightly luminescent dolomite cement that is Ca rich and Fe poor. Local calcitization of dolomite cores and this overgrowth is followed by precipitation of a vein-filling calcite spar, which is present in both formations and in the overlying Maastrichtian chalks. This late-calcite spar exhibits a uniform isotopic composition independent of its host formation. The early calcite associated with dedolomitization is enriched in Mg, and its $\delta^{18}O$ values lie intermediate between dolomite and late-calcite values. Because of similarities in their composition and fabric, it is suggested that both calcite and dolomite phases are related to a single diagenetic event. For example, carbonates of both formations were potentially affected by meteoric water derived at the post-Maastrichtian to pre-Eocene unconformity that has locally breached this sequence down to lower Cenomanian clastics.

The distinctive Fe^{2+} contents of dolomite crystal cores between these formations, however, suggest two temporally separate events of dolomitization, rather than a single diagenetic event. Elevated Fe and Na contents of El Heiz Formation dolomite and the association with coeval pyrite indicate a marine origin for initial dolomitization of this formation. In contrast, depleted $\delta^{18}O$ and trace-element values of dolomite from both formations and the similarity of compositional variation to co-occurring meteoric calcites argue for dolomitization within a mixed-water, meteoric-marine setting. Thus, it is suggested that a metastable precursor phase of dolomite formed in the El Heiz Formation from marine waters. This dolomite subsequently reacted with mixed-water fluids responsible for dolomitization of the El Hefhuf Formation. Such a process of dolomite by dolomite replacement is recorded in the mottled CL microfabrics of the El Heiz Formation dolomite crystal cores.

INTRODUCTION

Recent research in sedimentary geology has produced much data on the nature of dolomite rocks and various models of dolomitization (Morrow, 1982), especially in the recognition of sources of magnesium necessary for widespread dolomitization. Recently, Morrow (1982) and Land (1985) suggested that marine or marine-derived fluids are the only natural solutions that contain sufficient amounts of magnesium to form massive dolomite. In contrast, Hardie (1987) has suggested that most dolomite may, in fact, reflect the basin-wide circulation of meteoric-derived fluids. Whereas the former scenario implicates dolomitization at or near the sediment-water interface during or soon after deposition, the latter clearly suggests that dolomitization as a diagenetic process is more important at depth.

The relative timing and burial depths of dolomitization may constrain processes controlling the incorporation of other cations into the dolomite lattice and the relative timing of dedolomitization. Mossler (1971), Wong and Oldershaw (1981), and McHargue and Price (1982), for example, have attributed the formation of ferroan dolomite to epigenetic processes, with temperatures near 100°C, whereas Dickson and Coleman (1980) and M'Rabet (1981) suggest that ferroan dolomite forms at shallow burial and considerably lower temperatures. None of these studies, however, requires that ferroan dolomite formation is an either/or process. The common presence of ferroan dolomite in organic-rich argillaceous sediments indicates that ferroan dolomite may form at or near the sediment surface and continues to form with increasing burial depth (Curtis, 1978; Irwin, 1980; Taylor and Sibley, 1986). In such settings, a source of Fe^{2+} must be available and dominate its removal in pyrite accompanying sulfate reduction. It is clear from these studies that relations between depth, temperature, and the formation of stoichiometric and/or ferroan dolomite are poorly constrained.

Similarly, the calcitization of dolomite (dedolomitization) is often interpreted as a near-surface process, either reflecting near-surface alteration associated with subaerial exposure or late alteration during burial (Goldberg, 1967; Scholle, 1971; Mossler, 1971; Chafetz, 1972; Budai and others, 1984). As with dolomite formation, relations between depth, temperature, and calcitization of dolomite are also poorly constrained (Evamy, 1963; Folkman, 1969; Katz, 1971; Frank, 1981).

In this study, we describe the major textural and compositional features of a shallow burial-dolomite sequence that occurs within Upper Cretaceous shallow-water carbonates of northern Egypt near Bahariya Oasis. Petrographic and geochemical data on this sequence bear directly on the timing and diagenetic setting of these various dolomite-related alteration processes. In addition, most data on dolomitization and dedolomitization processes are from intensively studied sequences in North America and parts of Europe. Therefore, detailed study of textural and compositional trends in these Upper Cretaceous units not only sheds light on the origin of Upper Cenomanian-Campanian do-

Sedimentology and Geochemistry of Dolostones, SEPM Special Publication No. 43

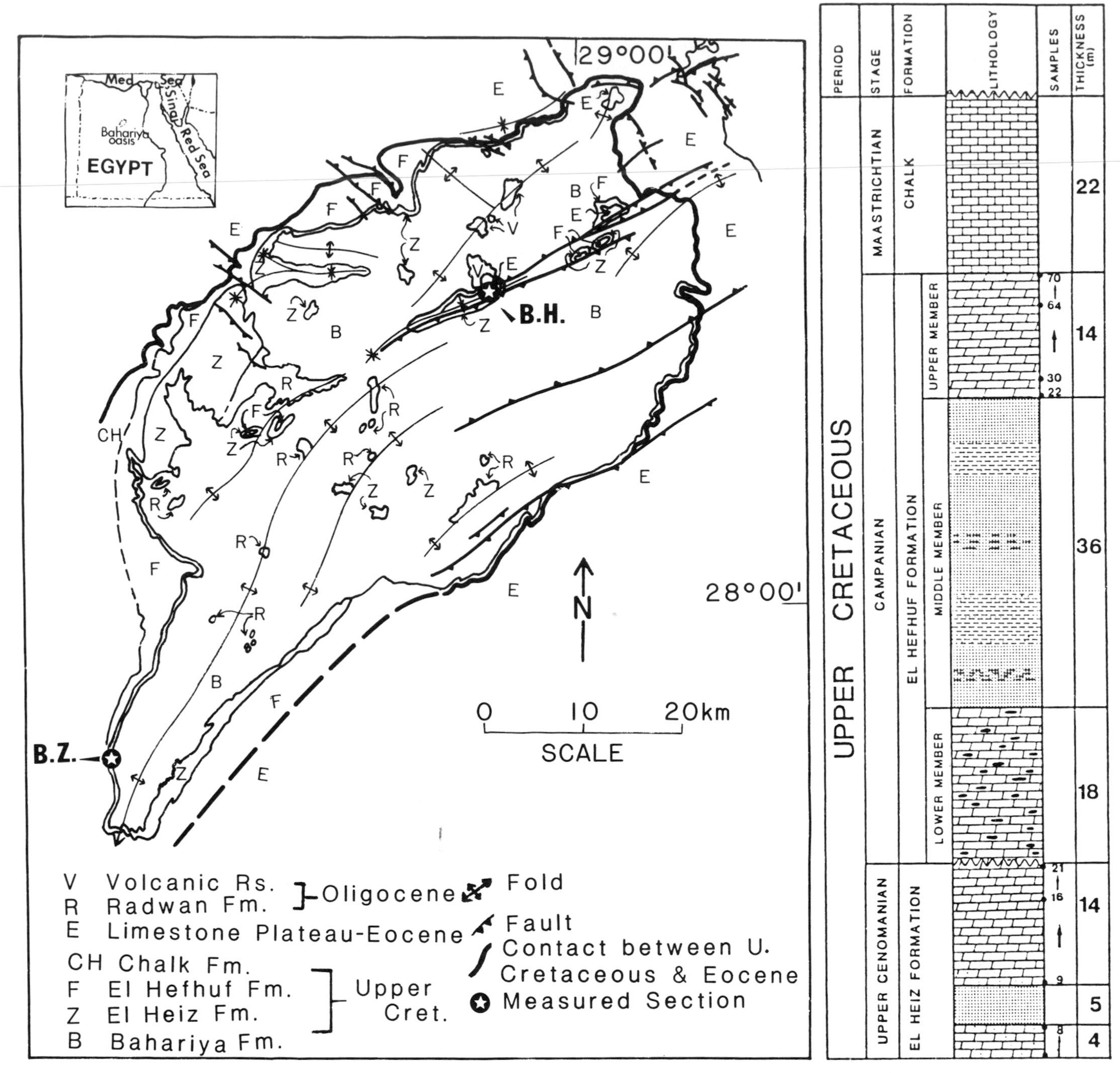

FIG. 1.—(A) Geologic map of Bahariya Oasis region. Areas labeled B.Z. and B.H. mark locations of El Heiz and El Hefhuf Formation sections sampled during this study. Modified from Said and Issawi (1964). (B) Stratigraphic section of the Upper Cretaceous sequence present at Baharyia Oasis.

lomites in this part of Egypt, but also adds important new data on ferroan-dolomite formation and calcitization processes at relatively shallow burial depths.

GEOLOGICAL SETTING

The Upper Cretaceous El Heiz and El Hefhuf formations are part of a predominantly carbonate sequence exposed along a northeast-trending anticlinal structure at Bahariya Oasis in the central part of Western Desert, Egypt (Fig. 1A). At this locality, the anticline is transected by a major axis-parallel fault zone. The geologic history of this area is similar to that of others within the Syrian Arc fold system, which extends from Syria and Israel to Egypt (El-Bassyony, 1978; Franks, 1982; Allam, 1986). Lithostratigraphically, the floor of the oasis is immediately underlain by an expanse of flat-lying Lower Cenomanian fluvial sandstone, whereas around the oasis margin the sandstone grades upward into terrestrial to marine terrigenous clastic units (Franks, 1982; Slaughter and Thurmond, 1974). The oldest marine unit exposed in this area is a thin, ferruginous dolomite that occurs near the base of carbonate beds in the Upper Cenomanian El Heiz Formation (Hermina, 1957; El-Akkad and Issawi, 1963; Dominik, 1984). This formation crops out in the southern and northwestern scarps of Bahariya Oasis. In the southern scarp, the El Heiz Formation is unconformably overlain by the Campanian El Hefhuf Formation, whereas in the northwestern scarp, it is unconformably overlain by the Middle Eocene Plateau Limestone (Said and Issawi, 1964; Bechmann and Hassan, 1968; Issawi, 1972). The El Heiz Formation is not present in the northern oasis scarp.

The El Hefhuf Formation is a dolomite and shale sequence that El-Akkad and Issawi (1963) divided into three members. The basal and upper members are largely dolomite, whereas the middle member consists of alternating beds of argillaceous and arenaceous terrigenous clastics. Like the underlying El Heiz Formation, this formation also is absent in some parts of the northern oasis, where it has been erosionally removed at a pre-Middle Eocene unconformity. To the southwest, this major regional unconformity separates a sequence of white chalky Maastrichtian limestones from the overlying Middle Eocene carbonates. The episode of subaerial exposure presumably had a profound effect on more than the Maastrichtian chalks, for within Bahariya Oasis, erosion at the pre-Eocene unconformity breached the crest of the anticline down to the lower Cenomanian clastics. Therefore, both the El Heiz and El Hefhuf formations have been variably affected by multiple events of subaerial exposure and its attendant diagenesis.

Dolomite samples were collected from the El Heiz anticline along the southern scarp of Bahariya Oasis at the type locality of El Heiz Formation (Hermina, 1957). Here, the formation consists of as much as 9 m of alternating sandy dolomite and sandstone overlain by as much as 14 m of brown to yellow dolomite immediately below the unconformity, where Turonian sediments have been removed by erosion. Dolomite was also collected frothe El Hefhuf area in the northeast part of Bahariya Oasis at the type locality of the El Hefhuf Formation (Said, 1962). The majority of samples examined during this study are from the uppermost strata immediately below the regional unconformity and/or exposure surfaces, especially in the El Hefhuf area (Fig. 1B). Here, samples were collected from the upper member of the El Hefhuf Formation, which consists of breccias of light-colored clasts in a darker matrix overlain by about 14 m of dolomite. These dolomites sometimes contain vugs and veinlets lined with crystalline calcite.

An important aspect of this sequence is its burial history and the implication of this history on maximum temperatures of formation for dolomitization and dedolomitization. Two principal episodes of exposure and erosion are recorded in the stratigraphic succession at Bahariya Oasis: (1) the pre-Campanian unconformity, which overlies the El Heiz Formation and reflects erosive loss of the Turonian strata, and (2) the Maastrichtian-pre-Middle Eocene regional unconformity. Figure 2 illustrates the progressive burial and periodic exposure of this sequence. It is important to note that, assuming a geothermal gradient of 30°C/km, the maximum temperature for the El Heiz Formation is limited to less than 40°C.

METHODS

Textural relations among component phases were examined in 60 polished and stained thin sections. Staining with potassium ferricyanide qualitatively indicated variations in iron contents, while microstructures were examined in thin section and by scanning electron microscopy (SEM) of broken surfaces.

Polished thin sections were also examined for cathodoluminescence fabrics to evaluate crystal growth zonation and alteration effects within the dolomites and calcites.

Representative powdered samples were analyzed by X-ray diffraction to determine carbonate mineralogy and dolomite stoichiometry. Calcium, magnesium, and iron contents were determined by electron probe microanalysis (EPMA). Operating conditions for EPMA include 15-KV accelerating voltage and 0.2-microamperes beam current with a 10-μm spot size. Trace-element (Sr, Na, Mn) signatures were determined by atomic absorption.

Various types of dolomite as well as calcite-filled veins and vugs were sampled for carbon- and oxygen-isotopic analysis employing a microscope-mounted drill assembly to extract 0.2 to 0.5 mg of powdered dolomite or calcite from polished slabs. These were roasted *in vacuo* at 380°C for 1 hr and then reacted with anhydrous phosphoric acid at 55°C. Dolomite samples were reacted off-line for 4 to 6 hr; calcite samples were reacted in an extraction line coupled directly to the inlet of a VG 602 E ratio mass spectrometer. All analyses were converted to PDB and corrected for ^{17}O in accordance with the procedure outlined by Craig (1957). No additional correction was applied for dolomite-phosphoric acid fractionation. Precision was monitored through daily analyses of NBS-20, which is better than 0.08 per mil for both oxygen and carbon determinations.

PETROGRAPHY

El Heiz Formation Dolomite

Noncarbonate components within the El Heiz Formation compose variable proportions of every sample; quartz and plagioclase feldspar contents increase downward in the section, composing as much as 30 percent at the base of the formation. In thin section, terrigenous constituents are present as individual grains or as grain aggregates. Where in contact with dolomite, no etching or corrosion is evident.

Dolomite consists primarily of medium-grained iron-rich crystals which, upon immersion in potassium ferricyanide, develop a typical bluish hue (Fig. 3A). Dolomite crystals compose sucrosic idiotopic masses with mud-supported to grain-supported textures, with few unidentifiable marine fossils. Dolomite occurs as two different textural phases: (1) dark rhombohedral cores surrounded by euhedral clear rims, and (2) dolomite cement overgrowths, which locally

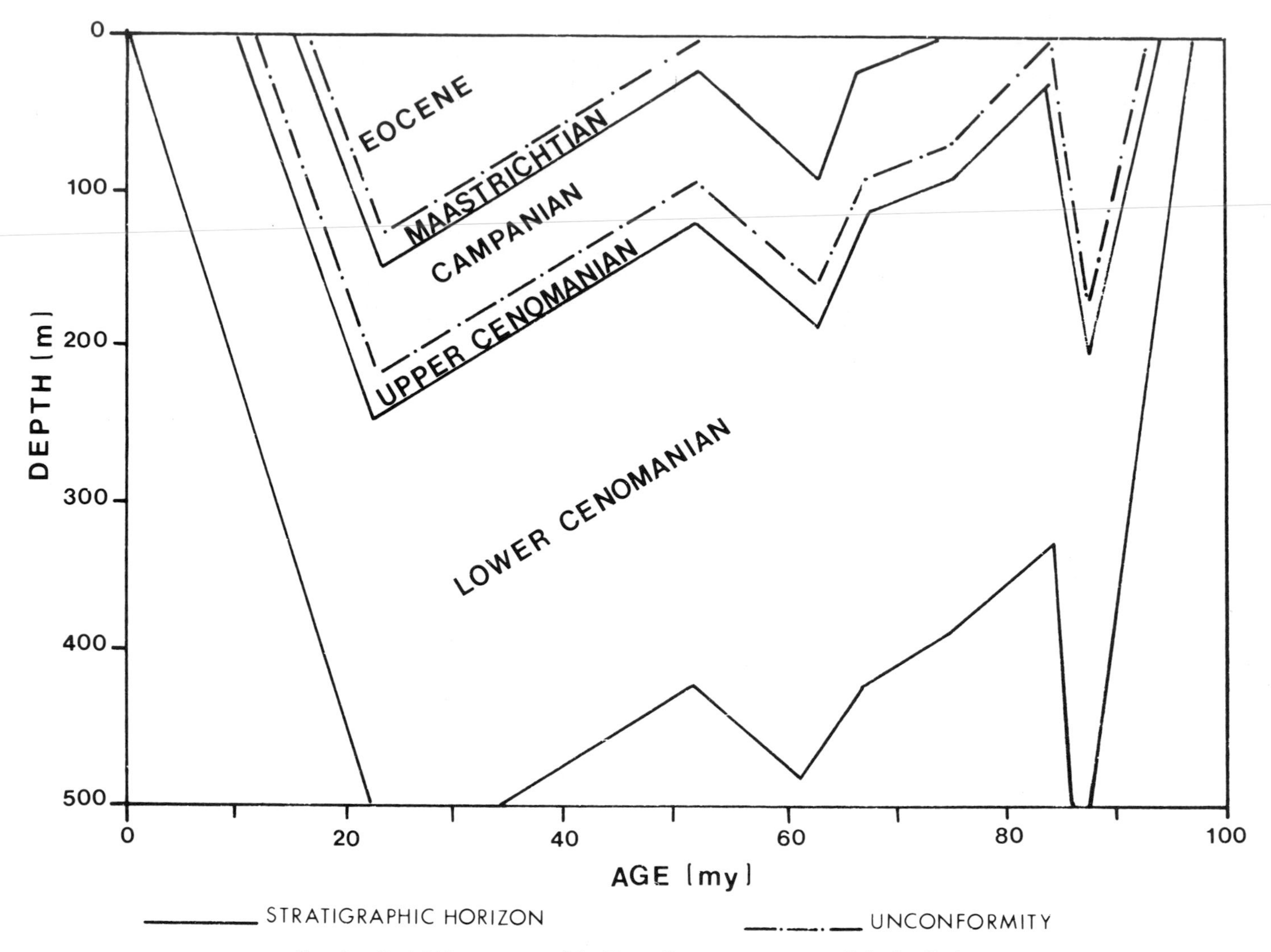

FIG. 2.—Burial-history curve of the Upper Cretaceous sequence, Bahariya Oasis.

infill pores between dolomite rhombs. Rare, 2- to 4-mm organic-rich stylolitic seams transect all types of dolomite (Fig. 3B). Most crystals are distinctly zoned, commonly containing a cloudy, iron-rich, rhombohedral center surrounded by clear overgrowth rims that progressively decrease in iron content (Fig. 3A, B).

Dolomite rhombs and overgrowth cements exhibit characteristic cathodoluminescent microfabrics. Dully luminescent rhomb cores exhibit extremely fine-scaled zonation and are overlain by a bright red luminescent overgrowth cement (Fig. 3C, D). This corresponds to the observed compositional trend of decreasing iron content from core to rim as suggested by staining.

When in close proximity to the unconformity, dolomite rhombs and cements have undergone extensive dissolution and calcitization (Fig. 4A, B). Dedolomitization is manifested by calcite cementation in dissolution voids of the iron-rich rhomb cores (Fig. 4C) and narrow zones within the rhombs (Fig. 4D). In other samples, however, dedolomitization was not zone-selective, and both entire rhombs and overgrowth dolomite cement have been completely replaced by calcite (Sibley, 1982). These dedolomite-related calcites, which exhibit alternating dull to bright cathodoluminescent zonation, occur both as a cement in inter-rhomb pores and as the replacive phase in optical continuity with dolomite substrates in dedolomite (Fig. 5B-D). Generally, replacive calcite crystals are anhedral, coarser, and exhibit more variable crystal size than precursor dolomite crystals (Fig. 5A). Sparry void-filling and veinlet-filling calcite (Fig. 6A, B) clearly postdates dedolomite-related calcite and exhibits a distinctive pattern of cathodoluminescent zonation. This calcite cement is largely nonluminescent but is marked by periodic, extremely thin, brightly luminescent bands.

El Hefhuf Formation Dolomite

Dolomite of the upper member of the El Hefhuf Formation is an iron-poor, medium- to coarse-grained dolomite (Fig. 7A). Dolomite in the lower 3 m of this member is brecciated and is composed of light yellow clasts in a darker

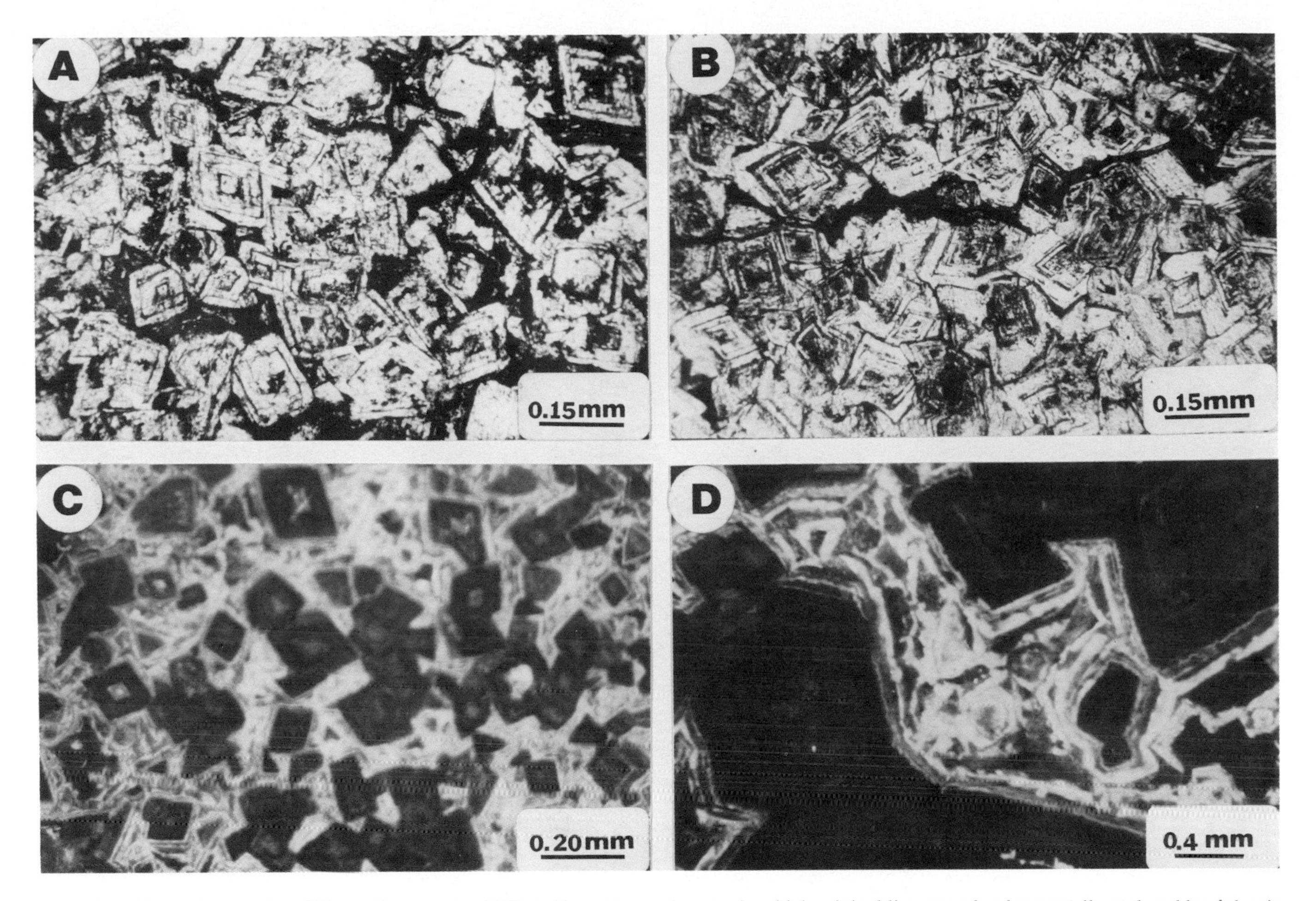

FIG. 3.—Photomicrographs of Upper Cenomanian El Heiz Formation carbonates in which original limestone has been totally replaced by dolomite. (A) Medium-grained, idiotopic, distinctly zoned, ferroan-dolomitc crystals (plane light). (B) Late stylolite transecting both dolomite rhomb cores and overgrowths. Note the tightly interlocking nature of this section, which consists of two different dolomite textures as euhedral rhombs and local inter-rhombic cement (crossed nicols). (C) Cathodoluminescent image showing alternating dull (dark) and bright (light) luminescent zones surrounding nonluminescent iron-rich dolomite crystal cores. (D) Cathodoluminescent image under high magnification shows dully luminescent cores rimmed by brightly luminescing overgrowths.

matrix. In thin section, this is marked by differences in crystallinity and concentration of iron oxides. Both clasts and matrix are pervasively dolomitized, with no preservation of precursor minerals or grains. The brecciated macrofabric of this unit is interpreted to reflect collapse structures related to the dissolution of interbedded evaporite lithologies. This is supported by locally silicified nodules that appear to have formerly been anhydrite.

Lighter clasts are irregularly distributed and exhibit equigranular hypidotopic to xenotopic textures (Fig. 7A). Rare iron oxide and organic grains are disseminated within clast rhombs, which are unzoned. In contrast, darker matrix dolomite exhibits a prophyrotopic fabric of zoned idiotopic rhombs embedded in iron oxide-stained microcrystalline dolomite crystals.

It is unlikely that such fine-scale variation from idiotopic dark matrix texture to hypidotopic and xenotopic light clast texture is the result of temperature differences during crystal growth. Whereas such a mechanism for producing dolomite textural variations has been suggested by Gregg and Sibley (1984), such an explanation seems untenable for the El Hefhuf dolomite because this sequence has never been deeply buried. More probably, such textural differences reflect different stages of dolomitization or result from differences in primary porosities of precursor sediments, rather than dolomite-dolomite neomorphism under elevated temperature conditions (e.g., Shukla, 1986).

The dolomite crystals in the upper member of the El Hefhuf Formation are sucrosic idiotopic to hypidotopic in texture (Fig. 7B) and contain no fossils. As in El Heiz Formation dolomite, inclusions are usually concentrated toward rhomb cores (Fig. 7C), giving sections a characteristic cloudy-center, clear-rimmed appearance (Fig. 7D). Sibley (1980) suggested that such cloudy centers arise because initial dolomitizing fluids were saturated with respect to calcite and, therefore, many crystals of calcite were included in early rhomb centers. Later, as diagenetic fluids became undersaturated with respect to calcite, dolomite zones were free of calcite inclusions. While such an explanation may well apply to dolomites of the El Heiz and El Hefhuf formatons,

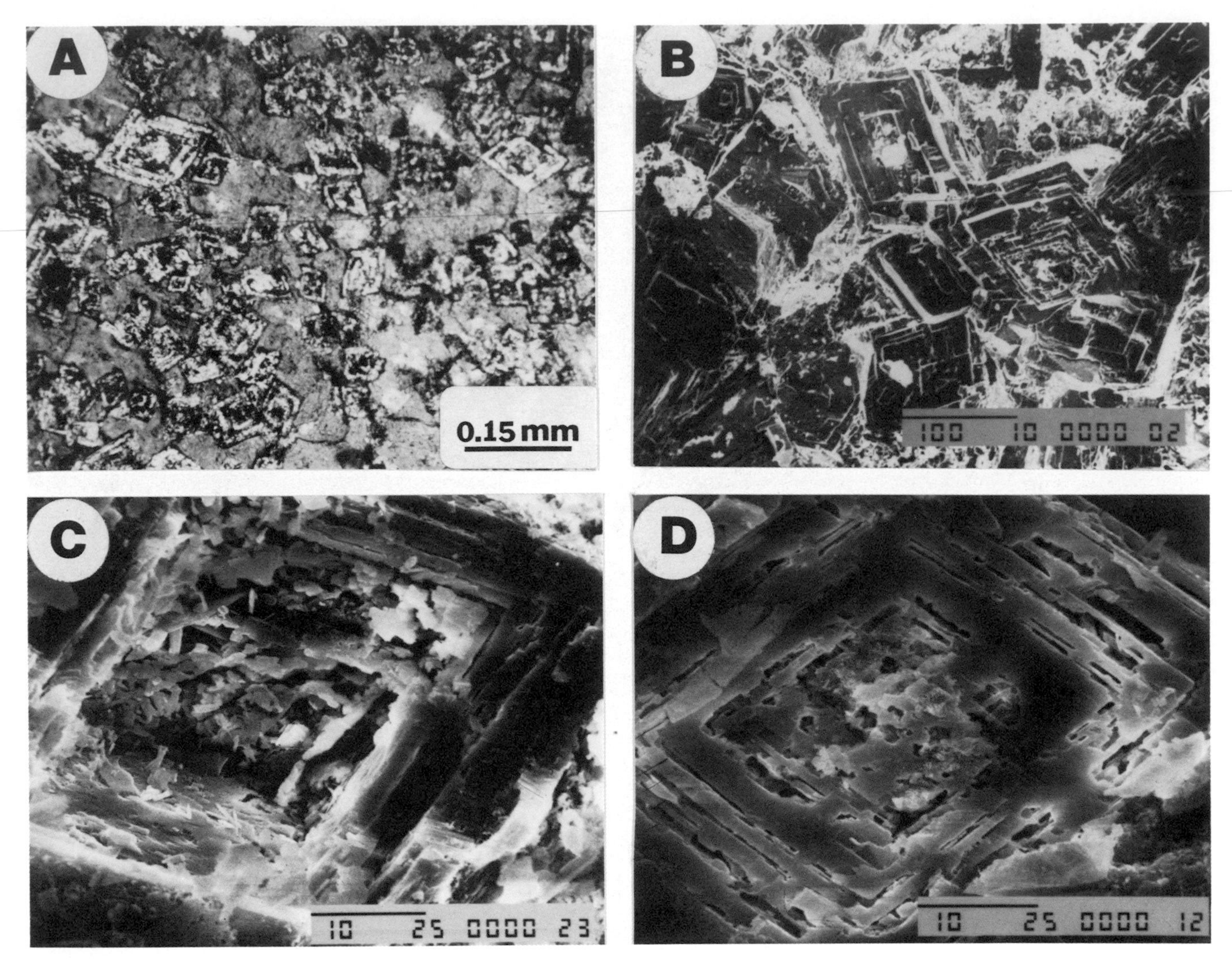

FIG. 4.—Photomicrographs of calcitized dolomite in the El Heiz Formation (A) Dolomite (light) and calcitized dolomite (dark) showing selective replacement of rhomb zones and rhomb cores. Such dedolomitization is typical of samples from the uppermost El Heiz Formation immediately below the unconformity (plane light). (B) SEM image of partially calcitized dolomite rhombs. Light areas are the iron-rich zones and cors of ferroan-dolomite crystals. (C) Enlargement of a corroded and partially dissolved dolomite crystal illustrating selective dissolution of the rhomb core. (D) Similar enlargement showing selective dissolution of zones within an individual rhomb.

most dolomite rhomb cores are stained by iron oxide, and it is difficult to determine if the inclusions are mainly calcite, or if they include a significant proportion of opaque minerals and/or organic accumulations.

Generally, dolomite in the El Hefhuf Formation exhibits two generations of crystal growth similar to those observed in the El Heiz Formation. These are cloudy-centered rhombs with clear euhedral rims and dolomite cement overgrowths (Fig. 8). Under cathodoluminescence, dolomite cores are largely nonluminescent, but unlike the El Heiz Formation, this does not reflect a high Fe^{2+} content. In contrast, clear rims and overgrowth cements are brightly luminescent and are similar to the late overgrowth rims of the El Heiz Formation.

Replacive diagenetic phases are developed in El Hefhuf Formation dolomites. Sparry calcite in the El Hefhuf Formation occurs both as replacive calcite precipitated during dedolomitization and as void-filling spar and is similar to calcites in the El Heiz Formation (Fig. 8C, D). In the upper part of the El Hefhuf Formation, calcitization has selectively affected zones and overgrowth rims during partial, and sometimes complete, dedolomitization (Fig. 9B). Quartz has locally replaced dolomite rhomb margins along pores now filled with quartz cement; contacts between replacive and/or void-filling quartz and dolomite are commonly corroded and irregular (Fig. 9A). Under cathodoluminescence, dolomites exhibit extensive reaction-replacement textures, whereby rhomb cores are partially to pervasively replaced by dolomite with a bright luminescence similar to that of the overlying dolomite cement. These outer-dolomite cement rims, which are clear and remain unaltered, are marked by a bright red luminescence similar to dolomite cements

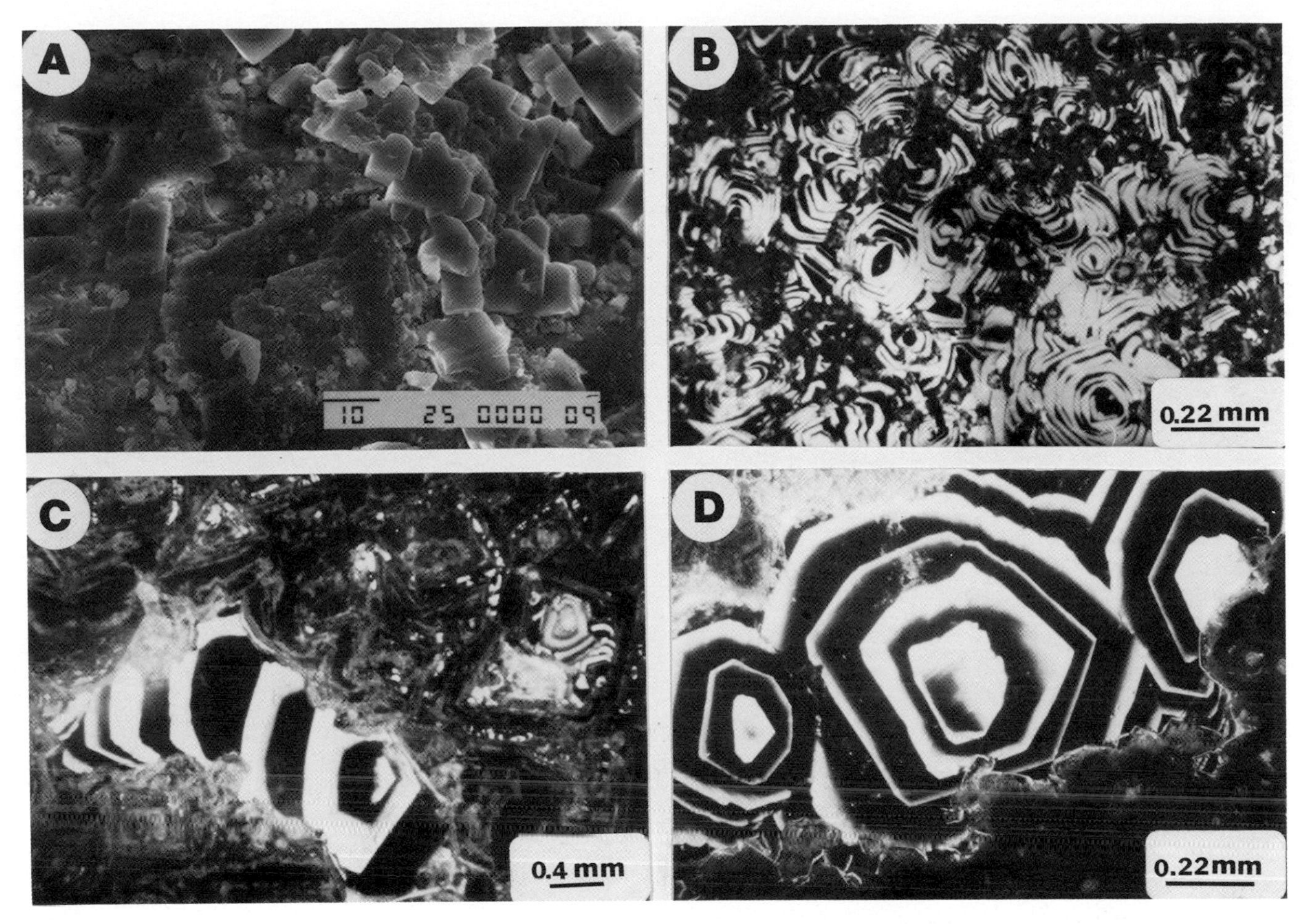

FIG. 5.—Early sparry calcite precipitated during dedolomitization of the El Heiz Formation. (A) SEM image showing the contact between calcite and dolomite. (B) Photomicrograph (cathodoluminescent light) showing strongly zoned early calcite and dully luminescent unreplaced dolomite crystals. (C) Closeup of replacive calcite spar. (D) Photomicrograph (cathodoluminescent light) showing that the calcite cement in the inter-rhomb pores and the calcite inside dolomite rhombs are in optical continuity with dolomite substrates.

observed throughout the Upper Cretaceous sequence in all formations (Fig. 8D).

GEOCHEMISTRY

X-ray diffraction analysis and elemental ratios from electron probe analysis demonstrate that El Heiz Formation dolomites are nonstoichiometric, with compositional zones ranging from 8.6 to 0.8 mole percent $FeCO_3$ (Table 1). In contrast, El Hefhuf Formation dolomites are nearly stoichiometric, and generally non-ferroan. These microprobe data also show that the dedolomite-related calcite and the late sparry calcite are chemically distinct in both formations. The early calcite is enriched in Mg (3.6 to 4.2 mole percent $MgCO_3$) relative to the late vein-filling calcite, which is depleted in Mg (0.9 to 1.3 mole percent $MgCO_3$).

Iron and Manganese

El Heiz Formation dolomites are ferroan with relatively high Mn^{2+} contents, whereas El Hefhuf Formation dolomites are non-ferroan and have relatively low Mn^{2+} contents. Of importance is that the dolomite overgrowth cements in the El Heiz Formation are similar in Fe^{2+} content to non-ferroan dolomites of the El Hefhuf Formation. Moreover, El Hefhuf dolomite samples are associated with authigenic quartz exhibiting high Mn^{2+} contents, averaging 1,150 ppm (Fig. 10A).

X-ray diffraction (XRD) analyses demonstrate that all dolomite is ordered and displays well-developed (211) and (111) superstructure reflections (Graf and Goldsmith; 1956; Gaines, 1977; Shimmield and Price, 1984); however, the (211) peak is shifted to mean values of 2.909 and 2.901 angstroms in El Heiz Formation and El Hefhuf Formation dolomites, respectively (compared with 2.886 angstroms for stoichiometric dolomite), reflecting replacement of Mg^{2+} by Fe^{2+} and excess Ca^{2+} in the structure (Lippmann, 1973; Runnells, 1974; Al-Hashimi and Hemingway, 1973).

Since Fe^{2+} and a portion of the Mn^{2+} ions occupy Mg^{2+} lattice sites in dolomite crystals, whereas remaining Mn^{2+} ions occupy Ca^{2+} lattice sites (Schindler and Ghose, 1970; Wildeman, 1970; Lumsden and Lloyd, 1984), elevated $(Fe^{2+} + Mn^{2+}):(Mg^{2+} + Ca^{2+})$ and $(Fe^{2+} + Mn^{2+}):(Mn^{2+} + Ca^{2+})$

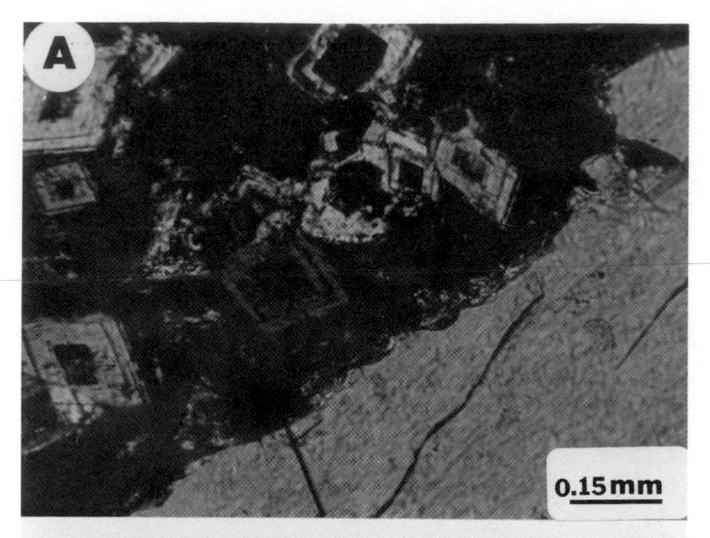

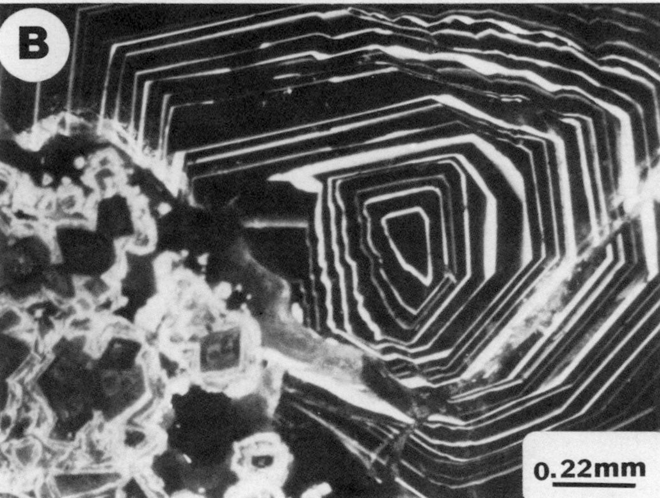

FIG. 6.—Late sparry void-filling calcite, which formed after dedolomitization. (A) Photomicrograph (crossed nicols) showing contact between calcite (lower right) and matrix dolomite (upper left), which consists of cloudy-cored rhombs. (B) Same view under cathodoluminescence showing extremely thin, yellow to nonluminescent bands of late replacive calcite. Dolomite crystals exhibit the same CL zones as those in Figure 3C.

ratios in dolomitizing solutions are necessary to form the ferroan dolomites with high Mn^{2+} contents, such as those in the El Heiz Formation. Such increases cannot be accompanied by significant decreases in Mg^{2+} activities, because such a change could inhibit dolomite formation (Katz, 1971). Thus, these cation ratio increases must be related to increasing Fe^{2+} and Mn^{2+} ion activities, suggesting that dolomitization occurred in the presence of reducing fluids (Aller, 1980; Klinkhammer, 1980; Wong and Oldershaw, 1981). Because Fe^{2+} substitutes for Mg^{2+} in dolomite crystals, a relation between $FeCO_3$ and $MgCO_3$ contents can be used to discriminate between El Hefhuf and El Heiz Formation dolomites (Fig. 10B). Finally, Morrow (1982) suggests that, in many dolomites, Fe^{2+} and Mn^{2+} increase in concentration toward the rims of dolomite crystals and exhibit a chemical zonation opposite to that of Sr^{2+} and Na^+. In contrast, our data indicate that cores of dolomite crystals are generally more iron-rich than outer rims. This reflects, however, a marked difference between the two phases, rather than a progressive trend within individual crystals.

Sodium and Strontium

Concentrations of Na^+ in El Heiz Formation ferroan dolomites are high (650±30 ppm) relative to El Hefhuf Formation dolomites (425±30 ppm). Concentrations of Sr^{2+} are similar, however, with averages of 120±25 ppm and 103±25 ppm, respectively. Dolomite from both formations shows good correlation between Na^+ and Sr^{2+} contents (Fig. 11A). Concentrations of Na^+ and Sr^{2+} in El Hefhuf Formation dolomites containing quartz cement are relatively low, averaging 265 and 60 ppm. Relations between Na^+ and Sr^{2+} values of these samples are opposite to those exhibited by Fe^{2+} and Mn^{2+} (Fig. 11B).

Sodium and strontium are among the most useful trace elements in evaluating the salinity of dolomitizing fluids owing to the nature of the partitioning between dolomite crystals and the waters from which they precipitate. Because of uncertainties, however, about the relative importance of bulk or surface equilibrium processes (Hardie, 1987) vs. kinetic processes (Lorens, 1981) in determining partitioning behavior and the possibility of later resetting of primary dolomite chemistries, interpretation of Na^+ and Sr^{2+} concentrations is less than straightforward. Like most ancient dolomites (Weber, 1964; Fritz and Katz, 1972; Land and others, 1975; Mattes and Mountjoy, 1980; Baum and others, 1985), El Heiz Formation and El Hefhuf Formation dolomites are depleted in Sr^{2+} relative to Holocene dolomite, which contains 500 to 700 ppm Sr^{2+} (Land, 1973; Land and Hoops, 1973). Whereas Na^+ contents of El Hefhuf Formation dolomites are in agreement with differences in Sr^{2+} content between ancient and Holocene dolomite, Na^+ values for El Heiz Formation ferroan dolomites are lower than those of Holocene counterparts, but are higher than those for most ancient dolomite. Because the difference in Na^+ concentration between the two studied dolomites is not that great, because Na^+ contents may reflect the presence of high-salinity fluid inclusions, and because both sequences yield nearly identical Sr^{2+} contents, dolomite in both formations may have formed from compositionally similar fluids that contained slightly different amounts of Fe^{2+} and Mn^{2+}.

Carbon and Oxygen Isotopes

Values of $\delta^{18}O$ for El Heiz Formation ferroan dolomites are lower (−1.4 to −3.3‰ PDB) than those of El Hefhuf Formation dolomites (−0.8 to 2.0‰ PDB), whereas $\delta^{13}C$ values of El Hefhuf Formation dolomites show a wide range (+0.0 to +5.2‰ PDB) relative to those of El Heiz Formation ferroan dolomites (+0.8 to +2.65‰ PDB; Figs. 12, 13).

The $\delta^{18}O$ values of dedolomite-related calcite (with high Mg^{2+} content) of the El Heiz Formation are similar in com-

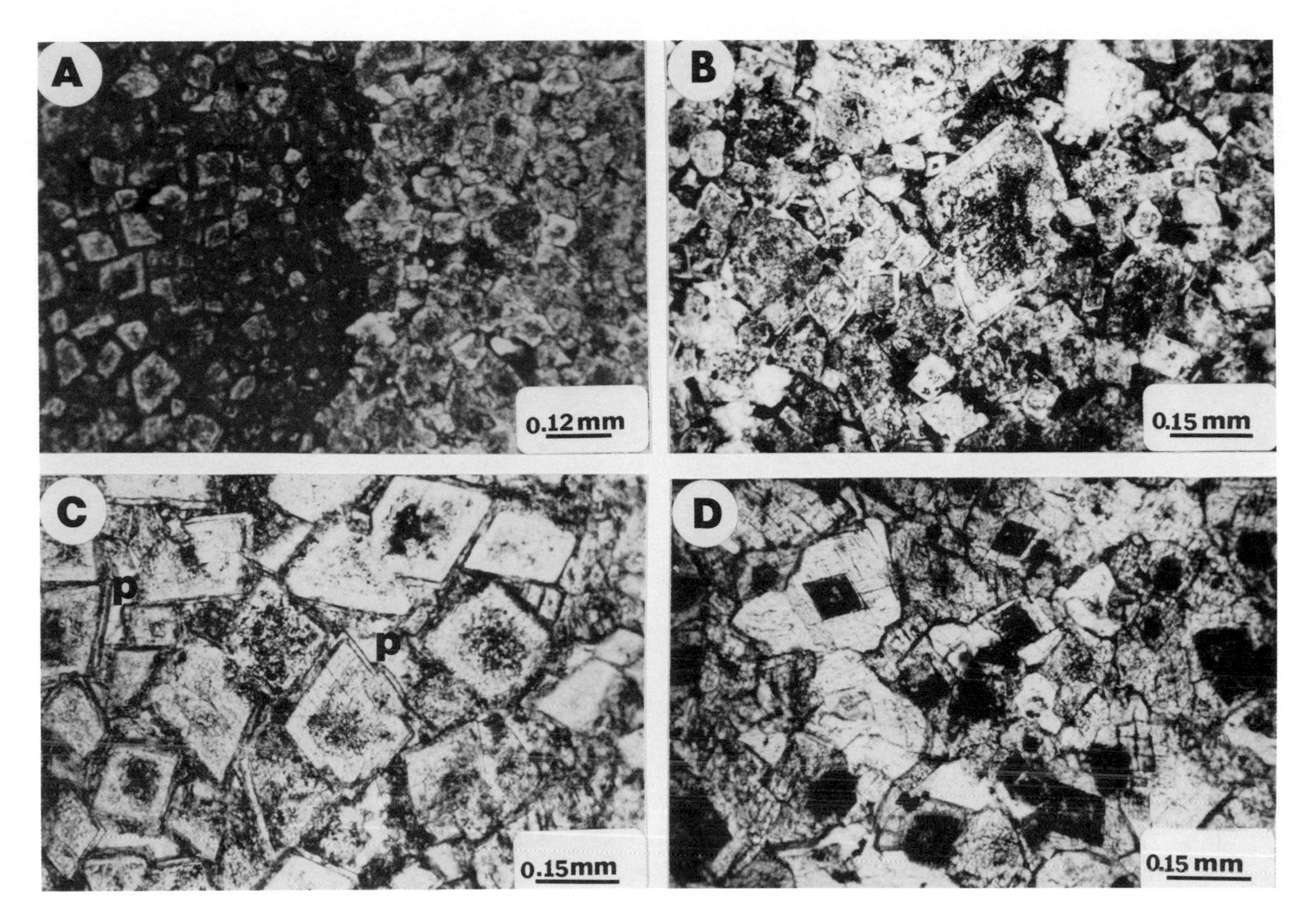

FIG. 7.—Photomicrographs of Campanian El Hefhuf Formation carbonates in which limestone has been totally replaced by dolomite (plane polarized light). (A) Brecciated dolomite composed of light clasts irregularly distributed in a darker, more iron-rich matrix. Lighter clasts consist of a mosaic of hypidotopic to xenotopic dolomite crystals. Matrix dolomite consists of idiotopic dolomite rhombs in a microcrystalline dolomite matrix stained by iron oxide. (B) Coarse- to medium-grained, zoned dolomite rhombs with cloudy centers. (C) Distinctly zoned dolomite rhombs with cloudy centers. Note pore spaces (p) between the dolomite crystals. (D) Tightly interlocking dolomite crystals consisting of euhedral cloudy cores and clear anhedral overgrowths. The rhombic shape of early crystals (dark) is clearly outlined by internal zoning.

position (−8.3 to −9.01‰ PDB) to those of El Hefhuf Formation replacive calcites (−6.5 to −7.6‰ PDB); these exhibit comparable $\delta^{13}C$ values in both formations. The $\delta^{18}O$ and $\delta^{13}C$ values of late-stage vein-filling calcites (with low Mg^{2+} contens) are identical for both the El Heiz and El Hefhuf Formations, ranging from −10.0 to −11.2‰ and from −3.8 to −7.8‰, respectively. Moreover, these are identical to the calcite veining within overlying Maastrichtian chalks, which are related to the pre-Middle Eocene erosional unconformity. Most important, the $\delta^{18}O$ and $\delta^{13}C$ values of dedolomite replacive calcites are intermediate between those of dolomite and the late vein-filling calcite.

Dolomites from both formations show a strong correlation between $MgCO_3$ content and $\delta^{18}O$ values (Fig. 14). Ferroan dolomites of the El Heiz Formation (average 41.3 mole percent $MgCO_3$) are isotopically lighter in $\delta^{18}O$ than the near-stoichiometric dolomites of the El Hefhuf Formation. This is particularly well demonstrated in the lower member of the El Hefhuf, where dolomite associated with silica replacement and cementation is notably depleted in Fe^{2+}, Mn^{2+}, Sr^{2+}, and Na^{+}, and possesses enriched $\delta^{18}O$ values.

The Upper Cretaceous $\delta^{18}O$ values are heavier than those of most ancient dolomites (Veizer and Hoefs, 1976; Badiozamani, 1973; Fritz and Katz, 1972) and are comparable to Middle Eocene dolomites reported by Land and others (1975); however, they are lighter than recent dolomites associated with evaporative conditions (Kinsman and Patterson, 1973; Kier, 1973). Regardless of the common association of older dolomite with supratidal sequences, most have negative $\delta^{18}O$ signatures (Sibley, 1980) that are comparable to, or lighter than, $\delta^{18}O$ values of El Heiz and El Hefhuf formations. This suggests that the isotope ratios of many ancient dolomites may have, at some stage, equilibrated in the presence of isotopically light water or elevated temperatures.

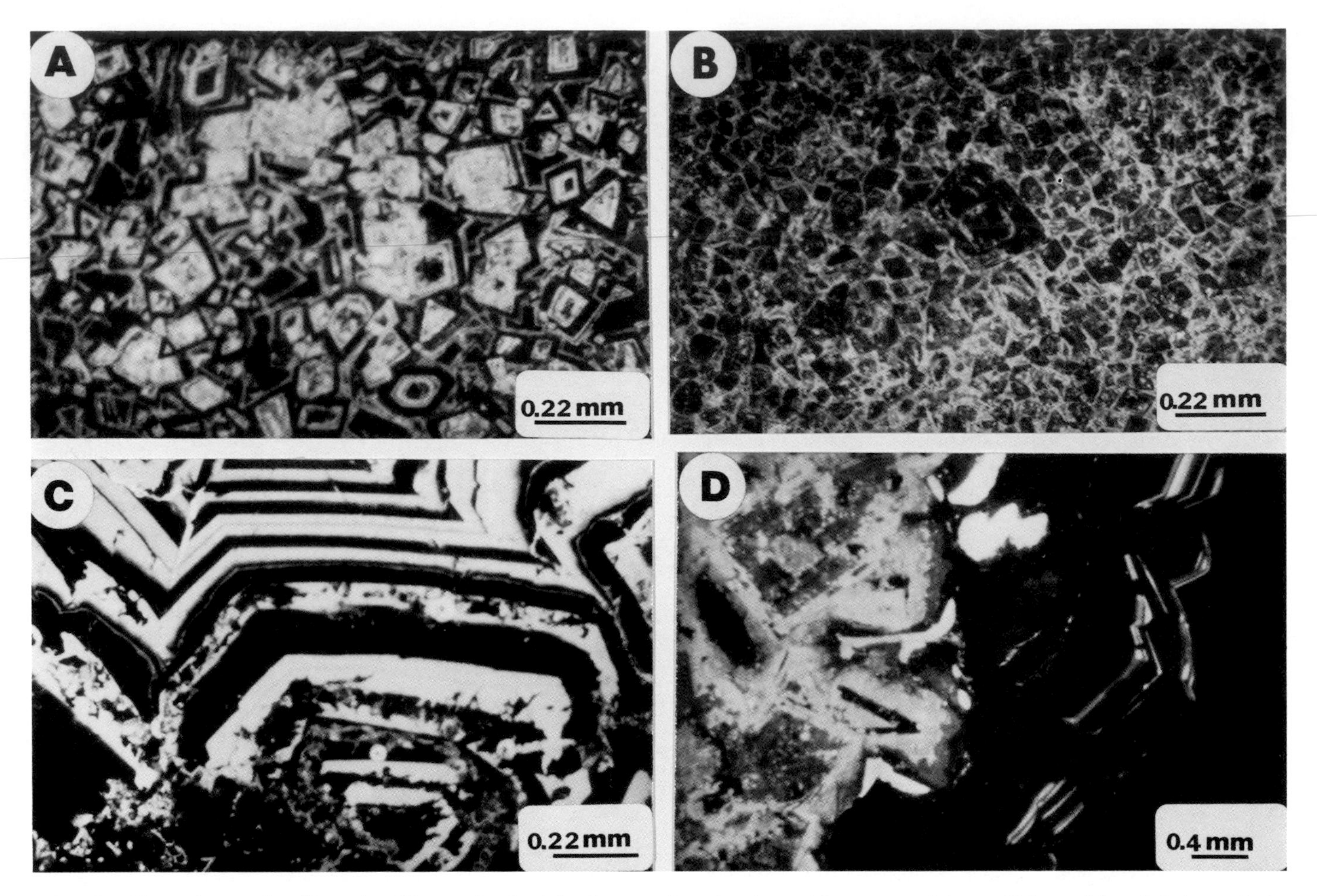

FIG. 8.—Photomicrographs, illustrating CL zonation in dolomite crystals and early- and late-stage calcites, which precipitated during and after dedolomitizatoin of El Hefhuf Formation. (A) Zoned dolomite crystals with clear centers. (B) Zoned dolomite crystals with cloudy centers. (C) Early calcite related with dedolomitization. Note the sharp boundaries between luminescent zones, indicating abrupt changes in porewater chemistry during precipitaton. (D) Late-stage vein-filling calcite with zoning similar to calcites of the El Heiz Formation.

DISCUSSION

Diagenetic processes by which carbonate sediments are modified, including those that have been proposed to explain dolomitization, are influenced both by depositional setting and burial history. Therefore, several settings of dolomitization must be considered: (1) synsedimentary hypersaline, (2) marine burial, and (3) mixed marine-meteoric environments. Petrographic relations among various carbonate phases in Upper Cretaceous dolomite units of El Heiz and El Hefhuf formations, in conjunction with geochemical data, will be combined to constrain diagenetic processes relative to the origin and timing of dolomite formation and dolomite calcitization. It will be shown that present isotopic and minor-element compositions of El Hefhuf and El Heiz dolomites and calcites suggest a mixed-water diagenetic setting for dolomitization. Reaction-replacement microfabrics of El Heiz dolomite indicates, however, that a precursor phase of dolomite had formed and subsequently has been chemically reset during dolomitization of the El Hefhuf. Various models follow.

Epigenetic Dolomitization

On the basis of burial history reconstructions, maximum burial temperatures could not have exceeded 40°C. Similarly, isotopic values for dolomites are too enriched to have formed under conditions of high-temperature subsurface fluids. Therefore, epigenetic dolomitization is unlikely and will not be considered.

Evaporative Hypersaline Dolomitization

Although primary evaporites are not present in either formation, brecciated dolomite textures in the lower part of the El Hefhuf Formation likely reflect the dissolution of evaporite beds. Local development of silicified nodules and abundant silica cement further support an association of evaporites, at least within this interval. The lack of such features within the majority of the sequence, however, cannot be easily examined. The majority of the El Heiz and El Hefhuf formations appears to represent open-marine facies prior to dolomitization. Moreover, the prevalence of

FIG. 9.—Photomicrographs of other diagenetic phases in the El Hefhuf Formation. (A) Partially silicified dolomite. Note irregular contacts between dolomite crystals and authigenic quartz (crossed nicols). (B) Early calcite spar precipitated during dedolomitization. Note that calcite has extensively replaced overgrowth dolomite and selectively replaced intra-rhomb zones, whereas most rhomb cores are still dolomite (plane light). These samples occur directly below the exposure surface.

ferroan dolomite within the El Heiz Formation is inconsistent with typical models for evaporite-related dolomitization.

Growing awareness of minor- and trace-element distributions within dolomite is shown by the numerous recent studies that incorporate such data into various dolomitization models. Both formations contain significantly lower Na^+ than would be expected from hypersaline dolomitizing fluids (Veizer, 1983). Similarly, Sr^{2+} contents are uniformly low in both the El Heiz and El Hefhuf. Oxygen-isotopic data for dolomite of both formations is roughly coincident with primary marine-calcite values estimated for the Upper Cretaceous (Scholle and Arthur, 1980). Expected values for primary hypersaline dolomite, however, should range from +2 to +5 per mil $\delta^{18}O$, which is significantly enriched over our observed values. On this basis, an evaporative origin for these dolomites seems untenable.

In contrast to an evaporative origin, Na^+ values for the El Hefhuf and El Heiz formations, 425 and 650 ppm, respectively, suggest involvement of marine waters during dolomitization. We cannot be certain, however, that this Na^+ is incorporated in crystal lattice sites, even though such enrichments should be matched by compatible concentrations of Sr^{2+}. The average concentration for both formations is approximately 100 ppm Sr^{2+}, considerably lower than the 300 to 500 ppm values anticipated for dolomites forming within a marine burial setting (Land, 1980; Murata and others, 1969; Veizer, 1983).

Marine Dolomitization

A feature that suggests dolomitization in a marine setting is the presence of pyrite and its relation to sulfate reduction. The coexistence of rare pyrite with dolomite suggests that some sulfate reduction occurred in association with dolomite formation, but not the levels proposed by Baker and Kastner (1981). The paucity of pyrite in these formations suggests that bacterial sulfate reduction did not play a major role in the origin of El Heiz and El Hefhuf Formation dolomites. The abundance of iron, especially within the El Heiz Formation, however, requires that diagenetic fluids were reducing and that a source of this iron was available at the time of dolomitization.

El Heiz Formation ferroan-dolomite crystals are composed of iron-rich centers surrounded by iron-rich to iron-poor rhombohedral zones. Similarly, the El Hefhuf Formation dolomite crystals have cloudy cores that are occasionally stained by iron oxides. The cloudy cores may reflect an early diagenetic origin for these dolomites (Sibley, 1980; Churnet and others, 1982). Clear rhombohedral zones form the outer part of dolomite crystals and are generally more stoichiometric than cloudy cores. The source of iron present within the El Heiz dolomites may have been derived from associated terrigenous clastics, which occur underlying and intercalated within the lower part of this formation (Kushnir and Kastner, 1984). Oxidation of marine organics could provide the mechanism for remobilizing Fe^{2+} from terrigenous phases. Moreover, because pyrite is virtually absent within this formation (less than 2 percent), it is apparent that either conditions were not sufficient for ongoing reduction of marine-derived sulfate, or that pore-water sulfate concentrations were depleted. If sulfate reduction were active, dissolved Fe^2 would have been sequestered as pyrite and would not have been available for incorporation into dolomites.

TABLE 1.—AVERAGE CHEMICAL COMPOSITIONS OF EL HEIZ AND EL HEFHUF DOLOMITES

Sample	$CaCO_3$	$MgCO_3$ (mole %)	$FeCO_3$	Sr	Na ppm	Mn	$\delta^{18}O$	$\delta^{13}C$
El Heiz	50.1–57.9	41.3	8.6–0.8	120	650	1900	−2.2	+1.7
El Hefhuf	51.0–52.5	47.1	1.9–0.4	103	425	720	−1.2	+2.9

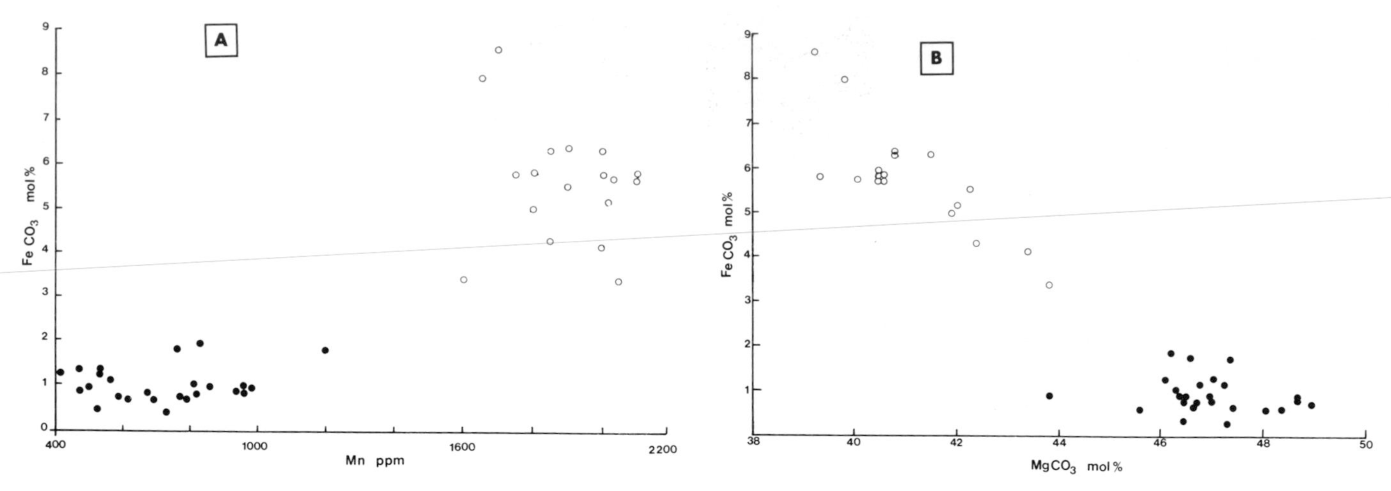

FIG. 10.—(A) Relation between Mn^{2+} content and mole percent $FeCO_3$ in El Heiz Formation ferroan dolomites (open circles) and El Hefhuf Formation dolomites (solid circles). (B) Relation between mole percent $MgCO_3$ and mole percent $FeCO_3$ in El Heiz (open circles) and El Hefhuf Formation (closed circles) dolomites.

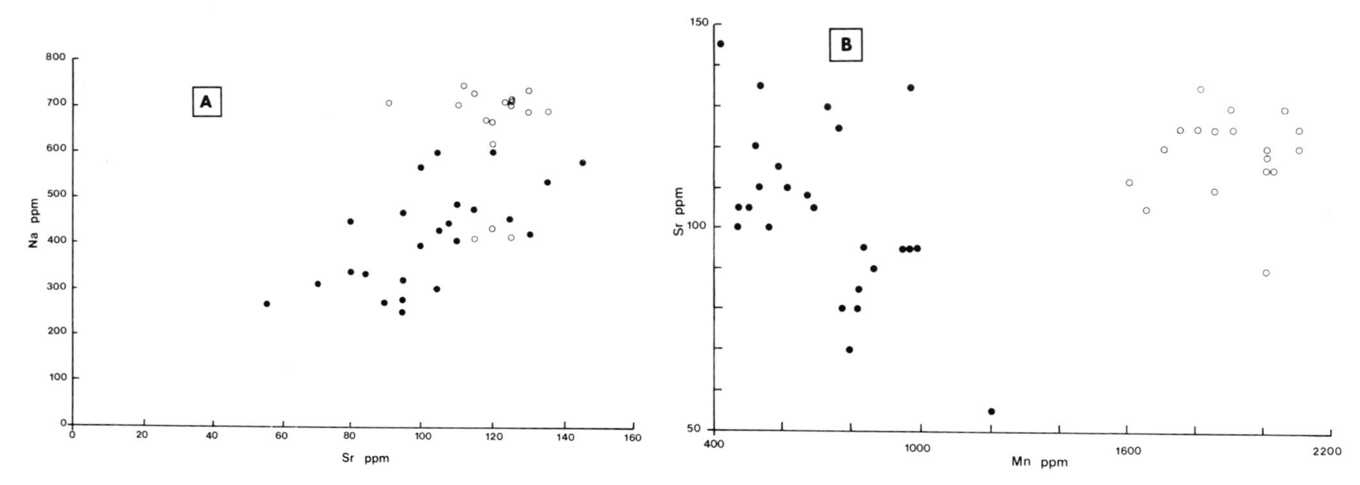

FIG. 11.—(A) Relation between Sr^{2+} and Na^{+} contents of El Heiz (open circles) and El Hefhuf Formation (closed circles) dolomites. (B) Relation between Mn^{2+} and Sr^{2+} contents of El Heiz (open circles) and El Hefhuf Formation (closed circles) dolomites.

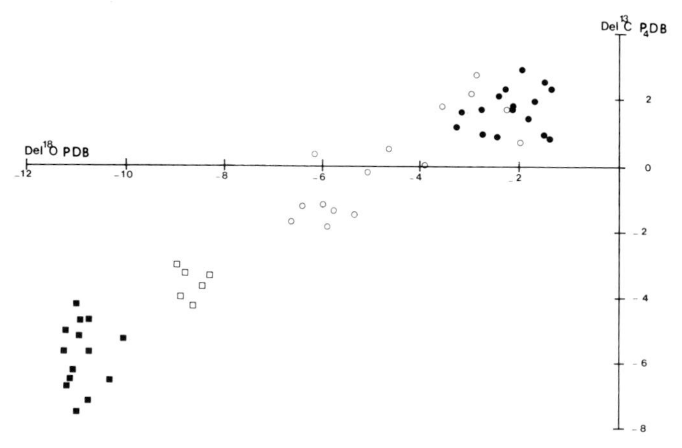

FIG. 12.—Isotopic composition of El Heiz Formation ferroan dolomites (solid circles), partially calcitized dolomites that contain a mixture of calcite and dolomite (open circles), early-stage calcite (open squares), and late-stage calcite spars (solid squares).

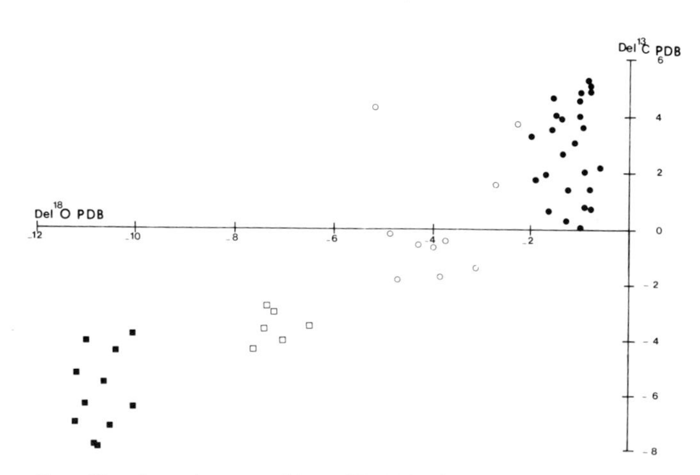

FIG. 13.—Isotopic composition of El Hefhuf Formation dolomites (solid circles), partially calcitized dolomites (open circles), early-stage calcite (open squares), and late-stage calcite spars (solid squares).

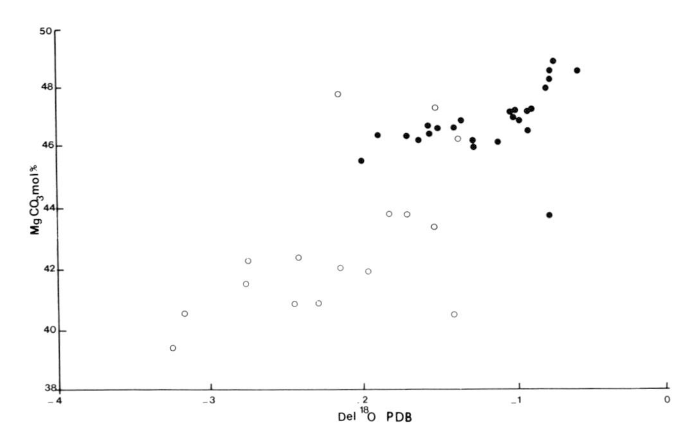

FIG. 14.—Relations between mole percent $MgCO_3$ and $\delta^{18}O$ values of El Heiz (open circles) and El Hefhuf (closed circles) dolomites.

Stable isotope evidence.—

In order to evaluate the consistency of stable isotope data with a marine dolomitization setting, it is necessary to estimate the composition of marine calcite that formed during Upper Cretaceous time. Based on estimates of Scholle and Arthur (1980), Cretaceous marine carbonates were depleted relative to present pelagic oozes with carbonate values of −2.0‰ $\delta^{18}O$ and about +2.5‰ $\delta^{13}C$. If the dolomites of this study reflect syndepositional marine formation and if estimates of dolomite-calcite mineral fractionation are correct, their $\delta^{18}O$ values should be about +2 to +5‰ enriched relative to coeval marine calcite. Given a marine calcite composition of −2‰, marine dolomite values would range from 0 to 3‰ $\delta^{18}O$. El Heiz and El Hefhuf Formation dolomites have $\delta^{18}O$ values that lie within the range of −3 to −1 per mil; however these values are significantly depleted relative to the expected marine-dolomite values for the Upper Cretaceous. Oxygen values for these dolomites are also depleted relative to Recent dolomites (Behrens and Land, 1972; Kier, 1973; Kinsman and Patterson, 1973; McKenzie, 1981) and suggest that present isotopic compositions could not reflect equilibrium with either hypersaline or marine waters. On the other hand, $\delta^{18}O$ values are very close to those for the Middle Eocene dolomites around Bahariya Oasis examined by Land and others (1975). They suggested that the Eocene dolomites formed by early replacement in a meteoric-marine diagenetic setting with further dolomite maturation in the presence of meteoric waters.

Trace-element evidence.—

It is important to note that minor- and trace-element chemistries and the oxygen-isotopic values are not independently variable. The oxygen-isotopic values of El Heiz and El Hefhuf Formation dolomites correlate with dolomite $MgCO_3$ contents, which in turn correlate with dolomite Sr^{2+} content. Moreover, nearly stoichiometric El Hefhuf Formation dolomites are enriched in ^{18}O and depleted in Sr^{2+} and Na^{+} relative to El Heiz Formation ferroan dolomites. Similar relations have been reported for other dolomitized sequences (Land, 1980; Swart and Dawans, 1984; Baker and Burns, 1985). This suggests that oxygen-isotopic trends in these components may not reflect primary values; rather, they may be controlled by processes that have defined the present stoichiometry of the dolomite phases. Therefore, interpretation of diagenetic setting based solely on the absolute values of oxygen enrichments may not be straightforward, at least for sequences in which differences in stoichiometry are present.

Mixed-Water Dolomitization

A significant aspect of the geologic history of this region was the development of two regional unconformities associated with subaerial exposure. The likelihood of meteoric-water infiltration during these events implicates mixed-water, meteoric-marine alteration as a candidate for dolomitization.

Such dolomitization may have developed during either or both documented events of subaerial exposure. If two events of mixed-water dolomitizaton developed, these would be related to the unconformity above the El Heiz and the other to the Maastrichtian to pre-Eocene exposure. Alternatively, mixed-water dolomitization related only to the latest unconformity may have pervasively altered carbonate strata of both horizons. Because this unconformity variably truncates the anticlinal structure, both formations were in close proximity to the exposure surface at pre-Middle Eocene time. Even though the geologic setting provides for meteoric-water interaction, specific evidence contained within the petrologic and geochemical record is necessary to demonstrate its potential for the diagenesis of these Upper Cretaceous carbonates.

The late vein-filling calcite spar, a distinctive CL-zoned phase, occurs throughout the stratigraphic sequence. Examination of this cement within the Maastrichtian chalks, which regionally overlie the El Hefhuf Formation (Holail and Lohmann, 1986), has demonstrated that this calcite is related to infiltration of meteoric waters during the regional erosional exposure coincident with the Maastrichtian-pre-Middle Eocene unconformity. The isotopic composition of this phase defines a distinctive field, characterized by an invariant $\delta^{18}O$ of −11 per mil and highly variable carbon, which does not vary among formations. Whereas the presence of meteoric waters in this diagenetic scenario need not implicate a mixed-water setting for dolomitization, the presence of the late-calcite cement throughout the stratigraphic sequence indicates that meteoric waters were involved with the late diagenesis of both the El Hefhuf and El Heiz dolomites. Numerous lines of evidence support initial dolomitization, as well as replacement of dolomite by dolomite in response to this diagenetic event.

Dolomite recrystallization.—

An additional component, the dolomite overgrowth cement that composes about 10 percent of the dolomites by volume, also occurs pervasively throughout the Upper Cretaceous sequence. This Fe^{2+}-free cement is calcian relative to dolomite core substrates, is brightly luminescent, and is commonly stained by iron oxides. Although precipitating pore waters must have been adequately reducing to allow for Mn^{2+} mobility, oxidation of dissolved iron is evidenced by associated iron oxides. Thus, this suggests that the dia-

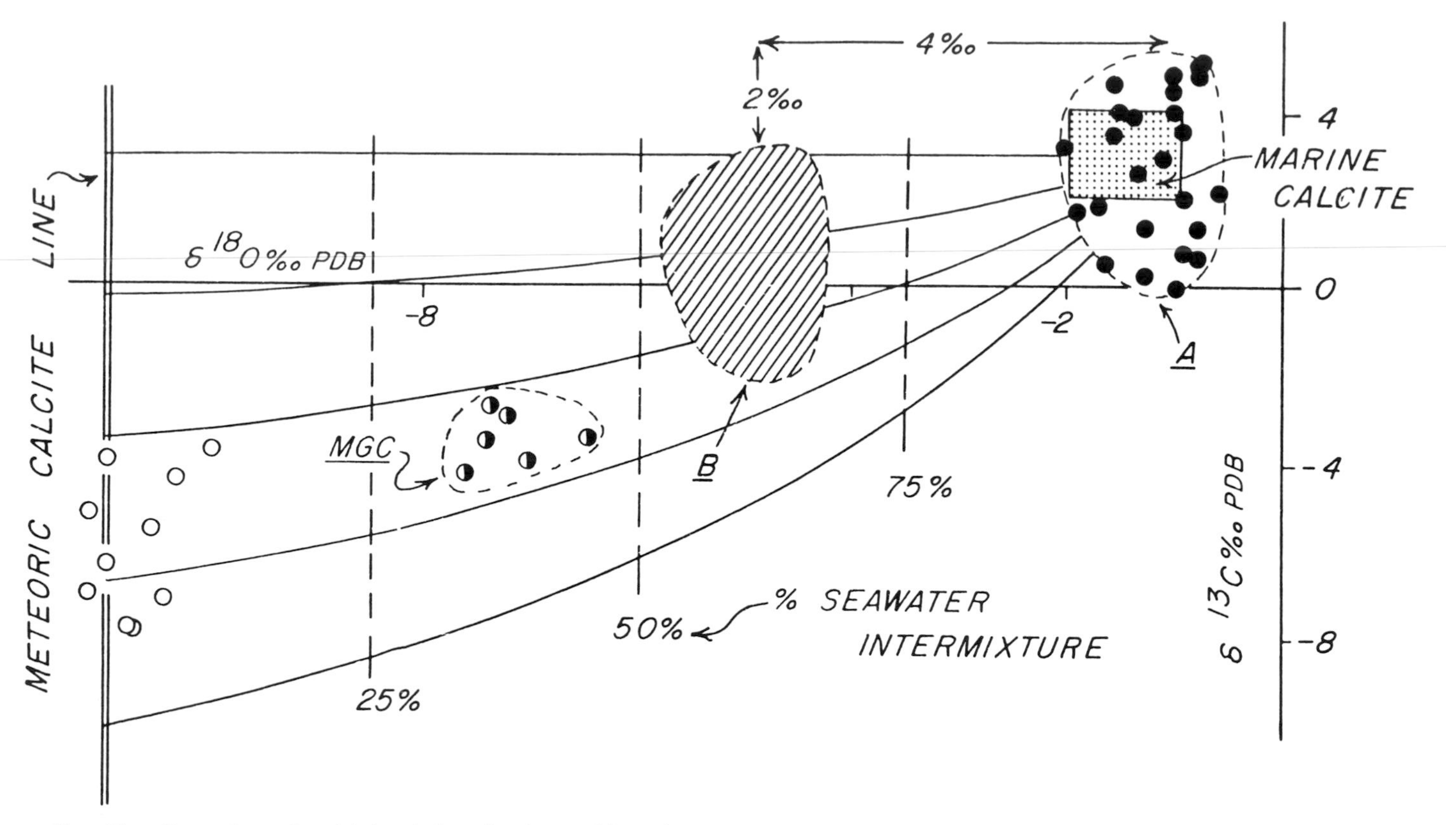

FIG. 15.—Comparison of modeled variation of carbon and isotopic composition of a mixed-water dolomitization setting with observed data of El Hefhuf Formation. Patterns resulting from fluid mixing are reflected by hyperbolic-mixing lines between marine- and meteoric-calcite end members. With increased intermixture of marine water, calcite may become undersaturated. The composition of dolomite precipitating within this region (field B) will, however, be displaced toward more enriched $\delta^{18}O$ and $\delta^{13}C$ values due to the larger dolomite-water mineral fractionation. Thus, measured values of dolomite forming at a seawater intermixture of field B will lie in range of 2 to 5 per mil enriched in $\delta^{18}O$ and 1 to 2 per mil enriched in $\delta^{13}C$ (field A). Calcite formed from mixed waters [MGC] possesses elevated Mg^{2+} contents and intermediate $\delta^{13}C$ and $\delta^{18}O$ values.

genetic fluids were at a transitional state between reducing and progressively more oxidizing conditions. Even though isotopic measurements of separates of this phase were not performed, the minor variation in $\delta^{18}O$ of the El Hefhuf dolomites suggests that its composition is similar to that of the dolomite cores.

It is important to note that this cement phase occurs uniformly throughout the sequence and is developed as the final phase of dolomite growth in both the El Heiz and El Hefhuf formations. The occurrence of this phase in both formations suggests precipitation during a single event and from an individual evolving pore water. The event that precipitated overgrowth cements may coincide with the pervasive dolomitization of the El Hefhuf Formation. In contrast, because of the distinctly higher Fe^{2+} contents of dolomite crystal cores, El Heiz dolomites may have developed prior to this late overgrowth. The similarity of isotopic compositions between these formations, however, suggests that this earlier formed dolomite of the El Heiz Formation subsequently underwent pervasive isotopic resetting in response to the fluid that dolomitized the El Hefhuf Formation. Whereas the geochemical constraints for mixed-water dolomitization of limestone have been amply presented, the potential for interaction of mixed waters with pre-existing dolomite, resulting in an effective dolomite to dolomite replacement, has received little notice (Land and others, 1975; Land, 1980; Frank and Lohmann, 1987).

Clearly, geochemical and petrographic evidence suggest alteration in response to a common fluid a synchronous event. Based on the geologic history of this region, this event was likely related to the regional exposure, which developed the erosional unconformity during post-Maastrichtian to pre-Eocene times. The variable truncation of the anticlinal structure has provided a setting in which each of the formations examined in this study was in close stratigraphic proximity to this erosional surface when examined on a regional scale. In this context, the proposed interaction with meteoric waters is not only possible, but also likely. The nature of this interaction, whether passive or active in modifying the sediment chemistry, can be evaluated by examining the consistency of observed geochemical patterns with trends predicted of a meteoric-marine, mixed water setting (Allan and Matthews, 1982; Lohmann, in prep.).

Range of mixing.—

Interpretation of data for the El Hefhuf Formation as reflecting a mixed-water origin is problematic within the context of existing models. It has been proposed that paired variation of $\delta^{18}O$ and $\delta^{13}C$ should define a straight-line mixture between marine and non-marine end-member fluids.

This, however, presumes that the total concentrations of carbon and oxygen are equivalent for marine and meteoric waters. In the case of oxygen, this is a reasonable assumption; in contrast, meteoric waters charged with soil gas CO_2 contain substantially more total dissolved carbon. Mixing fluids with such disparate carbon concentrations (Fig. 15) results in a hyperbolic mixing relationship, which deviates significantly from the simple straight-line model (Lohmann, in prep.). Temporal and spatial variation in the $\delta^{13}C$ composition of the meteoric end member in addition modifies the chemical trends. During the interaction of meteoric water with marine carbonate country rock, meteoric waters evolve in composition through rock-water interaction. The isotopic composition of carbon may range from highly depleted values coincident with soil gas CO_2 to values reflecting a dominance by carbon contained within the dissolving rock phases. With only slight variation in the degree of rock-water interaction, $\delta^{13}C$ variation can span the full range from marine (rock) values to meteoric-water values in equilibrium with surface-derived soil gas CO_2. Such rock-water reaction accompanying mixing, therefore, will induce the principal variation of isotopic compositions to be parallel to the *meteoric-calcite line,* the field of variable $\delta^{13}C$ and relatively invariant $\delta^{18}O$ that characterizes meteoric-water carbonate precipitates.

When considering the range of mixing that might produce dolomite saturation and calcite undersaturation, theoretical calculations of Plummer (1975) define a broad range of seawater intermixture, depending upon initial saturation of each solution and equilibrium pCO_2s. Thus, calcites that precipitate from intermixtures of sea water and meteoric water should reflect the increased availability of Mg, Sr, and Na associated with marine waters. Similarly, with an increased marine intermixture, calcite becomes undersaturated, while dolomite saturation increases.

When examined on a carbon-oxygen cross-plot, the isotopic composition of replacive dolomite may not lie directly within this predicted field because of differing dolomite-water and calcite-water fractionations. It has been proposed that dolomite co-precipitated with calcite will be enriched in $\delta^{18}O$ by 2 to 5 per mil; empirically, enrichment in $\delta^{13}C$ for dolomites co-occurring with calcites has been estimated within the range of +1 to 2 per mil (Budai and others, 1987). Given the uncertainty of this fractionation at low temperatures and the ambiguities introduced by kinetic fractionation effects and/or fractionation control by dolomite stoichiometry (Swart and Dawans, 1984), it is not possible to refine this estimate. Figure 15 schematically illustrates possible ranges of mixing in which dolomite mixed-water calcites might develop in association with the El Hefhuf Formation. Correction for dolomite-calcite mineral fractionation illustrates that present dolomite compositions coincide with the range where mixed-water dolomitization may develop.

The presence of Mg-enriched calcite, as dedolomite and cements, further supports a mixed-water setting. If the Mg contained within these calcites were derived solely from dissolving dolomite, mass balance calculations would require equilibration of $\delta^{13}C$ values to host rock dolomite values. In contrast, Mg^{2+}-enriched calcites possess depleted $\delta^{13}C$ and $\delta^{18}O$ values which coincide with a mixture of approximately 25% to 40% marine water.

The co-occurrence of the dolomite overgrowth cement, Mg^{2+}-enriched calcite, and late vein-filling calcite in both the El Hefhuf and El Heiz formations and the similarity of the isotopic compositions of all diagenetic phases strongly argue for alteration by a common fluid during a single event. In contrast, the elevated Fe^{2+} content El Heiz dolomite crystal cores and the association with pyrite suggest an earlier and chemically distinct origin. Dolomite of the El Heiz presumably pre-existed prior to the mixed-water setting that led to dolomitization of the El Hefhuf Formation. Cathodoluminescence microfabrics of El Heiz dolomites indicate dolomite-dolomite replacement. Such fabrics require reaction-replacement of pre-existing dolomite with a fluid saturated with respect to dolomite, and thus suggest that early formed dolomite of the El Heiz was chemically or mineralogically unstable. Present isotopic values of these dolomites, therefore, likely indicate a late chemical resetting of metastable dolomite of the El Heiz and primary dolomitization of carbonate in the El Hefhuf in response to the mixed-water setting, which pervasively affected the entire stratigraphic sequence.

CONCLUSIONS

Whereas the environmental setting of primary dolomitization of the El Heiz and El Hefhuf formations is not conclusive, the similarity of their present compositions and their correlation to the composition of late vein-filling calcite cement argue for alteration by a common diagenetic fluid related to the Maastrichtian to pre-Eocene regional unconformity. Isotopic variation of dolomites coincides with trends predicted for a mixed-water, meteoric-marine diagenetic environment. Meteoric-water composition is recorded in the late vein-filling calcite cement, which occurs in both formations and in the overlying Maastrichtian chalks. Replacive dolomites and Mg enriched dedolomite calcite cement define a pattern of decreasing marine water intermixture. The coherence of compositional variation among all diagenetic phases implicates a common origin within a single-fluid system that evolved from mixed, meteoric-marine to an exclusively meteoric composition. This suggests that the processes of mixed-water dolomitization, dolomite-dolomite replacement, and dolomite calcitization may occur coevally within an individual evolving diagenetic system.

ACKNOWLEDGMENTS

The senior author is particularly indebted to the numerous colleagues who shared their analytical and interpretive expertise during the course of this research. This notably includes Luis Gonzalez, Scott Carpenter, and David Dettman. Final completion of this project was aided by the constructive criticism and careful reading of the manuscript by Bruce Wilkinson. Moreover, we thank Ron Carlton and Vijai Shukla for their critical reviews. Support for the senior author was provided by the Ministry of Higher Education of Egypt. Extensive isotopic analytical study was supported by the Light Isotope Geochemistry Laboratory of the University of Michigaon.

REFERENCES

AL-HASHAMI, W. S., AND HEMINGWAY, J. E., 1973, Recent dedolomitization and the origin of rusty crusts of Northumberland: Journal of Sedimentary Petrology, v. 43, p. 82–91.

ALLAM, A. M., 1986, A regional and paleoenvironmental study on the Upper Cretaceous deposits of Bahariya Oasis, Libyan Desert, Egypt: Journal of African Earth Scinces, v. 5, p. 407–412.

ALLAN, J. R., AND MATTHEWS, R. K., 1982, Isotopic signature associated with early meteoric diagenesis: Sedimentology, v. 29, p. 797–817.

ALLER, R. C., 1980, Diagenetic processes near the sediment-water interface of Long Island Sound sediments, II–Fe and Mn: Advanced Geophysics, v. 22, p. 351–415.

BADIOZAMANI, KHOSROW, 1973, The Dorag dolomitization model–Application to the Middle Ordovician of Wisconsin: Journal of Sedimentary Petrology, v. 43, p. 965–984.

BAKER, P. A., AND BURNS, S. J., 1985, Occurrence and formation of dolomite in organic-rich continental margin sediments: American Association of Petroleum Geologists Bulletin, v. 69, p. 1917–1930.

———, AND KASTNER, MIRIAM, 1981, Constraints on the formation of sedimentary dolomite: Science, v. 213, p. 214–216.

BAUM, G. R., HARRIS, W. B., AND DREZ, P. E., 1985, Origin of dolomite in the Eocene Castle Hayne Limestone, North Carolina: Journal of Sedimentary Petrology, v. 55, p. 506–517.

BECHMANN, J. P., AND HASSAN, A. A., 1968, Stratigraphy of the northern Bahariya Oasis: Gulf of Suez Petroleum Company (GUPCO) Unpublished Report, Cairo, Egypt, 18 p.

BEHRENS, E. W., AND LAND, L. S., 1972, Subtidal Holocene dolomite, Baffin Bay, Texas: Journal of Sedimantary Petrology, v. 42, p. 155–161.

BUDAI, J. M., LOHMANN, K. C., AND OWEN, R. M., 1984, Burial dedolomitization in the Mississippian Madison Limestone, Wyoming and Utah thrust belt: Journal of Sedimentary Petrology, v. 54, p. 276–188.

———, ———, AND WILSON, J. L., 1987, Dolomitization of the Madison Group, Wyoming and Utah overthrust belt: American Association of Petroleum Geologists Bulletin, v. 71, p. 909–924.

CHAFETZ, H. S., 1972, Subsurface diagenesis of limestone: Journal of Sedimentary Petrology, v. 42, p. 325–329.

CHURNET, H. G., MISRA, K. C., AND WALKER, K. R., 1982, Deposition and dolomitization of upper Knox carbonate sediments, Copper Ridge District, east Tennessee: Geological Society of America Bulletin, v. 93, p. 76–86.

CRAIG, HARMON, 1957, Isotopic standards for carbon and oxygen correction factors for mass spectrometric analysis of carbon dioxide: Geochimica et Cosmochimica Acta, v. 12, p. 133–149.

CURTIS, C. D., 1978, Possible links between sandstone diagenesis and depth-related geochemical reactions occurring in enclosing mudstones: Journal of Geological Society of London, v. 135, p. 107–117.

DICKSON, J. A., AND COLEMAN, M. L., 1980, Changes in carbon and oxygen isotope composition during limestone diagenesis: Sedimentology, v. 27, p. 107–118.

DOMINIK, W. H., 1984, Sedimentologie, Lithostratigraphie and Palokologie der Bahariya-Formation (Cenoman) als mogliches Pendant der Kharaga und Baris-Formation des Dakhla-Beckens: Unpublished Ph.D. Dissertation, Technische Universitat, Berlin, 185 p.

EL-AKKAD, SALAH, AND ISSAWI, BAHAY, 1963, Geology and iron deposits of the Bahariya Oasis: Egypt Geological Survey Paper 18, 301 p.

EL-BASSYONY, A. A., 1978, Structure of the north-eastern plateau of the Bahariya Oasis, Western Desert: Geologie Mijnbouw, v. 57, p. 77–86.

EVAMY, B. D., 1963, The application of chemical staining technique to a study of dedolomitization: Sedimentology, v. 2, p. 164–170.

FOLKMAN, YEHOSHUA, 1969, Diagenetic dedolomitization in the Albian-Cenomanian Yagur Dolomite on Mount Carmel (Northern Israel): Journal of Sedimentary Petrology, v. 39, p. 380–385.

FRANK, J. R., 1981, Dedolomitization in the Taum Sauk Limestone (Upper Cambrian), southeast Missouri: Journal of Sedimentary Petrology, v. 51, p. 7–18.

FRANK, M. H., AND LOHMANN, K. C., 1987, Textural and chemical alteration of dolomite: Interaction of mineralizing fluids and host rock in a Mississippi Valley-type deposit, Bonneterre Formation, Viburnum Trend, *in* Hagni, R. D., ed., Process Mineralogy, VI. Applications to Precious Metals Deposits, Industrial Minerals, Coal, Liberation, Mineral Processing, Agglomeration, Metallurgical Products, and Refractories, with Special Emphasis on Cathodoluminescence Microscopy: The Mettalurgical Society, p. 103–115.

FRANKS, G. D., 1982, Stratigraphical modelling of Upper Cretaceous sediments of Bahariya Oasis: 6th Egyptian General Petroleum Corporation Exploration Seminar, Cairo, Egypt, 22 p.

FRITZ, PETER, AND KATZ, AMITAI, 1972, The sodium distribution of dolomite crystals: Chemical Geology, v. 10, p. 237–244.

GAINES, A. M., 1977, Protodolomite redefined: Journal of Sedimentary Petrology, v. 47, p. 543–546.

GOLDBERG, MOSHE, 1967, Supratidal dolomitization and dedolomtization in the Jurassic rocks of Hamaktesh Hagatan, Israel: Journal of Sedimentary Petrology, v. 37, p. 760–773.

GRAF, D. L., AND GOLDSMITH, J. R., 1956, Some hydrothermal syntheses of dolomite and protodolomite: Journal of Geology, v. 64, p. 173–186.

GREGG, J. M., AND SIBLEY, D. F., 1984, Epigenetic dolomitization and origin of xenotopic dolomite texture: Journal of Sedimentary Petrology, v. 54, p. 908–931.

HARDIE, L. A., 1987, Dolomitization: A critical view of some current views: Journal of Sedimentary Petrology, v. 57, p. 166–183.

HERMINA, M. H., 1957, Final geological report on geology of Bahariya Oasis: Sahara Petroleum Company Unpublished Internal Report 26, Cairo, Egypt, 35 p.

HOLAIL, HANAFY, AND LOHMANN, K. C., 1986, The role of early lithification on development of chalky porosity in calcitic micrite: Upper Cretaceous Chalks, Egypt: Society of Economic Paleontologists and Mineralogists Annual Midyear Meeting, Raleigh, North Carolina, Abstracts, p. 53.

IRWIN, HILARY, 1980, Early diagenetic precipitation and pore fluid migration in the Kimmeridge clay of Dorset, England: Sedimentology, v. 27, p. 577–591.

ISSAWI, BAHAY, 1972, Review of Upper Cretaceous-Lower Tertiary stratigraphy in central and southern Egypt: American Association of Petroleum Geologists Bulletin, v. 56, p. 1448–1463.

KATZ, AMITAI, 1971, Zoned dolomite crystals: Journal of Geology, v. 79, p. 38–51.

KIER, J. S., 1973, Primary subtidal dolomicrite from Baffin Bay, Texas (abst.): American Association of Petroleum Geologists Bulletin, v. 57, p. 788.

KINSMAN, D. J., AND PATTERSON, R. J., 1973, Dolomitization process in sabkha environment (abst.): American Association of Petroleum Geologists Bulletin, v. 57, p. 788–789.

KLINKHAMMER, G. P., 1980, Early diagenesis in sediments from the eastern equatorial Pacific II. Pore water metal results: Earth and Planetary Science letters, v. 49, p. 81–101.

KUSHNIR, J., AND KASTNER, MIRIAM, 1984, Two forms of dolomite in the Monterey Formation, Santa Maria and Santa Barbara areas, California, *in* Garrison, R. E., Kastner, M., and Zenger, D. H., eds., Dolomites of the Monterey Formation and Other Organic-Rich Units: Pacific Section, Society of Economic Paleontologists and Mineralogists, v. 41, p. 171–184.

LAND, L. S., 1973, Contemporaneous dolomitization of Middle Pleistocene reefs by meteoric water, North Jamaica: Marine Science Bulletin, v. 23, p. 64–92.

———, 1980, The isotopic and trace element geochemistry of dolomite: The state of the art, *in* Zenger, D. H., Dunham, J. B., and Ethington, R. L., eds., Concepts and Models of Dolomitization: Society of Economic Paleontologists and Mineralogists Special Publication 28, p. 87–100.

———, 1985, The origin of massive dolomite: Journal of Geological Education, v. 33, p. 112–125.

———, AND HOOPS, G. K., 1973, Sodium in carbonate sediments and rocks: A possible index to the salinity of diagenetic solutions: Journal of Sedimentary Petrology, v. 43, 614–617.

———, SALEM, M. R., AND MORROW, D. W., 1975, Paleohydrology of ancient dolomites, geochemical evidence: American Association of Petroleum Geologists Bulletin, v. 59, p. 1602–1625.

LIPPMANN, FRIEDRICH, 1973, Sedimentary Carbonates Minerals: Springer-Verlag, Berlin, 228 p.

LORENS, R. B., 1981, Sr, Cd, Mn, and Co distribution coefficients in calcite as a function of calcite precipitation rates: Geochimica et Cosmochimica Acta, v. 45, p. 533–561.

LUMSDEN, D. N., AND LLOYD, R. V., 1984, Mn(II) partitioning between calcium and magnesium sites in studies of dolomite origin: Geochimica et Cosmochimica Acta, v. 48, p. 1861–1865.

MATTES, B. W., AND MOUNTJOY, E. W., 1980, Burial dolomitization of the Upper Devonian Mitte buildup, Jasper National Park, Alberta, *in* Zenger, D. H., Dunham, J. B., and Ethington, R. L., eds., Concepts and Models of Dolomitization: Society of Economic Paleontologists and Mineralogists Special Publication 28, p. 259–297.

MCHARGUE, T. R., AND PRICE, R. C., 1982, Dolomite from clay in argillaceous or shale-associated marine carbonates: Journal of Sedimentary Petrology, v. 52, p. 873–886.

MCKENZIE, J. A., 1981, Holocene dolomitization of calcium carbonate sediments from the coastal sabkhas of Abu Dhabi, U.A.E.: A stable isotope study: Journal of Geology, v. 89, p. 185–198.

MORROW, D. W., 1982, Diagenesis 1 and 2. Dolomite. The chemistry of dolomitization and dolomite precipitation: Dolomitization models and ancient dolostones: Geoscience, v. 9, p. 5–13 and p. 95–107.

MOSSLER J. H., 1971, Diagenesis and dolomitization of Swope Formation (Upper Pennsylvanian), southeast Kansas: Journal of Sedimentary Petrology, v. 41, p. 962–970.

M'RABET, ALI, 1981, Differentiation of environments of dolomite formation, Lower Cretaceous of central Tunisia: Sedimentology, v. 28, p. 331–352.

MURATA, K. J., FRIEDMAN, IRVING, AND MADSEN, B. M., 1969, Isotopic Composition of Diagenetic Carbonates in Marine Miocene Formations of California and Orgeon: U.S. Geological Survey Professional Paper 614-B, 24 p.

PLUMMER, L. N., 1975, Mixing of seawater with calcium carbonate groundwater, *in* Whitten, E. H., ed., Quantitative Studies in the Geological Sciences: Geological Society of America Memoir, v. 142, p. 219–236.

RUNNELLS, D. D., 1974, Discussion of Recent dedolomitization and the origin of the rusty crusts of Northumberland: Journal of Sedimentary Petrology, v. 44, p. 270–271.

SAID, RUSHDI, 1962, The Geology of Egypt Elsevier, Amsterdam, 377, p.

———, AND ISSAWI, BAHAY, 1964, Geology of northern plateau, Bahariya Oasis, Egypt: Egypt Geological Survey Paper 19, 41 p.

SCHINDLER, PETER, AND GHOSE, SUBRATA, 1970, Electron paramagnetic resonance of Mn^{2+} in dolomite and magnesite and Mn^{2+} distribution in dolomites: American Mineralogist, v. 55, p. 1889–1896.

SCHOLLE, P. A., 1971, Diagenesis of deep-water carbonate turbidites, Upper Cretaceous Monte Antola Flysch, northern Apennines, Italy: Journal of Sedimentary Petrology, v. 41, p. 233–250.

———, AND ARTHUR, M. A., 1980, Carbon isotope fluctuations in Cretaceous pelagic limestones: Potential stratigraphic and petroleum exploration tool: American Association of Petroleum Geologists Bulletin, v. 64, p. 67–87.

SHIMMIELD, G. B., AND PRICE, N. B., 1984, Recent dolomite formation in hemipelagic sediments of Baja California, Mexico, *in* Garrison, R. E., Kastner, M., and Zenger, D. H., eds., Dolomites of the Monterey Formation and Other Organic-Rich Units: Pacific Section, Society of Economic Paleontologists and Mineralogists, v. 41, p. 5–18.

SHUKLA, VIJAI, 1986, Discussion of epigenetic dolomitization and origin of xenotopic dolomite texture: Journal of Sedimentary Petrology, v. 56, p. 733–736.

SIBLEY, D. F., 1980, Climatic control of dolomitization, Seroe Domi Formation (Pliocene), Bonaire, N. A., *in* Zenger, D. H., Dunham, J. B., and Ethington, R. L., eds., Concepts and Models of Dolomitization: Society of Economic Paleontologists and Mineralogists Special Publication 28, p. 247–258.

———, 1982, The origin of common dolomite fabrics: Clues from the Pliocene: Journal of Sedimentary Petrology, v. 52, p. 1087–1100.

SLAUGHTER, B. H., AND THURMOND, J. T., 1974, A Lower Cenomanian (Cretaceous) Ichthyofauna from the Bahariya Formation of Egypt: Annals, Geological Survey of Egypt, v. 4, p. 25–40.

SWART, P. K., AND DAWANS, J. M., 1984, Variations in the Mg/Ca ratio as a control on the distribution of Sr concentration and $\delta^{18}O$ in Late Tertiary dolomites from Bahamas (abst.): American Association of Petroleum Geologists Bulletin, v. 68, p. 533.

TAYLOR, T. R., AND SIBLEY, D. F., 1986, Petrographic and geochemical characteristics of dolomite types and the origin of ferroan dolomite in the Trenton Formation, Ordovician, Michigan Basin, U.S.A.: Sedimentology, v. 33, p. 61–86.

VEIZER, JAN, 1983, Chemical diagenesis of carbonates: Theory and application of trace element techniques, *in* Arthur, M. A., Anderson, T. F., Kaplan, I. R., Veizer, J., and Land, L. S., eds., Stable Isotopes in Sedimentary Geology: Society of Economic Paleontologists and Mineralogists Short Course No. 10, p. 3–1 to 3–100.

———, AND HOFFS, JOCHEN, 1976, The nature of O^{18}/O^{16} and C^{13}/C^{12} secular trends in sedimentary carbonate rocks: Geochimica et Cosmochimica Acta, v. 40, p. 1387–1395.

WEBER, J. N., 1964, Trace element composition of dolostones and dolomites and its bearing on the dolomite problem: Geochimica et Cosmochimica Acta, v. 28, p. 1817–1868.

WILDEMAN, T. R., 1970, The distribution of Mn^{2+} in some carbonates by electron paramagnetic resonance: Chemical Geology, v. 5, p. 167–177.

WONG, P. K., AND OLDERSHAW, A., 1981, Burial cementation in the Devonian Kaybob Reef complex, Alberta, Canada: Journal of Sedimentary Petrology, v. 51, p. 507–520.

DOLOMITE DIAGENESIS IN THE METALINE FORMATION, NORTHEASTERN WASHINGTON STATE

HOWARD J. FISCHER

Department of Geology and Geological Engineering, University of North Dakota, Grand Forks, North Dakota 58202

ABSTRACT: The middle member of the Cambrian Metaline Formation in northeastern Washington State is a 360-m-thick dolostone unit containing seven depositional and diagenetic lithotypes. Four major dolomite crystal fabrics (A–D) have been identified within these rocks. Cathodoluminescence was required to show subtle features in the dolomite crystal fabrics and helped separate discrete stages in the development of the four fabrics.

Fabric A displays bright orange luminescence and is composed of micritic dolomite. Fabric B is a hypidiotopic and xenotopic fabric that shows a red-blue luminescence that does not outline individual crystals. Fabric C is more coarsely crystalline than fabric B and displays the same luminescence as fabric B. Fabric D is composed of zoned, brightly colored luminescent crystals of white sparry and saddle dolomite. Zebra dolomite in the Metaline Formation is a special feature formed when fabric D crystals filled subparallel, linear vugs in fabrics B and C.

Each crystal fabric is related to one or more diagenetic process or diagenetic environment. The crystal fabrics can be interpreted for the Metaline section specifically and for dolostones in general. Fabrics, processes, and environments include: (1) fabric A (micritic dolomite), surface-controlled, penecontemporaneous, open systems, where surface-derived fluids cause dolomitization of original calcite/aragonite sediments and limestones; (2) early fabric B (idiotopic fabric), intraformational open systems, where surface-derived fluids or shallow-subsurface fluids cause dolomitization; (3) final form of fabrics B and C (hypidiotopic and xenotopic fabrics), intraformational closed systems, where subsurface fluids cause dolomitization diagenesis or neomorphism of earlier formed dolomites; and (4) fabric D (white sparry and saddle dolomite), deep-subsurface, secondarily opened systems that precipitate dolomite cements.

Because each environment and its associated process produces a particular dolomite crystal fabric, each major diagenetic environment determines an end-member fabric that may be formed and preserved. Relicts of prior fabrics may or may not remain in the rock, depending on the completeness with which later processes develop. Often, only cathodoluminescence can show the relicts of early processes. Complex dolostones, such as the middle member of the Metaline Formation, can thus be interpreted by analyzing their contained diagenetic-crystal fabrics. By matching fabric with process and diagenetic environment, the evolution of a dolostone can be traced.

INTRODUCTION

The purpose of this paper is to document multiple stages of dolomite diagenesis in the middle member of the Cambrian Metaline Formation. Four dolomite crystal fabrics described from the Metaline rocks have morphologies that suggest a specific diagenetic environment for their formation. The crystal types are used as the basis for a set of "end-member fabrics" that are applicable to dolomites and dolostones in general. Based on the relations seen in the middle member rocks, a general model for the formation of complex dolostones has been proposed.

The Metaline Formation, in northeastern Washington State (Fig. 1), was divided into four informal members. These include: lower, middle, upper, and Fish Creek members (Fischer, 1981; Fig. 2). Mapping by Dings and Whitebread (1965) showed a combined thickness of 1650 m for the formation and a map thickness of 360 m for the middle member (Park and Cannon, 1943). The middle member is a complex dolostone unit that lies disconformably above dark-colored, transgressive, subtidal limestones of the lower member, is overlain by gray subtidal limestones of the upper member, and is laterally equivalent to the slope deposits of the Fish Creek member (Fischer, 1984; Fig. 2).

Harbour (1978) studied numerous drill cores of the middle member from the Pend Oreille Mine in Pend Oreille County, and on the basis of that study, divided the member into six laterally discontinuous lithotypes. Cleveland (1982) proposed seven lithotypes for the middle member in the Clugston Creek area of Stevens County (Table 1). Both Harbour and Cleveland define a shallow peritidal complex that was part of a much larger peritidal mosaic extending at least as far east as northwestern Montana (Bush and others, 1980; Bush and Fischer, 1981) and merging with deeper water slope deposits of the Fish Creek member of the Metaline Formation on the west (Fischer, 1984). Aitken (1978) documents the northern extension of the same shoal complex in British Columbia.

During this study, samples were collected from each of the lithotypes listed in Table 1. Because of the extreme structural deformation and extensive glacial and forest cover in the study area, lateral relations between lithotypes are difficult to determine. Each lithotype does, however, show distinct, depositionally derived features that have been diagenetically altered to some degree. The least altered samples retain fine-grained, micritic fabric and contain well-preserved allochems. The more altered samples within any lithotype are composed of a hypidiotopic to xenotopic dolomite mosaic that contains ghosts of allochems or shows no retention of depositional features.

The gray massive dolostone lithotype (Table 1) contains the most altered rocks observed. It consists of featureless gray dolostone or gray dolostone containing vugs and pods of white sparry dolomite, areas of zebra dolomite, and rarely, relict allochems and textures similar to those found in the other lithotypes. Microscopic examination shows that most of the samples are composed of a medium crystalline, hypidiotopic to xenotopic, cloudy, dolomite mosaic. Micritic allochems and allochem ghosts within the crystalline dolomite mosaic are morphologically similar to allochems found in other lithotypes and indicate that the gray massive-dolostone lithotype is a diagenetic lithotype and is not, as suggested by Harbour (1978) and Cleveland (1982), a depositional lithotype (Fischer, 1981).

METHODS

The field aspects of this study were conducted during the summers of 1976–1978. The four main sample locations were in the Metaline Falls area of Pend Oreille County. The

Sedimentology and Geochemistry of Dolostones, SEPM Special Publication No. 43

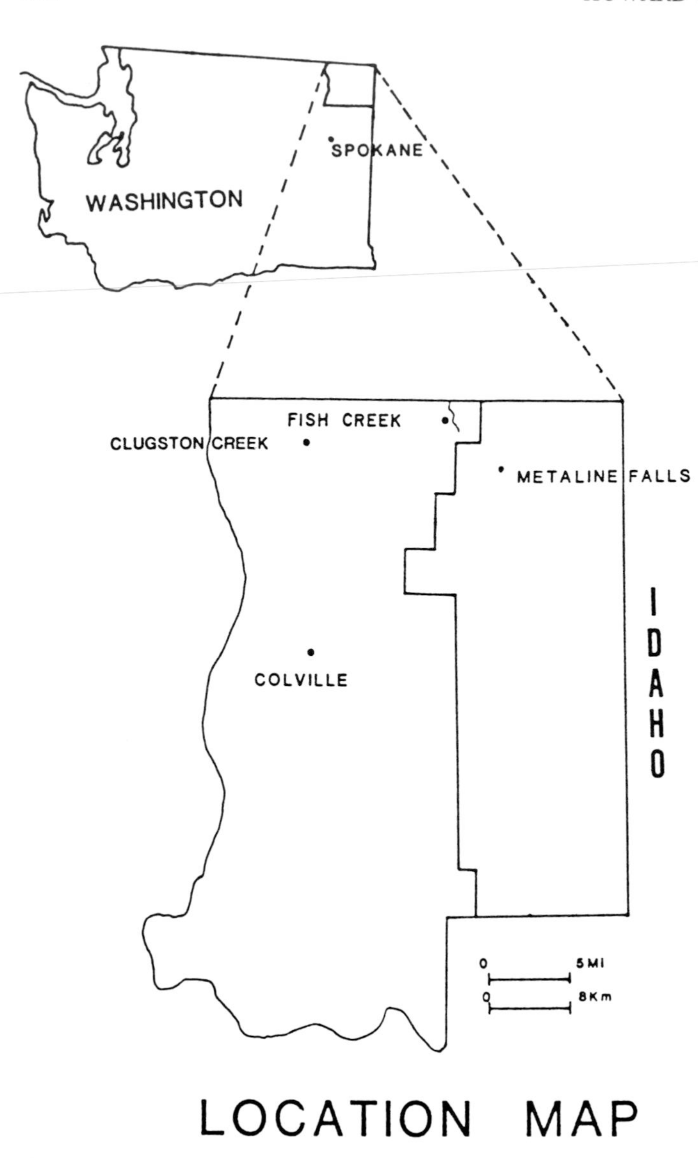

FIG. 1.—Location map of study area in northeastern Washington State.

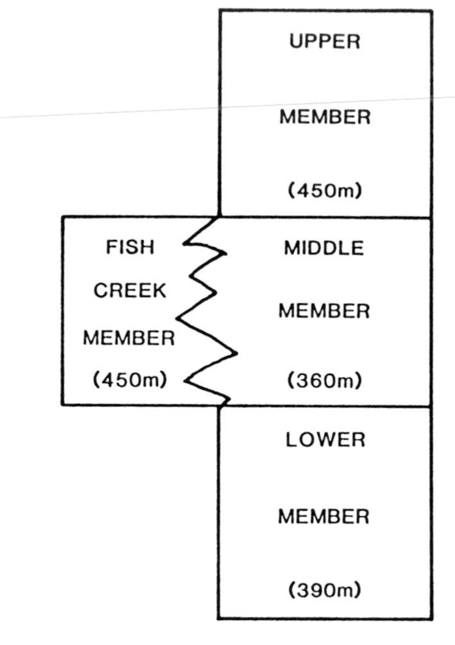

FIG. 2.—Generalized stratigraphic column of the Metaline Formation.

samples were selected in vertical succession from the four discontinuous outcrops at obvious changes in lithology. Because of the structural complications and covered intervals in the area, no reliable correlation between outcrops or features could be accomplished. Additional samples were taken from drill core supplied by the Bunker Hill Company from the Pend Oreille Mine, located at Metaline Falls, Washington.

The initial microscopic and cathodoluminescence examination of selected samples was done as a portion of the writer's Ph.D. dissertation, which was completed in 1981. Additional observations and analyses were made between 1981 and 1986.

The cathodoluminescence analyses and photographs were conducted on a Nuclide luminoscope using unstained, polished thin sections. The beam energy used was 10–20 KV at 100–200 microamperes of current.

DOLOMITE FABRICS

Four dolomite crystal fabrics were observed within the middle member rocks. The crystal fabrics are informally designated fabrics A–D for this discussion. The fabrics described below are found in varying concentrations within all of the lithotypes listed in Table 1. Most samples contain more than one of the fabrics described. Fine-grained fabrics become less common and coarse-grained fabrics become more common as samples become more complex.

Fabric A

Fabric A, the most finely crystalline dolomite fabric, is composed of equant crystals 3 to 15 mm in size (Fig. 3A). Fabric A occurs in most lithotypes and is most commonly within allochems (Fig. 3A B), but is also found as isolated patches of cloudy, dense, or carbon-stained crystals floating in more coarsely crystalline dolomite mosaics (Fig. 3A).

TABLE 1.—LITHOTYPES AND ASSOCIATED DEPOSITIONAL ENVIRONMENTS FOR THE MIDDLE MEMBER OF THE METALINE FORMATION

Lithotypes	Depositional Environment
1. Black birdseye dolomite	Supratidal
2. Cryptalgalaminate dolomite	Upper intertidal, lower supratidal
3. Dolomitized laminated intraclastic floatstone	Outer intertidal
4. Gray massive dolomite	Subtidal and intertidal
5. Dolomitized oncolitic floatstone	Shoals in upper subtidal lags in tidal channels
6. Lenticular-bedded dolomite	Subtidal
7. Breccia	Solution collapse

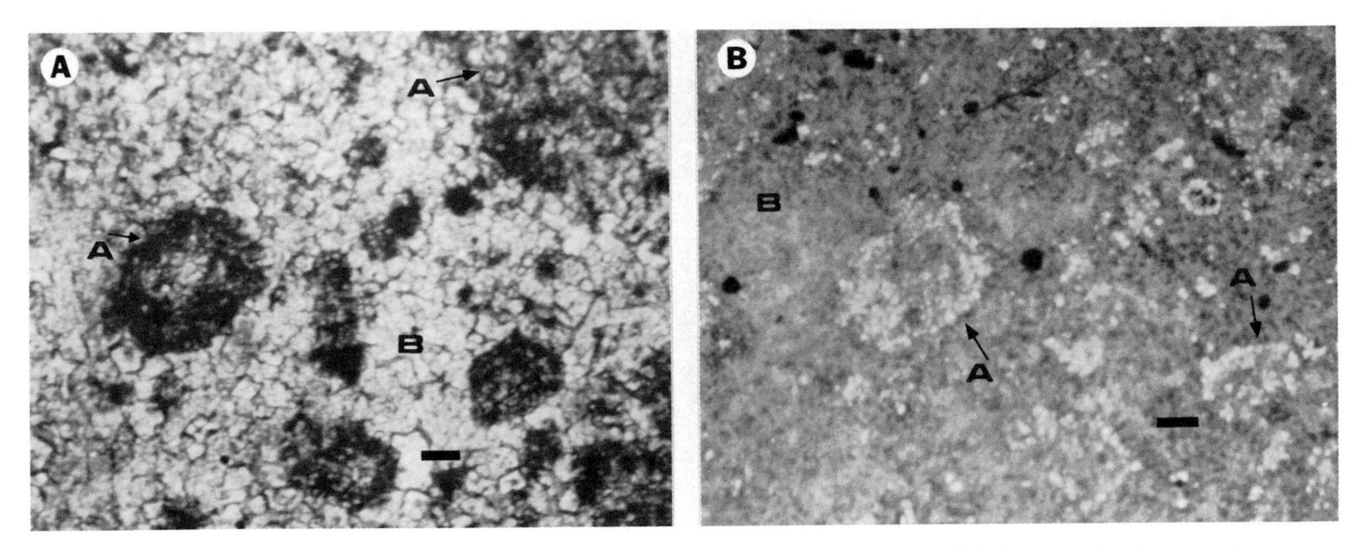

FIG. 3.—Allochems composed of fabric A in a mosaic of fabric B crystals. Scale bar 0.2 mm. (A) Micritic dolomite allochems A occur within fabric B mosaic, B. Plane polarized light. (B) Cathodoluminescence view of (A). Allochems, A, show bright luminescence compared to background response of fabric B, B.

Fabric A crystals may also occur as inclusions within larger dolomite crystals. Often, fabric A grades upward in size to merge with or be replaced by other, more coarsely crystalline fabrics (Fig. 4). Fabric A becomes less abundant as samples become more coarsely crystalline. Fabric A crystals are characterized by bright orange luminescence. Each crystal of fabric A is a single dot of luminescent dolomite (Fig. 3B), and this is consistent throughout the samples.

Fabric B

Fabric B is the most abundant crystal fabric seen in the rocks of the middle member. Two major sizes are found, a finely crystalline phase averaging 0.06 mm in diameter, grading into a more coarsely crystalline phase averaging 0.2 mm in diameter (Fig. 5A). In a given sample, crystals become larger and more anhedral with decreasing retention of depositional features. Usually, the smaller crystals show one or two planar crystal faces and have irregular, interpenetrating faces on the remainder of the crystal (Fig. 5A). More coarsely crystalline anhedral crystals have complex, interlocking grain margins (Fig. 5A).

Crystals of fabric B do not respond to Alizarin Red stain. They do not, therefore, contain any calcite. They are all cloudy crystals (Fig. 5A), although many show clear overgrowths that are usually 0.01 mm or less in width. Some crystals contain no distinct inclusions, whereas others contain single ghost euhedra within them. Other fabric B crystals may contain crystals of fabric A. Areas of fabric B mosaic may contain allochem ghosts composed of fabric A crystals (Fig. 4).

In samples containing well-preserved allochems, fabric B occurs mainly between allochems. Fabric B mosaics within the matrix may transect allochem margins and obliterate the allochem at that point (Fig. 4). Fabric B is initially matrix specific, and only in samples where most or all of the matrix has been observed to be converted to fabric B does fabric B become common within allochems (Fig. 4).

When viewed using cathodoluminescence, fabric B shows a patchy, red-blue luminescence that does not outline individual crystals (Fig. 5B). Fabric B crystals contain crystals of fabric A and euhedral ghosts in their cores. These dolomite inclusions retain the bright orange luminescence of fabric A (Fig. 3B).

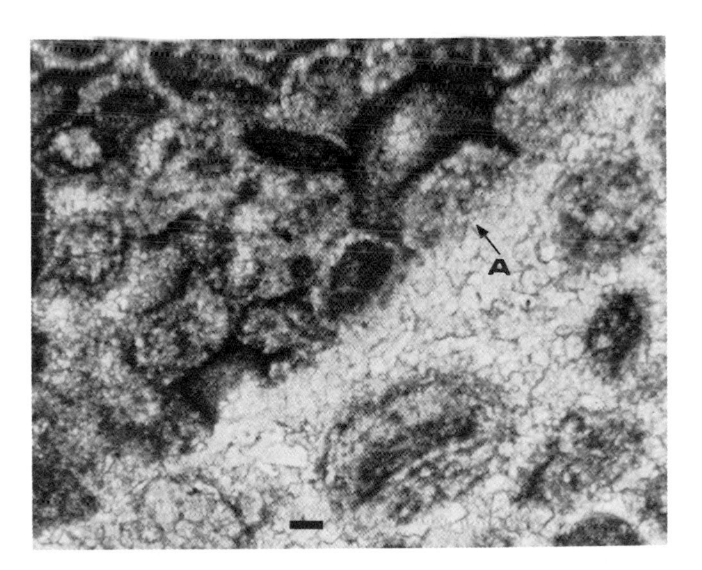

FIG. 4.—Allochems are well preserved at top of photomicrograph, but are preserved poorly or only as ghosts at bottom. Replacement of fabric A by fabric B is matrix specific at top, but affects allochems at bottom, where all matrix has been altered to fabric B. Note fabric B mosaic replacing allochems along diagonal line, A, crossing photo. Plane polarized light. Scale bar 0.2 mm.

Fabric C

Fabric C is composed of cloudy, mostly anhedral crystals (Fig. 5A). Fabric C contains two major crystal sizes, a finely crystalline phase composed of isolated or mosaic crystals

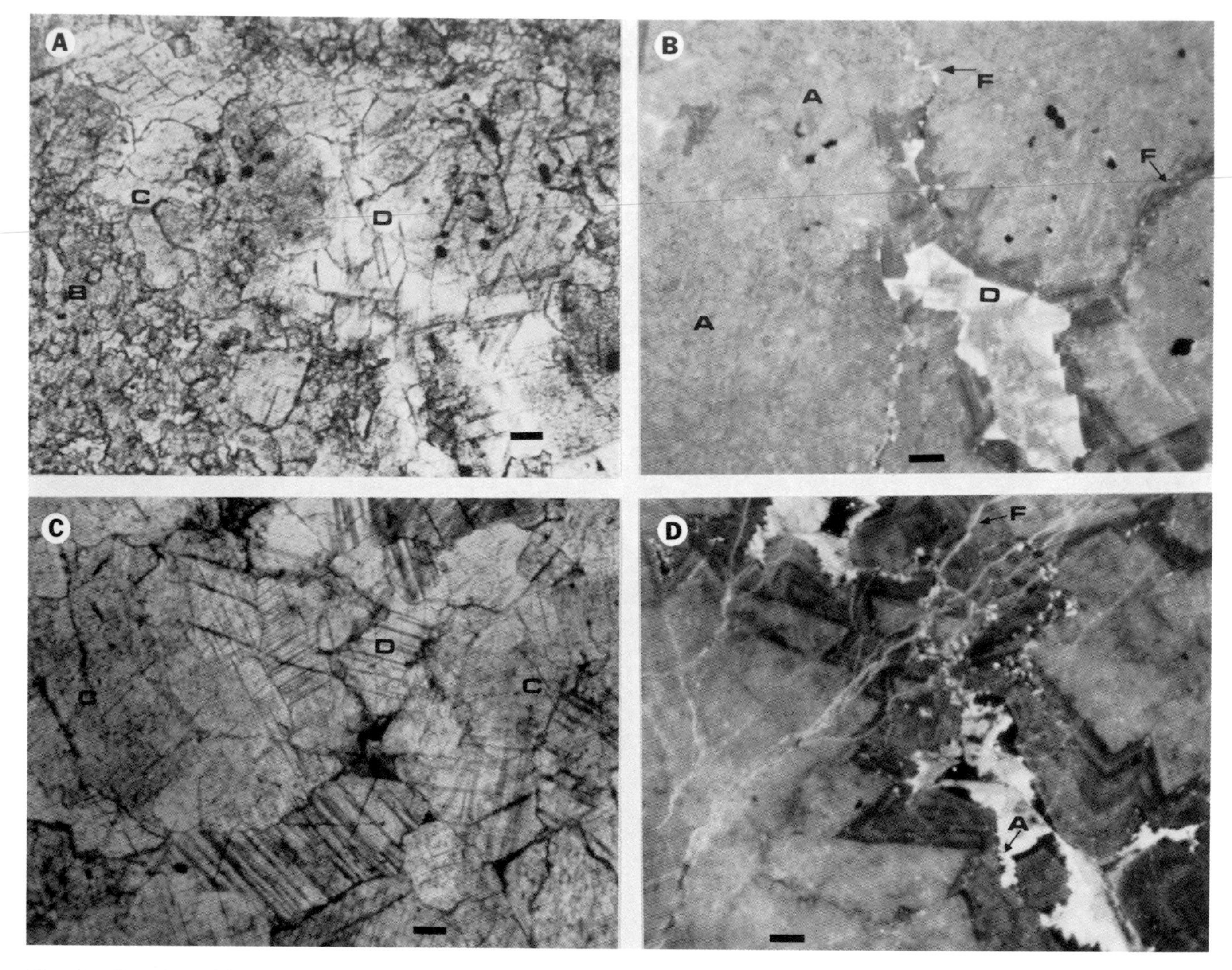

FIG. 5.—Two sizes of fabric B crystals and fabric C are gradational in size. Scale bar 0.2 mm. (A) Small, cloudy, anhedral crystals, B, of fabric B grade in size up to those of fabric C, C. Linear area of clearer crystals with cloudy internal bands in center of view is composed of fabric D, D. Plane polarized light. (B) Cathodoluminescence view of (A). Large area of non-specific luminescence is both fabrics B and C, A. Bright zoned areas are fabric D, D. Note that each fracture intersecting the vug is filled with a single zone of fabric D, F. (C) Fabric C mosaic composed of cloudy anhedral crystals, C. Clearer crystals are fabric D, D. Plane polarized light. (D) Cathodoluminescence view of (C). Large linear vug is filled with zoned fabric D crystals. Brightly luminescent fractures, F, crossing the vug contain luminescent dolomite of the same color as that composing one of the zones in the fabric D crystals. Note dissolution of early zones in fabric D crystals at A.

ranging from 0.2–0.8 mm in diameter, and a coarse phase, 0.6–1.2 mm in diameter. In most samples, fabric C is actually a coarse phase of fabric B, because, when both fabrics appear in the same sample, fabric C crystals grade in size into those of fabric B (Fig. 5). Fabric C shows the same luminescent response as fabric B, and when the two fabrics occur together in a sample, there is no visible difference in luminescent properties (Fig. 5B).

Fabric D

Fabric D is not as abundant as fabric B but is present in all of the samples studied. Fabric D occurs as optically continuous overgrowths on fabric B crystals in finely crystalline rocks and on fabric C crystals in coarsely crystalline rocks. Fabric D is composed of sparry dolomite crystals of various sizes (Fig. 5A). Growth of fabric D crystals was into vugs developed in fabric B and C mosaics. In most cases, fabric D crystals grew as overgrowths on seed crystals in the vug walls, until crystals from opposite walls met in the center of the vug and became intergrown. In some cases, vug centers are filled with calcite and, more rarely, quartz. Occasionally, vug centers are voids where fabric D growth ended before it could completely fill the vug.

Fabric D is a combination of white sparry dolomite (Beales and Hardy, 1980) and saddle dolomite (Radke and Mathis, 1980). In places, fabric D crystals contain cloudy layers that alternate with clear layers developed parallel to the faces of the crystal (Fig. 5A, C). Other crystals have undulatory extinction, and some have both undulatory extinction and saddle-shaped crystal form. Crystal size is a function of the size of the adjacent crystals of fabrics B or C and of the space available for the growth of the sparry fabric D crystals.

In both thin sections and hand specimens, concentrations of fabric D crystals are either pod-shaped or elongate (Fig. 5A). Where multiple elongate areas of fabric D occur, the subparallel layers of large, relatively clear fabric D crystals are the light-colored bands in zebra dolomite. The intervening areas of fabrics B or C are composed of smaller cloudy crystals and are the dark bands in the zebra pattern.

Cathodoluminescence shows that fabric D crystals are composed of two to five zones of luminescent dolomite, and at least one zone shows evidence of a dissolution event occurring prior to the precipitation of the next luminescent zone (Fig. 5B, D). The dolomite zones may be continuous or internally banded (Fig. 5B, D). There is a consistent order to the zones, and where they occur, the order of zones from the vug wall to its interior are: blue (dull), a dissolution event; black; bright yellow (very bright); pink (bright) or dark red (dull); and gold (dull) calcite. Each zone is an optically continuous overgrowth on the original vug wall seed crystal. The luminescent zones and the relations between fabrics C and D are best seen in Figure 5C and D.

In the more finely crystalline samples that contain relict allochems, only three zones occur within the overgrowths. The number of zones increases in the more coarsely crystalline samples. Vugs are largest in these samples and stylolites are common (Fig. 6A). Many of the larger vugs show breakage or brecciation of vug wall crystals or later overgrowth zones within the vugs themselves (Fig. 6B).

There are numerous fracture systems passing through the middle member rocks (Fig. 5B, D). These fractures are rarely visible without the use of cathodoluminescence. Each of the zones of dolomite found within vugs is associated with one of the fracture systems, each fracture system being lined with at least one distinct zone of luminescent dolomite (Fig. 5B, D). Larger fractures, developed late in the parageneses, are lined with two zones of luminescent dolomite and one of luminescent calcite, but do not contain material from earlier luminescent zones (Fig. 6C).

FABRIC INTERPRETATIONS

The four dolomite crystal fabrics described above occur together in some samples, as illustrated in Figure 7A and B. The fabric morphologies and the spatial relations be-

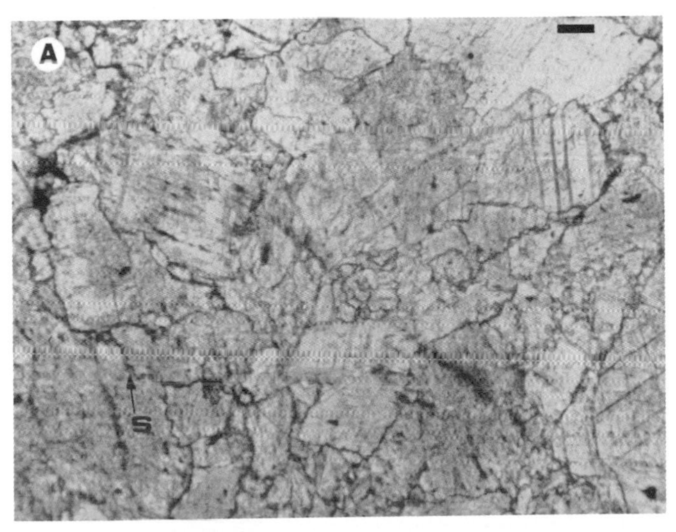

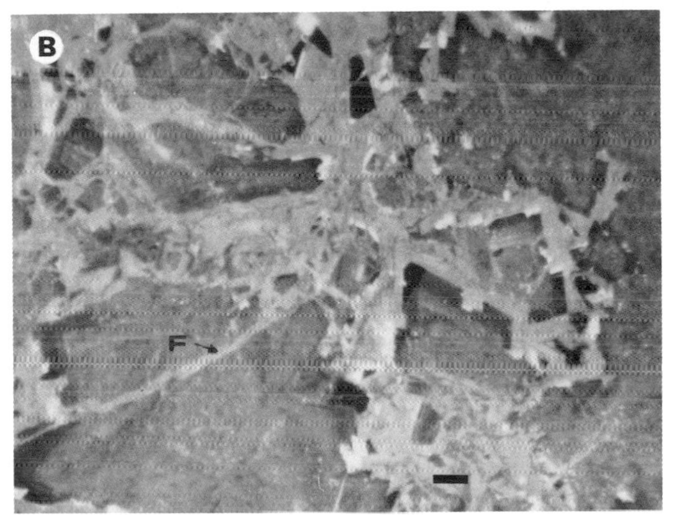

FIG. 6.—Fabrics C and D are difficult to separate in this view. Scale bar 0.2 mm. (A) Note small stylolite, S, crossing sample. Plane polarized light. (B) Cathodoluminescence view of (A). Primary zones of fabric D crystals have been broken during emplacement of later zones of dolomite through fractures, F, into same vug. (C) Cathodoluminescence view of large vug, V, filled with major zones of luminescent dolomite and a central zone of calcite, C. Late-stage fracture, F, contains only the final zone of luminescent dolomite and is filled with calcite.

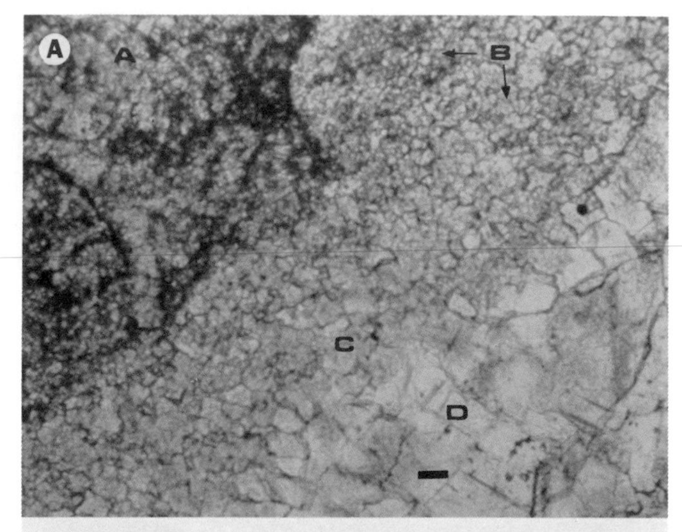

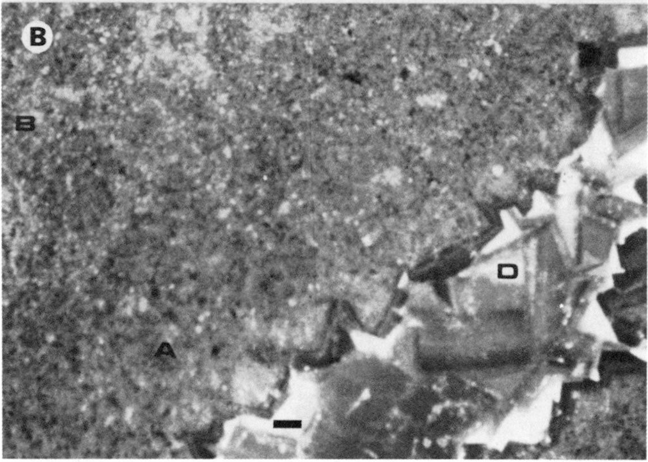

FIG. 7.—All four dolomite crystal fabrics occur together and illustrate spatial relations between them. Scale bar 0.2 mm. (A) Micritic fabric A preserved in allochems, A; fabric B partially replaces the allochems, B, and becomes a mosaic that grades into Fabric C, C. Fabric D crystals are optically continuous with those of fabrics C and B and form mainly clear, coarse crystals, D. Plane polarized light. (B) Cathodoluminescence view of (A) shows luminescent responses of the fabrics. Note similarity of fabrics B and C, A, and localization of fabric A in relict allochems, B. Fabric D is developed in the elongate vug at D.

tween the four fabrics can be seen in these illustrations. From these relations, the following interpretations were drawn, and these in turn were used to develop a model for the evolution of complex dolostones in general.

Fabric A

Because of the small crystal size of fabric A, its inclusion in primary features, such as grains and matrix, and its abundance in samples where depositional features are the most well preserved, fabric A is interpreted to be the product of early diagenetic, penecontemporaneous dolomitization. Because of its widespread occurrence in middle Metaline lithofacies, the processes that formed fabric A must have been responsible for the conversion of much of the original sediment into dolomite. Using the peritidal depositional environment interpreted by Harbour (1978) and Cleveland (1982), algal and evaporitic tidal flat processes are a reasonable model for the origin of fabric A.

Fabric B

The formation of fabric B is the result of dolomitization of original limestone and diagenesis of the early formed dolomites. The sequence of events responsible for the development of fabric B is complex and contains a number of steps. Early, shallow burial processes dolomitized any limestone remaining in the middle member. The euhedral ghosts, rarely preserved in the cores of some fabric B crystals, are the only evidence for this event. They are a product of dolomitization prior to dolomite diagenesis, and as such, represent a transition step between fabric A and the true fabric B. Because of the poor preservation of the products of the early portions of fabric B, the nature of the dolomitizing process is not clearly evident. Hurley (1980) suggested a schizohaline environment for the dolomitization of the middle member and, in the Clugston Creek area, Cleveland (1982) proposed a reflux model for middle member dolomitization. The present study finds little direct evidence to support either interpretation, since the products of this early event are rare and only poorly preserved when found.

The final form of fabric B was produced by alteration of the early formed portions of fabric B by (1) the overgrowth addition of dolomite onto micritic and euhedral crystals to produce subhedral crystals, (2) the closure of porosity and permeability by the overgrowths, and (3) neomorphic recrystallization of fabric A and the early formed portions of fabric B to produce the final, xenotopic mosaic.

Discussion.—

The continued evolution of fabric B beyond its initial dolomitization was in large part controlled by the evolution

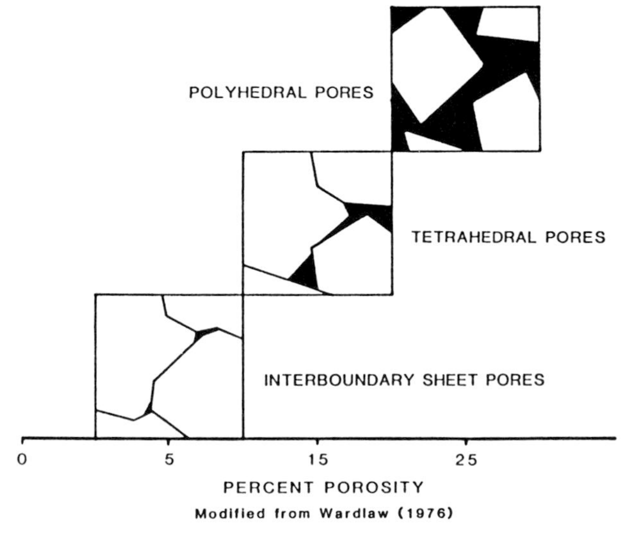

FIG. 8.—Diagram showing crystal and pore morphology changes and porosity reduction by addition of dolomite overgrowths in an idiotopic crystal fabric. Modified from Wardlaw (1976), with permission of the American Association of Petroleum Geologists.

of porosity within the dolomite mosaic. The evolution of dolomite porosity has been well described by Wardlaw (1976). His discussion shows how the shape, size, and distribution of crystals and pores in a dolostone may evolve, from an idiotopic mosaic containing polyhedral intercrystalline pores to a mosaic of intergrown anhedral crystals separated by interboundary sheet pores. This change reduces the porosity of the rock from an initial 20 to 25% to a low value between 3 and 5% (Fig. 8). Wardlaw's (1976) theoretical model can be used to explain the transformation of fabric A and the earliest portions of fabric B to a hypidiotopic fabric by the addition of dolomite overgrowths. The overgrowth stage is limited and is eventually self-destructive, since the pore system through which the fluids travel is progressively closed as the overgrowths develop.

Some of the overgrowths may have formed from fluids originating at the surface in either a reflux (Cleveland, 1982) or a mixed-water (Hurley, 1980) environment. Alternatively, the clear, euhedral nature of the overgrowths, the reduced permeability caused by the overgrowths themselves, and the abundance of stylolites associated with fabric B mosaics, suggests that some of the fluids responsible for the fabric B overgrowths may also have been derived intraformationally from pressure solution of other dolostones within the middle member. Since the rocks being dissolved along stylolites were dolostones, the fluids derived in this manner would be magnesium rich and would be capable of precipitating dolomite cement in optical continuity with the earlier formed dolomite. Due to the low permeability, supply of ions to nucleation sites would be slow, and few impurities would be included in the growing crystal. Therefore, clear euhedral overgrowths could form on earlier formed crystals by this method.

Once the overgrowth stage of fabric B was completed, the rocks became "tight" and were no longer effective conduits for the transmission of surface-derived or intraformational fluids. The remaining porosity would be in intercrystalline (interboundary sheet) pores, and as such, if the pores contained fluids, they could act as solution films for the transfer of ions by diffusion between adjacent grains (Wardlaw, 1976; Bathurst, 1975).

A similar process has been cited for the formation of neomorphic calcite by Bathurst (1975). The neomorphic fabric developed by these processes is morphologically the same as the xenotopic A fabric described by Gregg and Sibley (1984). In addition to the effects of temperature on the initiation of dolomite neomorphism discussed by Gregg and Sibley (1984) and Shukla (1986), this study suggests that there is also a level of dolomite diagenesis that must be reached before recrystallization can begin. Neomorphism is only initiated if (1) porosity is reduced to interboundary sheet pores, (2) permeability is effectively closed, and (3) interboundary sheet pores can become small enough to contain active solution films. If these criteria are met, then the conversion of an idiotopic dolomite mosaic to a xenotopic dolomite mosaic may begin.

An associated result of the neomorphic recrystallization of fabric B is the shift of porosity from intercrystalline pores to vugs. When the overgrowth stage ends, the effective porosity of the rocks is fixed at 3 to 5% (Wardlaw, 1976). With no new material being added, there is little chance to decrease that value further. Therefore, the porosity either remains in interboundary sheet pores and small, local vugs, or, during the neomorphic stage of internal size and shape rearrangement of the solid phase, the pores may also reform and coalesce, so that a portion of the 3 to 5% porosity contained in interboundary sheet pores becomes concentrated in vugs. The experimental process of sintering suggests a model for a similar redistribution of porosity. During sintering of metals and ceramics, a distinct rearrangement of pore size and shape can be seen as grain boundaries migrate (Lay, 1973, Barrett and Hardie, 1986).

Fabric C

The size gradation and luminescent properties of fabrics B and C suggests that both fabrics formed from similar diagenetic processes. Fabric C is primarily the product of dolomite neomorphism. Fabric C is also more xenotopic and shows more interlocking crystal boundaries than does fabric B. There is less preservation of the overgrowth stage within fabric C mosaics as a whole than there is in fabric B. Additional evidence for the neomorphic origin of fabric C is the increasing abundance of fabric C with increasing complexity of diagenetic alteration within a given sample. In many samples, the pattern of replacement of fabric B by fabric C is similar to the progressive replacement of microspar by pseudospar in limestone.

Fabric D

Fabric D is the product of void filling in vugs, channel pores, and fractures. The crystals of fabric D grew in optical continuity with crystals of dolomite that lined the walls of vugs and fractures developed in fabric B and C mosaics. The fractures intersect vugs, and because luminescent dolomite of a given zone is found within both the fractures and the vugs in which the fractures terminate, the fractures are interpreted to have been direct conduits for the dolomite composing a particular luminescent zone in that pore (Fig. 5B, D). Because the final set of fractures that developed does not contain the first three luminescent dolomite zones (Fig. 5C), those fractures must have formed after the precipitation of at least the first three zones. From this, it can be seen that at least two distinct precipitation and fracture events occurred in the evolution of fabric D. From the erosion of portions of the early zones seen in Figure 5D, there must also have been a dissolution event in the history of fabric D that followed the first major precipitation event and occurred before the second major fracture event.

The varied composition of the fabric D crystals and the cement stratigraphy indicate that a single fluid was not responsible for all of fabric D. The banding within luminescent zones suggests that the composition of the fluid changed during the precipitation of a particular zone of dolomite. The increasing number of fractures and associated luminescent zones indicates that progressively more stressed environments were responsible for the precipitation of the later luminescent zones in fabric D crystals. Since original permeability was essentially reduced to zero during the evolution of fabrics B and C, most of the movement of late diagenetic fluids in the middle member was along fractures.

Because the crystals of fabric D are late-stage additions to the rocks, because the rocks contain many stylolites, and because fabric D crystals are composed mainly of dolomite, a likely source of fluids for the precipitation of fabric D is from pressure solution of other dolomites within the middle member. The fractures that transported the fluids that formed fabric D are commonly associated with stylolites, and dolomitic solute generated by pressure solution late in the diagenetic history could easily have moved along the fractures and entered pores where precipitation could occur. Since the major zones of luminescent dolomite are each associated with a different fracture system, it is likely that each zone of dolomite was derived from a different source of dissolving dolomite within the formation. This would account for the variable trace-element composition of each zone implied by the varying luminescent characteristics shown by each zone. The majority of the cement stratigraphy seen in fabric D thus represents multiple sources for the fluids, most of which are intraformational, producing the individual zones composing the complex crystals. The final stage of fracture filling by calcite and occasionally quartz, however, suggests that at least some of the fluids passing through the rocks of the middle member were derived extraformationally, possibly from the dissolution of adjacent limestone units (lower and upper members) or from the diagenesis of black shales in the adjacent basin.

ZEBRA DOLOMITE

Introduction

In the middle member, zebra dolomite occurs most commonly in the rocks of the gray massive-dolomite lithotype. Petrographically, the zebra dolomite is composed of alternating bands of dark, medium-size, cloudy, hypidiotopic and xenotopic dolomite and light-colored, coarse, clear, sparry idiotopic dolomite. These compose the dark and light bands, respectively, of the zebra pattern. In the Metaline section, the two portions of the zebra pattern are composed of distinct dolomite fabrics. The darker bands contain the complex interlocking mosaics of fabrics B and C. The light-colored portion of the zebra dolomite contains fabric D crystals (Fig. 5A, B).

Discussion.—

Zebra dolomite is found in intensely altered carbonates, particularly in association with lead-zinc mineralization. Beales and Hardy (1980) suggest that zebra dolomite is the product of precipitation of white sparry dolomite in vugs formed by the dissolution of primary evaporite nodules. Fabric relations within the middle member of the Metaline Formation suggest that zebra dolomite may also be a product of advanced diagenetic alteration of a dolostone, regardless of whether or not there were nodular evaporites in its history. Both the middle member and the Fish Creek member of the Metaline Formation contain zebra dolomite. Although the middle member is the product of peritidal sedimentation (Harbour, 1978; Cleveland, 1982), the rocks show little or no evidence of nodular evaporites in their history. A few scattered dolomite pseudomorphs of gypsum were described by Cleveland (1982), but no widespread nodular facies was seen in this or any of the prior studies. Cleveland (1982) also suggested that his breccia facies was the result of solution collapse following the removal of former evaporites; however, this facies is aerially restricted and is rarely found associated with zebra dolomite.

The Fish Creek member is a deep-water slope deposit on the seaward edge of the prograding peritidal wedge that deposited the middle member (Fischer, 1984). The Fish Creek member shows no evidence of the presence of evaporites, nor has a model been suggested for the deposition of nodular evaporites in such a slope environment. The Fish Creek member, however, contains zebra dolomite. It can be shown, therefore, that at least for the Metaline Formation, in the absence of evidence supporting the occurrence of a significant evaporite facies, the zebra dolomite is the product of another process, namely, dolomite diagenesis.

The significant difference between this proposed origin for zebra dolomite and that of Beales and Hardy (1980) is that the origin of the vugs and that of the surrounding dolomite are related. Regardless of the origin of a dolostone unit, progressive diagenesis of the dolomites to an anhedral fabric with internal, subparallel vugs is required for the formation of zebra dolomite. In complex dolostones, whereas vugs may be formed early and retained through the neomorphic stage, they may also be formed during dolomite neomorphism by the porosity evolution discussed earlier for fabric B. The subparallel alignment of neomorphic pores can provide the substrate onto which the clear fabric D crystals can grow following introduction of dolomitic fluids after fracturing. The cementing fluids may be derived intraformationally during tectonic and burial deformation and enter the pores through tectonically induced fractures. An intermediate stage in the formation of zebra dolomite is shown in the hand sample photograph in Figure 9.

FIG. 9.—Hand sample showing incipient zebra dolostone. Dark areas of zebra pattern are fabric B; white areas are fabric D. Note small fractures at base of sample, F, that also contain white fabric D. Vugs in center and at top of sample are not closed; vug at top contains small, euhedral quartz crystals. Scale in millimeters.

DOLOMITE FABRIC EVOLUTION

Introduction

By identifying the crystal fabrics present in the middle member of the Metaline Formation, the evolution of that unit can be traced. Crystal fabrics in the middle member, and dolostones in general, can be separated into specific end-member types based on the extent of crystal fabric evolution they exhibit. The end members correspond to most of the classically described dolomite fabrics.

Diagenetic Stages

The fabric evolution of dolostones from the middle member of the Metaline Formation has been separated into four stages (I–IV) on the basis of diagenetic environment, processes acting in a given environment, end-member dolostones, and progressive increase in diagenetic intensity. During each stage, some portion of the fabric generated was preserved (end-member fabric) and, as diagenesis progressed past that stage, another fabric was generated that either replaced or modified the earlier fabric until another end-member fabric was reached. Partial or total destruction of earlier fabrics has occurred, and increasing levels of destruction of precursor fabrics makes interpretation of the early history of the middle member progressively more difficult. The stages, processes, and end members defined from the middle member can be applied to dolostone units in general and have been put in a flow chart diagram in Figure 10.

The overall progression from one dolomite fabric type or diagenetic stage to another is not always easy to document. Only the use of cathodoluminescence permitted the clear definition of the progression of diagenetic events for the middle Metaline. The exact progression of events in the evolution of the Metaline dolostones may be unique to that section; however, the relation between dolomite-fabric evo-

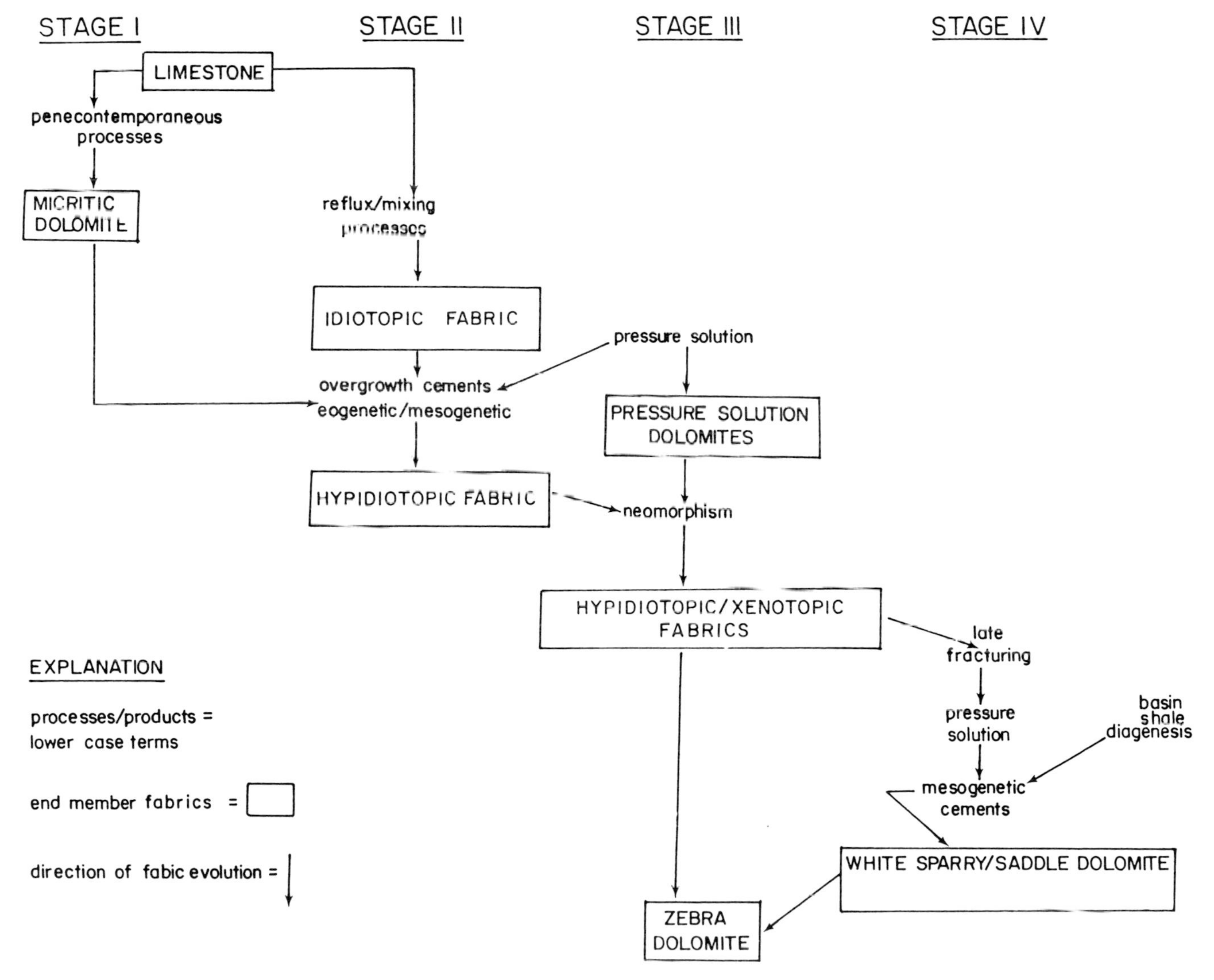

FIG. 10.—Flow chart diagram showing pathways, products, processes, end-member fabrics, and diagenetic stages in evolution of a complex dolostone. A carbonate can be altered to any one of the end-member dolomite fabrics and be preserved in that form. If the carbonate progresses to another diagenetic environment, shown by fabric evolution arrows, another diagenetic-crystal fabric can be formed and another end-member fabric can develop that often contains relicts of earlier fabrics.

lution and burial and tectonic history has a more universal application. The Metaline study suggests that many dolostones that have been poorly explained might be the products of multiple stages of dolomite diagenesis.

CONCLUSIONS

(1) The middle member of the Metaline Formation is a complex dolostone unit containing both depositional and diagenetic lithotypes.
(2) Four dolomite crystal fabrics (A–D) exist within the middle member lithotypes, and each fabric is the product of a different diagenetic event in the history of the rocks.
(3) Fabric A is composed of bright orange, luminescent micritic dolomite. It was formed by penecontemporaneous dolomitization processes.
(4) Fabric B is the most abundant and complex fabric in the middle member samples. Fabric B is composed of a hypidiotopic and xenotopic mosaic of red-blue luminescent dolomite. The luminescence does not outline individual crystals. It was formed by a series of surface-controlled and shallow-subsurface-controlled processes that converted original carbonate to dolomite and portions of fabric A to a hypidiotopic and xenotopic mosaic. The early processes produced euhedral rhombs. Other surface-derived fluids and pressure-solution-derived fluids formed overgrowths on the earlier crystals. Neomorphism of fabric A and the early portions of fabric B produced a xenotopic mosaic, which is the final form of fabric B.
(5) Fabric C is a coarse phase of fabric B. It shows the same luminescent response as fabric B and was produced by the continued neomorphism of fabric B.
(6) Fabric D is composed of zoned, brightly luminescent crystals of white sparry and saddle dolomite formed by the introduction of late-stage subsurface fluids through a set of tectonically induced fractures. Fabric D is found in vugs and fractures in fabric B and C mosaics.
(7) Zebra dolostone in the Metaline Formation was formed by complex diagenetic processes. Vug formation during dolomite neomorphism produced the substrate onto which late-stage, sparry-dolomite crystals could precipitate. The neomorphic dolomite (fabrics B and C) is the dark portion and the late-stage sparry dolomite (fabric D) is the light-colored portion of the zebra pattern.
(8) In complex dolostones, dolomite fabric evolution is a direct result of overlapping stages of dolomite diagenesis. By relating crystal fabrics to specific diagenetic environments, the history of a complex dolostone can be unraveled.

ACKNOWLEDGMENTS

The ideas and concepts expressed in this paper have developed over a number of years and were greatly aided by the guidance of John Bush at the University of Idaho. Others, at the University of Idaho, who contributed their ideas and discussions include: Scott Cleveland, Michael Janick, and William Dansart. Partial support for the project was provided by The Bunker Hill Company, Geological Society of America Penrose Grant No. 2190-77, and the Idaho Mining and Minerals Resource Research Institute. Mark Luther helped prepare the illustrations and Jean Hoff provided advice and editing during preparation of the manuscript. Reviews by John Kaldi and Vijai Shukla helped improve the manuscript significantly. Kris Reardon prepared the table and final version of the manuscript.

REFERENCES

AITKEN, J. D., 1978, Revised model for depositional grand cycles, Cambrian of the southern Rocky Mountains, Canada: Bulletin of Canadian Petroleum Geology, v. 26, p. 515–542.

BARRETT, M. L., AND HARDIE, L. A., 1986, Smackover (Jurassic) dolomite, Alabama: Evidence for dolomitization by upward-moving saline fluid: Society of Economic Paleontologists and Mineralogists Midyear Meeting, Raleigh, North Carolina, Abstracts, p. 6.

BATHURST, R. G. C., 1975, Carbonate Sediments and their Diagenesis, second edition: Elsevier, New York, 658 p.

———, AND HARDY, J. L., 1980, Criteria for the recognition of diverse dolomite types with an emphasis on studies on host rocks for Mississippi Valley-type ore deposits, *in* Zenger, D. H., Dunham, J. B., and Ethington, R. L., eds., Concepts and Models of Dolomitization: Society of Economic Paleontologists and Mineralogists Special Publication 28, p. 192–213.

BUSH, J. H., AND FISCHER, H. J., 1981, Stratigraphic and depositional summary of middle and upper Cambrian strata in northeastern Washington, northern Idaho, and northwestern Montana: U.S. Geological Survey Open-File Report 81-743, Second International Symposium on the Cambrian System, p. 42–45.

———, ———, AND AADLAND, K. R., 1980, The importance of a middle and upper Cambrian algal-shoal/tidal-flat barrier over northern Idaho: Geological Society of America, Abstracts with Programs, v. 12, p. 100.

CLEVELAND, S. R., 1982, Deposition and diagenesis of the middle member of the Metaline Formation, Clugston Creek, Stevens County, Washington: Unpublished M.S. Thesis, University of Idaho, Moscow, Idaho, 98 p.

DINGS, M. G., AND WHITEBREAD, D. H., 1965, Geology and ore deposits of the Metaline zinc-lead district, Pend Oreille County, Washington: U.S. Geological Survey Professional Paper 489, 109 p.

FISCHER, H. J., 1981, The lithology and diagenesis of the Metaline Formation, northeastern Washington: Unpublished Ph.D. Dissertation, University of Idaho, Moscow, Idaho, 175 p.

———, 1984, Pelagic and gravity sedimentation on the foreslope of a prograding Cambrian peritidal complex: Society of Economic Paleontologists and Mineralogists First Midyear Meeting, San Jose, California, Abstracts with Programs, v. 1, p. 32.

GREGG, J. M., AND SIBLEY, D. F., 1984, Epigenetic dolomitization and the origin of xenotopic dolomite texture: Journal of Sedimentary Petrology, v. 54, p. 908–931.

HARBOUR, J. L., 1978, The petrology and depositional setting of the middle dolomite unit, Metaline Formation, Metaline District, Washington: Unpublished M.S. Thesis, Eastern Washington State University, Cheney, Washington, 67 p.

HURLEY, B. W., 1980, The Metaline Formation-Ledbetter Slate contact in northeastern Washington: Unpublished Ph.D. Dissertation, Washington State University, Pullman, Washington, 141 p.

LAY, K. W., 1973, Grain Growth during Sintering, *in* Kuczynski, G. G., ed., Sintering & Related Phenomena: Proceedings, Third International Conference on Sintering and Related Phenomena: Plenum Press, New York, 457 p.

PARK, C. F., Jr., AND CANNON, R. S., Jr., 1943, Geology and ore deposits of the Metaline quadrangle, Washington: U.S. Geological Survey Professional Paper 202, 81 p.

RADKE, B. M., AND MATHIS, R. C., 1980, On the formation and occurrence of saddle dolomite: Journal of Sedimentary Petrology, v. 50, p. 1149–1168.

SHUKLA, VIJAI, 1986, Epigenetic dolomitization and origin of xenotopic dolomite texture: Discussion: Journal of Sedimentary Petrology, v. 56, p. 733–734.

WARDLAW, N. C., 1976, Pore geometry of carbonate rocks as revealed by pore casts and capillary pressure: American Association of Petroleum Geologists Bulletin, v. 60, p. 245–257.

SECTION VII
DOLOMITE ORIGINS: CASE HISTORIES

SECTION VII: INTRODUCTION
DOLOMITE ORIGINS: CASE HISTORIES

In this section three papers summarize data from contrasting locations and ages.

Sass and Bein present geochemical data from dolomites ranging in age from Permian to Neogene. In addition, the dolomites have widely differing characteristics, e.g., marine, gypsum- and halite-associated dolomites. These characteristics are explained by a geochemical model in which ionic ratios in dolomites are related to corresponding ratios in dolomitizing solutions. This paper should be compared and contrasted with those by Machel (Section IV) and Shukla (Section V).

Mullins and others describe Neogene dolomites from the Florida-Bahamas platform. This dolomite is calcium rich, euhedral, and occurs as pore fillings. Some dolomite precipitated in close association with dissolution of Sr-rich aragonite precursors, whereas other dolomite precipitated from normal marine water.

Ruppel and Cander present data from Permian dolomites of west Texas. Two dolomite phases are described. The older dolomite precipitated from marine waters, whereas the younger dolomite formed from evaporated seawater brines.

DOLOMITES AND SALINITY: A COMPARATIVE GEOCHEMICAL STUDY

EYTAN SASS
Department of Geology, The Hebrew University of Jerusalem, Jerusalem 91904, Israel
AND
AMOS BEIN
Geological Survey of Israel, Jerusalem 95501, Israel

ABSTRACT: A geochemical study of ancient (Permian to Neogene) dolomites that were formed under widely different salinities revealed certain distinct characteristics: (1) *marine (non-evaporitic) dolomites* vary in their calcium content from stoichiometric to calcian (57 mole percent, and their sodium content is 150–350 ppm; (2) *dolomites in association with gypsum* cover the same range of stoichiometry as the marine dolomites, but their sodium content reaches much higher values (as high as approximately 2,700 ppm). Furthermore, a negative correlation is present between excess calcium and sodium; and (3) *dolomites in association with halite* have an almost ideal stoichiometric composition and a sodium content overlapping the lower range of marine dolomites (150–270 ppm).

To explain these characteristics, a model is presented that states that ionic ratios in dolomites are related to the respective ratios in mother solutions. Those solutions are considered to have evolved through a combination of surface evaporation and modification in pore water, mainly through dolomitization. Accordingly, our model indicates that the compositional variations in the "evaporitic" dolomites correlate well with the expected Ca/Mg and Na/Ca ratios in modified evaporitic sea water. The "marine" dolomites, however, formed under varying degrees of modification, the two extremes being non-modified (open-system) solutions and fully modified (closed-system) solutions. In that context, the deviations from stoichiometric composition of the marine dolomites reflect different degrees of modification. Thus, advanced modification is expressed in higher Ca/Mg in solution and more calcian dolomite.

The distribution of $\delta^{18}O$ in the studied dolomites is consistent with anticipated variations, where $\delta^{18}O$ increases at the first stage of evaporation and tends to decrease beyond a certain high salinity.

The conclusions of the present study relate only to the main trends of changes that occur in dolomites forming under a wide range of salinity and ignore the fine details. They indicate, however, that the geochemical characteristics of ancient dolomites reflect the imprints of their depositional and diagenetic environments.

INTRODUCTION

The "dolomite problem" is still as hotly debated today as it was three decades ago (Fairbridge, 1957). The primary reasons for that are the scarcity of dolomite in Recent marine environments, as compared with the abundance of ancient dolomite, and the failure to synthesize dolomite in the laboratory under conditions simulating those prevailing in the past (Land, 1980). Attempts at utilizing geochemical tools in tackling this problem are likewise hampered by our flimsy knowledge of the mechanisms controlling the trace and bulk composition of dolomites. This state of confusion is linked to a widely held belief that ancient carbonates in general, and dolomites in particular, do not retain their original isotopic and elemental composition, so that geochemical data bear little or no evidence of the conditions prevailing at the time of dolomite formation (Land, 1980, 1985). Accordingly, it is often stated that recurrent processes of recrystallization and exchange cause calcian and poorly ordered dolomites to become stoichiometric and ordered, lead to depletion of their trace-elements content (such as Sr and Na), and lower their $\delta^{18}O$ and $\delta^{13}C$ signatures (Land and Hoops, 1973; Land, 1980, 1985; Sonnenfeld, 1986).

This notion is not shared by all researchers, as attested to by the multitude of papers (e.g., Fritz, 1971; Fritz and Katz, 1972; Möller and others, 1976; Morrow, 1978; Dunham and Olson, 1979; Dickson and Coleman, 1980; Bein and Land, 1982, 1983; Fairchild, 1985; Given and Lohmann, 1985; McKenzie, 1985; Renard, 1985; Veizer, 1985) dealing with ancient calcite and dolomite associations interpreted in terms of the conditions existing during their original formation.

The common occurrence of nonstoichiometric calcian dolomites in ancient rocks, some of them Paleozoic and Precambrian (Goldsmith and Graf, 1958; Atwood and Fry, 1967; Richter, 1974; Lumsden and Chimahusky, 1980; Fairchild, 1985; Land, 1985), indicates that metastable dolomite can resist stabilization for long geologic times, as also pointed out by Land (1985).

Sodium is sensitive to salinity variations in Recent sediments and skeletons (Gordon and others, 1970; Ishikawa and Ichikuni, 1984), and it was suggested (Fritz and Katz, 1972; Land and Hoops, 1973; Veizer and others, 1977) that it may also serve as a salinity indicator in ancient carbonates, including dolomites. Applying this supposition may meet with difficulties, partly because other factors aside from salinity seem to control the sodium content of dolomites (Sass and Katz, 1982).

One of the best ways to overcome these difficulties is to study the characteristics of dolomites that formed under different and constrained ranges of salinity. In these dolomites, the salinity-sensitive trends should be prominent, whereas other effects will appear as "noise" around the main trends. It is our aim to examine the potential of this approach. This is done by analyzing the trends of some geochemical parameters (stoichiometry, sodium content, $\delta^{18}O$) in ancient dolomites representing a wide range of salinity, from the marine range, through the low span of hypersalinity (dolomites associated with gypsum), to the range of higher salinity (dolomites associated with halite).

METHODS

The chemical analyses in the present study include previously reported data for the San Andres Formation (Bein and Land, 1982) and for the Judea Group (Sass and Katz, 1982), as well as new data for the Judea Group and Bira and Gesher formations. The laboratory procedures used in the various projects were designed to ensure the purity of

Sedimentology and Geochemistry of Dolostones, SEPM Special Publication No. 43

the analyzed dolomite, to avoid contamination of sodium (from soluble salts), and to ascertain that the reported sodium constitutes part of the dolomite crystals and is not derived from fluid inclusions.

The following specific methods were used in analyzing the Cretaceous and Neogene rocks. Finely ground samples were stirred in distilled water for 15 min in order to eliminate soluble salts and then centrifuged. Conductivity was monitored before deciding whether to repeat the dissolution procedure. Any existing calcite was thereafter selectively removed (as verified by X-ray diffraction) by stirring the samples in 5N acetic acid for about 10 min. The remaining dolomite fraction was dissolved in 1N nitric acid and filtered into a volumetric flask. To avoid contamination of sodium by sweating hands, latex gloves were used throughout.

Calcium and Mg were determined by using inductively coupled plasma (ICP) atomic emission spectrometry. The same solutions were further analyzed for Cl by colorimetry and for Na by atomic absorption spectrometry (AAS).

All the reported sodium data, excluding those from the Judean Hills (samples SKS), are calculated values of Na excess over Cl. Since Cl concentration in the Judea Group samples is negligible (as found in the Carmel samples and also in a few SKS samples), their Na data can be used with no further correction.

X-ray powder diffraction was routinely used for checking the purity of the treated dolomite samples and determining the various *d* spacings as a measure of the dolomite stoichiometry. Stable isotope ratios of oxygen in the dolomites were determined, using standard treatment techniques (McCrea, 1950).

GEOLOGIC BACKGROUND

This study is based on data gathered from three dolomite sequences differing in their locality, geologic age, and depositional environments.

The Judea Group—Cretaceous, Israel

The Judea Group includes thick sequences of dolomites and limestones as well as some chalks and marls. The dolomites formed on a shallow and extensive platform, separated from the open sea by a belt of barrier reefs (Bein, 1976; Sass and Bein, 1978, 1982). On the basis of field, petrographic, and geochemical data (Sass and Bein, 1982; Sass and Katz, 1982), it was concluded that these dolomites formed at an early diagenetic stage, from slightly evaporated sea water whose salinity fluctuated and only sporadically reached the range of gypsum and anhydrite precipitation. The studied dolomite samples are from the Judean Hills (Sass and Katz, 1982) and Mount Carmel.

The Bira and Gesher Formations—Neogene, Israel

The Bira and Gesher formations in the Jordan Valley consist of lagoonal sediments with intercalations of basalt flows and fluviatile conglomerates (Schulman, 1962). The sampled Bira Formation (in Wadi Bira, at the western margin of the Jordan rift valley) consists of alternating dolomite and thin argillaceous layers. A few kilometers to the east, it includes layers of gypsum and anhydrite and 10 km farther to the northeast, in the center of the rift valley, an equivalent sequence was found to contain anhydrite and halite layers (Marcus and Slager, 1985).

The Gesher Formation, which overlies the Bira Formation, consists mainly of limestones, dolomites, and some clays. Oolites are common in both the limestones and dolomites, and many ooids exhibit repeated alternations of dolomites and calcite laminae.

The lithologic associations of the Bira Formation and its lateral equivalents suggest that in the basin margin, the water salinity fluctuated mostly within the range of gypsum precipitation, whereas higher salinities were attained toward the basin center. The lack of evaporites in the overlying Gesher Formation is consistent with solutions having lower salinities, fluctuating between brackish and slightly hypersaline waters (Rosenfeld and others, 1981).

The San Andres Formation—Permian, Texas

The San Andres Formation constitutes the middle part of the Permian salt-bearing sequence found throughout northwest Texas and New Mexico. It covers the whole range, from thin dolomite layers interbedded with anhydrite and halite in the central evaporitic part of the basin, to a thick (over 500 m) dolomite sequence free of evaporite intercalations, marking the southward transition toward the open marine basin (Bein and Land, 1982, 1983). Geochemical data, as well as facies and rock sequence considerations, indicate that most of the dolomitizing solutions were brines formed through evaporation of sea water beyond halite precipitation, while some were diluted with normal sea water. The data used in the present study embrace only dolomites closely associated with anhydrite and halite, indicating deposition in highly saline waters (samples from cycles 5 and 6 in Swisher and Randall cores; Bein and Land, 1982, 1983).

The three dolomitic associations seem to represent different, although overlapping, ranges of salinity. The Judea Group dolomites formed at the lower range of salinity (from normal to slightly hypersaline sea water) and are referred to here as *marine dolomites*. The Bira and Gesher dolomites represent an intermediate range of salinity and are recognized as *dolomites associated with gypsum*. The San Andres dolomites formed at the higher range of salinity and are termed *dolomites associated with halite*.

It must be realized that the genetic implications of this terminology is appropriate for the majority of samples in each association, but not necessarily for all of them. For example, some of the Bira and Gesher dolomites probably formed in the marine rather than in the evaporitic range of salinity.

RESULTS

The geochemical data are presented in scatter diagrams (Figs. 1, 2). The three data sets, representing different depositional environments, form distinguishable clusters in these diagrams and define some general trends, as follows. (1) The Judea Group (marine) dolomites are mostly calcian, having a calcium content spanning the entire range from 50

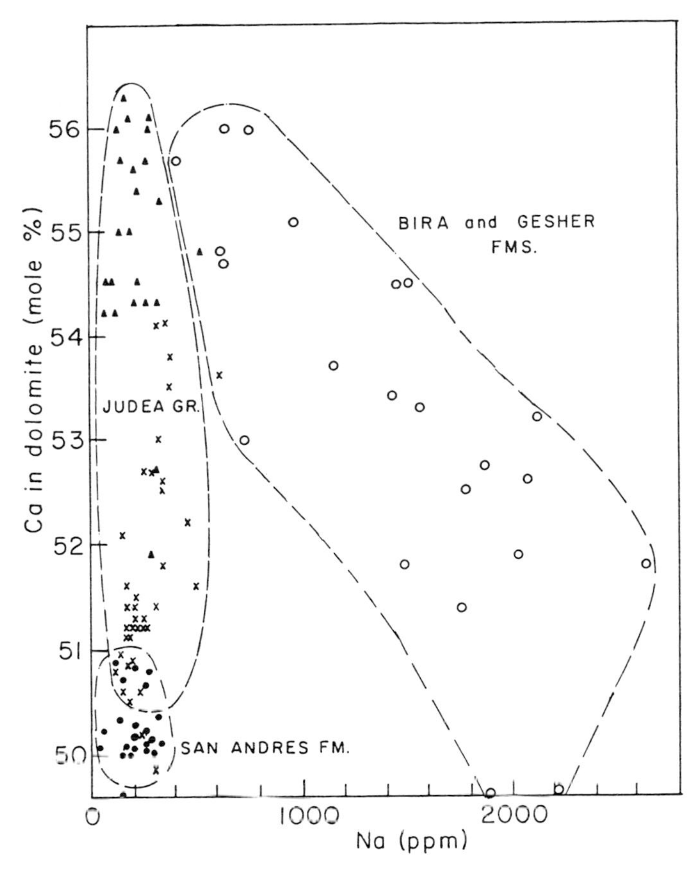

FIG. 1.—Distribution of mole percent calcium against sodium in the dolomite lattice of the three studied dolomite assemblages. Judea Group assemblage includes two sets of samples, one from Mount Carmel (triangles) and the other from Judean Hills (crosses).

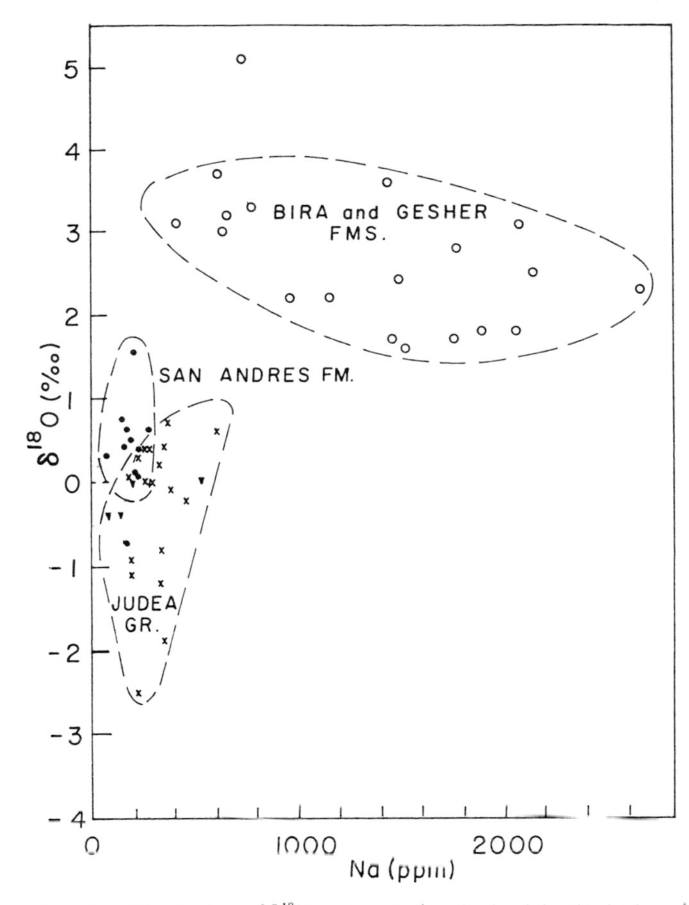

FIG. 2.—Distribution of $\delta^{18}O$ against sodium in the dolomite lattice of the three studied dolomite assemblages. Symbols are the same as in Figure 1.

to 57 mole percent. With few exceptions, the sodium content is rather low-ranging, between 150 and 300 ppm, and the $\delta^{18}O$ is within approximately 3‰ around 0. (2) The Bira and Gesher Formation dolomites, associated with gypsum, are also mostly calcian, covering the same range of calcium content as the marine dolomites. Their sodium values define a cluster across a wide range of 400 to 2,700 ppm, displaying a negative correlation with the calcium content. The $\delta^{18}O$ values are all positive, lying in the range of 1.5 to 4‰. (3) The San Andres dolomites, associated with halite, are characterized by their low-sodium content (mostly below 300 ppm) and nearly stoichiometric composition. The oxygen-isotopic composition field overlaps the upper part of the marine field, approaching a limit of +2‰.

DISCUSSION

The observed trends suggest that the properties of the studied dolomites directly reflect the conditions prevailing during their formation and are consistent with the notion that their original composition hardly changed with time. In order to explore the possible explanations for the observed trends, in terms of chemical parameters of the depositional environment, the changing ionic ratios in evaporating sea water should be compared with the respective ratios in the dolomites. In so doing, the various factors affecting ionic ratios in solution should be examined: (1) precipitation of evaporitic minerals, (2) mixing of solutions of different salinities, and (3) interaction between minerals and pore water.

Ionic Ratios in Evaporating Sea Water

The pathways of the chemical changes that follow evaporation of sea water are known from laboratory experiments (Zherebtsova and Volkova, 1966; Baseggio, 1974; Starinsky, 1974; Lazar and others, 1983) and from observations in evaporation ponds (Herrmann and others, 1973). Using these sources, the behavior of sodium and magnesium concentrations against calcium is plotted in Figures 3 and 4, and the Mg/Ca, Na/Ca, and Na/Cl ratios are plotted against Total Dissolved Solids (TDS) in Figure 5. The evaporation (solid) curves of the Mg/Ca and Na/Ca plots (Fig. 5a, b) allow us to distinguish three ranges of salinity defined by the break points *G* (beginning of gypsum precipitation) and *H* (beginning of halite precipitation).

Modified Ionic Ratios in Surface and Pore Waters

The ratios obtained along the evaporation curves are only limiting ones and represent the initial values for a specific

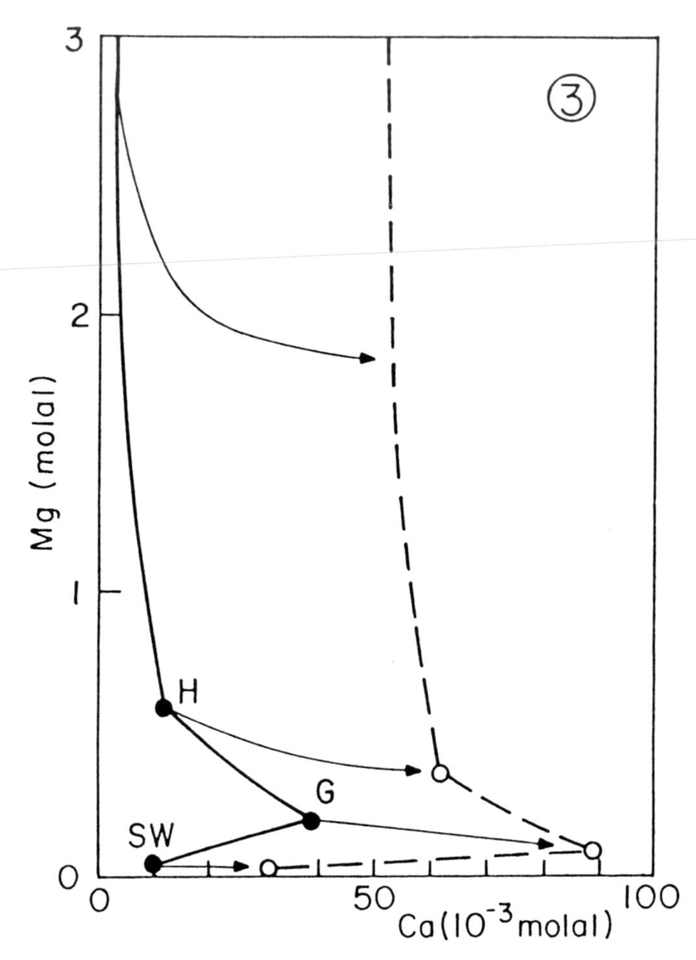

FIG. 3.—Variations of magnesium against calcium in evaporating sea-water-derived solutions. Solid curve shows variations through surface evaporation and evaporite precipitation alone. Dashed curve delineates probable limit of modification due to dolomitization in pore water. Critical points in the evolution of these solutions are designated *SW* (sea-water), *G* (beginning of gypsum precipitation), and *H* (beginning of halite precipitation). Tie lines between the two curves describe pathways of modification during dolomitization. Curved shape of tie lines at the higher range of salinity reflects constraint of equilibrium with calcium sulfate.

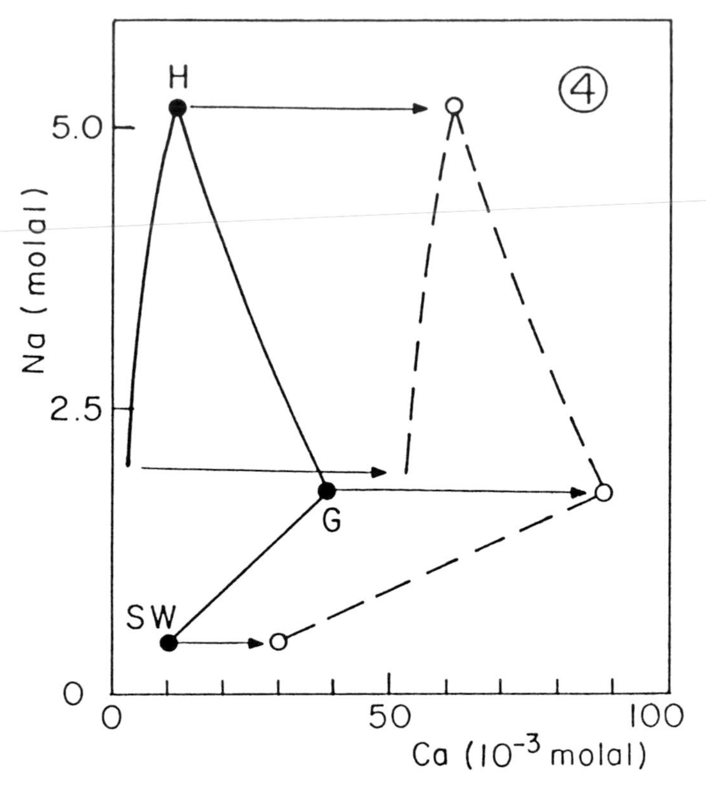

FIG. 4.—Variations of sodium against calcium in evaporating sea-water-derived solutions. Solid curve shows variations through surface evaporation and evaporite precipitation alone. Dashed curve delineates probable limit of modification due to dolomitization in pore water. Letter designation is the same as in Figure 3. Horizontal tie lines between the two curves reflect the fact that modification through dolomitization does not entail variations in sodium content in solution.

salinity. Further modifications can occur, however, due to homogeneous and heterogeneous reactions. Indeed, in marine and evaporitic environments, modified brines are more common than non-modified ones. This situation is clearly demonstrated in Figure 5a and b, where data points represent surface and pore water samples from the Persian Gulf lagoons and sabkhas (Butler, 1969; Butler's chemical analyses are incomplete and were recalculated for the purpose of the present study, as described in Appendix 1). It is evident that whereas some of the data points are very close to the evaporation curve, in particular at the lower range of salinity, many of them fall on both sides of this curve. Also, it appears that positive deviations (data points lying above the evaporation curve) are found mainly in lagoonal and intertidal environments, most of them surface waters, whereas negative deviations characterize the pore water of the supratidal environment. The main potential processes contributing to these modifications are the following:

Mixing.—

Figure 5 shows how a mixed solution, whose two end members belong to different sectors of the evaporation curve, will usually fall either above the evaporation curve, as for the Mg/Ca and Na/Ca ratios (Fig. 5a, c), or below it (Na/Cl ratio, Fig. 5c). Accordingly, the existing negative deviations of Na/Cl data points (Fig. 5c) are consistent with mixing, which could also account for the positive deviations of Mg/Ca and Na/Ca ratios (Fig. 5a, b).

FIG. 5.—Variations in ionic ratios in solutions as a function of Total Dissolved Solids (TDS) and salinity: (a) Mg/Ca; (b) Na/Ca; (c) Na/Cl. Solid curves describe path followed during surface evaporation of sea water; dashed curves outline the superimposed modifications due to dolomitization. Data points are from the Persian Gulf (Butler, 1969). Letter designation (SW, G, H) is the same as in Figure 3. Note that many of the data points deviate from the evaporation curves. Positive Mg/Ca and Na/Ca deviations occur mostly in lagoonal and intertidal solutions, whereas negative deviations in all three ratios characterize the supratidal pore waters. Possible mechanisms of deviation are illustrated by the "mixing" and "dolomitization" paths.

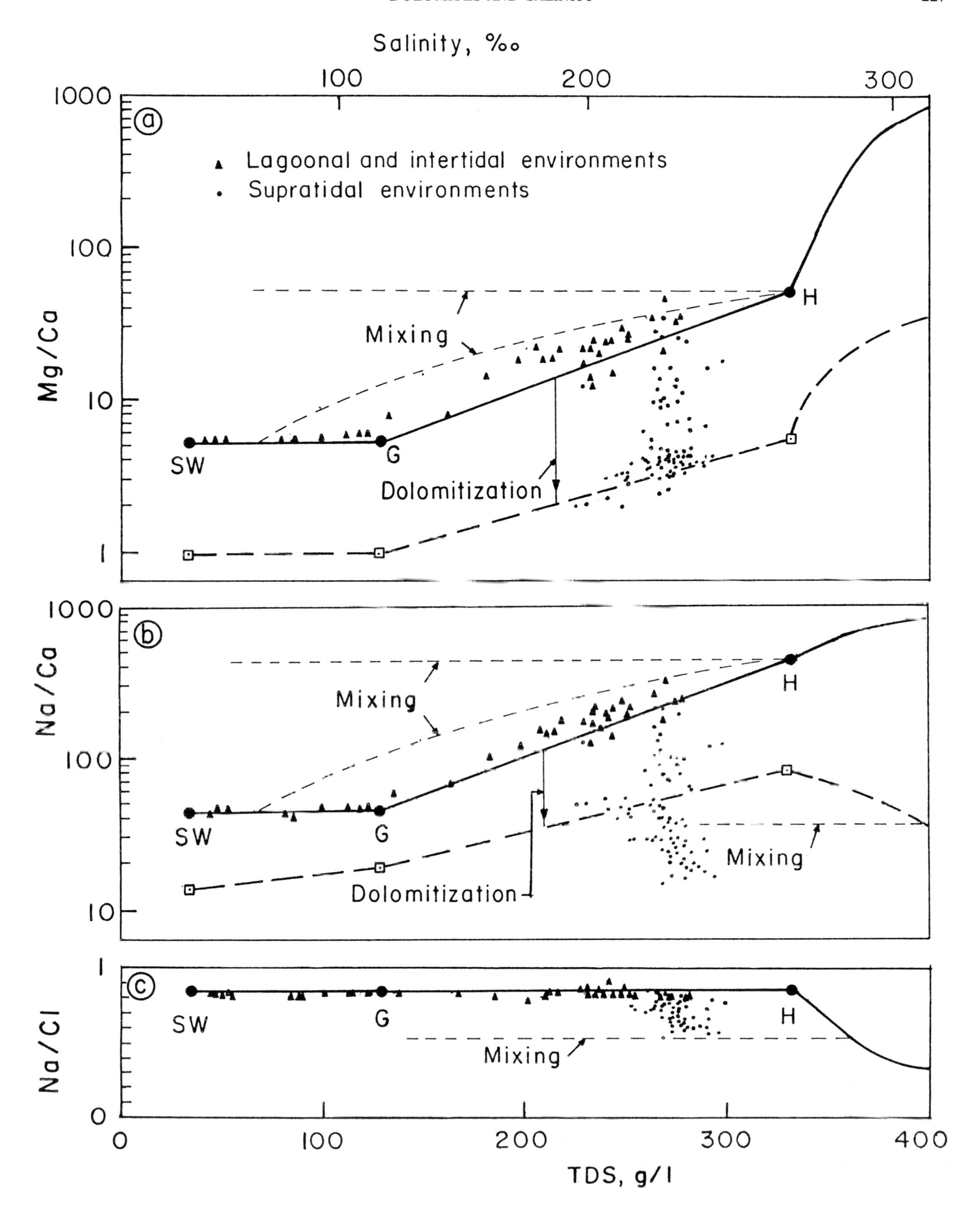
Salinity, ‰
100
200
300
1000
100
10
1
Mg/Ca
ⓐ
Lagoonal and intertidal environments
Supratidal environments
Mixing
H
SW
G
Dolomitization
1000
100
10
Na/Ca
ⓑ
Mixing
H
SW
G
Mixing
Dolomitization
1
0
Na/Cl
ⓒ
SW
G
H
Mixing
0
100
200
300
400
TDS, g/l

Dolomitization.—

This process entails both a decrease in magnesium concentration and an increase in calcium concentration, and its net effect is the *decrease* of the Na/Ca and Mg/Ca ratios. Dolomitization seems to be the only mechanism affecting both ratios in this way, and the negative deviations (Fig. 5a, b) of these ratios occurring at high salinities are therefore consistent with this explanation. The effect of dolomitization on the deviation of these two ratios is salinity-sensitive, as follows:

(1) At the lower range of salinity, the classic equation of dolomitization (equation 1) applies fairly well:

$$2CaCO_3 + Mg^{2+} \rightarrow CaMg(CO_3)_2 + Ca^{2+} \quad (1)$$

Under these conditions, the Na/Ca ratio decreases only through the increase of calcium (Fig. 4), whereas the Mg/Ca ratio further decreases due to removal of magnesium (Fig. 3). That decrease is limited and determined by the equilibrium of Mg/Ca ratio in solution between calcite and dolomite. If we accept the value of 1 as an approximation for the equilibrium ratio (Hanshaw and others, 1971; Folk and Land, 1975), then the limiting relative decrease of the Mg/Ca ratio is by a factor of about 5 (Fig. 5a), and it can easily be deduced that the limiting factor for the decrease of the Na/Ca ratio would only be about 3 (Fig. 5b). It should be noted, however, that the limiting (thermodynamic) deviations from the evaporation curves can only be achieved in closed systems. Dolomitization can be completed at some intermediate stage of deviation.

(2) At salinities approaching (and exceeding) the range of gypsum precipitation (point G of Figs. 3, 4, 5a, b), equation 1 no longer holds alone, since some of the calcium released by dolomitization combines with aqueous sulfate to form gypsum or anhydrite (Berner, 1971; Patterson and Kinsman, 1982). This situation is demonstrated by the curved tie lines (arrows, Fig. 3) depicting the covariation of Mg and Ca during dolomitization at the higher salinity. The absolute limit for a modified molar Mg/Ca ratio, which is around 1 at lower salinities, decreases markedly at higher salinities, due to decrease of the $\gamma Ca/\gamma Mg$ activity ratio (Carpenter, 1980). It is reasoned, however, that owing to the high initial Mg/Ca ratio in solution, the bulk of dolomitization would be completed before reaching the limiting modification. For comparative purposes, it is assumed that the limit of the net increase in calcium during dolomitization is the same throughout the range of gypsum precipitation. This is expressed graphically by the parallelism of the dashed and solid curves in Figure 4 and the near parallelism of curves in Figure 3. In that case, the Mg/Ca and Na/Ca ratios in some of the modified solutions would be similar to those of sea water or even lower (Fig. 5a, b, dashed curves).

(3) Beyond the onset of halite precipitation, diverging trends should occur for Na/Ca and Mg/Ca ratios in solutions modified by dolomitization and gypsum precipitation. Despite the similar behavior of the evaporation curves for the two ratios (Fig. 5a, b, solid curves), the distinct difference is that the magnesium concentration continues to rise with salinity (Fig. 3), whereas that of sodium drops (Fig. 4). If the assumption concerning a constant increase in calcium still applies in this range, it can readily be demonstrated (Figs. 4, 5b, dashed curves) that, starting with precipitation of halite (point H), the *modified* Na/Ca ratio reverses its previous trend and diminishes with the increase of salinity. In fact (depending on salinity and the degree of calcium modification in solution), there is a point beyond which the Na/Ca molar ratio becomes even smaller than that of sea water. By contrast, the Mg/Ca ratio continues its trend of increasing with salinity (Figs. 3, 5a).

We have demonstrated that deviation of Mg/Ca and Na/Cl ratios in solution from the evaporation curve is the rule, rather than the exception, in peritidal environments. Also, the agreement between the occurrence of dolomite in the supratidal areas of the Persian Gulf (Illing and others, 1965; Butler, 1969; Patterson and Kinsman, 1982) and the independent geochemical evidence for dolomitization is noteworthy. Of even more significance is the observation that most of the Mg/Ca data points from the supratidal environment fall between the "evaporation" and "modified" curves (Fig. 5a). This behavior is consistent with the notion of partial modification in sites of active dolomitization and gives credence to the tentatively drawn boundaries of modification.

The digression of Na/Ca data points from the expected field of modification (Fig. 5b) seems to disagree with the previous conclusion. This apparent contradiction can easily be explained by mixing, as illustrated in Figure 5b. Indeed, the negative deviations of the Na/Cl data points (Fig. 5c) provide independent support for a mixing process.

In conclusion, the observed trends of brine modification in the Persian Gulf–a modern environment of dolomite formation–are consistent with the prediction of a limited modification of pore water. The next step is to evaluate the effect of such modifications on the composition of the growing dolomite crystals.

Ionic Ratios in Dolomites and Solutions

We have discussed the effect of dolomitization on ionic composition in pore water. We will proceed by testing the hypothesis that a feedback mechanism operates, whereby the composition of dolomites is controlled by the composition of the solution. More specifically, we will examine whether the calcium and sodium contents in dolomites are directly related to the depicted Ca/Mg and Na/Ca ratios in non-modified and modified solutions at various salinities.

It is evident that the relation existing in the lower range of salinity is not simple. As can be seen in Figures 1 and 2, the calcium content of the marine dolomites (Judea Group) covers a wide range. By contrast, the Ca/Mg ratios (that is, the reciprocal values of Mg/Ca) along the range SW-G (Fig. 5a) are constant for non-modified (solid curve) and fully modified (dashed curve) solutions.

These observations could lead to the conclusion that the aforementioned hypothesis is incorrect. Alternatively, it is possible that, at the lower range of salinity, the composition of dolomites reflects intermediate states between non-modified and fully modified solutions. The results of a detailed geochemical study of some of the Judea Group dolomites (Sass and Katz, 1982) are indeed consistent with this idea.

The characteristic behavior of the more saline (Gesher, Bira, and San Andres) dolomites was demonstrated in a Ca vs. Na diagram (Fig. 1). A comparison of this diagram with the predicted trends of Ca/Mg vs. Na/Ca in solution can serve as a tool for identifying the causal relation between solution and rock composition. The predicted behavior is plotted for the two cases, namely, for solutions affected by evaporation only and for solutions that were modified also by dolomitization (Fig. 6a and b, respectively). This comparison indicates that the trend of variations in the "evaporitic" dolomites (Fig. 1) is strikingly similar to the portrayed changes in modified evaporitic solutions (Fig. 6b). The reversal (that is, the decrease) of the sodium content in dolomites at the halite stage makes this observation even more evident.

These deductions indicate that, in the evaporitic range of salinity, dolomitization mostly took place in modified solutions whose composition is approximated by the dashed curves in Figures 3, 4, 5a, and b. In the lower range of salinity preceding the onset of evaporite precipitation, however, the composition of dolomites seems to reflect varying degrees of modification in pore water, covering the entire range from open (non-modified) to closed (modified) systems.

At this point, the hypothesis concerning the dependence of dolomite composition upon that of the mother solution seems to be established. We will evaluate that dependence.

Stoichiometry.—

The implication of the above conclusion is that the Ca/Mg ratio in dolomite is directly related to the Ca/Mg of the solution, a notion shared by other investigators (Schmidt, 1965; Marschner, 1968; Katz, 1971; Patterson and Kinsman, 1974; Richter, 1974; Morrow, 1978; Sass and Katz, 1982; Ward and Halley, 1985). A concrete evidence for this hypothesis is concerned with abundant findings (e.g., Goldsmith and Graf, 1958; Marschner, 1968; Lumsden and Chimahusky, 1980) of stoichiometric dolomites (that is, dolomites with a relatively low Ca/Mg ratio) in association with evaporites (whose mother solutions are noted for their low Ca/Mg ratios).

The form of this dependence has hitherto never been formulated. We propose here the following distribution form:

$$(Ca_x/Mg)^D = D^D_{Cax} \cdot (mCa/mMg)^L \qquad (2)$$

where Ca_x is the molar fraction of calcium found in excess of the stoichiometric concentration (50%), and D^D_{Cax} is the molar partition coefficient of excess calcium in dolomite. The rationale behind this formulation is the following: calcium and magnesium ions occupy two different sites (known as sites *A* and *B*, respectively) in a stoichiometric, well-ordered dolomite (Lippman, 1973; Reeder, 1983). Any deviation from stoichiometry does not usually affect the *A* site which, ideally, is occupied by Ca ions only, but is rather expressed by excess calcium ions occupying site *B* at the expense of magnesium ions (Reeder, 1983). In this situa-

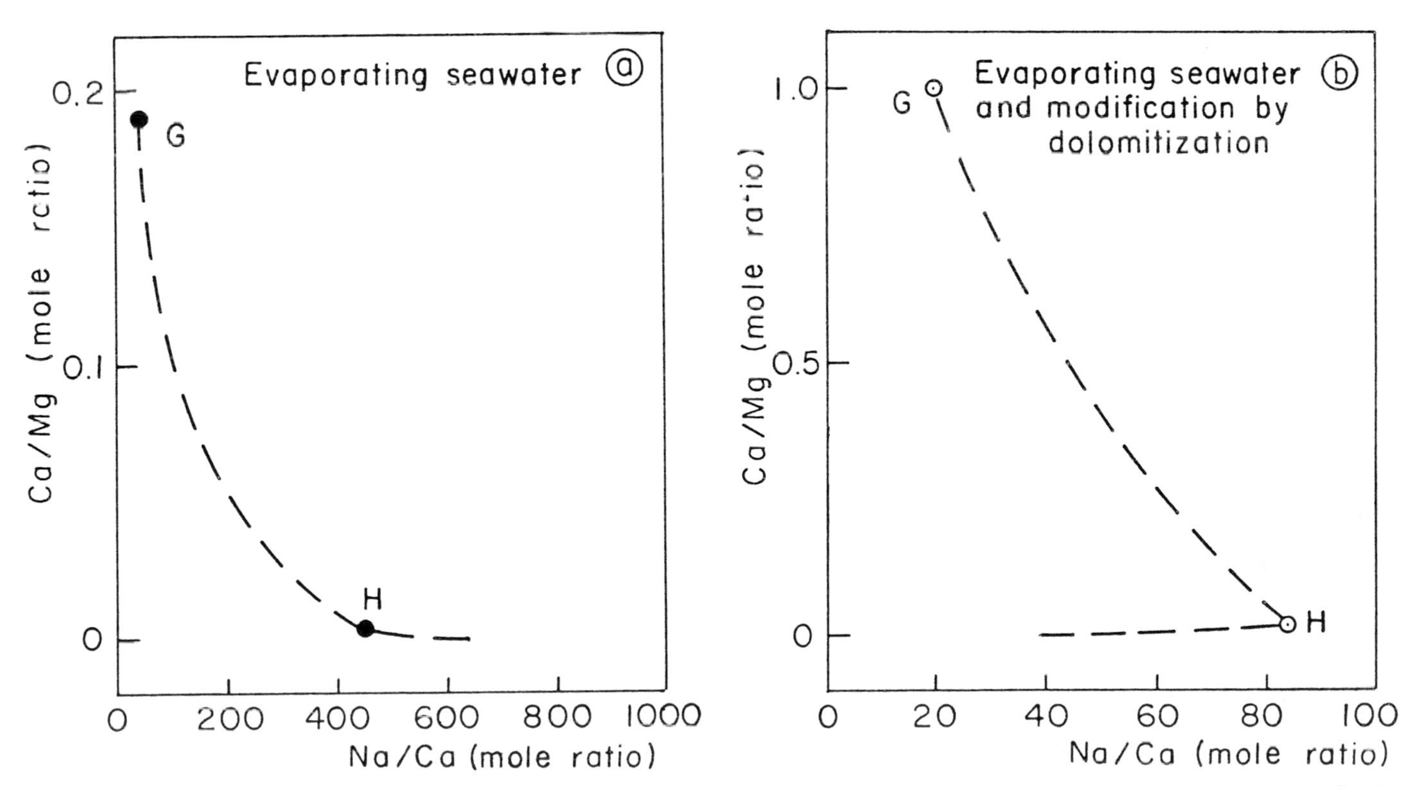

FIG. 6.—Ca/Mg vs. Na/Ca expected covariation in seawater-derived solutions due to evaporation alone (a) and to evaporation *and* modification through dolomitization (b). Letter designation (SW, G, H) is indicated in Figure 3. Note smooth and continuous increase of Na/Ca with decrease of Ca/Mg in (a), and sharp reversal in Na/Ca change beyond point *H* in (b).

tion, the structural formula of dolomite can be expressed as $Ca_{0.5}Mg_{0.5-x}Ca_xCO_3$, in which the *B* site Ca (Ca_x) behaves differently from the *A* site Ca, and can be considered as a distinct species, substituting for Mg.

We are aware that equation 2 can be an approximation only, being based on the assumption that magnesium does not substitute for calcium in the *A* site. The correlation existing between the expected trends in solution and the ones observed in dolomites suggests that equation 2 is approximately valid. In view of the uncertainties involved with partition mechanisms and constants in geochemistry, it seems that equation 2 can be used until further experimental and theoretical refinements are made.

An estimate of the partition coefficient of Ca_x in dolomite can be obtained for the limiting case of the highest Ca/Mg in solution (approximately 1) and highest Ca_x (approximately 6.5% or 0.065, Fig. 1). Substituting these values in equation 2 results in D^D_{Cax} of approximately 0.15.

The calculated value of the partition coefficient evidently depends on the chosen Ca/Mg equilibrium ratio between calcite and dolomite. As is shown by Carpenter (1980), this value increases markedly with the increase in salinity and temperature, which should be accompanied by a decrease in D^D_{Cax}. The derived *D* value can be tested by calculating the predicted Ca_x for dolomites forming under given Ca/Mg ratios in solution and comparing them with observed Ca_x values. This is illustrated for two cases: Ca/Mg ratio in solution of 0.18 (a non-modified seawater value). The predicted Ca_x is 1.3 mole percent, quite close to stoichiometry. This compares quite well with (1) the low range of Ca_x in the Judea Group dolomites (Fig. 1), interpreted as forming from "marine" open-system solutions, and (2) Ca/Mg ratio of 0.02, which applies to a modified solution in the halite range of salinity (Fig. 5a). The predicted Ca_x is 0.15 mole percent, which is even closer to stoichiometry than the first case. Indeed, the San Andres dolomites (in association with halite) are virtually stoichiometric (Fig. 1). Considering the approximations used in deriving the Ca_x partition coefficient, the conformity between the calculated and observed Ca_x values corroborates the presented model of bulk dolomite composition.

Sodium.—

The compositional trends observed in the studied dolomites suggest that the Na/Ca ratio in dolomites is roughly proportional to the Na/Ca ratio in solution. This behavior is consistent with the view that sodium substitutes for calcium in its lattice site. The distribution function can be described in a modified version of the Berthelot-Nernst law (McIntire, 1963) as follows:

$$(Na/Ca)^D = K'^D_{Na} \cdot (aNa/aCa)^L \quad (3)$$

in which the Na/Ca (molar or activity) ratio in dolomite is proportional to the activity ratio (aNa/aCa) in solution.

The use of a *prime* partition coefficient (K'^D_{Na}) is warranted due to the requirement for electrical neutrality, dictating that substitution of sodium ion (monovalent) for calcium (divalent) must be accompanied by a concomitant substitution of still another ion or by the creation of vacancies (McIntire, 1963). A single partition coefficient for a coupled substitution cannot, therefore, be a thermodynamic constant but may serve as an empirical one. Owing to our actual ignorance of the nature of the additional substitution, the prime partition coefficient cannot, as yet, be evaluated.

Instead of using an activity ratio and a quasi-thermodynamic constant, the distribution between dolomite and solution can be expressed in terms of molar values, as follows:

$$(Na/Ca)^D = D^D_{Na} \cdot (mNa/mCa)^L \quad (4)$$

where D^D_{Na} is the molar partition coefficient of sodium in dolomite. It can be demonstrated (Sass and Katz, 1982) that equations 3 and 4 are analogous, and this requires the following identity:

$$D^D_{Na} \equiv K'^D_{Na} \cdot (\gamma Na/\gamma Ca)^L \quad (5)$$

This relationship indicates that the activity coefficient ratio in solution [$(\gamma Na/\gamma Ca)^L$] plays an important role in determining the molar partition coefficient. Due to the sensitivity of this ratio to salinity variations, especially where altervalent ions are involved (the "solution interaction factor" of McIntire, 1963), D^D_{Na} is not expected to remain constant across the wide range of salinity deduced for the studied dolomites.

Empiric D^D_{Na} values can be estimated for the two limits of dolomite formation in association with gypsum (400 and 2,700 ppm Na) by resorting to the conclusion, derived at the beginning of this section, that those dolomites formed from modified solutions, and by assuming that the limiting values represent rather closely the *G* and *H* points (Figs. 4, 5b), respectively.

The Na/Ca ratios of the dolomites for those limits are $2.85 \cdot 10^{-3}$ and $21.6 \cdot 10^{-3}$. The Na/Ca ratios in solution (modified) of the respective points *G* and *H* are 20 and 85. By substituting these values in equation 4, the obtained estimates for the sodium partition coefficient in those dolomites are $1.4 \cdot 10^{-4}$ and $2.5 \cdot 10^{-4}$, respectively.

As shown in Figure 7 (based on Whitfield, 1979, and Krumgalz and Millero, 1982), the $(\gamma Na/\gamma Ca)^L$ ratio first increases with ionic strength, reaches a maximum at ionic strength of about 2, then decreases. It could be expected that D^DNa would change similarly, but the findings of Katz and others (1977) show that the sodium partition coefficient in aragonite keeps rising through the entire salinity range of gypsum (as much as TDS of 300 g/l). These findings, together with the D^DNa figures calculated above, suggest that the maximum of the *D* curve is located at a higher salinity than that of $(\gamma Na/\gamma Ca)^L$.

The anticipated decrease of both the Na/Ca ratio (Fig. 6b) and D^DNa at the salinity range of halite should result in a very low Na/Ca ratio in dolomite, in accordance with the observed low-sodium content of the San Andres ("associated with halite") dolomites (Fig. 1). This qualitative conclusion cannot as yet be quantified, because, as was shown above, D^D_{Na} is controlled by both $(\gamma Na/\gamma Ca)^L$ and K'^D_{Na} (equation 5), and the dependence of K'^D_{Na} upon salinity is still undefined.

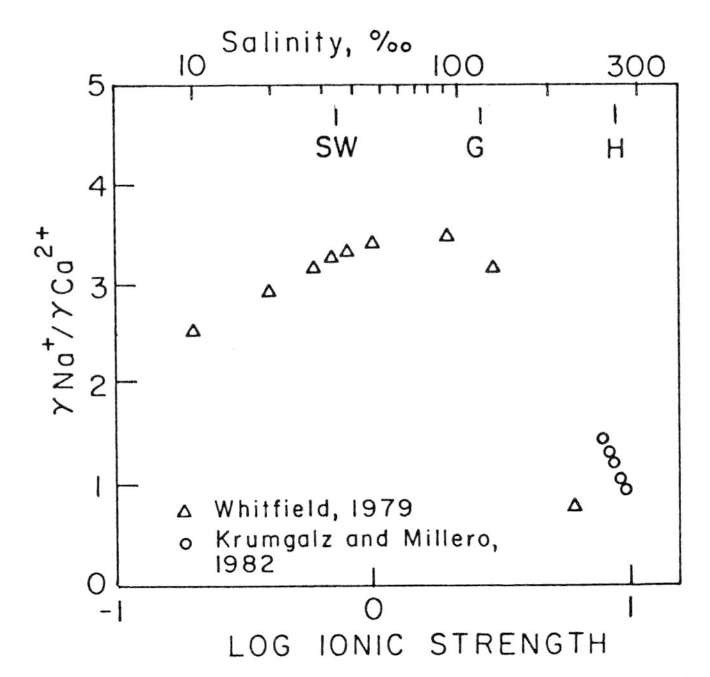

FIG. 7.—γNa/γCa activity coefficient ratio as a function of ionic strength and salinity. Note that the γNa/γCa ratio first increases with ionic strength, reaches a maximum at an ionic strength of about 2, and then decreases.

Stable Oxygen Isotopes

The $\delta^{18}O$ of carbonates reflects both the temperature and isotopic composition of the mother solution, and as such, is a very complicated function of salinity, hydrography, climate, and diagenesis. The nature of the available data in the present study does not justify a detailed examination of the subject, and we have restricted ourselves to evaluating the general trends only.

The $\delta^{18}O$ values of the gypsum-associated dolomites are significantly higher (by about 3‰ on the average) than those of the marine dolomites (Fig. 2). This observation agrees with the expected rise of $\delta^{18}O$ in solution with an increase of salinity (Lloyd, 1966). The $\delta^{18}O$ values of the halite-associated dolomites do not exhibit a continuation of this trend; instead, they are similar to those of the marine dolomites (Fig. 2).

This apparent deviation does not require the effect of fresher solutions. It has been repeatedly observed that $\delta^{18}O$ tends to decrease with an increase of salinity beyond a certain value of around 140‰ (Sofer and Gat, 1972, 1975; Nadler and Magaritz, 1980; Gonfiantini, 1986).

It appears, therefore, that the relatively low $\delta^{18}O$ values of the halite-associated dolomites reflect the reversal effect of $\delta^{18}O$ in evaporating brines, exhibiting consistency with the predicted behavior, as also suggested by Bein and Land (1982, 1983). In conclusion, the entire trend of $\delta^{18}O$ variations in the studied dolomites seems to conform to the expected behavior of $\delta^{18}O$ in evaporating seawater-derived solutions.

CONCLUSIONS

The geochemical trends revealed in the study of ancient dolomites representing widely different salinities can be explained in terms of original solution chemistry and distribution functions. The main conclusions drawn are the following:

(1) The stoichiometry and sodium content of the dolomites are, to a first approximation, the result of partitioning of ionic ratios (Ca/Mg, Na/Ca) between solid and liquid.

(2) The factors governing the ionic ratios in solution are a combination of two processes: (a) the degree of evaporation and evaporite precipitation, and (b) the modification of porewater chemistry through the process of dolomitization.

(3) The partition coefficient of sodium in dolomite is very sensitive to variations in salinity, most probably due to the strong dependence of the activity coefficient ratio, γNa/γCa, upon ionic strength.

(4) In the pre-evaporitic stage, the ionic ratios change very little during surface evaporation of sea water. Modification of pore water, therefore, plays the dominant role in controlling ionic ratios. The deviation of ionic ratios reflects the degree of isolation of the porewater system from the overlying water body. Within this range of salinity, it seems that dolomitization takes place under varying degrees of isolation, covering the entire range from open to closed systems. This explanation accounts for the wide spread of stoichiometry of the "marine" dolomites.

(5) Ionic ratios in hypersaline solutions are affected both by precipitation of evaporites and by modification in (nearly) closed systems. As a result, within the intermediate range of evaporation, at the stage of gypsum and anhydrite precipitation, the Ca/Mg ratio in solution drops with increase of salinity, and the dolomite composition changes gradually from calcian to stoichiometric. Concomitantly, both the Na/Ca and the γNa/γCa ratios in solution increase, as is reflected in the increase of sodium content in the dolomites.

(6) At the higher range of salinity, beyond the onset of halite precipitation, the Ca/Mg ratio in solution keeps decreasing and, consequently, the dolomites attain a nearly stoichiometric composition. The trends of Na/Ca and γNa/γCa ratios in solution, applying in the lower range of salinity, are reversed. This is reflected in the low values of sodium content characterizing the dolomites associated with halite.

(7) The observed trend of stable oxygen-isotopic composition of the dolomites is consistent with $\delta^{18}O$ variations in evaporating seawater-derived solutions. During evaporation and salinity increase of such solutions, the heavy-oxygen content of the water first increases, then levels off, and finally drops down to values approaching those of sea water.

ACKNOWLEDGMENTS

We thank L. Halicz, H. Eldad, and R. Binstock for performing the chemical analyses, and A. Pe'er for drawing the various versions of the figures. We are also grateful to Y. Druckman who cooperated in the field work in the Jordan Valley, and to J. Price, V. Shukla, and an anonymous reviewer for their invaluable constructive criticism and editorial comments.

The isotopic and XRD determinations were carried out

at the Department of Geology, the Hebrew University. The chemical analyses of the new data were performed at the geochemical laboratory of the Geological Survey of Israel.

REFERENCES

ATWOOD, D. K., AND FRY, H. M., 1967, Strontium and manganese content of some coexisting calcites and dolomites: American Mineralogist, v. 52, p. 1530–1535.

BASEGGIO, GINO, 1974, The composition of sea water and its concentrates, *in* Coogan, A. H., ed., Fourth Symposium on Salt: Northern Ohio Geological Society, Cleveland, Ohio, p. 351–358.

BEIN, AMOS, 1976, Rudistid fringing reefs of Cretaceous shallow carbonate platforms of Israel: American Association of Petroleum Geologists Bulletin, v. 60, p. 258–272.

———, AND LAND. L. S., 1982, San Andres Carbonates in the Texas Panhandle: Sedimentation and diagenesis associated with magnesium-calcium-chloride brines: Bureau of Economic Geology, The University of Texas at Austin, Report of Investigations No. 121, p. 48.

———, ———, 1983, Carbonate sedimentation and diagenesis associated with Mg-Ca chloride brines: The Permian San Andres Formation in the Texas Panhandle: Journal of Sedimentary Petrology, v. 53, p. 243-260.

BERNER, R. A., 1971, Principles of Chemical Sedimentology: McGraw-Hill, New York, 240 p.

BUTLER, G. P., 1969, Modern evaporite deposition and geochemistry of coexisting brines, the sabkha, Trucial Coast, Arabian Gulf: Journal of Sedimentary Petrology, v. 39, p. 70–89.

CARPENTER, A. B., 1980, The chemistry of dolomite formation. I: The stability of dolomite, *in* Zenger, D. H., Dunham, J. B., and Ethington, R. L., eds., Concepts and Models of Dolomitization: Society of Economic Paleontologists and Mineralogists Special Publication 28, p. 111–121.

DICKSON, J. A. D., AND COLEMAN, M. L., 1980, Changes in carbon and oxygen isotope composition during limestone diagenesis: Sedimentology, v. 27, p. 107–118.

DUNHAM, J. B., AND OLSON, E. R., 1979, Shallow subsurface dolomitization of subtidally deposited carbonate sediment in Hanson Creek Formation (Ordovician-Silurian), central Nevada–Evidence for groundwater mixing (abst.): American Association of Petroleum Geologists Bulletin, v. 63, p. 442.

FAIRBRIDGE, R. W., 1957, The dolomite question, *in* LeBlanc, R. J., and Breeding, J. G., eds., Regional Aspects of Carbonate Deposition: Society of Economic Paleontologists and Mineralogists Special Publication 5, p. 125–178.

FAIRCHILD, I. J., 1985, Petrography and carbonate chemistry of some Dalradian dolomitic metasediments: Preservation of diagenetic textures: Journal of the Geological Society of London, v. 142, p. 167–185.

FOLK, R. L., AND LAND, L. S., 1975, Mg/Ca ratio and salinity: Two controls over crystallization of dolomite: American Association of Petroleum Geologists Bulletin, v. 59, p. 60–68.

FRITZ, PETER, 1971, Geochemical characteristics of dolomites and the ^{18}O content of middle Devonian oceans: Earth and Planetary Science Letters, v. 11, p. 277–282.

———, AND KATZ, AMITAI, 1972, The sodium distribution of dolomite crystals: Chemical Geology, v. 10, p. 237–244.

GIVEN, R. K., AND LOHMANN, K. C., 1985, Derivation of the original isotopic composition of Permian marine cements: Journal of Sedimentary Petrology, v. 55, p. 430–439.

GOLDSMITH, J. R., AND GRAF, D. L., 1958, Structural and compositional variations in some natural dolomites: Journal of Geology, v. 66, p. 678–693.

GONFIANTINI, ROBERTO, 1986, Environmental isotopes in lake studies, *in* Fritz, Peter, and Fontes, J.Ch., eds., Handbook of Environmental Isotope Geochemistry: Elsevier, Amsterdam, p. 113–168.

GORDON, C. M., CARR, R. A., AND LARSON, R. E., 1970, The influence of environmental factors on the sodium and manganese content of barnacle shells: Limnology and Oceanography, v. 15, p. 461–466.

HANSHAW, B. B., BACK, WILLIAM, AND DEIKE, R. G., 1971, A geochemical hypothesis for dolomitization by groundwater: Economic Geology, v. 66, p. 710–724.

HERRMANN, A. G., KNAKE, DORIS, SCHNEIDER, JÜRGEN, AND PETERS, HEIDE, 1973, Geochemistry of modern seawater and brines from salt pans: Main components and bromine distribution: Contributions to Mineralogy and Petrology, v. 40, p. 1–24.

ILLING, L. V., WELLS, A. J., AND TAYLOR, J. C. M., 1965, Penecontemporaneous dolomite in the Persian Gulf, *in* Pray, L. C., and Murray, R. C., eds., Dolomitization and Limestone Diagenesis: Society of Economic Paleontologists and Mineralogists Special Publication 13, p. 89–111.

ISHIKAWA, M., AND ICHIKUNI, M., 1984, Uptake of sodium and potassium by calcite: Chemical Geology, v. 42, p. 137–146.

KATZ, AMITAI, 1971, Zoned dolomite crystals: Journal of Geology, v. 79, p. 38–51.

———, KOLODNY, YEHOSHUA, AND NISSENBAUM, ARIE, 1977, The geochemical evolution of the Pleistocene Lake Lisan-Dead Sea system: Geochimica et Cosmochimica Acta, v. 41, p. 1609–1626.

KRUMGALZ, B. S., AND MILLERO, F. J., 1982, Physico-chemical study of the Dead Sea waters. 1. Activity coefficients of major ions in Dead Sea water: Marine Chemistry, v. 11, p. 209–222.

LAND, L. S., 1980, The isotopic and trace element geochemistry of dolomite: The state of the art, *in* Zenger, D. H., Dunham, J. B., and Ethington, R. L., eds., Concepts and Models of Dolomitization: Society of Economic Paleontologists and Mineralogists Special Publication 28, p. 87–110.

———, 1985, The origin of massive dolomite: Journal of Geological Education, v. 33, p. 112–125.

———, AND HOOPS, G. K., 1973, Sodium in carbonate sediments and rocks: A possible index to the salinity of diagenetic solutions: Journal of Sedimentary Petrology, v. 43, p. 614–617.

LAZAR, N., STARINSKY, A., KATZ, A., SASS, E., AND BEN-YAAKOV, S., 1983, The carbonate system in hypersaline solutions: Alkalinity and $CaCO_3$ solubility of evaporated seawater: Limnology and Oceanography, v. 28, p. 978–986.

LIPPMAN, FRIEDRICH, 1973, Sedimentary Carbonate Minerals: Springer-Verlag, Berlin, 228 p.

LLOYD, R. M., 1966, Oxygen isotope enrichment of seawater by evaporation: Geochimica et Cosmochimica Acta, v. 30, p. 801–814.

LUMSDEN, D. N., AND CHIMAHUSKY, J. S., 1980, Relationship between dolomite nonstoichiometry and carbonate facies parameters, *in* Zenger, D. H., Dunham, J. B., and Ethington, R. L., eds., Concepts and Models of Dolomitization: Society of Economic Paleontologists and Mineralogists Special Publication 28, p. 123–137.

MARCUS, E., AND SLAGER, J., 1985, The sedimentary-magmatic sequence of the Zemah 1 well (Jordan-Dead Sea rift, Israel) and its emplacement in time and space: Israel Journal of Earth Sciences, v. 34, p. 1-10.

MARSCHNER, HANNELORE, 1968, Ca-Mg distribution in carbonates from the lower Keuper in northwest Germany, *in* Müller, German, and Friedman, G. M., eds., Recent Developments in Carbonate Sedimentology in Central Europe: Springer-Verlag, New York, p. 128–135.

MCCREA, J. M., 1950, On the isotopic chemistry of carbonates and a paleotemperature scale: Journal of Chemical Physics, v. 18, p. 849-857.

MCINTIRE, W. L., 1963, Trace element partition coefficients–A review of theory and application in geology: Geochimica et Cosmochimica Acta, v. 27, p. 1209–1264.

MCKENZIE, J. A., 1985, Stable isotope mapping in Messinian evaporative carbonates of central Sicily: Geology, v. 13, p. 851–854.

MÖLLER, PETER, RAJAGOPALAN, GOVINDARAJA, AND GERMANN, KLAUS, 1976, A geochemical model for dolomitization based on material balance, part II: Trace element distribution during dolomitization: Geologisches Jahrbuch, v. D20, p. 57–76.

MORROW, D. W., 1978, Dolomitization of Lower Paleozoic burrow fillings: Journal of Sedimentary Petrology, v. 48, p. 295–305.

NADLER, A., AND MAGARITZ, M., 1980, Studies of marine solution basins–Isotopic and compositional changes during evaporation, *in* Nissenbaum, A., ed., Hypersaline brines and evaporitic environments: Developments in Sedimentology 28: Elsevier, Amsterdam, p. 115–129.

PATTERSON, R. J., AND KINSMAN, D. J. J., 1974, Crystal chemistry and morphology of Recent dolomite from the Persian Gulf (abst.): EOS, American Geophysical Union, v. 55, p. 457.

———, ———, 1982, Formation of diagenetic dolomite in coastal sabkha along Arabian (Persian) Gulf: American Association of Petroleum Geologists Bulletin, v. 66, p. 28–43.

REEDER, R. J., 1983, Crystal chemistry of the rhombohedral carbonates, *in* Reeder, R. J., ed., Carbonates: Mineralogy and chemistry: Mineralogical Society of America, p. 1–47.
RENARD, MAURICE, 1985, Géochimie des carbonates pélagiques: Mise en évidence des fluctuations de la composition des eaux océaniques depuis 140 Ma. Essai de chimiostratigraphie: Mémoire Science Terre, Université Pierre et Marie Curie, Paris, v. 85-16, 650 p.
RICHTER, D. K., 1974, Origin and diagenesis of Devonian and Permotriassic dolomites in the Eifel Mountains (Germany), in German, with English summary: Contributions to Sedimentology, v. 2, p. 1–101.
ROSENFELD, A., SEGEV, A., AND HALBERGSBERG, E., 1981, Ostracod species and paleosalinities of the Pliocene Bira and Gesher Formations (Northwestern Jordan Valley): Israel Journal of Earth Sciences, v. 30, p. 113–119.
SASS, EYTAN, AND BEIN, AMOS, 1978, Platform carbonates and reefs in the Judean Hills, Carmel and Galilee: Tenth International Congress on Sedimentology, Guidebook, Part II, p. 239–274.
———, ———, 1982, The Cretaceous carbonate platform in Israel: Cretaceous Research, v. 3, p. 135–144.
———, AND KATZ, AMITAI, 1982, The origin of platform dolomites: New evidence: American Journal of Science, v. 282, p. 1184–1213.
SCHMIDT, VOLKMAR, 1965, Facies, diagenesis, and related reservoir properties in the Gigas Beds (Upper Jurassic), northwestern Germany, *in* Pray, L. C., and Murray, R. C., eds., Dolomitization and Limestone Diagenesis: Society of Economic Paleontologists and Mineralogists Special Publication 13, p. 124–168.
SCHULMAN, NACHMAN, 1962, The geology of the central Jordan Valley: Unpublished Ph.D. Dissertation, in Hebrew, with English summary, The Hebrew University, Jerusalem, 103 p.
SOFER, ZVI, AND GAT, J. R., 1972, Activities and concentrations of oxygen-18 in concentrated aqueous salt solutions: Analytical and geophysical implications: Earth and Planetary Science Letters, v. 15, p. 232–238.
———, ———, 1975, The isotope composition of evaporating brines: Effect on the isotope activity ratio in saline solutions: Earth and Planetary Science Letters, v. 26, p. 179–186.
SONNENFELD, PETER, 1986, Stable isotope mapping in Messinian evaporative carbonates of central Sicily: A comment: Geology, v. 14, p. 799.
STARINSKY, AVRAHAM, 1974, Relationship between Ca-Chloride brines and sedimentary rocks in Israel: Unpublished Ph.D. Dissertation, in Hebrew, with English summary, The Hebrew University, Jerusalem, 176 p.
VEIZER, JAN, 1985, Carbonates and ancient oceans: Isotopic and chemical record on time scales of 10^7–10^9 years, *in* Sundquist, E. T., and Broecker, W. S., eds., The Carbon Cycle and Atmospheric CO_2; Natural Variations, Archean to Present: Geophysical Monographs, p. 595–601.
———, LEMIEUX, JEAN, JONES, BRIAN, GIBLING, M. R., AND SAVELLE, JIM, 1977, Sodium: Paleosalinity indicator in ancient carbonate rocks: Geology, v. 5, p. 177–179.
WARD, W. C., AND HALLEY, R. B., 1985, Dolomitization in a mixing zone of near-seawater composition, Late Pleistocene, northeastern Yucatan Peninsula: Journal of Sedimentary Petrology, v. 55, p. 407–420.
WHITFIELD, MICHAEL, 1979, Activity coefficients in natural waters, *in* Pytkowicz, R. M., ed., Activity Coefficients in Electrolyte Solutions: CRC Press, Boca Raton, Florida, p. 153–299.
ZHEREBTSOVA, I. K., AND VOLKOVA, N. N., 1966, Experimental study of behaviour of trace elements in the process of natural solar evaporation of Black Sea water and Sasyk-Sivash brine: Geochemistry International, v. 3, p. 656–670.

APPENDIX 1

Butler's (1969) data on the chemical composition of solutions in the Persian Gulf are incomplete, including solely the determined values of specific gravity, chloride, sulfate, calcium, and magnesium. Potassium was determined in a small part only in the studied solutions.

The TDS and sodium values used in plotting Figure 5 are calculated as follows: TDS is empirically related to d (specific gravity) by TDS = $(1000 \cdot d - 997)/0.74$ (Starinsky, 1974). Sodium and potassium concentrations are calculated by solving the two simultaneous equations:

$$mNa^+ + mK^+ = 2mSO_4^{2-} + mCl^- - 2(mCa^{2+} + mMg^{2+}) \quad (A1)$$

$$mK^+ = 0.125(mCl^- - mNa^+) \quad (A2)$$

Equation A1 expresses the requirement for charge balance and is only approximate, since it does not include all existing ions, in particular, alkalinity.

Equation A2 is based on the assumptions that potassium is conservative and that precipitation of halite is the only heterogeneous process that affects sodium and chloride concentrations. The assumptions involved, as well as the computational method of obtaining the potassuium concentration through subtraction of two large values, must result in some error. A comparison of the calculated and analyzed values of potassium (where reported by Butler, 1969) indicates, however, that the discrepancy is usually very small and, by implication, the error involved with the calculated sodium concentration does not exceed 1 percent.

NEOGENE DEEP-WATER DOLOMITE FROM THE FLORIDA-BAHAMAS PLATFORM

HENRY T. MULLINS AND GEORGE R. DIX
Department of Geology, Heroy Geology Laboratory, Syracuse University, Syracuse, New York 13244
ANNE F. GARDULSKI
Department of Geology, Tufts University, Medford, Massachusetts 02155
AND
LYNTON S. LAND
Department of Geological Sciences, University of Texas, Austin, Texas 78713

ABSTRACT: Authigenic calcian dolomite is a common but rarely abundant (≤20%) component of Neogene deep-water (475–2,767 m) carbonates peripheral to the Florida-Bahamas Platform. Dolomite concentrations as high as 57% of the carbonate fraction occur in the Miocene of west Florida, however, and as much as 86% dolomite has been found in a hardground from the Bahamas.

Dolomite occurs principally as pore-filling euhedral rhombs (5–20 μm) that precipitated *in situ*, as well as by replacement of calcite at disconformities. Stable isotope ratios (oxygen and carbon) suggest dolomite precipitation from deep, cold, seawater-derived fluids, and trace-element (Sr) concentrations suggest strontium-rich aragonitic(?) precursors. Preliminary $^{87}Sr/^{86}Sr$ data suggest substantial lag times for dolomite precipitation and contamination by "old" strontium.

Because of the high diagenetic potential of periplatform carbonates, Bahamian deep-water dolomites appear to be a natural consequence of shallow subsurface (<60 m) burial diagenesis. In contrast, carbonate ramp slope sediments from west Florida, which have a much lower initial diagenetic potential, are punctuated by discrete concentrations of authigenic dolomite, which may represent paleoceanographically controlled dolomite "events." Overall, our data indicate that deep-water dolomite has had little difficulty in precipitating from normal marine-derived fluids.

INTRODUCTION

One of the long-standing difficulties in resolution of the "dolomite problem" has been the widely accepted premise that dolomite cannot precipitate from normal sea water at earth surface temperatures (Land, 1980, 1985). As a consequence, "special waters," such as those in mixing zones, on sabkhas, or within organic-rich sediments, have been sought as media for dolomitization (Machel and Mountjoy, 1986; Hardie, 1987). Recent discoveries of authigenic dolomite demonstrate, however, that dolomite can precipitate from sea water (Saller, 1984; Mullins and others, 1985; Carballo and others, 1987; Mitchell and others, 1987).

Relatively small percentages (5–15%) of dolomite are common in post-Jurassic deep-sea sediments throughout the world ocean sampled by the Deep Sea Drilling Project (DSDP; Lumsden, 1985). Most of these occurrences, however, are products of fluids derived from underlying evaporites, extensive bacterial degradation of organic-rich sediments, or submarine weathering of basalts (Garrison, 1981). Small quantities of dolomite have also been reported from deep-water periplatform carbonates that surround shallow-water platforms along continental margins (Zemmels and others, 1972; Worzel and others, 1973), but they were never further investigated, although Mitchum (1978) proposed that some west Florida slope dolomite might be authigenic. It was not until geochemical studies of Miocene-Pliocene deep-water periplatform dolomites north of Little Bahama Bank were undertaken (Mullins and others, 1984, 1985) that their authigenic origin became apparent. The regional significance of deep-water dolomites in the Bahamas was subsequently confirmed by preliminary shipboard X-ray diffraction data during Ocean Drilling Program (ODP) leg 101 (Austin and others, 1986). Shipboard analyses were on a limited number of samples, however, and no geochemical or scanning electron microscopic (SEM) data for the dolomites were generated.

The purpose of this paper is to report new data for authigenic dolomite in Neogene deep-water carbonates recovered from drill and piston cores peripheral to the Florida Bahamas platform and to discuss their significance toward resolution of the "dolomite problem."

METHODS

Sample spacing for Bahamian drill cores was about every 50 cm from 0 to 200 m subsurface and about every 1 m at greater subsurface depths. West Florida samples were taken every 20 cm from Exxon drill cores and every 10 cm from piston cores.

Quantitative carbonate mineralogy was determined by standard X-ray diffraction techniques corrected for the nonlinear relation of aragonite content with peak intensity (accuracy of 5 percent; Milliman, 1974). The mole percent $MgCO_3$ was evaluated by the shift of the d (211) peak from ideal dolomite, using the curve of Goldsmith and others (1961). Trace-element concentrations were determined by atomic absorption or inductively coupled plasma spectrometry.

Stable isotope ratios (relative to PDB) for oxygen ($^{18}O/^{16}O$) and carbon ($^{13}C/^{12}C$), as well as strontium ($^{87}Sr/^{86}Sr$) ratios, were determined by mass spectrometry on beneficiated dolomite. Dolomite was concentrated from bulk samples by leaching the fine fraction in 15% acetic acid for 2 hr. Complete removal of calcite was verified by X-ray diffraction and the dolomite concentrate analyzed. For strontium isotope ratios, samples were prepared in a manner similar to that of DePaolo (1986). Data have been normalized (by adding 0.00007 to determined values), so that modern marine skeletal carbonate yields values consistent with those of DePaolo and Ingram (1985).

Selected samples were observed in thin section and by SEM. A few Bahamian samples were also analyzed by microprobe and SEM backscatter imaging. Age estimates for

Sedimentology and Geochemistry of Dolostones, SEPM Special Publication No. 43

host sediments are based on planktonic foraminiferal and/or calcareous nannofossil biostratigraphy relative to the time scale of Berggren and others (1985).

RESULTS

Distribution of dolomite in Bahamian deep-sea drill cores.—

Dolomite occurs in relatively low concentrations (<20%) in all eight Bahamian deep-sea drill cores we have studied; the cores extend over a range of water depths from 807 m at ODP Site 630 to 2,767 m at DSDP Site 98 (Figs. 1, 2). Overall, dolomite is a common but rarely abundant component of deep-water periplatform carbonates in the Bahamas (Fig. 2). Dolomite has been found in fine-grained periplatform ooze, as well as in coarser grained turbidite layers and in limestones ranging in age from late Miocene to Pleistocene.

North of Little Bahama Bank, dolomite is present throughout most of ODP Site 630 (807 m) at concentrations of less than 13% (Fig. 2). At Site 628 (966 m), dolomite is restricted to periplatform facies within the upper 130 m of the drill core, as it is at Site 627 (1,027 m) between 12 and 76 m subsurface. There is also an unusual occurrence at Site 627 (249 m subsurface), where dolomite constitutes 86% of the carbonate fraction in a late Campanian hardground (Fig. 2).

In Exuma Sound, dolomite occurs throughout all three ODP drill cores below a certain subsurface depth, which decreases with increasing water depth: Site 631 (1,081 m), 64 m subsurface; Site 633 (1,681 m), 44 m; and Site 632 (1,996 m), 11 m. In all three drill cores, dolomite concentrations are always less than 20% and commonly less than 10% (Fig. 2).

Ocean Drilling Program Site 626 was plagued by poor recovery throughout (<37%), except for a mid-Miocene debris flow sequence between about 120 and 170 m subsurface. Because of this, the actual distribution of dolomite is impossible to depict. Dolomite was detected, however, at 46 m subsurface and within the Miocene debris flow, which has a maximum concentration of 23% at its disconformable top (Fig. 2).

At DSDP Site 98 (2,767 m) in Northeast Providence Channel (Hollister and others, 1972), dolomite is restricted to the upper 27 m of the drill core in late Miocene-Pleistocene sediment (Fig. 2). Dolomite concentrations are always less than 7%.

Microscopic characteristics of Bahamian deep-water dolomites.—

Our observations indicate that dolomite occurs as: (1) euhedral pore-filling rhombohedra 10–20 μm in size that occasionally contain embedded coccoliths (Fig. 3), indicating *in situ* precipitation, and (2) anhedral to subhedral microspar and pseudospar (commonly 5–50 μm in size, but ranging to 100 μm) that have formed by the replacement of calcite.

The replacement dolomite spar consists of clusters of crystals that form well-developed to vague outlines of much larger rhombohedra (Fig. 4). In finer crystalline clusters, dolomite crystals are inclusion rich with relict textures preserved, whereas in coarser crystalline spar, original textures

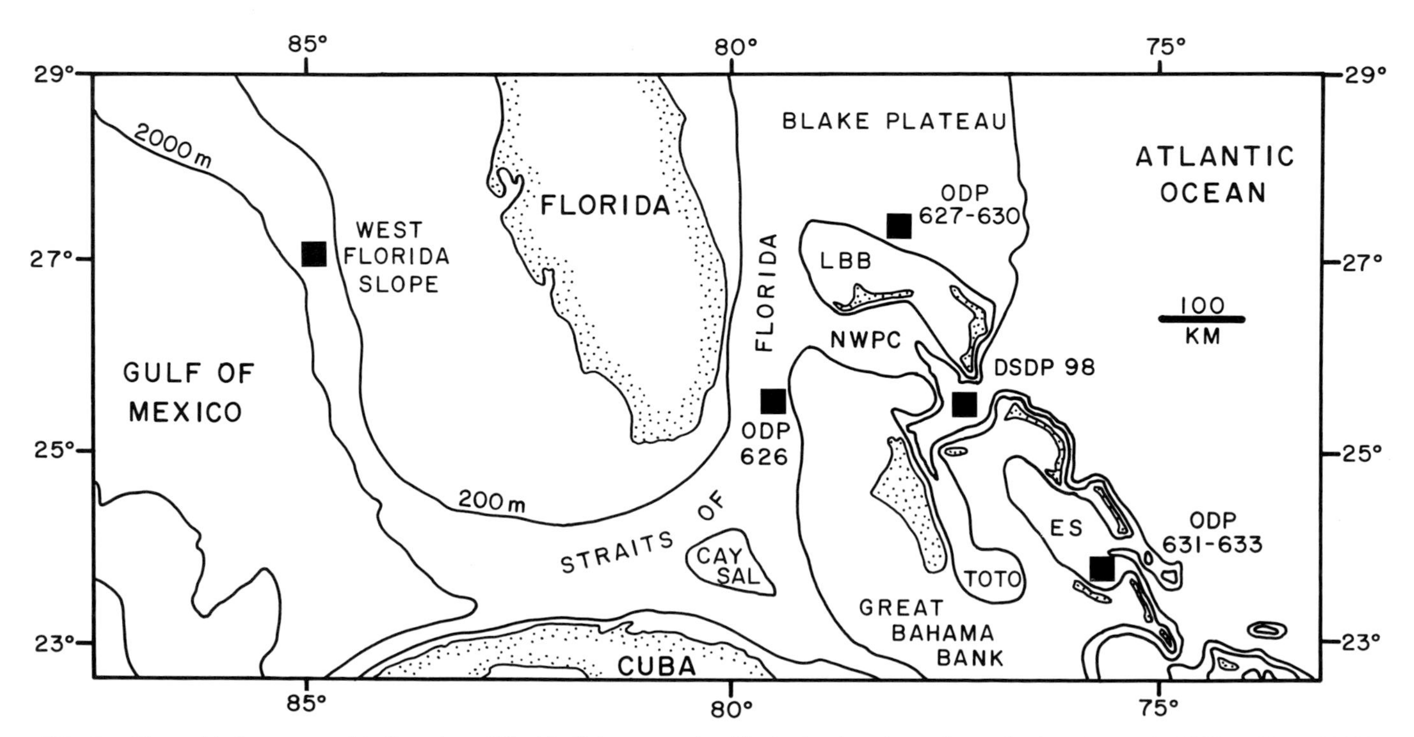

FIG. 1.—General index map and bathymetry of Florida-Bahamas region illustrating locations of sample sites (squares). ODP = Ocean Drilling Program; DSDP = Deep Sea Drilling Project; LBB = Little Bahama Bank; NWPC = Northwest Providence Channel; TOTO = Tongue of the Ocean; ES = Exuma Sound.

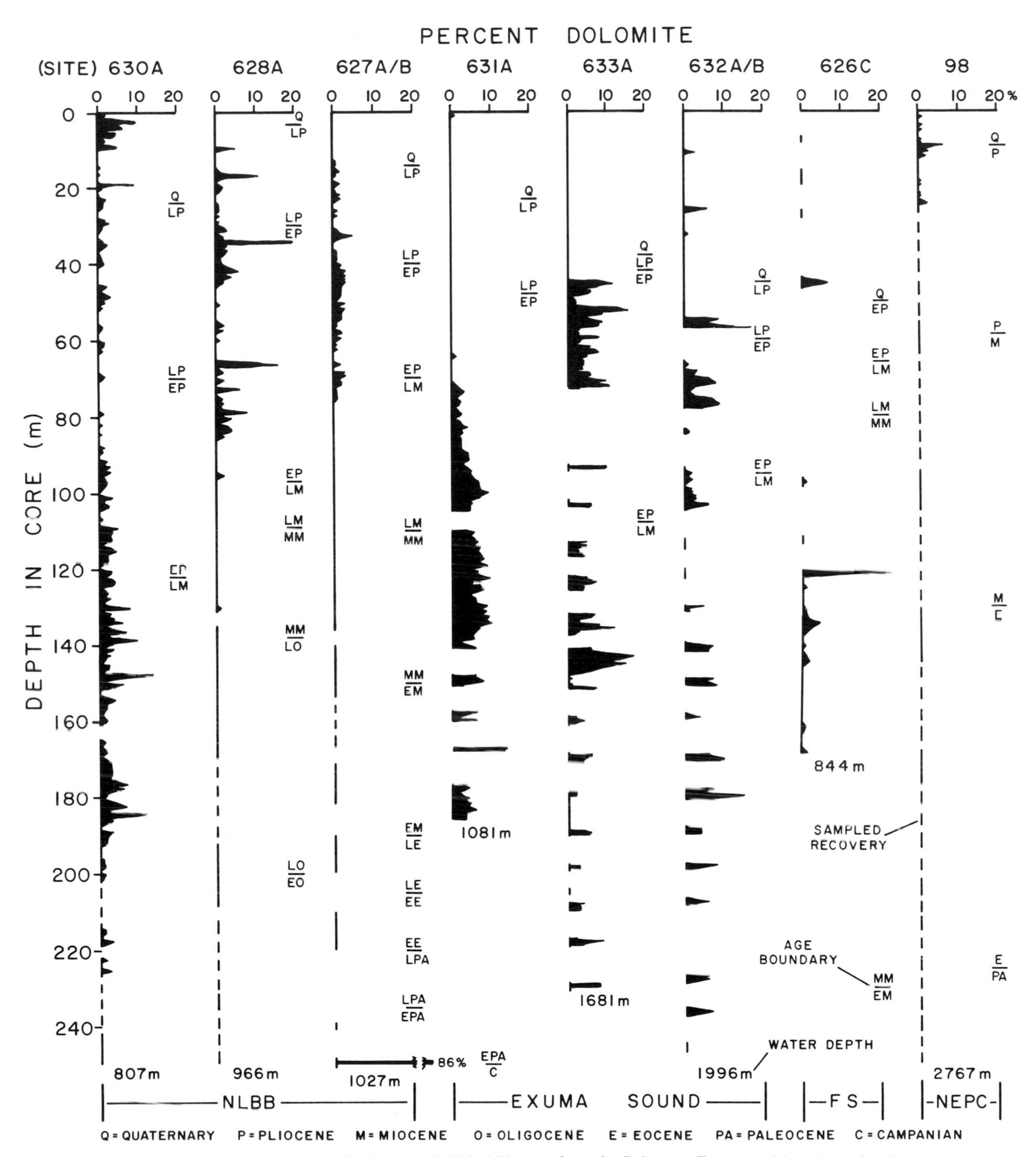

FIG. 2.—Downcore distribution of dolomite in ODP and DSDP drill cores from the Bahamas. For general locations of drill cores see Figure 1; for more detailed locations see Austin and others (1986) and Hollister and others (1972). Early (E), middle (M) and late (L) age divisions also shown; NLBB = northern Little Bahama Bank; FS = Florida Straits; NEPC = Northeast Providence Channel.

FIG. 3.—Scanning electron photomicrographs of deep-water dolomite. (A) Dolomite rhombs with embedded coccolith (arrow); early Pliocene, ODP Site 631 (83 m), northern Little Bahama Bank; scale bar = 5 μm. (B) Dolomite rhomb from 170 m subsurface in Exxon drill core CH-45 (Miocene) from west Florida slope with numerous embedded coccoliths; bar scale = 2 μm.

have been obliterated (Fig. 4). This complex replacement mode of dolomitization in deep-water periplatform carbonates is illustrated for the first time in SEM backscatter images (Fig. 5).

Mineralogy and geochemistry of Bahamian deep-water dolomites.—

X-ray diffraction data indicate that deep-water dolomite in the Bahamas is calcium rich, usually containing 57–59 mole percent $CaCO_3$; however, there is a range of values from near stoichiometric to 62 mole percent $CaCO_3$. Elemental analyses suggest a composition for these dolomites of 55–59 mole percent $CaCO_3$. Microprobe analyses of pore-fill dolomite from Site 626 indicate 54–56 mole percent $CaCO_3$, whereas replacement dolomite in the same sample has 59–62 mole percent $CaCO_3$.

Mullins and others (1985) have reported stable isotope ratios of oxygen and carbon for five dolomite samples of Miocene-Pliocene age from 600 m of water north of Little Bahama Bank. Values of $\delta^{18}O$ are heavy with a narrow range of +4.9 to +5.4‰. Two dolomite samples from Exuma Sound (Table 1) are also heavy, having $\delta^{18}O$ values of +6.4‰ and +6.7‰. Assuming: (1) an original sea water $\delta^{18}O$ value of 0‰ (Berger, 1979); (2) a $\Delta\delta^{18}O$ fractionation factor for dolomite of +3‰ relative to calcite (Land, 1980); and (3) using the following equation (Anderson and Arthur, 1983)

$$T°C = 16.9 - 4.2(\delta c - \delta w) + 0.13(\delta c - \delta w)^2 \quad (1)$$

where δc and δw represent values for calcite and original ambient sea water, respectively, we estimate paleotemperatures of dolomite formation that range from 1.8 to 9.4°C.

We interpret these oxygen isotope data as a consequence of dolomite precipitation in near equilibrium with cold, deep, normal marine waters. Maximum paleotemperature estimates (9.4°C) are for dolomites north of Little Bahama Bank, where water depth is only 600 m, whereas the lower paleotemperature estimates (1.8°C) are for dolomites in Exuma Sound, where water depths range from 1,081 to 1,996 m. Although we have no direct water temperature measurements from either northern Little Bahama Bank or Exuma Sound, Gulf of Mexico-Caribbean waters at a depth of 600 m are 5–10°C and less than 5°C at depths below about 1,000 m (Jones, 1973).

Stable carbon isotope ratios for Bahamian deep-water dolomites (Table 1) suggest that their carbon source has been from normal marine water. Six of our $\delta^{13}C$ values have a very narrow range of only +2.7 to +2.9‰ for samples from both northern Little Bahama Bank and Exuma Sound, with a seventh value of +3.6‰. We interpret these data as an indication of dolomitization from largely normal marine-derived pore fluids. An organic carbon source via complete sulphate reduction is unlikely on the basis of slightly positive $\delta^{13}C$ values (Irwin and others, 1977) and the presence of sulphate in interstitial waters (Austin and others, 1986).

Two dolomite samples from Exuma Sound were evaluated for their $^{87}Sr/^{86}Sr$ ratios to estimate the timing of dolomitization relative to the host sediment. A sample from Site 631 at 100 m subsurface has a strontium ratio of 0.70920 ± 5 and a sample from Site 632 at 180 m subsurface a value of 0.70896 ± 6. Using the data of DePaolo (1986) and assuming open-system dolomite precipitation, these ratios yield interpolated age estimates of dolomitization from modern to 2.5 ma and 8 to 13 ma, respectively.

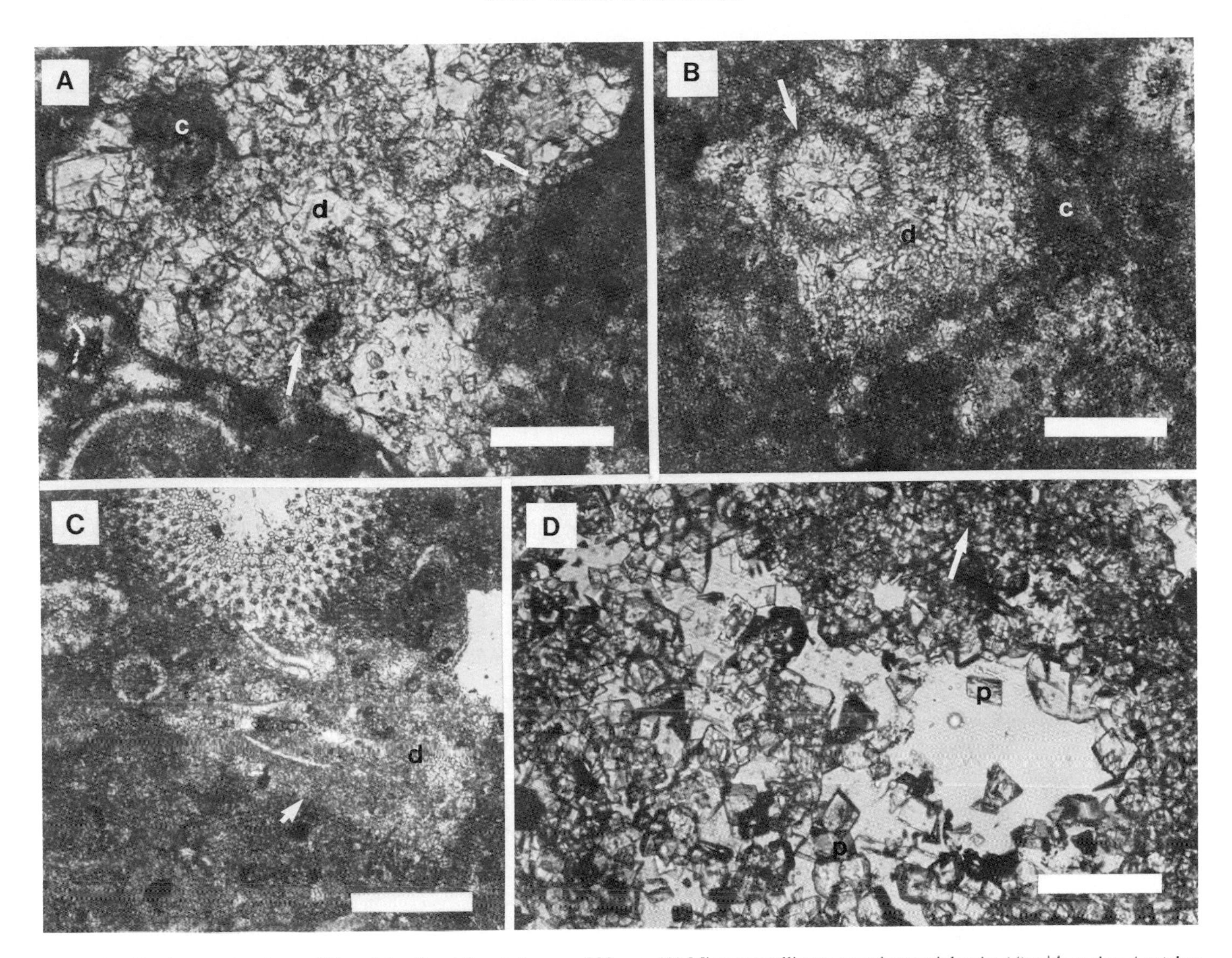

FIG. 4.—Replacement and pore-filling dolomite. All scale bars are 200 μm. (A) Microcrystalline to pseudospar dolomite (d) with enclosed patches of calcite (c and arrows). Microcrystalline dolomite is found typically asssociated with relic calcite fabric (upper arrow); middle Miocene, ODP Site 626C, 122 m subsurface. (B) Foraminifer, now dolomitized, within microcrystalline to pseudospar dolomite. Dolomite regions form rough outline of a rhombohedron; ODP Site 626C, 122 m. (C) Dolomite (d) replacing skeletal allochems, micrite, and part of a large foraminifer. A rectangular outline of the dolomite aggregate is present; ODP Site 627B, 82 m subsurface; late Miocene. (D) Dolomitic hardground, ODP Site 627B, 249 m, at a Paleocene-Campanian discontormity. Microcrystalline inclusion-rich dolomite (arrow) contains relic texture, whereas pore-filling dolomite (p) tends to form well-developed, clear crystals.

Biostratigraphically, the sample from 100 m at Site 631 is bound by nannofossil Zones NN12/13 and NN11 (4–8 ma) and by planktonic foraminiferal Zones N18/N19 (3–5 ma; Austin and others, 1986). Overlap of these zones suggests an early Pliocene age (4 ma) for the host sediment. Dolomitization at this site may therefore have occurred considerably later than sediment deposition by subsequent circulation of sea water. Such a time lag is supported by the fact that dolomite is virtually absent from the upper 70 m of sediment at Site 631 (Fig. 2).

For Site 632, sediment at 180 m subsurface is within nannofossil Zone NN11 (6–8 ma) and most likely near the top of this zone (Austin and others, 1986). The apparent older isotope age (8–13 ma) for dolomite relative to the sediment can be explained by either advective groundwater flow or diffusive loss of strontium, and by control of pore-fluid strontium by diagenesis of older carbonates. We wish to emphasize, however, the preliminary nature of these interpretations necessitated by our presently small data base and its inherent assumptions.

Distribution of deep-water dolomite on the west Florida slope.—

Deep-water dolomitic-carbonate sediments from the west Florida slope (Fig. 1) were sampled by two Exxon drill cores that extend 80 and 305 m into the subsurface, as well as by piston cores as long as 12 m. In Exxon drill core CH-45 from 1,477 m of water, early Miocene deep-water carbonates (argillaceous ooze) contain as much as 20% dolomite at a subsurface depth of 170 m (Fig. 6). At Exxon

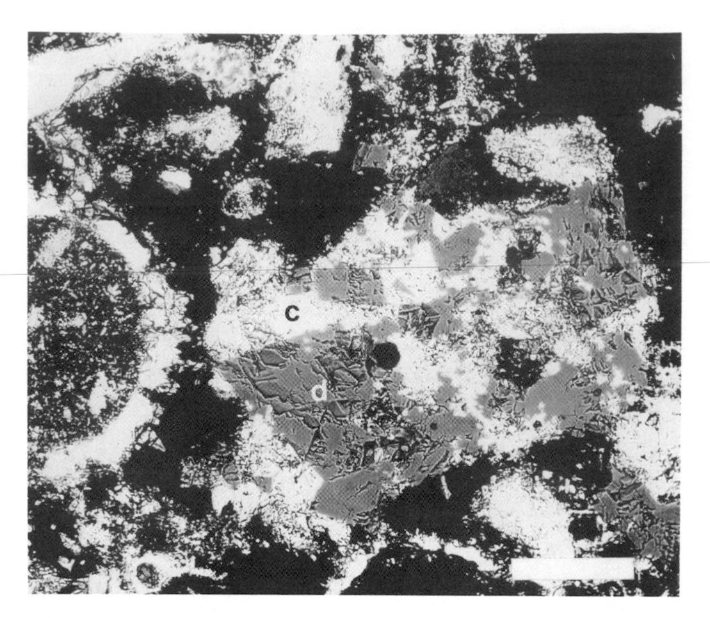

FIG. 5.—Scanning electron microscope backscatter photomicrograph illustrating complex replacement of calcite (white) by dolomite (gray). Note calcite test of planktonic foraminifer to left; black areas are pore space; scale bar = 100 μm; c = calcite; d = dolomite.

site CH-48 (495 m of water), dolomite occurs in mid-Miocene as well as Pliocene sediments. As much as 57% dolomite is present in the Miocene section and 13% in the Pliocene.

In piston cores, dolomite has been found in unlithified Pleistocene sediment (approximately 5%) and in chalk hardgrounds (as much as 10%; Gardulski, 1987). In oozes, dolomite is restricted to glacial intervals that are also relatively rich in magnesian calcite and pteropod-derived aragonite (Gardulski and others, 1986).

Under the scanning electron microscope, all the dolomites from west Florida appear as euhedral to subhedral rhombs as large as 10 μm, as well as larger aggregates. Embedded coccoliths (Fig. 3) are common, indicating *in situ* precipitation. X-ray diffraction results indicate that all dolomites analyzed from west Florida are calcium rich, containing 56–58 mole percent $CaCO_3$.

Geochemistry of west Florida deep-water dolomite.—

Three dolomite samples from Miocene sediment, three from Pliocene deposits, and a single sample from Pleistocene material have been analyzed for stable isotope ratios of oxygen and carbon as well as Sr 87/86 (Table 1). Values of $\delta^{18}O$ for dolomite from Miocene deep-water carbonates are consistently heavy, ranging from +4.5 to +6.3‰. Following the same assumptions as for our Bahamian samples for estimating paleotemperatures, we show that the west Florida isotope data suggest paleotemperatures of formation for Pliocene and Miocene dolomite of 10.9–11.7°C for site CH-48 (495 m) and 4.5°C for site CH-45 (1,477 m). These estimated paleotemperatures agree well with contemporary bottom water temperatures in the eastern Gulf of Mexico (Jones, 1973) of 8–9°C at 495 m and less than 5°C for water depths in excess of 1,000 m. We therefore believe that most of the $\delta^{18}O$ signal is largely temperature controlled, suggesting *in situ* precipitation of dolomite. The Pleistocene sample from 524 m of water depth, has a $\delta^{18}O$ value of +5.7‰ with an estimated paleotemperature of 6.5°C, very similar to contemporary bottom water temperatures of 7–8°C in the eastern Gulf (Jones, 1973). Values of $\delta^{13}C$ for deep-water dolomites in Miocene to Pleistocene west Florida sediments are narrow, ranging from +1.6 to +2.7‰, and are consistent with a normal marine source of carbon for dolomite precipitation.

Dolomite from early Miocene sediment at −170 m at site CH-45 has a $^{87}Sr/^{86}Sr$ ratio of 0.70863 ± 6, suggesting an interpolated age range of 18 ma (DePaolo, 1986) within the late part of the early Miocene, which implies synsedimentary dolomite precipitation. A second dolomite sample from middle Miocene sediment near the base of drill core CH-48 has a strontium ratio of 0.70916 ± 4, suggesting an interpolated age range of 2–4 ma. This is considerably younger than a planktonic foraminiferal biostratigraphic date of 13–15 ma (*Globorotalia fohsi fohsi* Zone; Exxon unpubl. data) for the sediment, which implies a considerable lag time for dolomitization.

A dolomite sample from mid-Pliocene sediment in drill core CH-48 at 43 m subsurface has a $^{87}Sr/^{86}Sr$ ratio of 0.70902 ± 5, suggesting a dolomitization date of about 7 ma, significantly older than the host sediment (3–4 ma).

TABLE 1.—SUMMARY OF FLORIDA-BAHAMAS DEEP-WATER DOLOMITE GEOCHEMICAL DATA

Sample No.	Location	Water Depth (m)	Sediment Age	Sample type	^{18}O (PDB)	^{13}C (PDB)	$^{87}Sr/^{86}Sr$	Sr (ppm)
630-148	NLBB	807	L. Mio-E. Plio	Chalk	—	—	0.70887 ± 8	340
631-100	Exuma Sound	1,081	L. Mio-E. Plio	Chalk	+6.7	+3.6	0.70920 ± 5	—
632-180	Exuma Sound	1,996	L. Miocene	Limestone	+6.4	+2.8	0.70896 ± 6	1,410
45-170	W. Florida	1,477	E. Miocene	Ooze	+6.3	+1.9	0.70863 ± 6	—
48-43.05	W. Florida	495	Pliocene	Sand	+4.3	+2.7	0.70902 ± 5	550
48-42.75	W. Florida	495	Pliocene	Sand	+5.1	+2.4	—	540
48-42.35	W. Florida	495	Pliocene	Sand	+4.9	+2.4	—	530
48-78.62	W. Florida	495	M. Miocene	Ooze	+5.1	+2.4	0.70916 ± 4	—
48-78.02	W. Florida	495	M. Miocene	Ooze	+4.5	+1.6	—	—
PC-87-2.4	W. Florida	524	Pleistocene	Ooze	—	—	0.70903 ± 5	1,060
PC-87-2.5	W. Florida	524	Pleistocene	Ooze	+5.7	+2.5	—	1,110
PC-81-0.6	W. Florida	482	Pleistocene	Chalk	—	—	—	1,640

NLBB = Northern Little Bahama Bank.

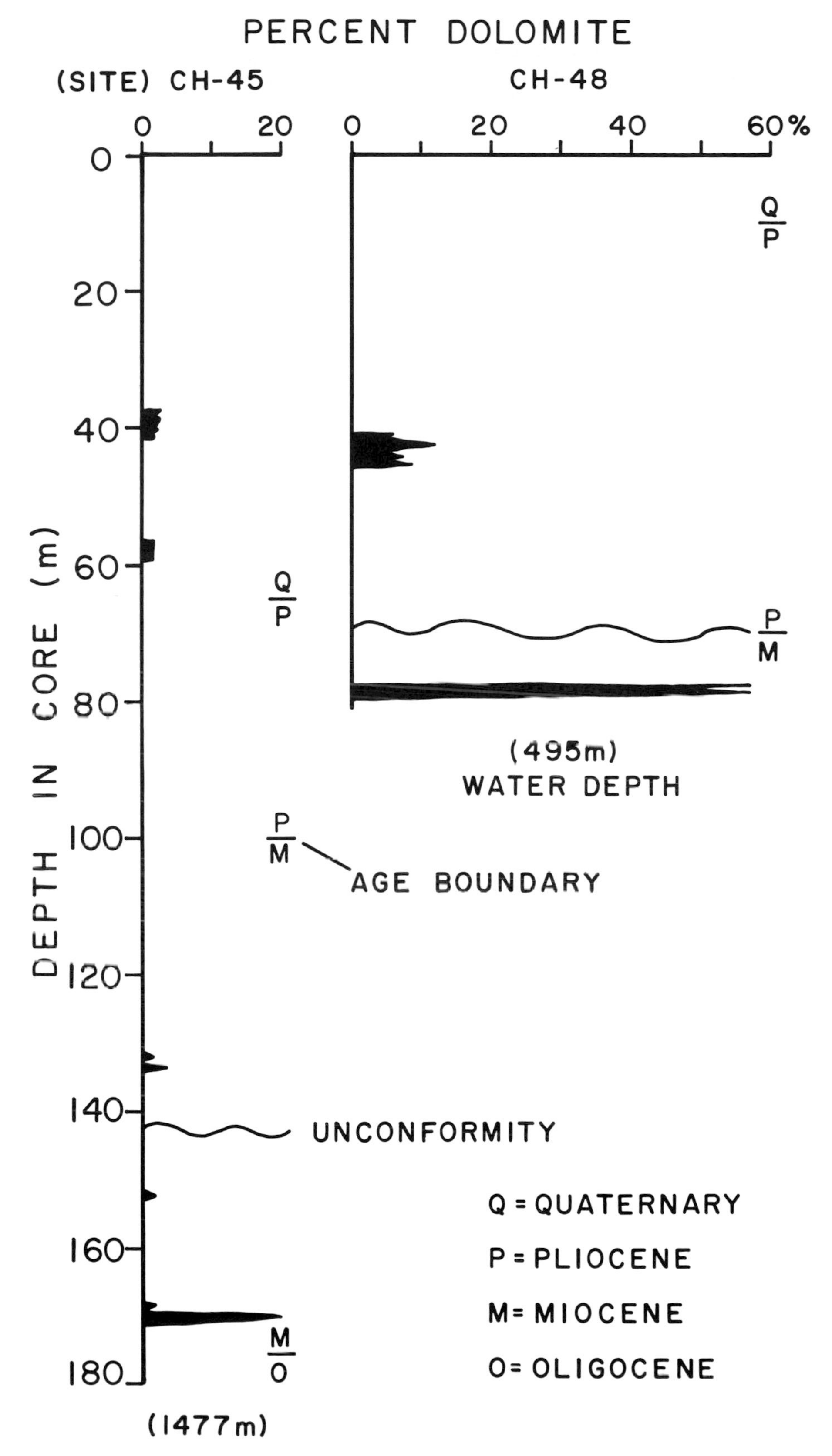

FIG. 6.—Downcore distribution of deep-water dolomite in Exxon drill cores from the west Florida slope.

This implies the incorporation of "old" strontium by discharge of interstitial fluids. Similar results characterize dolomites in Pleistocene sediments that have a $^{87}Sr/^{86}Sr$ value of 0.70903 ± 5, suggesting that the Mg for dolomitization in these mostly coarse-grained, high-energy deposits has been supplied by either diffusive or advective transport out of older sediment. Again, however, we emphasize the preliminary nature of these interpretations.

Selected dolomite samples were also evaluated for their Sr concentration. Dolomites contained within Miocene sediments from west Florida have Sr concentrations of 530–550 ppm, whereas dolomite from Plio-Pleistocene sediments contain as much as three times as much Sr (1,060–1,640) ppm). These data indicate that dolomite in Plio-Pleistocene sediment formed from very Sr-rich pore fluids.

DISCUSSION

Our results indicate that relatively modest amounts (<20%) of dolomite are common and occasionally abundant (57–86%) in largely Neogene deep-water carbonates peripheral to the Florida-Bahamas carbonate platform. Fresh euhedral rhombs of dolomite 5–20 μm across occur as pore-filling crystals scattered throughout these deep-water carbonate sediments and as replacements of calcite at disconformities. Stable isotope ratios of oxygen indicate *in situ* precipitation in near equilibrium with deep, cold, open-marine water, and carbon isotope data indicate a normal marine source. Strontium 87/86 ratios suggest that much of the dolomite formed after lag times, implying a possible kinetic control on precipitation. Other dolomites, however, apparently had Sr-rich precursors, which have resulted in apparent ages that are older than surrounding sediment.

Although the geochemical signatures of deep-water dolomites from both the Bahamas and west Florida are similar (Fig. 7), their downcore distributions are different. In the Bahamian drill cores, dolomite is more or less dispersed evenly throughout core sections of periplatform ooze that contained significant components of bank-derived metastable aragonite and magnesian calcite (Fig. 2), whereas the west Florida cores are punctuated by discrete intervals that are rich in dolomite (as much as 57%). West Florida slope sediment cores contain much less metastable aragonite/magnesian calcite than their Bahamian counterparts and, thus, initially have a much lower diagenetic potential.

Our interpretation of these data is that relatively small concentrations of dolomite are a natural consequence of early submarine, shallow-burial diagenesis of deep-water periplatform carbonates originally rich in metastable components. In contrast, carbonate ramp slope sediments, such as along the west coast of Florida, have a punctuated distribution of dolomite that may represent "dolomite–events," controlled paleoceanographically by the introduction of different water masses and/or circulation changes over time. Brunner and Keigwin (1981), Gartner and others, (1983) and Mullins and others (1987) have all concluded that the Miocene was a time of changing oceanographic conditions in the eastern Gulf of Mexico. Episodic influxes of corrosive waters (relative to calcite) and periodic fluctuations in current strengths are trends that have persisted throughout the Neogene.

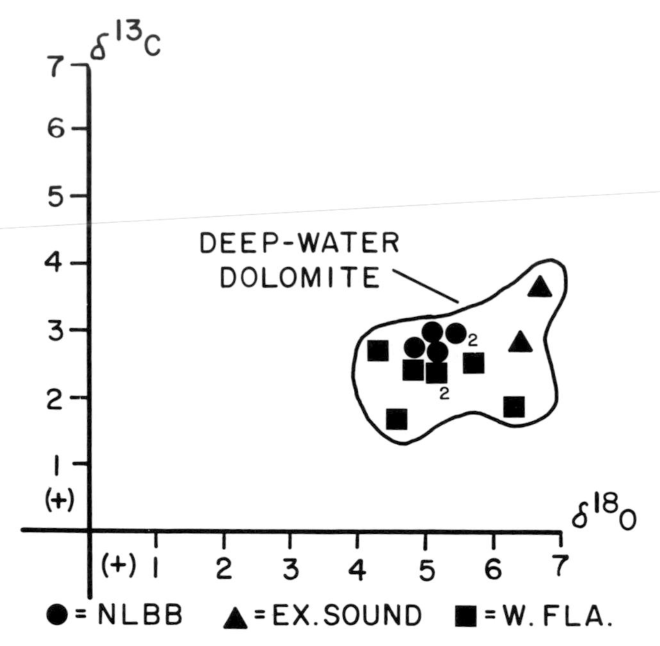

FIG. 7.—Stable isotope cross-plot (relative to PDB) of deep-water dolomites from the Bahamas and west Florida. Small numbers '2' indicate identical isotopic values for two samples.

In the Bahamas, dolomite occurs in consistently higher concentrations in drill core samples from Exuma Sound, a closed seaway, than the open-ocean slope north of Little Bahama Bank (Fig. 2). Geochemical analyses of pore waters (Austin and others, 1986) indicate that interstitial fluids are geochemically distinct between these different deep-water carbonate settings, which appear to be related to rates of sediment accumulation. At ODP Site 632, sedimentation rates are as high as 210 m/million years vs. a maximum of 62 m/million years for Site 630 north of Little Bahama Bank (Austin and others, 1986). Ocean Drilling Program porewater chemistry data also indicate that, although there is ample generation of H_2S, sulphate has not been depleted, thus casting doubts on the Baker and Kastner (1981) model of SO_4 inhibition of dolomite precipitation from sea water. In fact, there was sufficient SO_4 available along with aragonite-derived Sr for the precipitation of celestite ($SrSO_4$) as fracture fill at site 632.

SUMMARY AND CONCLUSIONS

(1) Calcium-rich dolomite is a common (<20%) but rarely abundant authigenic component of deep-water periplatform carbonates that contained metastable aragonite and magnesian calcite.
(2) Deep-water dolomite occurs mostly as scattered euhedral rhombs, 5–20 μm across, precipitated as pore fillings.
(3) Along discontinuities, deep-water dolomite can constitute as much as 86% of the carbonate fraction in hardgrounds occurring as both pore fill and as calcite replacement.
(4) Stable oxygen isotope ratios are enriched in ^{18}O and consistent with precipitation from deep, cold, ma-

rine-derived pore fluids. Stable carbon isotopes have a very narrow range, indicating a largely normal marine carbon source.

(5) Preliminary Sr 87/86 ratios suggest that dolomite precipitation occurred after lag times of a few million years.

(6) Trace-element concentrations of Sr indicate that some dolomites precipitated in close association with the dissolution of Sr-rich aragonite(?) precursors.

(7) Dolomite precipitation in deep-water carbonate ramp slope sediments along west Florida occurs as discrete, punctuated, paleoceanographically controlled "dolomite events" that can result in dolomite concentrations in excess of 50%.

ACKNOWLEDGMENTS

Data for this paper were partially derived from dissertation research projects by Dix (Bahamas) and Gardulski (Florida). This research was supported by National Science Foundation Grants OCE-83-08168 and OCE-85-17622, as well as Joint Oceanographic Institute Grants 62-84 (subcontract from University of Texas) and 1892-001 to H. T. Mullins at Syracuse University. Bahamian samples were collected during Leg 101 of the Ocean Drilling Program. West Florida samples were obtained during cruises CH-1-84 and CH-4-85 of the R/V CAPE HATTERAS as well as from Exxon, whose gracious assistance is gratefully acknowledged.

REFERENCES

ANDERSON, T. F., AND ARTHUR, M. A., 1983, Stable isotopes of oxygen and carbon and their application to sedimentological and paleoenvironmental problems, *in* Arthur, M. A., Anderson, T. F., Kaplan, I. R., Veizer, J., and Land, L. S., eds., Stable Isotopes in Sedimentary Geology: Society of Economic Paleontologists and Mineralogists Short Course Notes No. 10, p. 1-1 to 1-151.

AUSTIN, J. A., SCHLAGER, W., AND PALMER, A. A., AND 19 OTHERS, 1986, Proceedings, Initial Reports, part A, Leg 101, Ocean Drilling Program: Texas A & M University, College Station, Texas, 569 p.

BAKER, P. A., AND KASTNER, M., 1981, Constraints on the formation of sedimentary dolomite: Science, v. 213, p. 214–216.

BERGER, W. H., 1979, Stable isotopes in foraminifera, *in* Lipps, J. H., ed., Foraminiferal Ecology and Paleoecology: Society of Economic Paleontologists and Mineralogists Short Course Notes No. 6, p. 156–198.

BERGGREN, W. A., KENT, D. V., FLYNN, J. J., AND VAN COUVERING, J. A., 1985, Cenozoic geochronology: Geological Society of America Bulletin, v. 96, p. 1407–1418.

BRUNNER, C. A., AND KEIGWIN, L. D., 1981, Late Neogene biostratigraphy and stable isotope stratigraphy of a drilled core from the Gulf of Mexico: Marine Micropaleontology, v. 6, p. 397–418.

CARBALLO, J. D., LAND, L. S., AND MISER, D. E., 1987, Holocene dolomitization of supratidal sediments by active tidal pumping, Sugarloaf Key, Florida: Journal of Sedimentary Petrology, v. 57, 153–165.

DEPAOLO, D. J., 1986, Detailed record of the Neogene Sr isotopic evolution of sea-water from DSDP Site 590B: Geology, v. 14, p. 103–106.

———, AND INGRAM, B. L., 1985, High-resolution stratigraphy with strontium isotopes: Science, v. 227, p. 938–941.

GARDULSKI, A. F., 1987, Climatic and oceanographic controls on the Neogene sedimentary framework of the outer west Florida carbonate ramp: unpublished Ph.D. Dissertation, Syracuse University, Syracuse, New York, 227 p.

———, MULLINS, H. T., OLDFIELD, B., APPLEGATE, J., AND WISE, S. W., 1986, Carbonate mineral cycles in ramp slope sediment: Eastern Gulf of Mexico: Paleoceanography, v. 1, p. 555–565.

GARRISON, R. E., 1981, Diagenesis of oceanic carbonate sediment–A review of the DSDP perspective, *in* Warme, J. E., Douglas, R. G., and Winterer, J. L., eds., The Deep Sea Drilling Project: A Decade of Progress: Society of Economic Paleontologists and Mineralogists Special Publication 32, p. 181–207.

GARTNER, S., CHEN, M. P., AND STANTON, R. J., 1983, Late Neogene nannofossil stratigraphy and paleoceanography of the northeastern Gulf of Mexico and adjacent areas: Marine Micropaleontology, v. 8, p. 17–50.

GOLDSMITH, J. R., GRAF, D. L., AND HEARD, H. C., 1961, Lattice constants of the calcium-magnesium carbonates: American Mineralogist, v. 46, p. 453–546.

HARDIE, L. A., 1987, Dolomitization: A critical view of some current views: Journal of Sedimentary Petrology, v. 57, p. 166–183.

HOLLISTER, C. D., EWING, J., AND 9 OTHERS, 1972, Initial Reports of the Deep Sea Drilling Project, v. 11: U. S. Government Printing Office, Washington, D. C., 1077 p.

IRWIN, H., CURTIS C., AND COLEMAN, M., 1977, Isotopic evidence for source of diagenetic carbonates formed during burial of organic-rich sediments: Nature, v. 269, p. 209–213.

JONES, J. I., 1973, Physical oceanography of the northeastern Gulf of Mexico and Florida continental shelf, *in* Summary of Knowledge, Eastern Gulf of Mexico: Florida State University, Tallahassee, Florida, p. IIB-1 to IIB-41.

LAND, L. S., 1980, The isotopic and trace element geochemistry of dolomite, *in* Zenger, D. H., Dunham, J. B., and Ethington, R. L., eds., Concepts and Models of Dolomitization: Society of Economic Paleontologists and Mineralogists Special Publication 28, p. 87–110.

———, 1985, The origin of massive dolomite: Journal of Geological Education, v. 33, p. 112–125.

LUMSDEN, D. N., 1985, Secular variations in dolomite abundance in deep marine sediments: Geology, v. 13, p. 766–769.

MACHEL, H. G., AND MOUNTJOY, E. W., 1986, Chemistry and environments of dolomitization – A reappraisal: Earth Science Reviews, v. 23, p. 175–222.

MILLIMAN, J. D., 1974, Marine Carbonates: Springer-Verlag, New York, 375 p.

MITCHELL, J. T., LAND, L. S., AND MISER, D. E., 1987, Modern marine dolomite cement in a north Jamaican fringing reef: Geology, v. 15, p. 557–560.

MITCHUM, R. M., 1978, Seismic stratigraphic investigation of west Florida slope, Gulf of Mexico, *in* Bouma, A. H., Moore, G. T., and Coleman, J. M., Framework, Facies and Oil Trapping Characteristics of the Upper Continental Margin: American Association of Petroleum Geologists, Studies in Geology No. 7, p. 193–223.

MULLINS, H. T., HEATH, K. C., VAN BUREN, H. M., AND NEWTON, C. R., 1984, Anatomy of a modern open-ocean carbonate slope: Northern Little Bahama Bank: Sedimentology, v. 31, p. 141–168.

———, WISE, S. W., LAND, L. S., SIEGEL, D. I., MASTERS, P. M., HINCHEY, E. J., AND PRICE, K. R., 1985, Authigenic dolomite in Bahamian periplatform slope sediment: Geology, v. 13, p. 292–295.

———, GARDULSKI, A. F., WISE, S. W., AND APPLEGATE, J., 1987, Middle Miocene Oceanographic event in the eastern Gulf of Mexico: Implications for seismic stratigraphic succession and Loop Current/Gulf Stream circulation: Geological Society of America Bulletin, v. 98, p. 702–713.

SALLER, A. H., 1984, Petrologic and geochemical constraints on the origin of subsurface dolomite, Enewetak Atoll: An example of dolomitization by normal seawater: Geology, v. 12, p. 217–220.

WORZEL, J. L., BRYANT, W., AND 7 OTHERS, 1973, Initial Reports of the Deep Sea Drilling Project, v. 10: U. S. Government Printing Office, Washington, D. C., 748 p.

ZEMMELS, I., COOK, H. E., AND HATHAWAY, J. C., 1972, X-ray mineralogy studies–Leg XI, *in* Hollister, C. D., Ewing, J., and 9 others, Initial Reports of the Deep Sea Drilling Project, v. 11: U. S. Government Printing Office, Washington, D. C., p. 729–790.

DOLOMITIZATION OF SHALLOW-WATER PLATFORM CARBONATES BY SEA WATER AND SEAWATER-DERIVED BRINES: SAN ANDRES FORMATION (GUADALUPIAN), WEST TEXAS

STEPHEN C. RUPPEL AND HARRIS S. CANDER

Bureau of Economic Geology, The University of Texas at Austin, University Station, Box X, Austin, Texas 78713-7508

ABSTRACT: The Upper Permian San Andres Formation is composed of a shallowing-upward sequence of shallow-water carbonates that were deposited on a gently sloping ramp during the middle Guadalupian. Normal marine fusulinid packstone and wackestone (open-platform facies) are progressively overlain by skeletal grainstone (sand shoal facies), burrowed skeletal mudstone and wackestone (restricted inner-platform facies), and pisolitic, fenestral, and cryptalgal mudstone (supratidal facies).

Two major stages of dolomitization are recorded in these deposits. Early dolomite (dolomite 1) occurs at and below a regionally extensive marker horizon that is rich in siliciclastics and organic matter and that records a slowing of deposition followed by a possible major fall in sea level. Stable isotope data from this red-luminescent dolomite (average $\delta^{13}C = +4.7‰$ PDB; average $\delta^{18}O = +2.3‰$ PDB) are consistent with contemporaneous precipitation from near-normal marine waters in a subtidal setting or, alternatively, during exposure following relative sea-level fall. This dolomitization event may have been extensive, but dolomite 1 is preserved only in low-permeability rocks.

Younger, matrix-replacive, brown-luminescent dolomite (dolomite 2) constitutes the bulk of the dolomite in the San Andres Formation. Fluids responsible for precipitation or stabilization of this phase were derived from evaporated seawater brines. The isotopic composition of dolomite 2 (average $\delta^{13}C = +4.9‰$ PDB; average $\delta^{18}O = +3.7‰$ PDB) suggests evaporation to near calcium sulfate saturation.

Strontium isotopes constrain the age of the fluids responsible for stabilization of dolomite 2. Least radiogenic $^{87}Sr/^{86}Sr$ ratios in dolomite 2 are coincident with minimum $^{87}Sr/^{86}Sr$ values recorded for sea water during the Phanerozoic (0.7067). This indicates that dolomite 2 stabilized in fluids generated during this seawater minimum in the late Guadalupian. Such an interpretation is consistent with the development of major complexes of evaporitic environments in the area at this time.

Decreasing-downward trends in Fe concentration and $^{87}Sr/^{86}Sr$ ratios in dolomite 2 subjacent to siliciclastic-rich beds indicate that these beds served as local sources of Fe and radiogenic strontium. These trends also suggest that dolomitizing fluids had at least a minor downward component of flow or mixing through the San Andres section.

INTRODUCTION

Dolostones of the San Andres Formation constitute a major hydrocarbon reservoir in West Texas. Although several recent studies (Bebout, 1986; Garber and Harris, 1986; Ruppel, 1986) have documented a marked influence of original depositional texture on porosity and permeability, it is also apparent that diagenetic change, chiefly dolomitization and sulfate emplacement, has played a significant role in reservoir development. Development of predictive models of lateral and vertical heterogeneity within carbonate reservoirs thus rests heavily on the ability to document variations in diagenetic as well as depositional facies. This, in turn, depends on recognition of the nature and origin of diagenetic fluids responsible for the alteration and modification of original textural fabrics. Equally important is an understanding of the timing and sequence of diagenetic events that have affected the reservoir section.

A number of tools are available for studying carbonate diagenesis, including cathodoluminescence (CL) petrography, major- and trace-element geochemistry, and stable and radiogenic-isotope geochemistry. The application of these techniques to processes of dolomitization, however, is fraught with difficulties (see for example reviews by Land, 1980; Machel, 1985; Hardie, 1987). The chemical composition of dolomite is a product of both the dolomitizing fluids and the modification of these fluids by host sediment. In order to deduce the origin of dolomitizing fluids from textural and chemical evidence in dolomite, potential fluid interactions with mineralogic components in the formation must be considered. Thus, as the mineralogy of the carbonate sequence varies, so will modification of the dolomitizing fluids. Furthermore, massive dolomitization can be a multistep, episodic process, wherein early formed dolomite can be recrystallized or overgrown repeatedly. Detailed petrographic techniques are often the only method of recognizing episodes of dolomitization. Origins of dolomitizing fluids based on chemical composition of massive dolomite can thus only be correctly interpreted within the framework of detailed facies and petrographic studies.

In this paper we document (1) the depositional fabrics of San Andes platform carbonates and the conditions under which these deposits formed, and (2) the sequence and timing of diagenetic events that these deposits have subsequently undergone. Then, combining a variety of petrographic and geochemical techniques with lithologic data, we document the origins of fluids responsible for the diagenetic alteration of these rocks and their modification during passage through the stratigraphic section.

METHODS

Three cores in Emma Field were supplemented by borehole logs from approximately 170 wells. More than 150 thin sections were examined under normal transmitted light and cathodoluminescence (CL); CL analysis was performed on a Technosyn Model 8200 MK II system mounted on a Nikon Labophot binocular microscope; operating conditions were: 18–22 kv, 300- to 600-μA beam current, less than 0.01 Torr vacuum.

Concentrations of Ca, Mg, Fe, Mn, and Sr in dolomite were determined by electron microprobe analysis (more than 300 points) using a Cameca MBX. Operating conditions were 15-kv accelerating potential, 15-nA beam current, 2-μm spot size, and 120-sec maximum counting time. Average calculated detection limits at 95% confidence level are as follows: Fe = 120 ppm; Mn = 100 ppm; Sr = 90 ppm.

Sedimentology and Geochemistry of Dolostones, SEPM Special Publication No. 43

Samples for stable isotope analysis of dolomite were collected from thin section chips using a 0.5-mm carbide dental burr. Cathodoluminescence petrography of polished chips and corresponding thin sections was used to select samples that contained greater than 90 percent one type of dolomite; calcite was negligible in all samples based on X-ray diffraction study. Powders were vacuum roasted at 375°C for 1 hr to remove volatile organics; CO_2 gas was liberated by reaction with 100% phosphoric acid at 60°C and analyzed with a fully automated VG SIRA-24 mass spectrometer. Average precision, based on duplicate analyses (13), is $\pm 0.27‰\ \delta^{13}C$ and $\pm 0.47‰\ \delta^{18}O$.

Eleven dolomite and 10 anhydrite samples were analyzed for $^{87}Sr/^{86}Sr$. As with stable isotope samples, CL was used to select relatively pure samples of each dolomite type. Chemical separation of dolomite from the noncarbonate fraction was achieved by: (1) dissolution of anhydrite in doubly distilled water, (2) leaching of loosely bound Sr from clays with 1.0 N ammonium acetate, and (3) dissolution of dolomite in 0.2 N distilled HCl. Rubidium/strontium concentrations were measured in three samples. Assuming a closed system with respect to Rb in the dolomites, measured Rb/Sr ratios (<0.027) indicate that $^{87}Sr/^{86}Sr$ values are not significantly affected by Rb content.

Anhydrite samples for $^{87}Sr/^{86}Sr$ analysis were powdered or microdrilled and separated from the solid by selective dissolution in NaCl saturated solution. Comparison of duplicate analyses indicates that precision is better than 0.0001 $^{87}Sr/^{86}Sr$ (0.01%). All values are corrected to NBS 987 = 0.71015.

STRUCTURAL AND STRATIGRAPHIC SETTING OF THE EMMA FIELD

Emma Field, located in south-central Andrews County, Texas (Fig. 1), is one of several major oil fields developed in the San Andres and Grayburg formations (Guadalupian) along the eastern margin of the Central Basin Platform (CBP), an elongate carbonate platform that developed during the Permian over structurally positive areas formed during earlier (Carboniferous) deformation (Galley, 1958; Ward and others, 1986). During most of the Permian, the CBP was a topographic high that separated the Delaware Basin to the west and the Midland Basin to the east (Fig. 1). Predominantly shallow-water platform carbonates accumulated on the CBP, while deeper water siliciclastics and carbonates were deposited in the basins.

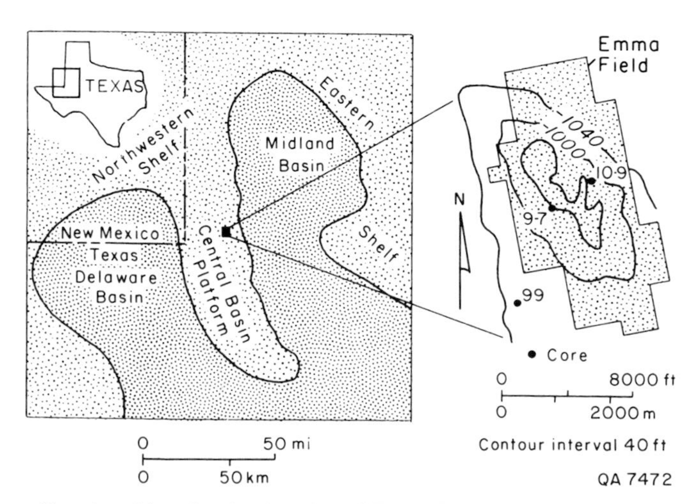

FIG. 1.—Map showing location of Emma Field and location of cored wells and structure contour map of the San Andres Formation (base of A marker, see Fig. 2).

The San Andres Formation in Emma Field is predominantly dolomite (Fig. 2). Anhydrite is ubiquitous in dolomitized intervals, generally constituting 20–30% of the rock. Thin beds of fine-grained, arkosic, quartz sandstone and siltstone are locally common in the uppermost San Andres; these form correlatable markers and at least locally are isochronous (Bebout, 1986; Ruppel, 1986). Both anhydrite and siliciclastics are more abundant in overlying formations (Fig. 2) and updip. To the north in the Texas Panhandle, the San Andres is primarily composed of halite and bedded anhydrite with lesser amounts of limestone, dolomite, and redbeds deposited in cyclic sequences (Meissner, 1969; Fracasso and Hovorka, 1986). A thick evaporite section also overlies the San Andres/Grayburg interval in Emma Field. Both the Queen and Seven Rivers formations (Fig. 2) are characterized by abundant anhydrite/gypsum and minor interbedded halite and terrigenous clastics. The Upper Permian (Ochoan) Salado and Rustler formations are predominantly halite.

The age of the San Andres Formation is imprecisely known. Most workers assign the San Andres to the Guadalupian on the basis of fusulinids recovered from outcrops and the subsurface (Silver and Todd, 1969; Wilde, 1986). It is probable, however, that the age varies regionally, as suggested by Todd (1976), due to the migration of depositional environments through time. In the Delaware and Midland basins and adjacent CBP areas, the San Andres contains early to middle Guadalupian fusulinid faunas. Zaaza (1978) and Vogt (1986) identified the fusulinid *Parafusulina deliciasensis* from the upper San Andres in Gaines County about 65 km north of Emma Field, suggesting that upper San Andres rocks in this area are middle Guadalupian (Wilde, 1975).

The topography of the CBP platform margin is incompletely known. Carbonate buildups have been documented along the western margin of the CBP for much of the middle Permian sequence (Silver and Todd, 1969). During development of the Capitan Formation buildup (late Guadalupian), platform-to-basin relief may have reached 600 m. No equivalent buildups are known from either side of the CBP during San Andres deposition (early to middle Guadalupian), however, indicating that relief was probably much less (Silver and Todd, 1969; Ward and others, 1986). Seismic data indicate that the eastern margin of the CBP was characterized by a gently eastward-dipping carbonate ramp throughout much of San Andres deposition (Sarg, pers. commun., 1986).

FACIES AND DEPOSITIONAL ENVIRONMENTS

The upper San Andres Formation in Emma Field is composed of nine intergradational carbonate lithofacies and one siliciclastic lithofacies (Figs. 2, 3) that represent four major paleoenvironments: open platform, sand shoal, restricted inner platform, and supratidal.

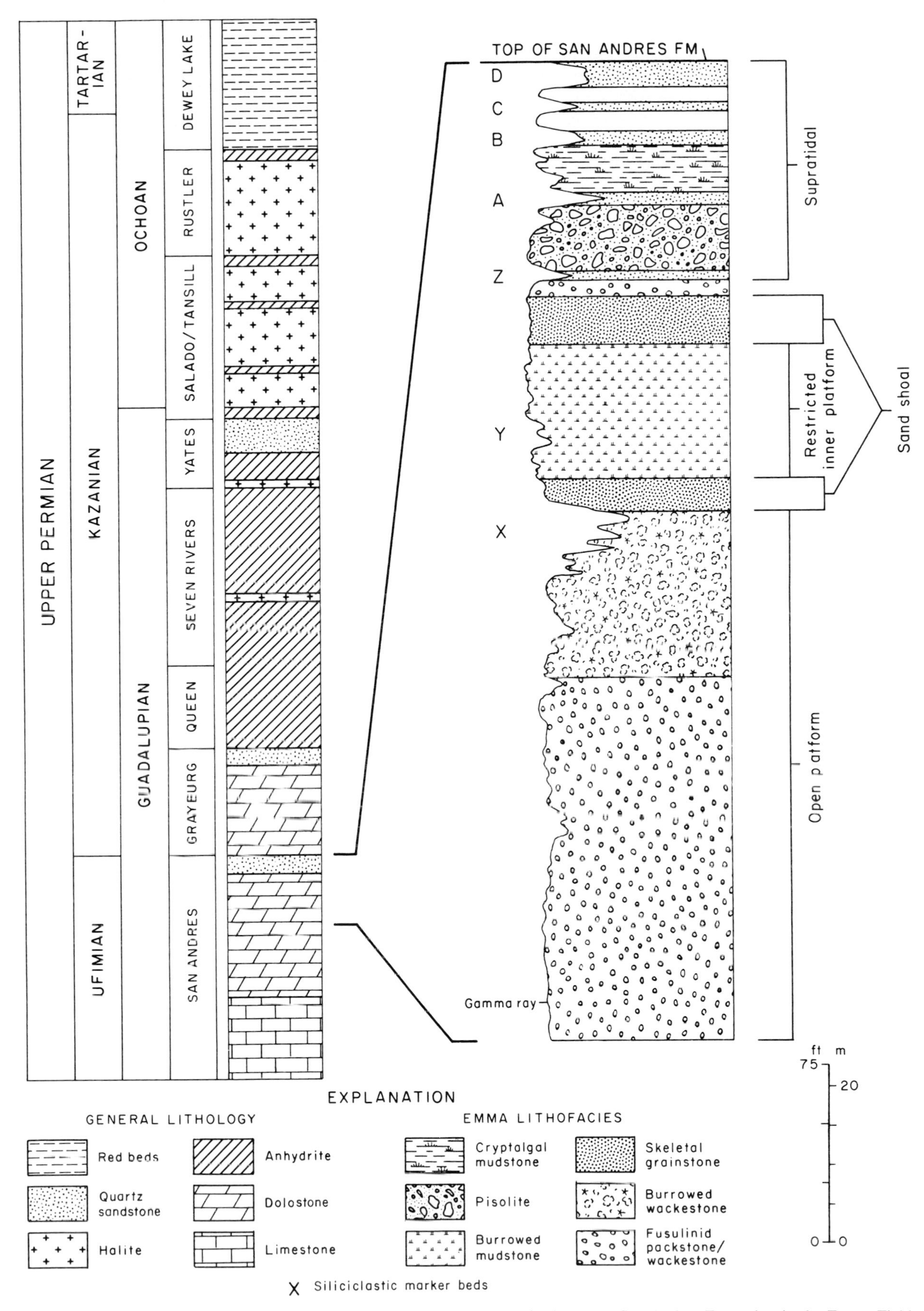

FIG. 2.—General Upper Permian stratigraphy and vertical facies sequence in the upper San Andres Formation in the Emma Field area.

Open platform.—

Rocks interpreted to represent deposition in an open-platform setting include fusulinid/echinoderm packstone, fusulinid packstone/wackestone, and burrowed wackestone (Figs. 2, 3). These extensive and easily correlatable deposits form the lower half of the upper San Andres reservoir interval in Emma Field (Fig. 3). Fusulinids are common throughout this sequence but are especially abundant in the fusulinid packstone/wackestone facies, where they are primarily preserved as open or anhydrite-filled molds (Fig. 4). Fusulinid molds combine with intercrystalline pores in these rocks to create significant reservoir porosity. Where fusulinids have not been leached to produce molds, pervasive dolomitization has so obscured original fabrics that CL petrography is necessary for their recognition. (Fig. 5).

Open-platform facies rocks exhibit progressive vertical and lateral changes in abundance of allochems and lime mud; allochem-rich fusulinid/echinoderm packstone is present in the lower part of the section and to the east, whereas more mud-rich wackestone characterizes the upper part of the sequence (Fig. 3). A persistent high gamma ray anomaly (X marker) occurs near the top of these deposits.

FIG. 4.—Photomicrograph of fusulinid packstone/wackestone (open-platform facies). Fusulinids are preserved as open and anhydrite-filled molds (shown here). Core 9-7, 1,319 m.

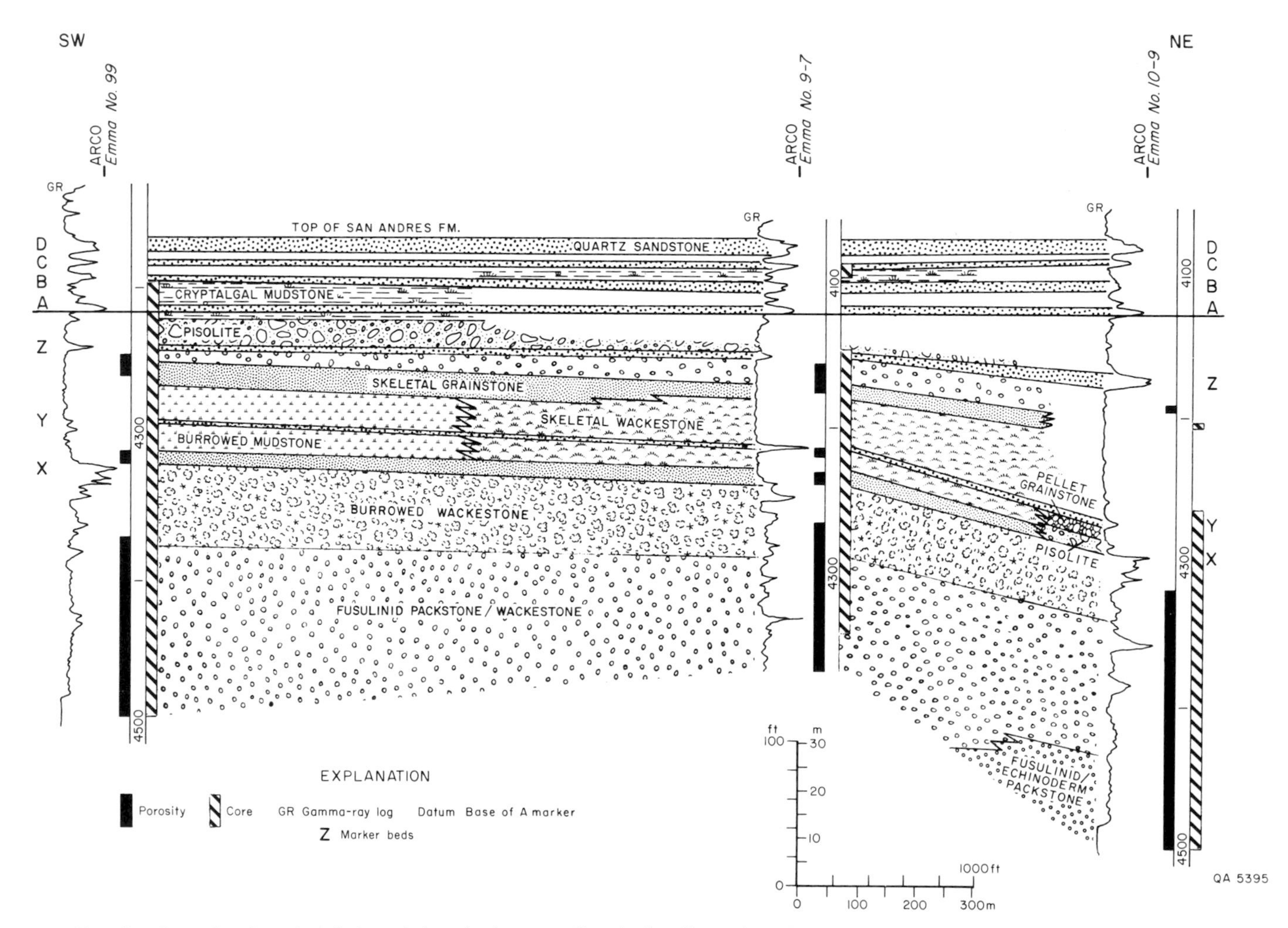

FIG. 3.—Lateral and vertical facies relations in the upper San Andres Formation. Emma Field. Line of section is shown in Figure 1.

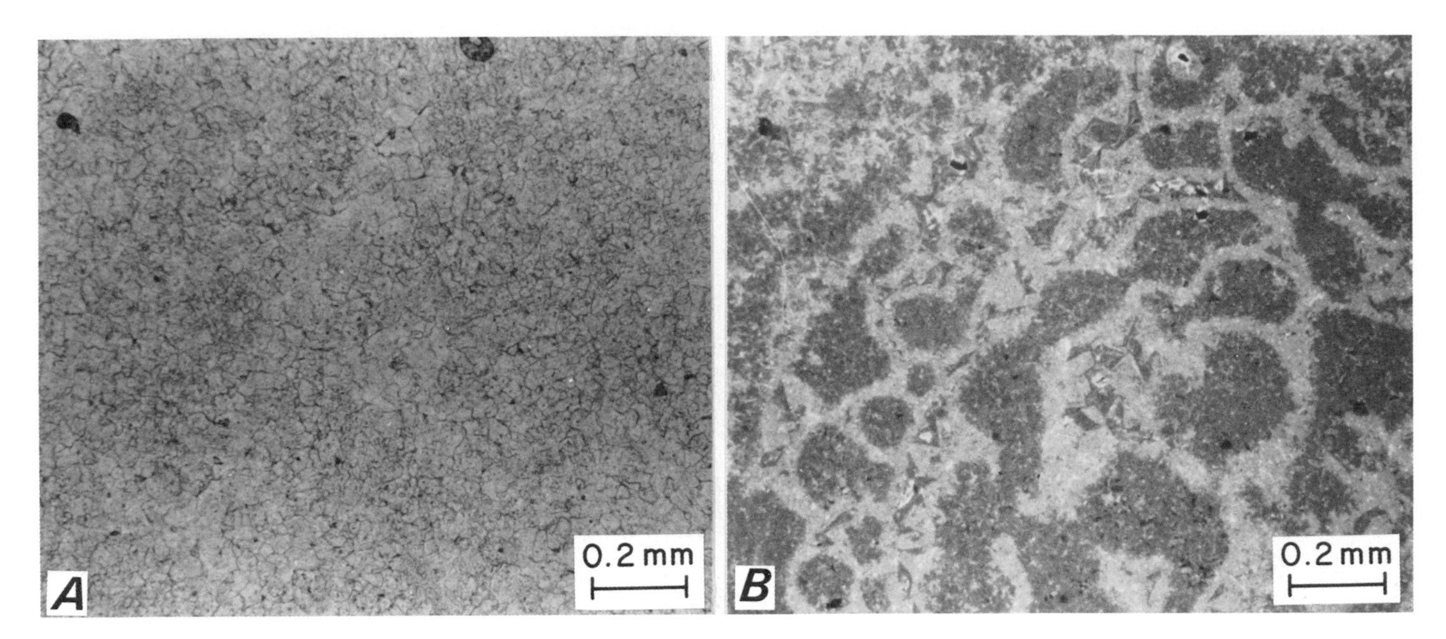

FIG. 5.—Paired photomicrographs of the open-platform facies (fusulinid packstone/wackestone). (A) Due to replacive dolomitization, plane light view reveals no aspects of the original depositional fabric, only interlocking dolomite crystals. (B) Under CL, the identical field of view reveals a cross section through a fusulinid. Cathodoluminescence is useful is delineating otherwise obliterated fabrics throughout the San Andres. Core 9-7, 1,326 m.

The X marker consists of laminated dolomite that contains relatively high amounts of terrigenous clastics (silt and clay) and organic carbon. Insoluble residues and total organic carbon (average 2.2% and 0.2%, respectively) are an order of magnitude greater in this interval than in carbonates elsewhere in the section (average 0.3% and 0.02%, respectively). The top of the X marker forms the sharp upper contact of the open-platform facies, which is easily recognizable in well logs (Fig. 2) and in core.

Open-platform rocks are interpreted to represent deposition in an open-marine setting on the basis of their normal marine faunas. Vertical and lateral changes in fauna and lime mud content apparently reflect lower energy conditions in the upper part of the sequence and to the west, which may be the result of the development of progressively shallower water conditions due to aggradation and progradation. The presence of pisolites immediately above the X marker in parts of the field (Fig. 3) suggests that a significant break in sedimentation may have occurred following deposition of the open-platform facies. Development of a major hiatus at this point in the sequence is consistent with interpretations of a rapid fall in relative sea level, based on study of outcrops and seismic sections in West Texas (Sarg, 1986).

Sand shoals.—

In the central part of the field, open-platform deposits are sharply overlain by relatively thin (3–6 m), laterally discontinuous intervals of skeletal grainstone (Figs. 2, 3). These deposits, which are composed of abundant calcareous algae and fusulinids, contain well-developed intergranular pore space (Fig. 6) and constitute the primary reservoir interval in the field.

Skeletal grainstone deposits are interpreted to have accumulated in a migrating complex of skeletal-sand shoals. Associated muddier rocks probably reflect lateral migration of shoals and deposition of lime mud in slack-water areas developed on and around the shoal complex.

Restricted inner platform.—

The restricted inner-platform facies is composed of intergradational burrowed mudstone and skeletal wackestone, which are interbedded with skeletal grainstone of the sand shoal facies (Figs. 2, 3). Skeletal debris, principally molluscs, decreases in abundance to the west (Fig. 3).

The abundance of carbonate mud suggests that these rocks were deposited in a low-energy, shallow subtidal to intertidal setting (James, 1984). The decrease in epifauna and concommitent increase in burrowing infauna to the west suggest increasing current restriction to the south and west. Isolated zones of fenestral carbonate, carbonate crusts, and pisolites document periodic exposure.

Supratidal.—

Supratidal-facies rocks are composed of pisolite mudstone to grainstone, cryptalgal mudstone, and quartz sandstone and form the uppermost part of the San Andres in Emma Field (Fig. 3). Pisolitic rocks are composed of large (2 to 10 mm), irregularly coated grains, fossil debris (molluscs and fusulinids), peloids, lithoclasts, and irregular and laminar birdeyes. Tepee structures containing fractures filled with internal sediment in their upper parts and anhydrite in their lower parts (Fig. 7A) are associated with pisolite rocks.

Cryptalgal mudstone overlies pisolite grainstone at the top of the cored intervals (Fig. 3), where it is interbedded with quartz sandstone. These deposits contain rare scattered

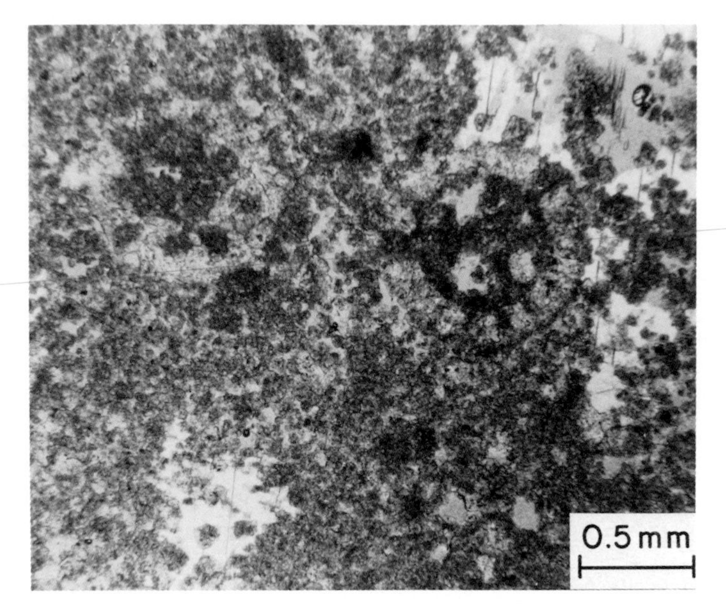

FIG. 6.—Skeletal grainstone of the sand shoal facies containing the dasyclad alga *Mizzia*. Anhydrite replacement of dolomite and subsequent leaching of anhydrite has enhanced the original intergranular porosity of this grainstone, which constitutes the most important reservoir interval in the Emma San Andres. Core 99, 1,315 m.

algal debris in horizontally laminated (algal?) carbonate mud that locally contains mudcracks (Fig. 7B) and rip-up clasts.

Arkosic quartz sandstone is most common as persistent beds (A, B, C, and D) in the uppermost San Andres Formation (Figs. 2, 3). These rocks are composed of sand- and silt-size quartz and feldspar that is commonly leached. Sandstone is also present, although in much lower quantities, in the Z marker and locally in the Y marker; the X marker contains only trace amounts of quartz sand but contains significant quantities of silt and clay.

Supratidal-facies rocks represent the culmination of the shallowing-upward trend observed in the upper San Andres in the Emma Field area. Cryptalgal and pisolitic carbonate deposits accumulated during periods of alternating wetting and drying in a predominantly supratidal setting. On the basis of lateral extent and association with supratidal facies, arkosic quartz sandstone is interpreted to have been deposited by aeolian processes active during periods of exposure. Such an origin has also been suggested for late Guadalupian sandstones (Queen, Tansill, and Yates formations) in the Guadalupe Mountains of West Texas (Mazzullo and Hedrick, 1985). No source for these sandstones has yet been documented. The nearest exposed land masses that may have acted as sources during late San Andres time are the Pedernale, Sierra Grande, and Uncompaghre uplifts in northern

FIG. 7.—Supratidal facies. (A) Interlayered pisolites, fenestral fabric, and mudstone with sheet cracks are common in the supratidal facies at the top of the San Andres. Vertical cracks, such as that shown here, are typically associated with tepee structures (note vertical offset) produced by dessication. Core is 9 cm wide. Core 99, 1,303 m. (B) Cryptalgally laminated mudstone is most common in the uppermost San Andres. These deposits are interbedded with fenestral mudstone, pisolite grainstone, and less commonly, pellet grainstone. Core is 9 cm wide. Core 9-7, 1,246 m.

New Mexico (Dixon, 1967; Oriel and others, 1967). These rocks are composed primarily of Precambrian granitic intrusives and metasediments (Gonzalez and Woodward, 1972; Grambling, 1979).

PALEOGEOGRAPHIC SETTING

It is apparent from the vertical-facies sequence documented earlier that the upper San Andres in Emma Field constitutes a shallowing-upward trend of shallow subtidal to peritidal and supratidal deposits. The vertical sequence of lithofacies observed in Emma Field is typical of sequences produced during the progradation of shallow-water environments across a shallow-water carbonate ramp (Ahr, 1973; Read, 1982). Lateral-facies relations indicate that this ramp sloped gently eastward across the Emma Field area (Fig. 8). A sudden eastward shift in paleoenvironments (offlap) appears to have taken place at the time represented by the X marker. Similar facies changes have been recognized in San Andres outcrop sections (Guadalupe Mountains, New Mexico), where they have been interpreted to represent rapid falls in relative sea level (Sarg. 1986). It is not possible, at present, to determine whether the X marker in Emma Field represents a major sequence boundary (Vail and others, 1984), such as that recognized by Sarg (1986). Such a conclusion requires seismic data and/or more extensive platform-to-basin core control than are presently available; however, the rapid shift in depositional conditions and local evidence of subaerial exposure associated with the X marker in Emma Field is consistent with rapid sea-level fall.

PETROGRAPHY

Dolomite.—

Four types of dolomite can be distinguished in the upper San Andres of Emma Field on the basis of crystallinity, abundance of inclusions, and CL (Table 1). The bulk of the dolomite is inclusion rich, subhedral, matrix replacive, and ranges in size from about 10 to 100 μm. Inclusion-rich dolomite (dolomite 1 and 2) constitutes as much as 95% of the total carbonate in the section. Throughout most of the upper San Andres, inclusion-rich dolomite is dark to light brown in CL (dolomite 2; Fig. 9A, B). In the X marker at the top of the open-platform facies (Fig. 3), however, in-

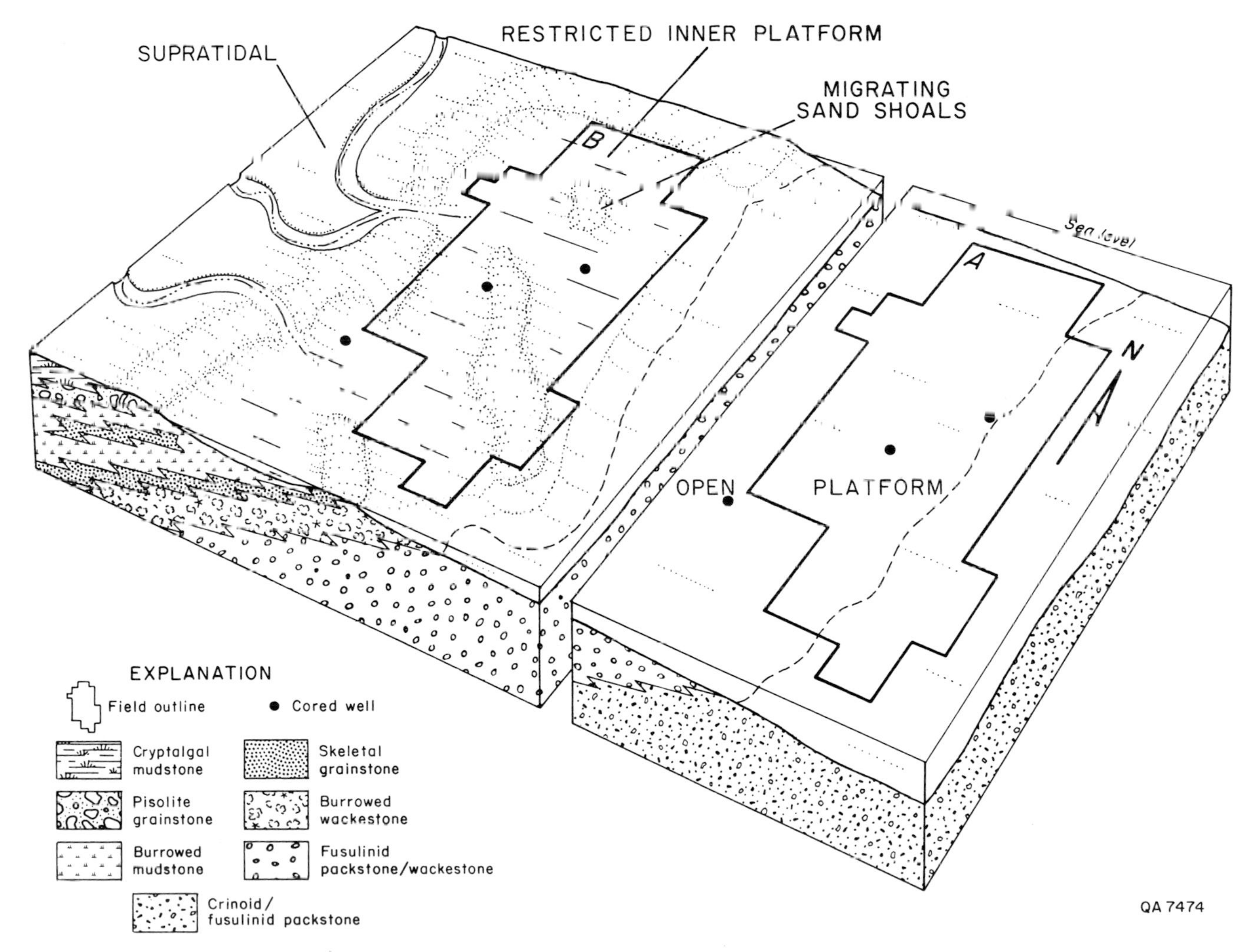

FIG. 8.—Paleoenvironmental reconstruction of area during late San Andres (middle Guadalupian) time. Field outlines indicate relative position of Emma Field on the paleoenvironmental transect at two different times. Time A represents the lower half (below the X marker bed) of the sequence; time B the upper half (above the X marker bed).

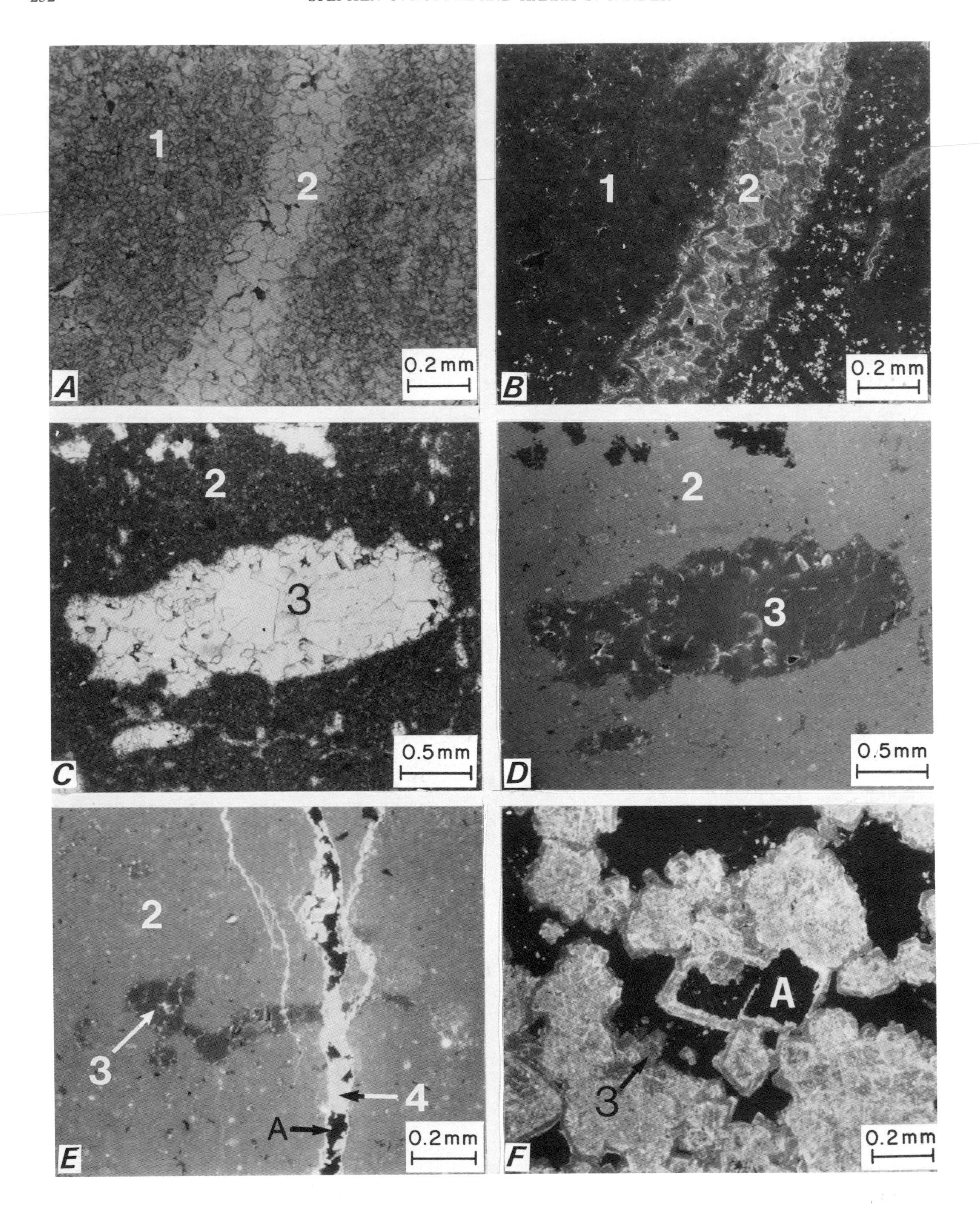
1
2
0.2 mm
A
1
2
0.2 mm
B
2
3
0.5mm
C
2
3
0.5mm
D
2
3
4
A
0.2mm
E
A
3
0.2mm
F

Table 1.—San Andres Dolomite Petrography and Geochemistry

Dolomite Type	Plain Light Features	Dominant Occurrence	Cathodoluminescence	Trace Elements*	
				FE	MN
1	inclusion rich subhedral	matrix replacement at and below X-marker	red	600	70/130
2	inclusion rich subhedral	matrix replacement throughout section	tan brown	470	50/130
3	inclusion poor euhedral	cement and overgrowth	red	2600	100/160
4	inclusion poor euhedral	overgrowth and fracture fill	bright yellow	40/130	410

*Average concentration in ppm. Two average values were calculated where some analyses were below detection limits: the smaller number is based on the assumption that these analyses represent 0 ppm, whereas the larger number considers them equivalent to detection limits (Fe: 120 ppm: Mn: 100 ppm). These numbers, thus, represent maximum and minimum averages.

clusion-rich, matrix-replacive dolomite luminesces red (dolomite 1; Fig. 9C, D). Red-luminescent dolomite 1 is also finer grained (average: 10 μm) than brown-luminescent dolomite 2. Dolomite 1 is locally overgrown by dolomite 2, indicating that red-luminescent, inclusion-rich dolomite formed first (Fig. 9B).

Inclusion-poor dolomite postdates inclusion-rich dolomite and is much less significant volumetrically in the Emma San Andres (Table 1). Two types of inclusion-poor dolomite can be distinguished. Dolomite 3 is red-luminescent and unzoned and most commonly forms 1 to 10-μm-thick syntaxial overgrowths on inclusion-rich dolomite (primarily dolomite 2) or coarse crystalline pore space linings and fillings as much as 400 μm in diameter (Fig. 9B, C). Brightly luminescent, inclusion-poor dolomite 4 forms thin syntaxial overgrowths on dolomite 3 and lines the walls of fractures that crosscut dolomites 2 and 3 (Fig. 9E). These four types of dolomite exhibit a consistent sequence stratigraphy across Emma Field, indicating that dolomitization events were relatively widespread.

Anhydrite.—

Anhydrite constitutes 20 to 30% of the whole rock in the San Andres Formation in Emma Field and exhibits four distinct habits: (1) poikilotopic crystal masses, (2) nodules, (3) allochem mold fillings, and (4) fracture fillings. All four varieties are ubiquitous.

Although it is difficult to discern in core, poikilotopic anhydrite is the most abundant form. Anhydrite nodules, which are much more obvious, are particularly common in the subtidal part of the upper San Andres, although they are also found in the supratidal and peritidal facies. Nodules vary from a few to several centimeters in diameter and usually consist of lath-shaped or fibrous crystals with or without blocky crystals on the outer rim. Mold-filling anhydrite is common in fusulinid molds, where it exhibits a blocky- or bladed-crystal morphology.

The following observations indicate that anhydrite emplacement postdates dolomitization in the Emma San Andres: (1) anhydrite fills cores of hollow dolomite rhombs (both dolomite 2 and dolomite 3; Fig. 9F); (2) inclusions of dolomites 2 and 3 are common in poikilotopic, mold-filling, and nodular anhydrite crystals; such inclusions are in some cases euhedral but usually have etched faces; (3) inclusions of dolomite in mold-filling anhydrite often define skeletal architecture (Fig. 10); and (4) anhydrite fills fractures that crosscut all dolomite types, including dolomite 4 (the youngest dolomite generation), which in turn crosscuts older dolomites (Fig. 9E).

The presence of compactional margins and indications of displacive growth of sulfate around some nodules, however, suggest that many may have formed early in the diagenetic history of the sequence, presumably prior to dolomitization. These nodules may have precipitated first as gypsum. During a second stage of sulfate (possibly anhydrite) emplacement that postdated all dolomitization, most nodules were partially recrystallized, especially along their margins where, in many cases, evidence of replacive growth is present. Similar recrystallization of sulfate nodules has been observed elsewhere in the San Andres (Sarg, 1981; Vogt, 1986).

GEOCHEMISTRY

Major and Trace Elements.—

Table 1 lists the Fe and Mn concentrations of dolomites 1, 2, 3, and 4. All dolomites are nearly stoichiometric; Mg/Ca ratios, determined from microprobe analysis, range from 49:51 to 50:50. There is, however, nearly an order of magnitude difference in Fe concentration between dolomite 2

Fig. 9.—Paired plane light (A) and CL (B) photomicrographs showing inclusion-rich, matrix-replacive dolomite 1 (1) that occurs at and below the X marker. Brown-luminescent dolomite 2 (2) fills void space within dolomite 1 interval, thus constraining the relative timing of these two dolomites. Core 9-7, 1,297 m. Paired plane light (C) and CL (D) photomicrographs showing two major types of dolomite present in the upper San Andres in the Emma Field. Matrix-replacive, inclusion-rich, brown-luminescent dolomite 2 (2) constitutes more than 90% of the dolomite in the sequence. Inclusion-free, red-luminescent dolomite 3 (3) is present as a void-filling cement. Core 99, 1,289 m. (E) CL photomicrograph showing cross cutting relations that constrain the relative timing of three of the types of dolomite present in the Emma San Andres. Brown-luminescent dolomite 2 (2) is crosscut by red-luminescent dolomite 3 (3), which is, in turn, crosscut by fracture-filling, brightly luminescent dolomite 4 (4). Anhydrite (A) fills post-dolomitization porosity. Core 99, 1,289 m. (F) Cathodoluminescence photomicrograph showing anhydrite filling rhomb molds (A). Most molds are lined with red-luminescent dolomite 3 (3), which has overgrown brown-luminescent dolomite 2 (2). Core 9-7, 1,274 m.

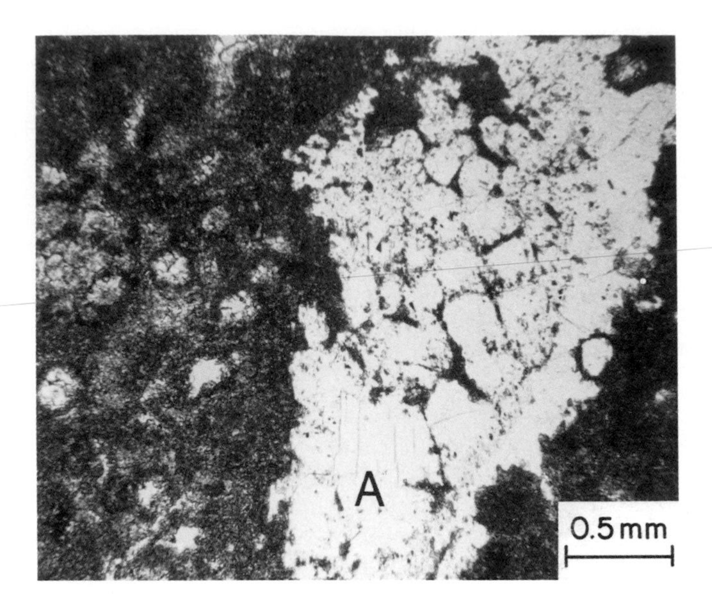

FIG. 10.—Plane light photomicrograph illustrating anhydrite replacement of dolomite. Residual dolomite and dolomite inclusions in the anhydrite (A) define the original skeletal structure of a nearly completely replaced calcareous alga. Note unreplaced skeleton at left. Core 99, 1,296 m.

(average 470 ppm) and dolomite 3 (average 2,600 ppm). Iron concentration also varies with depth; Fe decreases by an order of magnitude downsection in both dolomite 2 and dolomite 3 (Fig. 11). A similar downsection decrease in dolomite Fe concentration has been observed in the overlying Grayburg Formation and in each case is directly related to the abundance of beds of silicilcastics in the sequence (Fig. 11). Manganese is generally below electron microprobe detection limits (100 ppm) in dolomites 1, 2, and 3 (Table 1). Dolomite 4 is relatively Mn rich and Fe poor (Table 1).

The positive correlation between Fe concentration in dolomites 2 and 3 and beds of terrigenous siliciclastics suggests that the clastics were a source of Fe during dolomitization. The vertical distribution of Fe in dolomite 2 constrains fluid transport directions during dolomitization. Assuming $D_{Fe} > 1$ in dolomite (Veizer, 1983) and a single source of Fe (the siliciclastics), dolomitizing fluids must have had at least a minor downward component of flow in order to cause downward depletion in Fe. An upward component of fluid flow would, conversely, have resulted in an upward depletion in Fe. The restriction of high-Fe concentrations in dolomite 2 to the upper part of the San Andres

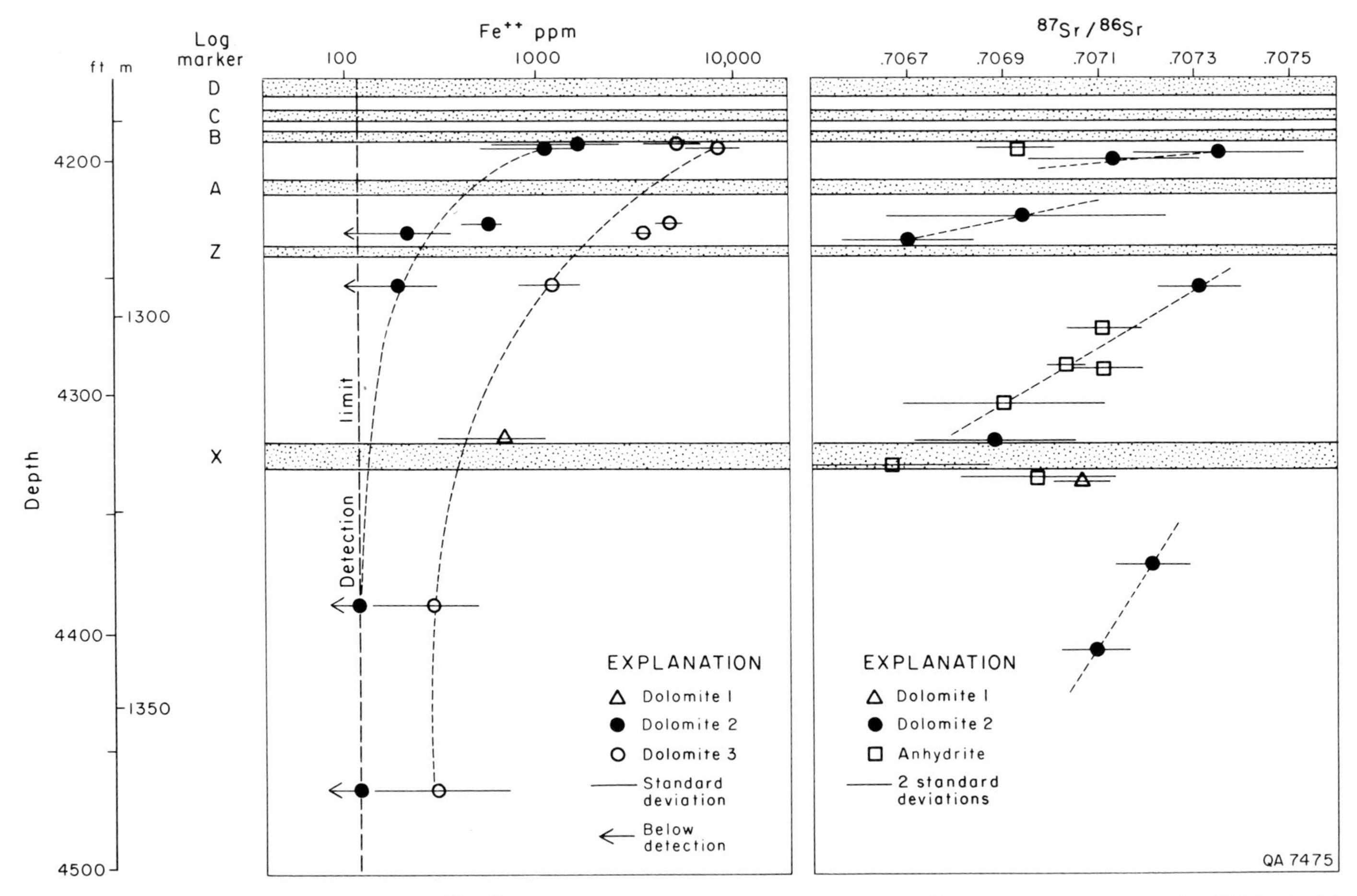

FIG. 11.—Vertical distribution of Fe^{++} and $^{87}Sr/^{86}Sr$ with depth in the Arco Emma #99 core. Fe^{++} concentration is greatest in siliciclastic-rich beds in the upper part of the San Andres and decreases downsection, indicating (1) a local source of Fe^{++} in the siliciclastic beds, and (2) a downward flow component associated with the movement of dolomitizing fluids through the section. Variations in $^{87}Sr/^{86}Sr$ parallel those delineated by Fe^{++} and indicate that siliciclastic beds in the upper part of the section were also significant sources of radiogenic strontium. Note that decreases in $^{87}Sr/^{86}Sr$ are punctuated by siliciclastic beds.

suggests rapid uptake of Fe from siliciclastics and limited downward mobility of Fe.

Processes accounting for the distribution of Fe in dolomite 3 are equivocal. Since dolomite 3 replaced or overgrew dolomite 2 rhombs, dolomite 3 may have derived some or all of its Fe from dolomite 2. Again, assuming rapid uptake of local Fe, the trend of Fe in dolomite 3 should parallel that of dolomite 2, as it does. Alternatively, identical processes of Fe mobilization may have occurred during two episodes of dolomitization. In either case, it appears that Fe was mobile on the scale of only several meters during dolomitization of the San Andres Formation in Emma Field. The enrichment in Fe of dolomite 3 relative to dolomite 2, along with the more euhedral morphology of the former, may reflect slower rates of precipitation. Distribution coefficients greater than one, such as D_{Fe} in dolomite, increase with decreasing rates of precipitation (review by Veizer, 1983).

Several sources of Fe in clastic beds were identified by X-ray diffractograms, including smectite clays, pyrite, iron oxides, and hydroxides (hematite and goethite). Since the pH of the dolomitizing fluid was carbonate-buffered (pH >5; Garrels and Christ, 1965), Fe was mobilized as a divalent species, which is consistent with substitution for divalent Ca or Mg in the dolomite lattice. Derivation of Fe from the aforementioned sources may have been accomplished via: (1) reduction of iron oxides and hydroxides. (2) oxidation of sulfide in pyrite, or (3) cation exchange in smectites. Flux of cations from clastics to dolomitizing fluids has also been proposed by Bein and Land (1983) for San Andres rocks in the Texas Panhandle.

Strontium isotopes.—

Ten samples of anhydrite, including nodular, poikilotopic, and fracture-filling types, 10 dolomite samples containing greater than 95 percent dolomite 2, and one sample containing greater than 95 percent dolomite 1 were analyzed. Combined dolomite and anhydrite data define repeated downsection decreases in $^{87}Sr/^{86}Sr$ punctuated by increases associated with siliciclastic marker beds (Fig. 11). The most radiogenic values (0.7070–0.7074) are from samples in the upper part of the section and, more specifically, from samples at and immediately below beds of siliciclastics. The least radiogenic values (0.7067–0.7069) are from samples downsection, generally just above siliciclastic beds. Two sequences of decreasing $^{87}Sr/^{86}Sr$ are apparent between siliciclastic beds (Fig. 11).

Comparison of San Andres $^{87}Sr/^{86}Sr$ data with reported secular trends (Burke and others, 1982; Popp and others, 1986a) illustrates that the lowest San Andres values recorded for both dolomite and anhydrite (0.7067–0.7069) are equivalent to the minimum $^{87}Sr/^{86}Sr$ estimated for sea water during the Phanerozoic (minimum ratios of 0.7067–0.7068 are recorded during the late Guadalupian and Late Jurassic). Significantly, these least radiogenic $^{87}Sr/^{86}Sr$ compositions are lower than estimates of middle Guadalupian sea water (0.7071; Burke and others, 1982), the depositional age of the Emma San Andres. It is unlikely that the low $^{87}Sr/^{86}Sr$ values recorded in San Andres dolomite and anhydrite are the result of fluids modified locally by nonradiogenic Sr, since exposed crystalline rocks in central New Mexico–the presumed sources of the siliciclastics in the upper San Andres Formation–are primarily sialic rocks (Gonzalez and Woodward, 1972; Nielsen and Scott, 1979; Grambling, 1979) that are typically sources of radiogenic Sr. These low $^{87}Sr/^{86}Sr$ values were probably derived from late Guadalupian sea water. (Fluids derived from sea water of any other age, except the Late Jurassic, would possess higher $^{87}Sr/^{86}Sr$ values. For a number of reasons, it is highly unlikely that dolomitization of the San Andres took place during the Jurassic. Perhaps most significantly, during the Jurassic, San Andres carbonates were buried beneath more than 1,300 m of sediment, nearly 700 m of which are evaporites of extremely low permeability.) We therefore conclude that San Andres carbonates were dolomitized by late Guadalupian sea water. Significant overprinting of precursor carbonate $^{87}Sr/^{86}Sr$ compositions by dolomitizing fluids suggests high water:rock ratios or an open diagenetic system. Depleted Sr concentrations in dolomite 2 (<300 ppm) relative to presumed carbonate precursors support this interpretation by indicating an open diagenetic system with respect to Sr and loss of most original rock Sr.

The vertical trends in $^{87}Sr/^{86}Sr$ values are somewhat similar to those observed in the distribution of Fe in dolomite (Fig. 11). As in the case of Fe, this suggests that siliciclastic beds were sources of radiogenic Sr. The vertical trend in $^{87}Sr/^{86}Sr$ also supports the theory that dolomitizing fluids had at least a minor downward component of flow. Radiogenic Sr was apparently leached from clays (smectites) or feldspars and incorporated into dolomite and anhydrite precipitating near these source beds. Dolomite and anhydrite that precipitated distal to siliciclastic beds acquired the least radiogenic values. We suggest that more radiogenic $^{87}Sr/^{86}Sr$ compositions resulted from mixing of relatively unaltered late Guadalupian seawater-derived brines with brines of similar origin modified by rock-water interaction in the siliciclastic-rich strata. Because $^{87}Sr/^{86}Sr$ compositions of dolomite and anhydrite decrease below siliciclastic-rich strata and not above, the relative proportion of altered brines mixing with unaltered brines must have decreased downward, away from siliciclastic-rich strata. In order to effect this trend, downward mixing of radiogenic fluids with relatively unaltered late Guadalupian seawater brines must have occurred.

Stable isotopes.—

Samples chosen for analysis of carbon and oxygen isotopes consist of 32 samples that contained greater than 90% dolomite 2 (in most cases >95%) and five samples consisting of greater than 90% dolomite 1. Dolomite 1 and dolomite 2 stable isotopes define two distinct populations (Fig. 12). Dolomite 2 samples exhibit uniformly heavy isotopic signatures for $\delta^{13}C$ and $\delta^{18}O$; no significant differences in isotope composition are apparent in samples from different wells. Dolomite 1 samples are depleted in $\delta^{18}O$ and slightly depleted in $\delta^{13}C$ relative to dolomite 2. This relation is illustrated in Fig. 13, which shows variation in $\delta^{13}C$ and $\delta^{18}O$ with depth. Oxygen and carbon isotope trends are somewhat covariant and exhibit correlative anomalies. Isotope values decrease sharply in the X marker (dolomite

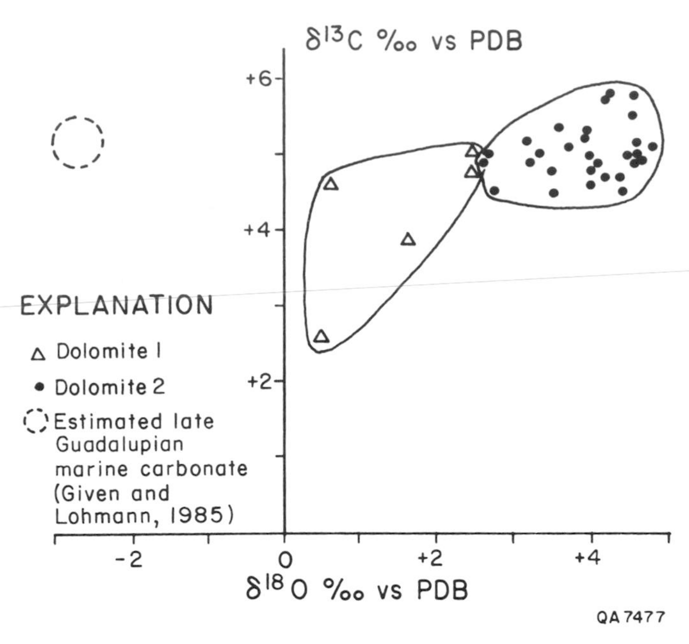

FIG. 12.—Combined stable isotope data from all cores in the Emma Field. Note that samples of dolomites 1 and 2 form distinct populations.

1) and remain depleted in $\delta^{18}O$ in dolomite 2 samples several meters below (Fig. 13).

Isotopic compositions of replacive dolomites are only interpretable when the isotopic composition of the marine precursor can be estimated. Due to obliterative dolomitization and anhydrite emplacement in Emma Field reservoir rocks, unaltered marine components are not available for analysis. Thus, the original isotopic composition of marine calcium carbonate in these rocks must be estimated. Given and Lohmann (1985) determined the stable isotope composition of late Guadalupian (Ufimian to early Kazanian) marine cements ($\delta^{13}C = +5.2$‰ PDB, $\delta^{18}O = -2.8$‰ PDB) in the Capitan reef sequence outcrop of the Guadalupe Mountains. Texas, some 225 km west of Emma Field. Although the significance of these data relative to global secular changes in seawater isotope chemistry has been questioned (Popp and others, 1986b), this estimate of marine carbonate composition is the best currently available for the Late Permian inland sea.

We interpret dolomite 2 as resulting from dolomitization of calcium carbonate from evaporatively concentrated sea water. Both stable isotope data and regional geology support this interpretation. Dolomite 2 is 6–7‰ enriched in $\delta^{18}O$ relative to the estimated value of the marine carbonate precursor (−2.8‰). Estimates of calcite-dolomite oxygen isotope fractionation predict dolomite to be approximately 3‰ enriched in ^{18}O relative to calcite (Land, 1980). An additional 3–4‰ $\delta^{18}O$ enrichment is predicted by the nonequilibrium evaporation of sea water to near $CaSO_4$ saturation (Lloyd, 1966; Holser, 1979). Dolomite 2 exhibits a $\delta^{13}C$ signature that is essentially the same as that of the estimated late Guadalupian marine carbonate (Given and Lohman, 1985). This interpretation is consistent with regional geology; late Guadalupian (Queen and Seven Rivers formations) and early Ochoan (Salado Formation) evaporite deposits, including bedded anhydrite and halite, overlie the San Andres in the study area as well as throughout the region (Fig. 2). These deposits document local generation of large volumes of evaporatively concentrated sea water.

Dolomite 1 contains depleted $\delta^{18}O$ and $\delta^{13}C$ values relative to dolomite 2 (Fig. 12). We suggest that this dolomite, which occurs in or subjacent to an organic-rich siliciclastic-rich marker bed, precipitated from fluids of near-seawater composition very early in the diagenetic history of the reservoir. Dolomite 1 samples associated with these siliciclastics are enriched in $\delta^{18}O$ (approximately 3–4.5‰) and depleted in $\delta^{13}C$ (1–3‰) relative to estimated marine calcium carbonate. The enrichment in $\delta^{18}O$ is accounted for by calcite-dolomite fractionation (Land, 1980). The slight depletion in $\delta^{13}C$ requires input of light carbon (^{12}C), such as soil gas carbon or carbon liberated during oxidation of organic matter. There is no unequivocal evidence of soil formation in the San Andres study interval. The siliciclastic-rich beds, however, are as much as an order of magnitude richer in total organic carbon than siliciclastic-free strata. Oxidation of some of this organic matter would introduce isotopically depleted CO_2 into dolomitizing fluids (review by Anderson and Arthur, 1983). Dolomitization from fluids of near-seawater composition suggests early or penecontemporaneous processes. This scenario is consistent with petrographic data, which indicate that dolomite 1 is the earliest diagenetic phase in the study interval.

In summary, stable isotope data from San Andres dolomites indicate localized early dolomitization by fluids of near-seawater composition (dolomite 1), followed by pervasive dolomitization by sea water near gypsum saturation (dolomite 2). Dolomites 3 and 4 represent very minor precipitates from fluids that probably did not significantly alter the chemistry of the bulk rock. Although early dolomitization appears to have been restricted to siliciclastic-rich strata or strata immediately subjacent to siliciclastics, it is possible that much more of the San Andres interval was dolomitized by this contemporaneous process. Present isotope compositions reflect only the latest recrystallization or stabilization event. Therefore, the two major isotope populations may reflect differences in dolomite recrystallization and diagenetic overprinting by late Guadalupian hypersaline fluids. Since isotopically depleted dolomite 1 occurs only in strata of very low porosity and permeability (X marker), the degree of diagenetic overprinting by hypersaline dolomite 2 fluids may have been controlled by variations in San Andres permeability after dolomite 1 formation. In other words, it is possible that dolomite 2 fluids caused recrystallization and re-equilibration of contemporaneous dolomite in more permeable strata, but did not re-equilibrate dolomite in the low-permeability siliciclastic-rich beds.

ORIGIN AND TIMING OF DOLOMITIZING FLUIDS

Interpretations of the petrographic, trace-element, and isotopic data of San Andres dolomite and anhydrite can be coupled with the regional geology and stratigraphy to model

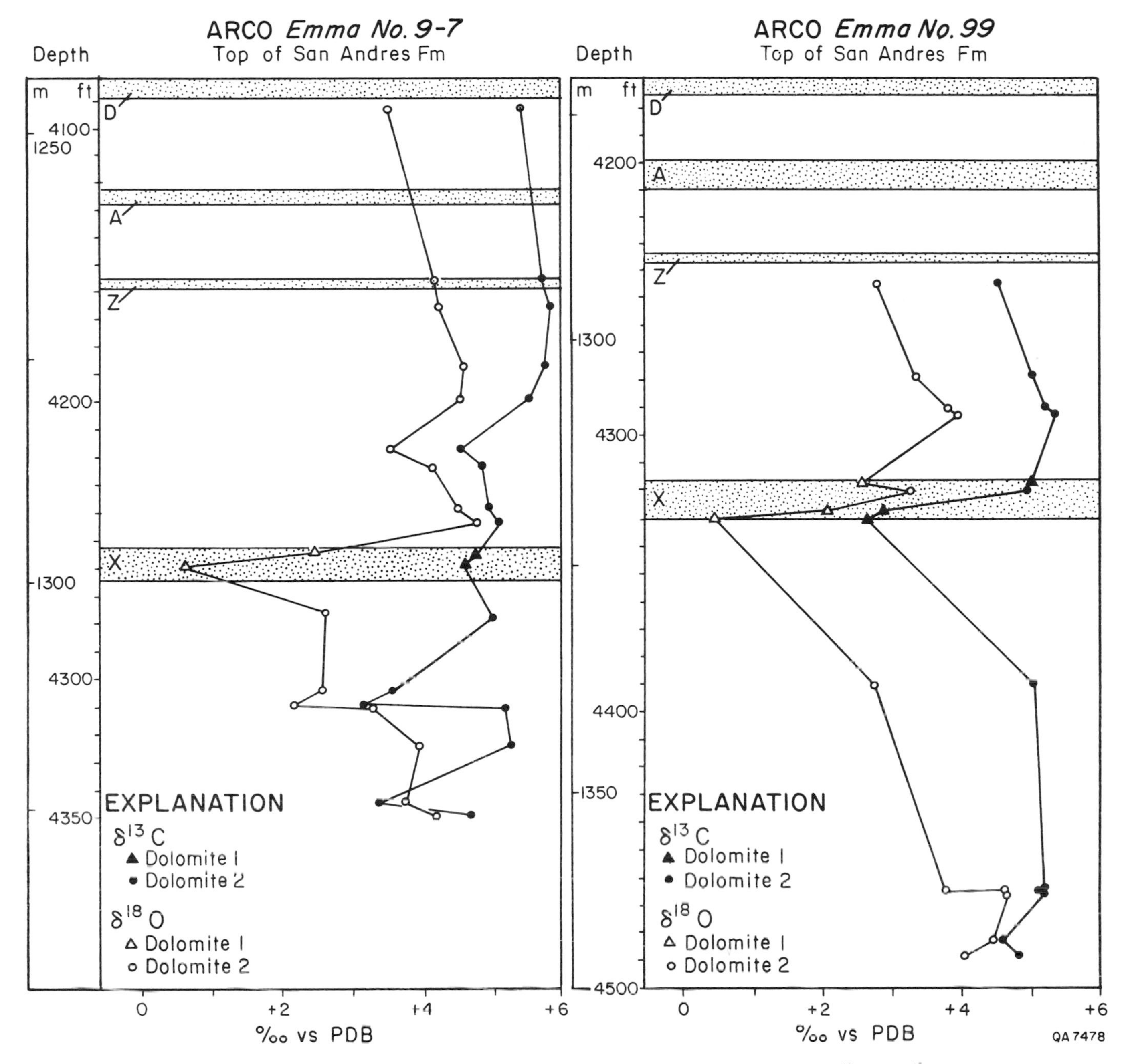

FIG. 13.—Plots of stable isotope data with depth in the Arco Emma #9-7 and #99 cores. Note that both $\delta^{13}C$ and $\delta^{18}O$ are strongly depleted in samples of dolomite 1 in and immediately below the X marker. A similar degree of depletion in the upper part of the #99 core may indicate the presence of dolomites formed by related processes.

the chemistry and movement of dolomitizing fluids (Figs. 14, 15).

Dolomite 1.—

On the basis of petrographic relations, dolomite 1 represents the first episode of dolomitization in the Emma San Andres. The timing of this event, however, is not precisely constrained. We can state only that it took place prior to formation of dolomite 2, during the middle to late Guadalupian. The pervasiveness of this early phase of dolomitization is also poorly known, because dolomite 1 chemistry is only retained in low-permeability, siliciclastic-rich strata. We propose two alternative models for dolomite 1 formation (Fig. 14): (1) syndepositional subtidal dolomitization by sea water (Fig. 14A); or (2) post-depositional, penecontemporaneous supratidal dolomitization by sea water during subaerial exposure associated with periods of siliciclastic deposition (Fig. 14B).

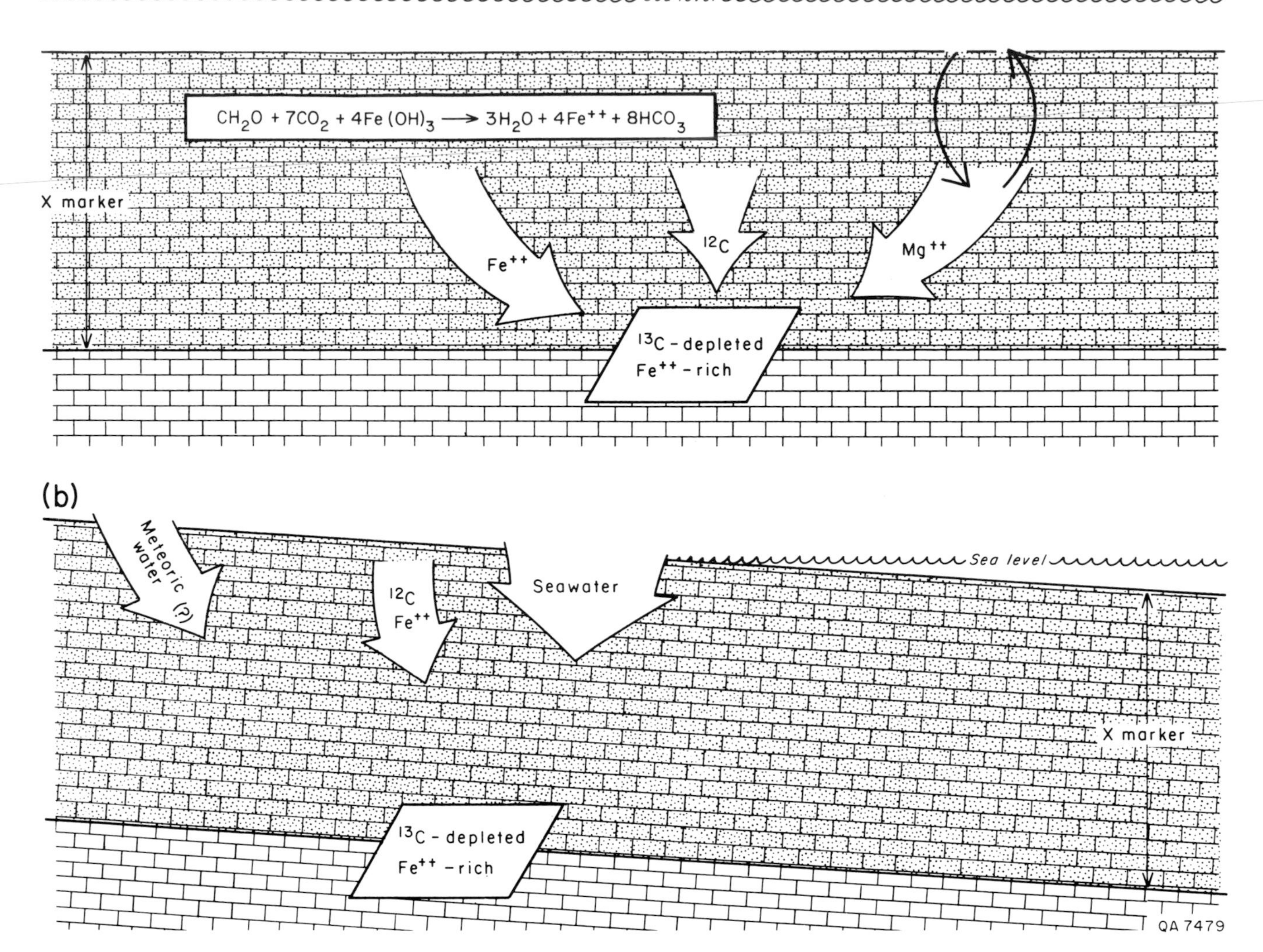

FIG. 14.—Proposed alternative models for formation of dolomite 1. (A) Syndepositional dolomitization by normal marine sea water in a shallow subtidal setting. (B) Early, post-depositional dolomitization by sea water or mixed sea water/meteoric water during subaerial exposure. In either case, bacterially catalyzed reduction of iron oxides by organic matter, both of which are abundant in the X marker interval, liberates divalent iron and results in depleted $\delta^{13}C$.

Subtidal dolomitization is suggested by the stratigraphically confined occurrence of dolomite 1. Dolomites reported by Behrens and Land (1972) from Baffin Bay, Texas, may represent a modern analog of dolomite 1. Like dolomite 1, Baffin Bay dolomites are fine-grained, anhedral, occur in terrigenous muds, and both are 3‰ enriched in $\delta^{18}O$ and slightly $\delta^{13}C$ depleted relative to estimated contemporaneous marine calcium carbonate. Behrens and Land (1972) interpreted dolomitization to have been early, entirely subtidal process produced by slightly hypersaline sea water (salinity approximately 45–65‰). Simultaneous Fe^{2+} enrichment and $\delta^{13}C$ depletion in dolomite 1 relative to seawater carbonate could have resulted from the bacterially catalyzed reduction of Fe hydroxides and oxides by organic matter:

$$CH_2O + 7CO_2 + 4Fe(OH)_3 = 4Fe^{2+} + 8HCO^{3-} + 3H_2O$$

This reaction liberates both divalent Fe and ^{13}C depleted bicarbonate. Woronick and Land (1985) proposed a similar reaction to account for compositions of baroque dolomite in Lower Cretaceous limestones of south Texas; Behrens and Land (1972) proposed that organically derived ^{12}C from terrigenous muds caused $\delta^{13}C$ depletion in Baffin Bay dolomites. Land (1982) has subsequently distinguished Holocene Baffin Bay dolomite from any ancient counterpart on the basis of its highly disordered crystallography and nonstoichiometry. Although dolomite 1 is clearly a more stable phase that Baffin Bay dolomite, the hydrodynamics and fluid chemistry of the dolomitization process may have been similar. Dolomite 1 may be the end result of nearly iso-

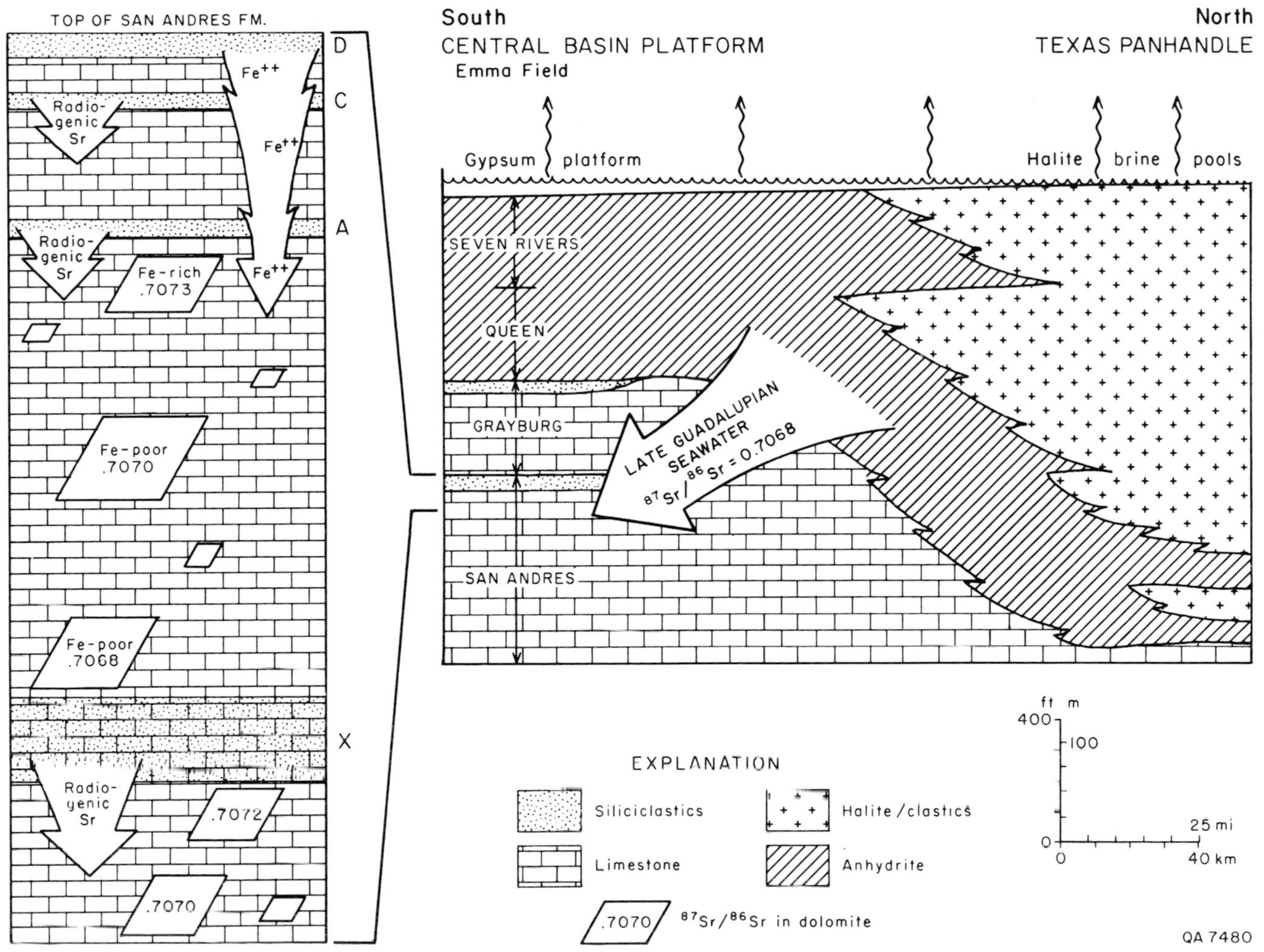

FIG. 15 —Proposed model for formation of dolomite 2. Dolomite 2 is the product of pervasive, matrix-replacive, post-depositional dolomitization of San Andres carbonates by hypersaline brines generated upsection and updip in gypsum-precipitating environments. These fluids, which are constrained by $^{87}Sr/^{86}Sr$ data and stratigraphic relations to have been derived from late Gaudalupian sea water, were modified by siliciclastic rich beds that acted as local sources of Fe^{++} and radiogenic strontium.

chemical recrystallization of a highly disordered, nonstoichiometric precursor similar to Baffin Bay dolomite. Subtidal dolomitization by near-normal sea water circulating through shallow-subbottom sediments has been documented by Mullins and others (1985). Normal seawater dolomitization has also been proposed by Saller (1984) in Enewetak Atoll, Carballo and others (1987) in Sugarloaf Key, Florida, Mazzullo and others (1987) in Belize, and Mitchell and others (1987) in Jamaican reefs. Land (1980, 1982, 1985) has concluded that sea water is the most likely naturally occurring dolomitizing fluid, since it is an inexhaustible source of Mg. The precise reason for contemporaneous selective dolomitization of carbonate at the X marker is unclear. Although unaltered sea water is capable of dolomitization, it is possible that the decrease in sedimentation rate associated with the X marker allowed for prolonged water-rock interaction between sea water and siliciclastic-rich carbonate sediments. It is also possible that the increased smectite/calcium carbonate ratio in the X marker bed resulted in interstitial marine pore waters with higher Mg/Ca ratios than interstitial pore waters in siliciclastic-free beds.

Alternatively, dolomitization may have been initiated from exposure associated with periods of siliciclastic influx at the X marker and overlying marker beds. As discussed previously, the top of the X marker is associated with a sudden vertical change in depositional environment that may represent a sequence boundary formed by a rapid fall in sea level. At least locally within Emma Field, there is evidence of subaerial exposure immediately above this horizon. The apparent restriction of dolomite 1 to carbonate at and below the X marker is consistent with this interpretation. Dolomitization may have been facilitated by storm recharge and/or tidal pumping. Evidence of exposure at the X marker suggests that meteoric-water infiltration could have occurred during dolomite 1 precipitation. There is, however, no preserved evidence of meteoric diagenesis in the study interval. As in the case with previously proposed subtidal dolomitization, depleted $\delta^{13}C$ and enriched Fe compositions in dolomite 1 resulted from the bacterially catalyzed reduction of Fe hydroxides by organic matter.

Dolomite 2.—

Radiogenic and stable isotope data indicate that dolomite 2 precipitated or stabilized in late Guadalupian sea water evaporated to near $CaSO_4$ saturation. The presence of bedded anhydrite, halite, and gypsum in late Guadalupian strata throughout northern and western Texas and southeastern New Mexico is consistent with this interpretation, in that it documents generation of large volumes of concentrated seawater brines at this time. Evaporites formed north of Emma Field and prograded generally southward. Since brines were generated in strata overlying Emma Field and to the north, there existed a potential for vertical, cross-formational flow as well as horizontal flow. Dolomitization of West Texas Permian carbonates by downward flux of evaporatively concentrated sea water has been proposed by several authors (Adams and Rhodes, 1960; Zaaza, 1978; Leary, 1984; Vogt, 1986). The efficiency of dolomitization by hypersaline sea water to dilute (mixing-zone) waters has been determined quantitatively by Sears and Lucia (1980) and Land (1982). Brine reflux has been theoretically modeled on regional scales by Simms (1984). Independent evidence for at least a minor downward component of flow or mixing of dolomitizing fluids in Emma Field comes from the downward decrease in Fe and $^{87}Sr/^{86}Sr$ in dolomite 2. Timing of dolomite 2 is further constrained by the $^{87}Sr/^{86}Sr$ composition of later anhydrite phases, which, based on secular trends, is most compatible with precipitation from late Guadalupian sea water (Fig. 15).

Studies of San Andres dolomite in West Texas by Leary (1984), Vogt (1986), and Leary and Vogt (1986) document similar CL, $\delta^{18}O$, $\delta^{13}C$, and $^{87}Sr/^{86}Sr$ compositions as observed in dolomite 2. These studies also concluded that dolomitization occurred via reflux of gypsum-saturated brines. These authors concluded, however, that brine reflux was contemporaneous with deposition. This interpretation requires that the seawater $^{87}Sr/^{86}Sr$ minimum (0.7067) occurred during the middle Guadalupian rather than the late Guadalupian, as reported by Burke and others (1982) and Popp and others (1986a). Our interpretation of the timing of dolomite 2 is consistent with both the reported secular variations in seawater $^{87}Sr/^{86}Sr$ and the generation of significant volumes of hypersaline brines on the Central Basin Platform during late Guadalupian time. We believe that only dolomite 1, which predates dolomite 2, formed contemporaneously with deposition and that it precipitated from near-normal sea water, as opposed to sea water evaporated to near-gypsum saturation.

Dolomites 3 and 4 and anhydrite.—

Subsequent precipitation of dolomites 3 and 4 need not have involved large mass transfer. Both types of dolomite are volumetrically very minor. Since dolomite 3 partially replaces dolomite 2 in many instances, an allochthonous source of cations is not required. It is conceivable that dolomites 3 and 4 precipitated from minor fluctuations in San Andres pore waters that were otherwise in near equilibrium with dolomite 2. Relatively enriched Fe concentrations in dolomite 3 may reflect slower rates of precipitation, resulting in increasing effective partitioning of Fe into the solid phase. In other words, dolomites 3 and 4 may have resulted from the nearly isochemical process of stabilization of dolomite 2.

The $^{87}Sr/^{86}Sr$ compositions of dolomite 2 and anhydrite suggest that both timing and origin of these phases were similar. As previously discussed, the $^{87}Sr/^{86}Sr$ signature of anhydrite is most compatible with precipitation from concentrated late Guadalupian sea water. This suggests that anhydrite emplacement quickly followed dolomitization, and therefore most of the San Andres diagenesis occurred shortly after deposition and entirely within the Permian.

SUMMARY AND CONCLUSIONS

The San Andres Formation in the Emma Field area is composed of a shallowing-upward sequence of shallow-platform carbonates deposited on an east-dipping ramp during the early Late Permian (middle Guadalupian). Open-platform packstone and wackestone are overlain by progressively shallower water deposits, including skeletal grainstone deposited in carbonate sand shoals, and an uppermost sequence of supratidal pisolitic, fenestral, and cryptalgal mudstone. Beds of terrigenous siliciclastics, presumably of windblown origin, are commonly associated with evidence of exposure. These beds are most common in the uppermost, supratidal part of the San Andres sequence but are also present lower in the section, where they may also be indicative of subaerial exposure.

On the basis of petrographic and geochemical evidence, dolomitization of these shallow-platform carbonates is constrained to have taken place during two major events. Early, matrix-replacive, red-luminescent dolomite 1 occurs at and immediately below a relatively thin interval within the section that is rich in terrigenous silt and organic carbon. Precipitation of this dolomite, which is 3–4.5‰ enriched in $\delta^{18}O$ but slightly depleted in $\delta^{13}C$ relative to estimated San Andres marine calcium carbonate, occurred in near-normal middle Guadalupian sea water in either shallow subtidal conditions or in a supratidal setting formed during a lowstand in sea level. Sedimentologic data and regional studies document the development of exposure at this time but do not rule out the possibility of subtidal dolomitization prior to this event. In either case, dolomitization may have been promoted, in part, by an increased Mg/Ca ratio of interstitial waters of siliciclastic-rich sediments. Siliciclastics were a local source of Fe and ^{12}C, which were liberated in a redox reaction of organic matter and Fe hydroxides. This early dolomitization may have been extensive but is only preserved in low-permeability strata.

The bulk of the dolomite in the San Andres (dolomite 2) was precipitated or stabilized subsequent to San Andres deposition by evaporatively concentrated sea water. Strontium isotope compositions suggest that this sea water was late Guadalupian; trends in $^{87}Sr/^{86}Sr$ ratios as well as in trace elements document a downward-flow component to the movement of these diagenetic fluids. The stable isotope signature of dolomite 2 (greatly enriched in $\delta^{18}O$ but equivalent in $\delta^{13}C$ relative to Permian marine carbonate) supports evaporation of seawater brine to gypsum saturation. Large volumes of evaporites developed updip from the study area during the late Guadalupian.

Following dolomitization, anhydrite was emplaced under similar hydrologic and geochemical conditions by late Guadalupian seawater brines. Anhydrite occluded much of the remaining porosity and locally replaced dolomite. Chemistry of $^{87}Sr/^{86}Sr$ indicates that anhydrite was precipitated from a fluid similar to that responsible for dolomite 2 formation. Trends in $^{87}Sr/^{86}Sr$ are similar to those observed in dolomite and indicate a downward component to fluid migration and leaching of radiogenic Sr from siliciclastics.

Trends in the distribution of Fe in dolomite and strontium isotopes in dolomite and anhydrite reveal that the composition of diagenetic fluids was modified by variations in the rock chemistry of the stratigraphic section. Both Fe and $^{87}Sr/^{86}Sr$ decrease gradually downsection below thin beds of siliciclastics in the uppermost San Andres, indicating that these sediments provided local sources of these elements to migrating fluids. The recognition of these geochemical variations and of the fact that they are produced by local water-rock interactions has been extremely valuable in this study in documenting fluid flow migration directions. Lithologic heterogeneities of the sort responsible for these local variations in geochemistry are probably common in carbonate sequences. It is thus extremely important to recognize and characterize such heterogeneities and resultant variations in dolomite geochemistry in attempting to reconstruct accurately the origin and chemistry of dolomitizing fluids in dolostone sequences.

ACKNOWLEDGMENTS

This work was funded by the University of Texas System. We have benefited from valuable discussions with numerous individuals, including D. G. Bebout, C. Kerans, L. S. Land, F. J. Lucia, R. P. Major, H. H. Posey, and Z. Sofer. Various drafts of the manuscript were reviewed by A. Bein, W. C. Dawson, P. M. Harris, R. P. Major, H. H. Posey, and V. Shukla.

Stable isotope analyses were performed at the Benedum Stable Isotope Laboratory of Brown University. All $^{87}Sr/^{86}Sr$ ratios were determined at Geochron Laboratories (Kreuger Enterprises, Inc., Cambridge, Massachusetts) and at the University of Texas at Austin.

REFERENCES

ADAMS, J. E., AND RHODES, M. L., 1960, Dolomitization by seepage refluxion: American Association of Petroleum Geologists Bulletin, v. 44, 912–920.

AHR, W. M., 1973, The carbonate ramp: An alternative to the shelf model: Gulf Coast Association of Geological Societies, Transactions, v. 23, p. 221–225.

ANDERSON, T. F., AND ARTHUR, M. A., 1983, Stable isotopes of oxygen and carbon and their application to sedimentologic and paleoenvironmental problems, *in* Stable Isotopes in Sedimentary Geology: Society of Economic Paleontologists and Mineralogists Short Course No. 10, p. 1-1 to 1-51.

BEBOUT, D. G., 1986, Facies control of porosity in the Grayburg Formation–Dune Field, Crane County, Texas, *in* Bebout, D. G., and Harris, P. M., eds., Hydrocarbon Reservoir Studies, San Andres/Grayburg Formations, Permian Basin: Permian Basin Section, Society of Economic Paleontologists and Mineralogists Publication No. 86-26, p. 107–111.

BEHRENS, E. W., AND LAND, L. S., 1972, Subtidal Holocene dolomite, Baffin Bay, Texas: Journal of Sedimentary Petrology, v. 42, 155–161.

BEIN, AMOS, AND LAND, L. S., 1983, Carbonate sedimentation and diagenesis associated with Mg-Ca-Cl brines: The Permian San Andres Formation in the Texas Panhandle: Journal of Sedimentary Petrology, v. 53, p. 243–260.

BURKE, W. H., DENISON, R. E., HETHERINGTON, E. A., KOEPNICK, R. B., NELSON, H. F., AND OTTO, J. B., 1982. Variation of seawater $^{87}Sr/^{86}Sr$ throughout Phanerozoic time: Geology, v. 10, p. 516–519.

CARBALLO, J. D., LAND, L. S., AND MISER, D. E., 1987, Holocene dolomitization of supratidal sediments by active tidal pumping, Sugarloaf Key, Florida: Journal of Sedimentary Petrology, v. 57, 153–165.

DIXON, G. H., 1967, Paleotectonic Investigations of the Permian System in the United States, *in* Northeastern New Mexico and Texas-Oklahoma Panhandles: U.S. Geological Survey Professional Paper 515-D, p. 61–80.

FRACASSO, M. A., AND HOVORKA, S. D., 1986, Cyclicity in the middle Permian San Andres Formation, Palo Duro Basin, Texas Panhandle: Bureau of Economic Geology Report of Investigations No. 156, The University of Texas at Austin, 48 p.

GALLEY, J. E., 1958, Oil and geology in the Permian Basin of Texas and New Mexico, *in* Habitat of Oil–A Symposium: American Association of Petroleum Geologists, Tulsa, Oklahoma, p. 395–446.

GARBER, R. A., AND HARRIS, P. M., 1986, Depositional facies of the Grayburg/San Andres dolomite reservoirs–Central Basin Platform, Permian Basin, *in* Bebout, D. G., and Harris, P. M., eds., Hydrocarbon Reservoir Studies, San Andres/Grayburg Formations, Permian Basin: Permian Basin Section, Society of Economic Paleontologists and Mineralogists Publication No. 86-26, p. 61–66.

GARRELS, R. M., AND CHRIST, C. L., 1965, Solutions, Minerals, and Equilibria: Freeman, Cooper and Company, San Francisco, 450 p.

GIVEN, R. K., AND LOHMANN, K. C., 1985, Derivation of the original isotopic composition of Permian marine cements: Journal of Sedimentary Petrology, v. 55, p. 430–439.

GONZALEZ, R. A., AND WOODWARD, L. A., 1972, Petrology and structure of Precambrian rocks of the Pedernal Hills, New Mexico: New Mexico Geological Society Guidebook, 23rd Field Conference, p. 144–147.

GRAMBLING, J. A., 1979, Precambrian geology of the truchas Peaks region, north-central New Mexico, and some regional implications: New Mexico Geological Society Guidebook, 30th Field Conference, p. 135–143.

HARDIE, L. A., 1987, Dolomitization: A critical review of some current views: Journal of Sedimentary Petrology, v. 57, 166–183.

HOLSER, W. T., 1979, Trace elements and isotopes in evaporites, *in* Burns, R. G., ed., Marine Minerals: Mineralogical Society of America, Reviews in Mineralogy, v. 6, p. 295–346.

JAMES, N. P., 1984, Shallowing-upward sequences in carbonates, *in* Walker, R. G., ed., Facies Models, second edition: Geoscience Canada Reprint Series 1, p. 213–228.

LAND, L. S., 1980, The isotopic and trace element chemistry of dolomite: The state of the art, *in* Zenger, D. H., Dunham, J. G., and Ethington, R. L., eds., Concepts and Models of Dolomitization: Society of Economic Paleontologists and Mineralogists Special Publication 28, p. 87–110.

———, 1982, Dolomitization: American Association of Petroleum Geologists Educational Short Course Notes No. 24, 20 p.

———, 1985, The origin of massive dolomite: Journal of Geological Education, v. 33, p. 112–125.

LEARY, D. A., 1984, Diagenesis of the Permian (Guadalupian) San Andres and Grayburg Formations, Central Basin Platform, West Texas: Unpublished M.S. Thesis, The University of Texas at Austin, Texas, 121 p.

———, AND VOGT, J. N., 1986, Diagenesis of Permian (Guadalupian) San Andres Formation, Central Basin Platform *in* Bebout, D. G., and Harris, P. M., eds., Hydrocarbon Reservoir Studies, San Andres/Grayburg Formations, Permian Basin: Permian Basin Section, Society of Economic Paleontologists and Mineralogists Publication No. 86-26, p. 67–68.

LLOYD, R. M., 1966, Oxygen isotope enrichment of sea water by evaporation: Geochimica et Cosmochimica Acta, v. 30, p. 801–814.

MACHEL, H.-G., 1985, Cathodoluminescence in calcite and dolomite and its chemical interpretation: Geoscience Canada, v. 12, p. 139–147.

MAZZULLO, S. J., AND HEDRICK, C. L., 1985, Lithofacies and depositional models of the back reef Guadalupian section (Queen, Seven Rivers, Yates, and Tansill formations), *in* Permian Carbonate/Clastic Sedimentology, Guadalupe Mountains: Analogs for Shelf and Basin Reservoirs: Road Log and Locality Guide: Permian Basin Section, Society of Economic Paleontologists and Mineralogists Annual Field Trip, p. 1–80.

———, REID, A. M., AND GREGG, J. M., 1987, Dolomitization of Holocene Mg-calcite supratidal deposits, Ambergis Cay, Belize: Geological Society of America Bulletin, v. 98, p. 224–231.

MEISSNER, F. F., 1969, Cyclic sedimentation in middle Permian strata of the Permian Basin, West Texas and New Mexico, *in* Elam, J. G., and Chuber, S., eds., Cyclic Sedimentation in the Permian Basin: West Texas Geological Society, p. 135–151.

MITCHELL, J. T., LAND, L. S., AND MISER, D. N., 1987, Modern marine dolomite cement in a north Jamaican fringing reef: Geology, v. 15, p. 557–560.

MULLINS, H. T., WISE, S. W., LAND, L. S., SIEGEL, D. I., MASTERS, P. M., HINCHEY, E. J., AND PRICE, K. R., 1985, Authigenic dolomite in Bahamian periplatform slope sediment: Geology, v. 13, p. 292–295.

NIELSEN, K. C., AND SCOTT, T. E., JR., 1979, Precambrian deformational history of the Picuris Mountains, New Mexico: New Mexico Geological Society Guidebook, 30th Field Conference, p. 113–120.

ORIEL, S. S., MYERS, D. A., AND CROSBY, E. J., 1967, Paleotectonic Investigations of the Permian System in the United States *in* West Texas Permian Basin Region, U.S. Geological Survey Professional Paper 515-C, p. 17–60.

POPP, B. N., PODOSEK, F. A., BRANNON, J. C., ANDERSON, T. F., AND PIER, JEAN, 1986a, $^{87}Sr/^{86}Sr$ ratios in Permo-Carboniferous sea water from the analyses of well-preserved brachipod shells: Geochimica et Cosmochimica Acta, v. 50, p. 1321–1328.

———, ANDERSON, T. A., AND SANDBURG, P. A., 1986b, Brachiopods as indicators of original isotopic compositions in some Paleozoic limestones: Geological Society of America Bulletin, v. 97, p. 1262–1269.

READ, J. F., 1982, Carbonate platforms of passive (extensional) continental margins: Types, characteristics, and evolution: Tectonophysics, v. 81, p. 195–212.

RUPPEL, S. C., 1986, San Andres facies and porosity distribution–Emma Field, Andrews County, Texas, *in* Bebout, D. G., and Harris, P. M., eds., Hydrocarbon Reservoir Studies, San Andres/Grayburg Formations, Permian Basin: Permian Basin Section, Society of Economic Paleontologists and Mineralogists Publication No. 86-26, p. 99–103.

SALLER, A. H., 1984, Petrologic and geochemical constraints on the origin of subsurface dolomite, Enewetak Atoll: An example of dolomitization by normal seawater: Geology, v. 12, p. 217–220.

SARG, J. F., 1981, Petrology of the carbonate-evaporite facies transition of the Seven Rivers Formation (Guadalupian, Permian), southeast New Mexico: Journal of Sedimentary Petrology, v. 51, p. 73–96.

———, 1986, Facies and stratigraphy of upper San Andres basin margins and lower Grayburg inner shelf, *in* Moore, G. E., and Wilde, G. L., eds., Lower and Middle Guadalupian Facies, Stratigraphy, and Reservoir Geometries–San Andres/Grayburg Formations–Guadalupe Mountains, New Mexico and Texas: Permian Basin Section, Society of Economic Paleontologists and Mineralogists Publication No. 86-25, p. 83–105.

SEARS, S. O., AND LUCIA, F. J., 1980, Dolomitization of northern Michigan Niagara reefs by brine refluxion and freshwater/seawater mixing, *in* Zenger, D. H., Dunham, J. B., and Ethington, R. L., eds., Concepts and Models of Dolomitization: Society of Economic Paleontologists and Mineralogists Special Publication 28, p. 215–235.

SILVER, B. A., AND TODD, R. G., 1969, Permian cyclic strata, northern Midland and Delaware Basins, West Texas and southeastern New Mexico: American Association of Petroleum Geologists Bulletin, v. 53, p. 2223–2251.

SIMMS, M. S., 1984, Dolomitization by groundwater flow systems in carbonate platforms: Gulf Coast Association of Geological Societies, Transactions, v. 34, p. 411–420.

TODD, R. G., 1976, Oolite-bar progradation, San Andres Formation, Midland Basin, Texas: American Association of Petroleum Geologists Bulletin, v. 60, p. 907–925.

VAIL, P. R., HARDENBOL, J., AND TODD, R. G., 1984, Jurassic unconformities, chronostratigraphy, and sea-level changes from seismic stratigraphy and biostratigraphy: American Association of Petroleum Geologists Memoir 36, p. 129–144.

VEIZER, JAN, 1983, Chemical diagenesis of carbonates: Theory and application of trace element technique, *in* Stable Isotopes in Sedimentary Geology: Society of Economic Paleontologists and Mineralogists Short Course No. 10, p. 3-1 to 3-100.

VOGT, J. N., 1986, Dolomitization and anhydrite diagenesis of the San Andres (Permian) Formation, Gaines County, Texas: Unpublished M.S. Thesis, The University of Texas at Austin, Texas, 202 p.

WARD, R. F., KENDALL, C. G. ST. C., AND HARRIS, P. M., 1986, Upper Permian (Guadalupian) facies and their association with hydrocarbons–Permian Basin, West Texas and New Mexico: American Association of Petroleum Geologists Bulletin, v. 70, p. 239–262.

WILDE, G. L., 1975, Fusulinid-defined Permian stages, *in* Permian Exploration, Boundaries, and Stratigraphy: Symposium and Field Trip: Permian Basin Section, Society of Economic Paleontologists and Mineralogists Publication No. 75-65, p. 67–83.

———, 1986, Stratigraphic relationship of the San Andres and Cutoff Formations, northern Guadalupe Mountains, New Mexico and Texas, *in* Moore, G. E., and Wilde, G. L., eds., Lower and Middle Guadalupian Facies, Stratigraphy, and Reservoir Geometries, San Andres/Grayburg Formations, Guadalupe Mountains, New Mexico and Texas: Permian Basin Section, Society of Economic Paleontologists and Mineralogists Publication No. 86-25, p. 49–63.

WORONICK, R. E., AND LAND, L. S., 1985, Late burial diagenesis, Lower Cretaceous Pearsall and Lower Glen Rose Formations, south Texas, *in* Schneidermann, Nahum, and Harris, P. M., eds., Carbonate Cements: Society of Economic Paleontologists and Mineralogists Special Publication 36, p. 265–276.

ZAAZA, M. W., 1978, The depositional facies, diagenesis, and reservoir heterogeneity of the upper San Andres: Unpublished Ph.D. Dissertation, The University of Tulsa, Oklahoma, 182 p.

SUBJECT INDEX